INTEGRATING HISTORY AND PHILOSOPHY OF SCIENCE

BOSTON STUDIES IN THE PHILOSOPHY OF SCIENCE

VOLUME 263

For further volumes:
http://www.springer.com/series/5710

INTEGRATING HISTORY AND PHILOSOPHY OF SCIENCE

Problems and Prospects

Edited by

SEYMOUR MAUSKOPF
Duke University, Durham, NC, USA

TAD SCHMALTZ
University of Michigan, Ann Arbor, MI, USA

Springer

Editors
Seymour Mauskopf
Duke University
History
Durham
USA
shmaus@duke.edu

Tad Schmaltz
University of Michigan Philosophy
South State St. 435
48109-1003 Ann Arbor
Michigan
USA
tschmalt@umich.edu

ISSN 0068-0346
ISBN 978-94-007-3807-2 ISBN 978-94-007-1745-9 (eBook)
DOI 10.1007/978-94-007-1745-9
Springer Dordrecht Heidelberg London New York

Printed on acid-free paper

Springer is part of Springer Science+Business Media (www.springer.com)

Contents

Contributors

Theodore Arabatzis Department of Philosophy and History of Science, University of Athens, University Campus, Ano Ilisia, 157 71 Athens, Greece, tarabatz@phs.uoa.gr

Kenneth L. Caneva University of North Carolina at Greensboro, Greensboro, NC, USA, klcaneva@uncg.edu

Hasok Chang University of Cambridge, Cambridge, UK, hc372@cam.ac.uk

Peter Dear Cornell University, Ithaca, NY, USA, prd3@cornell.edu

Melinda Bonnie Fagan Rice University, Houston, TX, USA, mbf2@rice.edu

Ronald N. Giere University of Minnesota, Minnesota, MN, USA, giere@umn.edu

Jan Golinski University of New Hampshire, Durham, NH, USA, jan.golinski@unh.edu

Jane Maienschein Arizona State University, Tempe, AZ, USA, maienschein@asu.edu

Seymour Mauskopf Duke University, Durham, NC, USA, shmaus@duke.edu

David Marshall Miller Duke University, Durham, NC, USA, david.m.miller@duke.edu

Wolfgang Pietsch Carl von Linde-Akademie, Technische Universität München, München, Germany, pietsch@cvl-a.tum.de

Dean Rickles University of Sydney, Sydney, NSW, Australia, dean.rickles@sydney.edu.au

Jutta Schickore Indiana University, Bloomington, IN, USA, jschicko@indiana.edu

Tad Schmaltz University of Michigan, Ann Arbor, MI, USA, tschmalt@umich.edu

About the Authors

Theodore Arabatzis is Assistant Professor of Philosophy at the National and Kapodistrian University of Athens. He is the author of *Representing Electrons: A Biographical Approach to Theoretical Entities* (Chicago, 2006).

Kenneth L. Caneva is Professor of History at the University of North Carolina, Greensboro. He is the author of *Possible Kuhns in the History of Science of Science: Anomolies of Incommensurable Paradigms* (Studies in History and Philosophy of Science, 2000).

Hasok Chang is Professor of Philosophy of Science at University College London. He is the author of *Inventing Temperature: Measurement and Scientific Progress* (Oxford, 2006).

Peter Dear is Professor of History at Cornell University. He is the author of *The Intelligibility of Nature: How Science Makes Sense of the World* (Chicago, 2006).

Melinda Bonnie Fagan is Assistant Professor of Philosophy at Rice University. She is the author of *The Search for the Hematopoetic Stem Cell: Social Interaction and Epistemic Success in Immunology* (Studies in History and Philosophy of Biological and Biomedical Sciences, 2007).

Ronald N. Giere is Professor Emeritus of Philosophy at the University of Minnesota. He is the author of *Scientific Perspectivism* (Chicago, 2006).

Jan Golinski is Professor of History and Humanities at the University of New Hampshire. He is the author of *Making Natural Knowledge: Constructivism and the History of Science* (Chicago, 2005).

Jane Maienschein is Regent's Professor in the School of Life Sciences at Arizona State University. She is the author of *Whose View of Life? Embryos, Cloning and Stem Cells* (Harvard, 2003).

Seymour Mauskopf is Professor Emeritus in the Department of History at Duke University (specialty, history of science). His most recent publication is "The Historiography of Science and Technology" (with Alex Roland), *The Oxford History of Historical Writing, Volume 5: Historical Writing Since 1945* (Oxford, 2011).

David Marshall Miller is Visiting Assistant Professor of Philosophy at Duke University. He is the author of *O Male Factum: Rectilinearity and Kepler's Discovery of the Ellipse* (Journal of the History of Astronomy, 2008).

Wolfgang Pietsch is Wissenschaftlicher Mitarbeiter at the Technical University Munich. He is the author of *On conceptual problems in classical electro-dynamics: Prospects and problems for an action-at-a-distance interpretation* (forthcoming in Studies in History and Philosophy of Modern Physics).

Dean Rickles is Professor in the Unit for the History and Philosophy of Science at the University of Sydney. He is the author of *The Ashgate Companion to Contemporary Philosophy of Physics* (Ashgate, 2008).

Jutta Schickore is Assistant Professor in the Department of History and Philosophy of Science at Indiana University, Bloomington. She is the author of *The Microscope and the Eye: A History of Reflections, 1740–1870* (Chicago, 2007).

Tad Schmaltz is Professor in the Department of Philosophy at the University of Michigan, Ann Arbor. He is the author of *Descartes on Causation* (Oxford, 2008).

Chapter 1
Introduction

Seymour Mauskopf and Tad Schmaltz

Just over 50 years ago, the National Science Foundation began to support "history and philosophy of science."[1] In 1960, programs in history and philosophy of science were begun at Princeton University and Indiana University. The phrase—and its acronym, "HPS"—soon became current. The History of Science Society and the Philosophy of Science Association began to hold joint annual meetings. All of these developments certainly suggested the coming to interdisciplinary fruition of a natural affinity between the two fields.

And yet, inspection of the interactions between the relatively new discipline of the history of science and its more established philosophical partner belies, for the most part, both the natural affinity and the interdisciplinary fruition. Reflecting on his days as a history of science graduate student in the Princeton program in the late 1960s, Kenneth Caneva flatly asserts, in his contribution to this volume: "I neither knew nor cared where the philosophers were." From the philosophical side, Ronald Giere, who had been a faculty member of the Indiana program, famously characterized the yoking of history and philosophy of science as "a marriage of convenience."

Nevertheless, there remains a sense, perhaps nostalgic but perhaps also programmatic, that there are natural affinities and promises of fruitful interactions between these disciplines. The purpose of this volume is to explore the current relationships between them. The fact that both Caneva and Giere are contributors to our discussion is, itself, indicative of the interest among historians and philosophers of science to pursue exploration of these relationships.

Fifty years ago, the history of science was coming into discernable existence as a discipline both in this country and in the UK. There was, perhaps, a greater urgency among historians of science for disciplinary identity than for philosophers of science. After all, from the time of the *Prior* and *Posterior Analytics* of Aristotle, issues

[1] See Rossiter (1984) for the interesting and complex history of the genesis of NSF funding programs.

S. Mauskopf (✉)
Duke University, Durham, NC, USA
e-mail: shmaus@duke.edu

S. Mauskopf, T. Schmaltz (eds.), *Integrating History and Philosophy of Science*,
Boston Studies in the Philosophy of Science 263, DOI 10.1007/978-94-007-1745-9_1,

in philosophy of science (formerly conceived as "natural philosophy") have always central to philosophical analysis. Before World War II, philosophers associated with the Vienna Circle had already established a clearly delineated philosophical perspective in "logical positivism." This emphasized the importance of the task of constructing a framework drawn from mathematical logic that would serve to advance current empirical research in the hard physical sciences, especially physics. Their framework entailed the analysis of completed systems of scientific knowledge, not the genesis of such systems. Consequently, there was no reference to—or particular interest in—history of science in this perspective.

If philosophy of science was comparatively well housed in philosophy, the same could not be said for history of science vis-à-vis history. History of science had virtually no purchase among historians. It had largely been pursued by scientists and by philosophers (e.g. Ernst Cassirer), not historians. Even Alexandre Koyré, the author of the *Études galiléennes* (Koyré 1939), which became the foundational work in the modern historiography of science, was professionally a philosopher. While not sundering ties to the historical profession, historians of science felt some necessity to establish a distinct professional and institutional identity. For example, the HPS Program at Princeton arose largely from Charles Gillispie's desire to provide advanced training to students with a real commitment to history of science, difficult to attract just through the history department at Princeton. On the other hand, Gillispie's initiative had little if anything to do with a felt need on his part to associate with philosophers of science. Nor, in Gillispie's account of the origins of this program, did the distinguished Vienna Circle philosopher of science, Carl G. Hempel, play any role in its formation (Gillispie 1999).[2]

Conversely, at Indiana University, the impetus from an HPS program did come from a philosopher of science, but a rather dissident one, N. R. Hanson. Although an American, Hanson came to Indiana after getting his D. Phil at Oxford and teaching philosophy of science at Cambridge, where he and the historian of science A. R. Hall constituted the teaching core of that HPS program in the early 1950s (Hall 1984). Unlike the dominant logical positivists, Hanson *was* concerned with origins and genesis of systems of scientific knowledge and, hence, with the history of science. But Hanson was unusual in this respect among philosophers of science (Hanson 1958).[3]

Yet, if only in hindsight, some general parallels in perspectives between historians and philosophers of science ca. 1960 can be perceived that might have made the union of them not unreasonable. One was an "intellectualist" approach (labeled "internalist" among historians of science). This was undoubtedly due to the impact of Koyré's studies of the Scientific Revolution on historians of science. Perhaps reflecting his philosophical training, Koyré viewed this watershed episode

[2]Gillispie's account is contained in a supplement of Isis with the title: "Catching up with the Vision: Essays on the Occasion of the 75th Anniversary of the Founding of the History of Science Society."

[3]Hanson brought A. R. Hall and his wife, M. B. Hall to Indiana in 1961. He and they left Indiana in 1963 but the program had been well launched and survived these losses.

in strongly intellectualist terms, delineating it as "intellectual mutation," in which even experimentation played little or no role, much less social and cultural contexts. This intellectualist approach was coupled with a strong guiding belief in scientific progress, perhaps expressed with the most sophistication in Charles Gillispie's *The Edge of Objectivity* (Gillispie 1960). Intellectualist and progressivist perspectives of science were certainly something historians and philosophers of science had in common and, in the case of the formation of the Indiana program, the intellectualist commonality, at least, was a positive factor.[4] Moreover, in this volume, Jan Golinski gives the intellectualist and progressivist perspectives contemporaneous political contexts in anti-Nazism and anti-Communism.

And then came Thomas Kuhn's *Structure of Scientific Revolutions* (Kuhn 1962). This work attracted the attention of philosophers of science in a way that no previous work in the history of science had been able to do. Part of Kuhn's attention from philosophers of science was due to the fact that contemporaries in philosophy had been developing somewhat similar approaches and perspectives to his own. Hanson was one; Stephen Toulmin was another.[5] But it was *Structure of Scientific Revolutions* that brought philosophers of science to consider seriously the history of science. There was now the thought in the work of philosophers such as Imre Lakatos—which one cannot find in the work either of the logical positivists or of their early philosophical critics—that historical studies could be used as a resource for honing philosophical analyses of scientific change.

In his final extended interview, Kuhn claimed that he thought *Structure* "when I got to it finally, as being a book for philosophers."[6] Although this retrospective assessment may have more to do with Kuhn's own later self-refashioning as a philosopher, certainly an important strand of Kuhn's perspective in *Structure* was the intellectualism of Koyré: Koyré's name was in fact the first to be mentioned in the book. The intellectualist strand in so ambitious a work as this no doubt recommended it to at least some philosophers of science.

And, yet, there were two other strands that came to undermine the commonality (however implicit) in perspective that had existed between historians and philosophers of science. The more important for this discussion was Kuhn's turn against the progressivist perspective.[7] In the *Introduction*, he strongly set forth his anti-progressivist perspective—perhap even more forcefully than he intended:

> Aristotelian dynamics, phlogistic chemistry, or caloric thermodynamics . . . were, as a whole, neither less scientific nor more the product of human idiosyncrasy than those current today.[8]

---

[4]See Grau (1999, S318).

[5]See Toulmin (1961).

[6]Zammito (2004, 53). In the same interview, Kuhn made an even more astonishing repudiation of history of science: "philosophers and scientists are much closer to one another, because they all come in being concerned about what's right and wrong—not about what happened" (181).

[7]Kuhn's consideration social and professional contexts was the second important component of his historical view of scientific change. Under the influence of the hitherto neglected Polish scientist, Ludwig Fleck, Kuhn gave important—if historically very general—place to scientific *communities*.

[8]Kuhn (1996, 2).

As Kuhn made clear in the body of the text, scientific change had to be seen as consisting not in smooth advance, but rather in epochs of paradigm-guided "normal science" punctuated by episodes of revolutionary paradigm change. Theory and practice in a science in its post-revolutionary state were "incommensurable" with the pre-revolutionary aspects of the same science. Given this new account of science, it was impossible to delineate a steady, cumulative "advance" over time for a science that went through a revolution. At the end of *Structure*, Kuhn pushed the argument further, arguing that it was impossible to define scientific "progress" in terms of closer and closer approximation to natural "truth."

With the notable exception of Paul Feyerabend, philosophers who initially embraced Kuhn's view that the history of science could alter the philosophical image of science nonetheless rebelled against the anti-progressivist implications of *Structure*. Indeed, Kuhn himself subsequently equivocated over such implications.

In contrast, historians of science were taken by the anti-progressivist rhetoric of *Structure* even as they tended to ignore Kuhn's abstract schema of paradigms, normal science and revolutions in their research. Part of their reaction may have been influenced by the more general contemporaneous cultural turn from a positive to a critical valuation of science in the 1960s and 1970s. Part was also a symptom of the maturation of history of science as an historical discipline.[9] Historians of science had become more and more focused on the understanding passė scientific enterprises *for their own sake and in their own terms* with as little concern as possible with how they related—much less contributed—to contemporary scientific understanding. Much of this had little or nothing to do with the impact of Kuhn's *Structure*.[10] To employ a distinction used by a number of our authors, history of science became irrevocably *descriptive* (and historicist), whereas philosophy of science, even of those reacting favorably to Kuhn, remained *normative* (and progressivist).

Kuhn's anti-progressivist rhetoric had a more direct and overt influence on the work of a group of sociologists, mainly British, who in the early 1970s introduced a program they called "sociology of scientific knowledge" (SSK). SSK had eclectic roots—the philosophy of Wittgenstein and sociology, anthropology and perhaps Marxist historiography, among others. But it was the impact of Kuhn's *Structure* that proved critical. The Edinburgh formulation of SSK, termed the 'Strong Programme' by David Bloor and Barry Barnes, was grounded upon a number of methodological principles. Most fundamental was the prescription that science be studied like any other aspect of human culture, without regard to its truth value. Associated with this was a second prescription, the "symmetry postulate." Particularly important

[9]This, despite feelings of marginalization within the historical discipline. It should be pointed out that the historical discipline itself changed profoundly in the 1960s and 1970s and did so in directions that undercut its relationship to history of science and traditional intellectual history. The major changes were away from elitist and Eurocentric history.

[10]Koyré's approach had been very much in this vein and it was his careful attention to intellectual context in late sixteenth century thought to the work of Galileo and Descartes that made his work "foundational" for historians of science.

methodologically in the analysis of scientific controversy, this postulate prescribed that neither side of the controversy be assumed to have intrinsically superior scientific merit or epistemic privilege.

These prescriptions could be seen—and were seen by their formulators—as conclusions developed from Kuhn's perspective on scientific change. What was particularly influential in Kuhn was the view of science as involving the imprinting of "paradigms" on the student through scientific training. In SSK this view was transmuted into scientific "practices," carried on and passed on by scientific "subcultures," with no one having methodological or epistemic superiority. Moreover, the old talk of scientific "discovery" was to be replaced by talk of natural knowledge "construction." Historians of science increasingly saw their concerns as naturally linked to the project in SSK of offering normatively neutral contextual explanations of past and present scientific constructs.

If philosophers reacted negatively to Kuhn's anti-progressivist stance, they continued to be influenced by *Structure* in others ways. For example, since the 1970s there has been an increasing emphasis among philosophers of science on attention to the details of work in specific scientific disciplines. In this way, philosophers were taking seriously Kuhn's injunctions against focusing on abstract scientific frameworks that are disconnected from actual scientific practice. Moreover, there was a turn away from the positivistic procedure of using physics as a model for a unified account of science, and a concern to be sensitive to the fact that disciplines such as chemistry and biology, as well as the social sciences, proceed in ways that differ from physics.

One might think that these developments in philosophy of science would bring this field closer to a history of science that stresses the value of narratives concerning practices in particular sciences. However, there remain significant tensions between the disciplines, as indicated by the emphasis among proponents of the Strong Programme that their explanations in terms of social factors do not merely differ from but also are intended to replace the normatively-charged evaluations in philosophy of science. This sort of emphasis has prompted the suspicion among many philosophers of science that any history of science informed by the Strong Programme has little to contribute to their research.

Even so, there have been some signs that historians and philosophers of science have become dissatisfied with this mutual estrangement of the disciplines. There have been attempts to explore a rapprochement. For instance, a recent issue of *Isis* devoted a group of essays to this topic from prominent scholars.[11] Moreover, there has been a series of workshops—announced with the new acronym '&HPS'—intended to bring historians and philosophers of science together to discuss integrative strategies for studying science. These workshops are described as guided by the conviction that "good history and philosophy of science is not just history of science into which some philosophy of science may enter, or philosophy of science into which some history of science may enter," but rather "work that is both

[11] "Focus: Changing Directions in the History and Philosophy of Science," *Isis* 99 (2007): 88–134.

historical and philosophical at the same time."[12] Several chapters of this volume are in fact drawn from &HPS presentations (viz., those of Pietsch, Schickore, Rickles and Fagan). This is not yet to declare success in the attempt to fully integrate history and philosophy of science. In his contribution to this volume, for instance, David Miller claims that the programs of the &HPS meetings and other similar conferences generally reflect rather than bridge the disciplinary divide between intellectual and social studies of science. However, there is at least a serious attempt to explore whether this divide can be overcome.

To be sure, disciplinary differences, as well as the history of strife between history of science and philosophy of science, make integrated discussion difficult. However, these differences are to a considerable extent artifacts of our practices, and if the practices can change, then so too perhaps can the conception of the intellectual landscape. The question is whether work can be produced that changes practices in a way that makes disciplinary room for research that truly integrates history and philosophy of science. Perhaps the best way to address this question is in the terms in which Alasdair MacIntyre addressed the question of whether a work can be written that shows that the achievements of philosophy must be judged in terms of its history: "The only way to answer that question is by trying to write it and either failing or succeeding" (Rorty et al. 1984, 47).

We propose in this volume to explore the problems and prospects for integrating history and philosophy of science. The volume is divided into two main parts. The first part comprises six general reflections on the development, present state and future possibilities of an integrated HPS, whereas the second part includes six studies of the history and philosophy of particular scientific disciplines. Here the intent is to provide concrete cases for testing the mutual relevance of the history of science and the philosophy of science.

Jan Golinski opens the first part of this volume, on general reflections, with an historicized (and externalist) diagnosis of the problems for integrated history and philosophy of science. In particular, his proposal is that we are to understand the different and conflicting appropriations of Kuhn's work among historians and philosophers in light of the political ideology of campaigns against Nazism and Communism. The emphasis here is on the problems of overcoming the divisions that Kuhn's text has prompted between those who take epistemological considerations to be central to the study of science and those who want to bracket such considerations altogether.

As previously indicated, David Miller claims that there is currently a fundamental divide in "science history" between the "intellectual history of science," which focuses on scientific ideas and argument, and "social history of science," which focuses rather on the socio-cultural conditions of science. He argues that this disciplinary division is rooted in the historical fact that practitioners of science history have had to decide whether to rely for institutional support on philosophy or history,

[12]From the website for the meeting of &HPS3 at the University of Notre Dame, at http://www.nd.edu/~andhps/about.html.

and thus have been forced to choose between an intellectual or a social perspective on science. Drawing on this history of science history, Miller offers the conclusion that science history can be unified only if it manages to break free of the hold of its cognate disciplines.

Kenneth Caneva stresses the differences between philosophy of science and the history of science, but not in terms of Miller's intellectual/social dichotomy. Rather, Caneva's dichotomy is between the historians' "time-bound particulars" and the philosophers' abstract and universal "timeless truths." Moreover, he makes the prima facie paradoxical argument that it is only the historians' perspective that can provide "epistemic warrant" for universalist claims of scientific objectivity. Caneva ends with a critique of Kuhn's later attempt to free philosophy of science from an entanglement with the idiosyncratic details of the history of science.

In the fifth chapter, Ronald Giere offers his reflections on the 1973 review article in which he emphasized the notion of HPS as involving merely a "marriage of convenience." He notes that he was then concerned with a perceived dichotomy between the two fields, similar to Caneva's: the descriptive practice of history of science and the normative aspirations of philosophy of science. He has since come to hold that there is the possibility of a "naturalized" philosophy of science that draws normative conclusions from actual (and historical) scientific practice. But though this may seem to allow for a more intimate relationship between history of science and philosophy of science than a mere marriage of convenience, Giere himself cautions that even naturalized philosophy of science tends to be presentist in its orientation, and so not readily assimilated with any history of science that lacks such an orientation.

In the sixth chapter, Peter Dear emphasizes the role of SSK in fostering tensions between the history of science and the philosophy of science. However, he suggests that there are ways forward for HPS that focus on accounts of the history of philosophy of science and on the project of "epistemography," i.e., providing empirical accounts of knowledge-practices. Dear also argues that both philosophy (history of philosophy as well as philosophy of science) and history of science have a role to play in constructing a genealogy that yields a better understanding of current scientific practice.

In the final chapter of the first part, Wolfgang Pietsch attempts to illustrate the importance of an historical perspective in the philosophy of science by considering the thesis of the underdetermination of scientific theory. He contrasts the two versions of this thesis in the work of Pierre Duhem and W. V. O. Quine, and argues that the superiority of Duhem's version derives from the fact that it is drawn from the history of science rather than from an ahistorical consideration of issues in the philosophy of logic and language. In light of Miller's discussion, one question Pietsch's argument raises is whether philosophy of science needs only intellectual history of science, or whether a more social history of science has something to contribute as well. Moreover, there is room to wonder whether Pietsch's argument is for an integrated history and philosophy of science, or whether Giere's proposal for collaboration without assimilation would suffice.

The second part of this volume, on particular case studies of the possibility of an integrated history and philosophy of science, begins appropriately enough with

Hasok Chang's discussion of problems with the use of case studies. In particular, Chang proposes that one reason for the difficulty of integrating history and philosophy of science is that considerations of case studies have prompted either hasty philosophical generalizations that distort the history or local histories devoid of general philosophical import. In order to offer a way forward, Chang considers what two episodes in the history of science can teach us about one proper methodology for the use of case studies. The lesson of the first episode, from the history of temperature measurement, is that scientific progress involves a kind of "epistemic iteration," that is, a process of starting from imperfect assumptions and then using results to refine and correct those assumptions. The second episode, from the Chemical Revolution, teaches us that standard explanations of scientific progress in terms of simplicity or predictive success fail, and that a new sort of explanation is required. The alternative explanation Chang tentatively sketches involves the relative success and coherence of a set of practices that Lavoisier initiated. Chang takes the investigation of the two episodes to illustrate a procedure in which one uses puzzles that the episodes raise for an existing philosophical framework to develop a new framework that allows for a better historical understanding of the episodes.

Theodore Arabatzis investigates the prospects of an integrated history and philosophy of science by examining various cases of the scientific postulation of "hidden entities" from the eighteenth to the twentieth centuries. In effect, Arabatzis attempts a synthesis of the historical descriptive perspective, here deployed around the issue of how theoretical entities get established in scientific practice and become objects of scientific conviction, and the philosophical normative perspective regarding the reality of such entities. In Abaratzis's view, historical considerations tell against a philosophical argument for realism concerning hidden entities that emphasizes simply their manipulability in experimental situations, whereas philosophical considerations tell against defining such entities simply in terms of a particular systematic theory concerning their nature. Arabatzis thus proposes an investigative program that deploys these different perspectives interactively to hone both historical and philosophical understanding of scientific practice.

Jutta Schickore explores a seldom-studied aspect of scientific writing about experiments: methods accounts. These are scientists' own writings on information and arguments concerning experimental techniques and procedures. She examines in detail an historical case, the methods accounts of the nineteenth-century American neurologist S. Weir Mitchell on his experiments dealing with poisonous snake venom. Schickore argues that Mitchell's quite extensive writings on his experimental methods had an historical context in the hostility of many in the contemporaneous American medical community to medical experimentation, then coming into ascendancy in nineteenth-century Europe. Mitchell was particularly concerned with experimental repetition (as contrasted with replication by other experimenters) and with accounting for discrepancies that inevitably occurred between his repetitions.

Schickore's perspective on history and philosophy of science is sharply different from, say, Abaratzis's. Arguing that the fragmented and ever shifting natures of these fields preclude clear boundaries between them, Schickore is decidedly skeptical about the prospects of "combining history and philosophy." Rather, she appeals to a more general, multi-perspective term, "metascientific analysis", and utilizes ideas from Peter Galison, Harry Collins, Jim Bogen and Alan Franklin to frame and test her own analysis of Mitchell's experimental accounts.

Dean Rickles also notes that there are deep differences between the perspectives and projects of philosophers, historians and sociologists of science. The principal difference is that philosophers tend to analyze either textbook accounts of "the finished products of science" or at best make rational reconstructions of how these came to be, whereas historians and sociologists are concerned to investigate the actual diachronic genesis and development of science, in which contextual and contingent factors are at play. Rickles favors the latter approach but is also interested in integrating it with philosophical analysis.

He argues that a good candidate for integrating philosophy of science with history and sociology of science is found in the investigating the development of theories of quantum gravity during the first two thirds of the twentieth century. Theories of quantum gravity, themselves attempting to integrate the two fundamental yet disparate domains of modern physics, general relativity and quantum mechanics, are characterized by a number of interesting features. Firstly, there has been a variety of different approaches to constructing such theories, the principal ones stemming from the geometrical perspective of general relativity and from the perspective of quantum mechanics and, later, particle physics. Moreover, these theories are basically untestable; therefore, theory evaluation must rely on other considerations. In his own analysis, Rickles focuses on "constraints" (Galison), which, in this case, are necessarily theoretical. But they may also have historical and sociological dimensions; different traditions and research communities will have different criteria regarding proper constraints. Finally, there has as yet been no resolution to this development through emergence of a definitive theory of quantum gravity. Therefore, the study of its development is very much a study of "science in action" (Latour) rather than of "finished products of science."

The final two chapters focus on the more recent science of regenerative medicine, and in particular on stem cell research. In her contribution, Jane Maienschein is concerned to use historical study and philosophical analysis (a) to demonstrate the very complex and tangled attempts to investigate, understand and utilize cellular regeneration and (b) to clarify concepts and perspectives and highlight possible promising connections between them. In this way, she proposes that history and philosophy of science can make "regenerative medicine research better." Perhaps the most explicit illustration of this improvement emerges from her philosophical challenge to the traditional reductionist (and deterministic) genetic "programming" metaphor for stem cell transformation.

Whereas Maienschein is not especially concerned with the tensions between a descriptive history of science and a normative philosophy of science, such tensions are central to Melinda Fagan's discussion of stem cell research. Using a strategy that relates closely to that of Giere's naturalized philosophy of science, Fagan proposes a bridging of the descriptive/normative dichotomy by the deployment of what she terms a "social epistemology of scientific inquiry." Such an epistemology is grounded in complex and highly contingent social negotiations, but it avoids mere "descriptivism" insofar as it employs a general notion of instrumental rationality that concerns the "fit" of means to ends. Fagan attempts to illustrate the nature of this social epistemology by focusing on the particular case study of the search for the hematopoietic stem cell (HSC). She finds in the outcome of this episode two components of "success": an improved model (of the development and function of the immune system) and new interfaces with research disciplines previously working independently (immunology and hematology). The nature of the success will depend on contingent facts concerning the goals of the participants in a certain scientific practice and the nature of their social negotiations. However, Fagan insists that insofar as there is a shared goal and negotiated agreement among the scientists concerned, one can speak of there being "objective knowledge" in that case.

References

Gillispie, Charles C. 1960. *The Edge of Objectivity*. Princeton, NJ: Princeton University Press.

Gillispie, Charles C. 1999. "Apologia pro Vita Sua." *Isis* 90: S84–94.

Grau, Kevin T. 1999. "Force and Nature: The Department of History and Philosophy of Science at Indiana University, 1960–1998." *Isis* 90: S295–318.

Hall, A. Rupert. 1984. "Beginnings at Cambridge." *Isis* 75: 22–25.

Hanson, N.R. 1958. *Patterns of Discovery: An Inquiry into the Conception Foundations of Science*. Cambridge: Cambridge University Press.

Koyré, Alexandre. 1939. *Études galiléennes*, 3 vols. Paris: Hermann.

Kuhn, Thomas. 1962. *Structure of Scientific Revolutions*. Chicago, IL: University of Chicago Press.

Kuhn, Thomas. 1996. *The Structure of Scientific Revolutions*. 3rd edition. Chicago, IL: University of Chicago Press.

Rorty, Richard, Jerome B. Schneewind, and Quentin Skinner, eds. 1984. *Philosophy in History: Essays on the Historiography of Philosophy*. Cambridge: Cambridge University Press.

Rossiter, Margaret W. 1984. "The History and Philosophy of Science Program at the National Science Foundation." *Isis* 75: 95–104.

Toulmin, Stephen. 1961. *Foresight and Understanding: An Enquiry into the Aims of Science*. Bloomington, IN: Indiana University Press.

Zammito, John H. 2004. *A Nice Derangement of Epistemes: Post-Positivism in the Study of Science from Quine to Latour*. Chicago, IL: University of Chicago Press.

Part I
General Reflections

Chapter 2
Thomas Kuhn and Interdisciplinary Conversation: Why Historians and Philosophers of Science Stopped Talking to One Another

Jan Golinski

Paradigm was a perfectly good word until I messed it up
– Thomas S. Kuhn

Thomas Kuhn's work continues to reverberate through the fields of science studies more than a decade after his death. Nearly half a century since the appearance of his seminal work *The Structure of Scientific Revolutions* (Kuhn 1962/1970), his name still features in the titles and abstracts of papers by philosophers, sociologists, and historians. We don't have a comprehensive history of Kuhn's influence in the many academic disciplines that were touched by his work, though writings by Steve Fuller, John Zammito, and others have painted parts of the picture.[1] I am going to draw on this work to explore an aspect of Kuhn's legacy, namely his contribution to the process whereby the history and the philosophy of science have largely gone their separate ways in recent decades. As has been noted, historians and philosophers of science began to speak of their fields as "married" to one another in the 1960s, at just about the point when it began to appear that they were heading for a divorce.[2] The divorce has in fact been announced many times since, though the couple continues to cohabit in certain institutional locations. I see Kuhn's role in this process as a crucial but rather paradoxical one. On the one hand, *Structure* (as it is conventionally abbreviated) was an explicitly interdisciplinary work, which did in fact form a common reference point for historians, philosophers, and many others in the humanities and the sciences. Many cross-disciplinary conversations took it as their point of departure. On the other hand, however, scholars of different disciplines pursued its implications in quite distinct directions. In this respect, Kuhn's

[1] Fuller (2000) and Zammito (2004) together provide the nearest thing available to an intellectual history of Kuhn's work and its influence on science studies. Fuller also provides some valuable biographical information, and the autobiographical comments in "A Discussion with Thomas S, Kuhn," in Kuhn (2000, 255–323), are also valuable. An illuminating article, Andresen (1999), indicates the possibility for more work along these lines. Philosophical studies of Kuhn include: Bird (2000) and Hoyningen-Huene (1993).

[2] On the "marriage" and prospects for divorce, see Zammito (2004, 95–96).

J. Golinski (✉)
University of New Hampshire, Durham, NH, USA
e-mail: jan.golinski@unh.edu

S. Mauskopf, T. Schmaltz (eds.), *Integrating History and Philosophy of Science*,
Boston Studies in the Philosophy of Science 263, DOI 10.1007/978-94-007-1745-9_2,

book encouraged historians and philosophers of science to stop talking to one another, because it lent itself to quite divergent interpretations in the two disciplinary communities.

What makes it particularly interesting to consider Kuhn in this connection is that he had of course a great deal to say about discourse within—and across the boundaries between—disciplinary communities. His own analyses of how his work was read show him reflexively applying some of his leading ideas of incommensurability, paradigm-shifts, and so on. His theory of scientific change highlighted the importance of the extra-disciplinary interloper who often shook things up, a role Kuhn himself played. There were deep ironies here, and even a dimension of personal tragedy. Kuhn failed to win complete acceptance from the philosophers he wanted to appeal to, he found himself popular with postmodern cultural theorists and sociologists whom he despised, and he felt himself increasingly isolated from historians as they made a generational shift of the kind he should have been able to recognize from his own historical work. Nobody could blame Kuhn for feeling bewildered and perhaps rather resentful about this. From an outsider's point of view, however, it could be that Kuhn's own model of intellectual change does offer us tools for analyzing the reception of his work among historians and philosophers of science. I suggest that this is in fact so, provided we modify his model to acknowledge the importance of context in shaping the concerns of disciplinary communities. In other words, we can understand Kuhn's legacy if we adopt the framework of the sociologists of scientific knowledge who appropriated—and in his view distorted—Kuhn's ideas. Using this approach, we can see why *Structure* did not attain paradigmatic authority within a single discipline, but rather became a site of contested interpretations in various academic communities, its meaning pulled in different directions by readers who approached it with different interests.

My claim then is that the context in which *Structure* appeared led different groups of readers to emphasize different aspects of the text, hence amplifying the disparities between them. For philosophers, the most crucial issue was that of relativism, raised by Kuhn's suggestion that different paradigms were "incommensurable." Philosophers' preoccupation with relativism—and issues such as truth and rationality that were bound up with it—determined the interpretation of Kuhn in philosophical circles. Historians, on the other hand, were more concerned with the question of historical causality, specifically the matter of "internal" versus "external" causes of scientific change. The divergent responses of the two groups can be understood in terms of their different disciplinary interests, but they also shared a common intellectual context in debates at the time about the politics of scientific knowledge. Kuhn's work emerged from a Cold War context in which the whole notion of a social dimension of science was highly politicized.[3] This common

[3]The importance of the Cold War context in Kuhn's work is emphasized by Fuller (2000, esp. 1–37). Although I have learned a lot from Fuller, readers will be able to discern that I do not share his overall interpretation of Kuhn, which portrays him as an ideological foot soldier in a Cold-War campaign to shield science from critical examination. Instead, I see aspects of the ideological context of the times as having shaped the readings of Kuhn's work in sometimes contradictory and paradoxical ways, and often against his own inclination. The various contemporary readings

context informed the readings of Kuhn by both philosophers and historians, different though those readings were from one another. Such a situation has many historical precedents. Historians are familiar with other cases of rich and original texts that offered themselves to widely divergent interpretations in the intellectual contexts in which they were read. Consider Newton's "General Scholium," read by some in the eighteenth century as Anglican orthodoxy and by others as Arian heresy; or Darwin's *Origin of Species*, which some readers interpreted as theistic and others as atheistic in its implications for divine design; or Einstein's 1905 paper on special relativity, which seemed to some readers perfectly consistent with the idea of an electromagnetic ether and to others to have disposed conclusively of that notion. In each of these cases, the ambiguities of the text were amplified by its serving as the nexus of different interest groups, each of which was inclined to interpret it in accordance with its own aims.[4]

This was also the situation with Kuhn's *Structure*. Among the many ideas that the text offered to its different constituencies of readers, the one that particularly stirred philosophers was the supposed "incommensurability" between paradigms. This raised the specter of relativism, which had been associated since the 1930s with totalitarianism of the fascist or communist varieties. Kuhn was accused of reviving the threat of science's subjection to political domination or irrational "mob psychology." For Karl Popper especially—the presiding genius of postwar philosophy of science—this was anathema. Popper emerged in the postwar period as a leading anti-Marxist philosopher; he was particularly concerned by the threat posed by totalitarianism to the freedom of scientific inquiry. This made his ideas popular in the climate of Cold War thought, in which scientific individualism was prized as an aspect of Western freedom.[5] In this climate, relativism was seen as a tool of totalitarianism, and Popper's antipathy to totalitarian regimes sharpened his response to what he perceived as Kuhn's dangerous flirtation with relativism. As David Hollinger explains, "Popper spoke for a generation that saw the world almost lost to a regime that distinguished Aryan physics from Jewish physics, and almost saved by a regime that distinguished proletarian science from bourgeois science" (Hollinger 1995, 452). To Popper, Kuhn's doctrine of incommensurability was an instance of "the myth of the framework," which held that beliefs could not be criticized from a standpoint outside the framework within which they occurred.

of Kuhn's *Structure* reflect the multiple interests of its readers, and do not require us to suppose, as Fuller does, that Kuhn's work was largely vacuous and unworthy of serious attention. Nor do I share Fuller's view that Kuhn's influence constituted a malign force shaping the whole field of science studies. As I shall indicate, I see Kuhn as having little positive influence on philosophers and almost none (directly) on historians. His most significant influence within science studies was mediated by sociologists, whose reading of his work he specifically repudiated.

[4]On Newton, see Stewart (1996), Snobelen (2001). On Darwin, see Young (1985), Moore (1979). On Einstein, see Warwick (1992, 1993).

[5]The intellectual historian Peter Novick has written: "one cannot reduce Popper's philosophy to his struggle against Marxism, but that concern permeated all his work"—including, I would suggest, his response to Kuhn (Novick 1998, 298). See also Hacohen (2000, 530–34).

This he denounced as, "in our time, the central bulwark of irrationalism" (Popper 1970, 56). To reduce scientific theories to merely one cultural framework among others was to "psychologize" or "sociologize" them, in Popper's view, thereby surrendering the freedom of objective scientific reason. Kuhn's vision of the history of science as a succession of incommensurable paradigms threatened to insinuate the ominous Marxist doctrine that scientific beliefs were products of fundamental social or political interests.

Popper's criticisms of Kuhn dominated the conference organized by his colleague Imre Lakatos in London in 1965. Lakatos had quite a different relationship to Marxism than Popper's, as a recent book by John Kadvany has shown (Kadvany 2001). In fact, Lakatos's debt to Hegel's thought—mediated by Hungarian Marxist philosophy—made him much more willing than Popper to draw philosophical implications from the history of science. Lakatos was quite receptive to the historicist character of Kuhn's work, and went on to incorporate elements of a historical perspective in his own methodology of scientific research programs. He recognized the justice of Kuhn's claim, in the opening lines of *Structure*, that the history of science posed a fundamental challenge to philosophical accounts of scientific rationality. However, it is also telling that Lakatos vehemently rejected Kuhn's supposed relativism, which he claimed reduced scientific debate to the level of "mob psychology." He labeled as "irrationalism" the claim that paradigms could not be compared from an independent standpoint. Lakatos charged that Kuhn had endorsed, albeit unwittingly, the "*credo* of contemporary religious maniacs ('student revolutionaries')" (Lakatos 1970, 178, 93).

A similar concern with the threat of relativism was manifested at the second major conference on philosophical implications of Kuhn's work, at Urbana, Illinois in 1969. That conference, and the volume that emerged from it, consolidated a shift of the debate onto the grounds of the semantics of scientific language. Kuhn was interpreted, as Frederick Suppe explained, as a philosopher of worldviews or *Weltanschauungen* (Suppe 1977, 135–51). His suggestion that the different worldviews arising in the course of scientific development were incommensurable was taken to mean that terms used in one framework could not be fully translated into those used in another. Several philosophers took up this question of the meanings of the language used in different theoretical frameworks and the possibility of accurate translation between them. As Zammito explains, "the philosophical reception shifted everything Kuhn had been saying into the key of philosophy of language" (Zammito 2004, 66). It was in these terms that the issue of relativism was couched in the subsequent debate between Kuhn and his philosophical critics.

The underlying anxiety about the dangers of relativism had deep roots in the philosophical tradition; it had arisen in the twentieth century in connection with arguments about history and anthropology, as well as those about science.[6] Popper had earlier attacked its manifestation in the Copenhagen Interpretation of quantum

---

[6]On the previous history of arguments over relativism, see Herbert (2001), Smith (2006, 18–45), Novick (1988, 133–67, 281–319).

mechanics, targeting especially Niels Bohr's supposedly "subjectivist" account of the reality of physical phenomena. But, in the Cold War era, relativism was particularly associated with political totalitarianism, paradoxical though that connection might seem. The link began to be made in the 1930s, when many liberal intellectuals accused political extremists of the Left and the Right of employing a relativist or subjectivist model of knowledge. Liberal observers charged that extremist regimes had denigrated the autonomy of science in order to insinuate their own ideologically biased theories, such as Nazi racial eugenics or Soviet Lysenkoism. In 1938, the sociologist Robert K. Merton wrote that, "totalitarian theorists have adopted the radical relativistic doctrines of *Wissenssoziologie* as a political expedient for discrediting 'liberal' or 'bourgeois' or 'non-Aryan' science."[7] In Merton's view, the work of the Hungarian sociologist of knowledge Karl Mannheim had lent itself to this totalitarian appropriation. The suggestion was often reiterated in the years after World War II, in part under the influence of Popper's widely-read work, *The Open Society and Its Enemies* (1945), which traced the totalitarian tendencies of Marxist thought to its ancient roots in the philosophy of Plato. Popper criticized Marx for his "*historical relativism* in the field of ethics," the claim that moral standards were not absolute but dependent on historical circumstances. For Marx, the inevitability of the coming socialist revolution provided the only foundations necessary for moral judgment. Similarly, according to Popper, epistemological relativism in the field of sociology of knowledge was licensed by Marxist claims that future historical change could be predicted.[8] George Orwell was one of those who seized upon the link between authoritarian politics and the deployment of moral and epistemological relativism. In 1946, he attacked the Marxist crystallographer J. D. Bernal for making moral judgments on the sole basis of loyalty to Stalin's Soviet Union, so that "we must alter our conception of right and wrong from year to year, and if necessary from minute to minute" (Orwell 1970, 4:186).[9] In Orwell's novel *Nineteen Eighty-Four* (1949), the totalitarian regime displays its relativist attitude by shamelessly manipulating historical and other facts without regard for their truth. The tendency of Communist governments to suborn objective science to political ends continued to be widely commented upon in the postwar debate over Lysenkoism. Western observers during the Cold War years noted that relativism was a convenient doctrine for totalitarian regimes that aimed to subjugate individual scientists to specific social or political priorities.

This context appears to have shaped philosophical criticisms of Kuhn by Popper and his followers. Kuhn's suggestion that different paradigms could be incommensurable with one another roused anxieties about relativism that had been heightened by the political disputes of the preceding decades. In this situation, the whole issue

---

[7]Merton, "Science and the Social Order" (1938), quoted in Novick (1988, 289). On Merton's interpretation of Mannheim, see Kaiser (1998).

[8]Popper (1945, 2:187–211). On Popper's work in this connection, see Hacohen (2000, 512–13), Novick (1988, 298–99), Anderson (1992, 60–65, 71–72).

[9]On Bernal, see Brown (2005).

of the social dimension of science was philosophically and politically contentious. As Hollinger has shown, even the notion that science was the product of a community was resisted in the United States from the 1930s through the 1950s, because it was associated with Marxist and socialist demands for planning of scientific research (Hollinger 1990). Leading American advocates of public investment in science, such as Vannevar Bush (vice-president of MIT and author of *Science—The Endless Frontier* [1945]), argued that the best rate of progress would follow from individual scientists being given the maximum freedom to pursue their research. After World War II, the freedom of science—understood as the entitlement of individual researchers to support without governmental or ideological direction—was upheld as part of anti-Communist propaganda. The theme was articulated, for example, by the Congress for Cultural Freedom, which was supported with funds from the CIA. In July 1953, the Congress sponsored a conference on freedom in science in Hamburg, with concluding remarks delivered by the Hungarian physical chemist Michael Polanyi. Polanyi, who was emerging as a leading public philosopher and anti-Marxist intellectual at this time, stressed that science must be allowed to flourish without any sort of planning or managerial direction.[10] Kuhn in fact incorporated some of Polanyi's points about scientists' tacit skills that could not be managed by outsiders. That being so, it no doubt surprised Kuhn to find himself criticized by philosophers for opening the door to relativism and "mob psychology." It was one thing, apparently, to argue that politicians and bureaucrats could not understand science well enough to direct it; it was quite another to propose that different groups of scientists could sometimes find themselves unable to understand one another. The latter claim aroused the ire of the philosophers, much to Kuhn's chagrin. He spent many subsequent years refining and qualifying his doctrine of incommensurability to try to address their concerns.[11] Toward the end of his life, he voiced his revulsion at prevailing relativistic attitudes that had appropriated and (in his view) misread his work.[12] The image of Kuhn in his last years was of an unwitting pioneer of postmodernism, a reluctant prophet embarrassed by the honors paid to him by the unashamed relativists who followed in his wake.

Among historians, Kuhn's work was received in the context of a rather differently configured debate. The central questions concerned historical causation, specifically

[10]Polanyi had criticized Bernal and other Marxists since the late 1930s for subordinating science to social ends and denying its autonomy as a search for truth. See Scott and Moleski (2005, 174–75). For Polanyi's influence on Kuhn, see Fuller (2000, 139–49).

[11]Kuhn's most important statement on the topic after *Structure* was "Commensurability, Comparability, Communicability" (1983), in Kuhn (2000, 33–57). See also Zammito (2004, 52–89), Hoyningen-Huene (1993, 206–22). Zammito notes that Kuhn's acceptance that the problem of incommensurability was basically a linguistic issue, as asserted by such philosophers as Dudley Shapere and Israel Sheffler, was ultimately self-frustrating. It is hard to avoid the reflection that the debate that resulted was an instance of sustained mutual incomprehension, if not of ultimate linguistic incommensurability, between Kuhn and his philosophical interlocutors. His continued struggles to resolve the issues at stake suggest, however, that Kuhn was reluctant to acknowledge this.

[12]See especially Kuhn's lecture at Harvard University in 1991: "The Trouble with Historical Philosophy of Science," in Kuhn (2000, 105–20).

whether scientific developments were understood to be influenced only by factors in the intellectual realm or whether broader social forces had an effect. This also was a highly politicized question in the period from the 1930s to the 1960s because of the challenge posed by Marxist theories in which the intellectual superstructure was determined by the economic base of society. Historians of science had become aware of the Marxist approach to their subject since the appearance of a Soviet delegation, including Boris Hessen and Nikolai Bukharin, at an international conference in London in 1931.[13] The paper Hessen delivered on that occasion, which purported to expose the "social and economic roots" of Newton's *Principia*, became a touchstone of the Marxist outlook, hailed by some as a pioneering effort in the new social history of science while being roundly denounced by many established scholars. On this issue, Kuhn trod a careful line that largely circumvented criticism from the historians who were hostile to Hessen's approach. He took the precaution of specifically discounting "external" social influences on the content of science. Such influences, he remarked, could not outweigh the technical factors that led to crisis and subsequent revolution in a paradigm. In the "mature" sciences, external factors had only a very marginal influence, and, Kuhn declared, "issues of that sort are out of bounds for this essay" (Kuhn 1962/1970, 69).

Kuhn's use of the vocabulary of "internal" and "external" factors relied upon an already longstanding tradition in history and sociology of science, which has been surveyed by Steven Shapin (Shapin 1992). The terminology seems to have been introduced by Merton, who was working on the sociology of science since the 1930s and well aware of its political implications. Merton shared the general antipathy to relativism among Western liberals, linking it (as I have mentioned) directly to totalitarianism. But he also found a way to conceptualize the social dimension of science while saving it from the taint of Marxism. He did this by articulating the distinction between internal and external factors in scientific development. External factors were allowed to influence the rate of progress but not its direction or the content of scientific ideas. In 1942, Merton proposed his famous four norms, supposedly the overarching values prevalent in the scientific community: universalism, communism, disinterestedness, and organized skepticism. The suggestion drew upon ideas about the scientific "code" or "ethos" that had been circulating in anti-fascist polemics from the late 1930s.[14] Although one of these norms was initially labeled "communism" (with a small *c*)—later diplomatically renamed "communalism"—his proposal kept at bay the suggestion, by Marxists such as Hessen and Bernal, that scientific ideas were determined by social or economic conditions. Merton frequently emphasized that totalitarian societies did not nurture the norms that ensured the growth of science. But external factors were allowed to be influential only to a limited extent: ethical or religious values could encourage

---

[13]On the London conference and its impact, see Werskey (1988, 138–49), Mayer (2004).

[14]Robert K. Merton, "The Normative Structure of Science" (1942), in Merton (1973, 267–78). See also Mendelsohn (1989), Hollinger (1983). Peter Novick has noted that Merton's original explanation of the coincidence between scientific norms and the values of a democratic society specifically criticized only Nazi Germany. When the exposition was revised in 1949, criticisms of the Soviet Union were added. See Novick (1988, 296–97 fn. 28).

or discourage the growth of scientific knowledge along its predetermined path, but could not affect the content of scientific theories. To admit that would be to indulge the outlook that Merton saw as characteristic of totalitarianism.

Merton's model of science was reflected in Kuhn's *Structure*, in the image of a scientific community that was singularly autonomous from more general social influences. In a "Postscript" added to the second edition of his book in 1969, Kuhn attempted to resolve some of the ambiguities of his core notion of paradigm. He differentiated what he called "philosophical" from "sociological" meanings of the term, distinguishing the exemplary achievements that guided scientific work from the communities whose property they were. Merton and his followers, he said, had catalogued the sociological markers of scientific identity, but these were quite independent of the actual scientific knowledge produced by members of the community. Science was necessarily an autonomous social enterprise, and its proper history was thus largely an internal one.[15] This conception reflected the influences that shaped Kuhn's own training in the history of science, including those of James B. Conant, the Harvard president who invited him to teach the subject, of Alexandre Koyré, the Russian émigré whose antipathy to Marxism was expressed in his foundational work in the discipline in the 1940s, and of the philosophically-minded historians whose work Kuhn read, including Hélène Metzger and Émile Meyerson.[16] From his wide and largely unguided reading, Kuhn derived his model of science as developing under the auspices of a series of worldviews or *Weltanschauungen*. He grafted onto this a model of the social dimension of science, largely derived from Merton, in which scientific communities were believed to be substantially insulated from the forces in society at large.

In excluding external factors, Kuhn's model was broadly consistent with the prevailing conception of the history of science in the Cold War years, especially in the United States and Britain. In the US, an indicator of the prevalence of internalism was Charles Gillispie's influential survey, *The Edge of Objectivity* (1960).[17] Internalism in the history of science went along with an emphasis on individualism in contemporary science, in what Hollinger calls the *Kulturkämpfe* of the postwar period.[18] Gillispie, for example, criticized the interpretation of Chinese science by the historian Joseph Needham, which tied its historical development to the

---

[15] Kuhn (1962/1970, 174–210). See also Kuhn (2000, 286–87), where Kuhn records that he saw his own work as "pretty straight internalist" and was surprised when Alexandre Koyré congratulated him for having brought together internalist and externalist approaches.

[16] On Conant, see Fuller (2000, 150–226), Kuhn (2000, 275–76, 282–89). See also Fuller (2000, 60–70 (on Koyré), 392–97 (on Meyerson)), Chimisso (2001), Dennis (2003), Porter (1986).

[17] Introducing a new edition of his text in 1990, Gillispie acknowledged that the discipline of history of science had shifted in the intervening years to embrace a more externalist outlook, a shift that he ascribed to the leftist cultural movements of the 1960s and 1970s. See Gillispie (1990, xiv–xv, xvi–xvii, xx–xxi).

[18] Hollinger (1995). See also Dennis (2003), who points out that Henry Guerlac at Cornell represented an alternative approach, open to consideration of external influences on scientific development and anticipating some of what followed after Kuhn.

prevailing social conditions. To Gillipsie, Needham's Marxist sympathies had led him to "an abject betrayal of the autonomy of science, and a surrender of the measure of independence which it has won for scholarship and thought" (Gillispie 1957, 176). Likewise, Conant, discussing the contemporary situation, hailed scientists as exemplars of the value of individual freedom and said nothing about inequalities of resources or authority within their communities. A similar outlook prevailed among historians of science in Britain, where Anna Mayer has shown how, in the late 1940s, "Anti-Marxism formed a defining feature of the process by which the image of scientific work as a disinterested journey of the mind came to be institutionalized" (Mayer 2000, 41). In Cambridge, for example, under Herbert Butterfield's leadership, the social vision associated with Hessen and Bernal before the war, and represented locally by Needham, was excluded from the new Department of History and Philosophy of Science.

In this respect, Kuhn's synthesis did not challenge historians' prejudices, but it did present them with a rather indigestible mix of ideas. His general picture of scientific development was not resisted by historians as fiercely as it was by philosophers, but his scheme of historical periodization found very few disciples. To use Kuhn's own terms of analysis, one could say that at least some philosophers were provoked to a crisis by the revolutionary implications of his work, but that historians who might have been looking for a new research paradigm failed to find one. Having sketched a very large-scale narrative of the development of science, he failed to convert scholars who were generally moving toward smaller-scale historical studies. They were familiar with the idea that revolutionary change sometimes occurred in the sciences, but "*The* Scientific Revolution" of the sixteenth and seventeenth centuries—which Koyré had helped establish as a foundation stone of the discipline—did not seem to fit the Kuhnian model.[19] Notwithstanding the pedagogical roots of Kuhn's own historiographical awakening in his teaching for the General Education program at Harvard, no one attempted to write a Kuhnian textbook of the history of science. It probably did not help his standing among historians that Kuhn largely turned his attention to interactions with philosophers for the remainder of his professional career. Toward the end of his life, he said that he had never really seen himself as a historian of science, notwithstanding his distinguished monographs on Copernicanism and on the early history of quantum physics (Kuhn 2000, 276).

There was, nonetheless, much that was historiographically innovative in Kuhn's *Structure*. Most readers understood that he had presented scientists not as autonomous individuals but as subject to what M. D. King called "a system of traditional authority" (King 1980, 103). This opened the door to consideration of the operations of power within scientific communities, a potential realized in the work of the Edinburgh school in the 1970s and early 1980s. The philosopher David Bloor and the sociologist Barry Barnes articulated a new approach to the sociology of science on the basis of Kuhn's work, and this in turn proved influential among historians. Bloor and Barnes picked up particularly on the Wittgensteinian elements

[19] Porter (1986).

in Kuhn's account of paradigms. Kuhn had talked of paradigms as irreducible to sets of propositions or rules, crediting the point to a suggestion by the philosopher Stanley Cavell, his colleague in the Harvard Society of Fellows and then at Berkeley.[20] Paradigms were envisioned as exemplary problem solutions, models for ways of looking at the world that did not contain specific instructions for how they should be applied. This made them consistent with Wittgenstein's understanding of language, in which words do not have their meanings inherent within them. As the meaning of language is determined only by its use in particular settings, so the application of paradigms is worked out only in the course of the process Kuhn had called "normal science." To the Edinburgh school, it was Kuhn's notion of normal science that seemed to be the revolutionary feature of his work, rather than the theory of revolutions as such. Normal science was thought to be governed by models of practice in which theoretical concepts, methods, techniques, and so on, were implicit but not fully explicated. The models were applied to new situations not by a process of logical deduction valid in all times and places but by a kind of judgment that was tied to local circumstances and what Wittgenstein called a "form of life." Science thus appeared less as the paragon expression of human reason, and more as a form of culture comparable to many others.[21]

The Edinburgh school's recovery of the Wittgensteinian aspects of Kuhn had a further payoff. Viewing paradigms along the lines of Wittgenstein's "language-games" or forms of life suggested that the relevant social units of science were relatively small groups that shared a particular form of practice. This offered a way to conceive of the social dimension of science without invoking the notorious "external" factors. Bloor and Barnes were intrigued by Kuhn's interest in the groups that formed around a particular experiment or instrument, for example the eighteenth-century "electricians" who explored the phenomena of the Leiden Jar.[22] Such groups would be expected to be narrower than scientific disciplines, and certainly narrower than the population of all professional scientists studied by Merton and his followers. Kuhn was suggesting the importance of studying smaller-scale groups whose social identity was bound up with allegiance to a specific model of practice. These subcultures were not to be defined in the traditional "externalist" terms of scientific institutions or disciplines, but in terms of their particular way of doing science. With this kind of analysis, the members of the Edinburgh school hoped to show that social relations penetrate to the very core of scientific practice.

Bloor and Barnes presented themselves as building a new kind of sociology of scientific knowledge on the foundations supplied by Kuhn. Their interpretation of Kuhn reflected the considerable attention accorded to Wittgenstein in British academic philosophy, sociology, and anthropology at the time. Under the philosopher's posthumous influence, the social sciences were being encouraged to free themselves

---

[20]On Cavell, see Kuhn (2000, 297).

[21]Barnes (1982, 1985), Bloor (1983, 1991).

[22]Kuhn (1962/1970, 18–19). See also Barnes (1982, 120–26).

from the dominance of the natural sciences and configure themselves as independent interpretive disciplines.[23] In the work of the Edinburgh school, sociology took on the ambitious agenda of studying science itself; and, although Kuhn himself found the Edinburgh appropriation of his work distasteful, it issued in important work by sociologists and subsequently by historians. The focus on scientific subcultures paid off in the controversy studies, initiated by the sociologist Harry Collins in the mid 1970s, and triumphantly applied to historical episodes by Steven Shapin and Simon Schaffer in *Leviathan and the Air-Pump* (1985), and by Martin Rudwick in *The Great Devonian Controversy* (also 1985). These studies exploited Kuhn's Wittgensteinian insight that, in disputes between paradigms, "incompatible modes of community life" were at stake.[24] The controversy studies showed that fundamental, usually implicit, values and assumptions come to the surface in prolonged disputes. Participants in controversies were found to articulate normally unspoken stipulations about method, about the expertise or reliability of their colleagues, about the propriety of their ways of collaborating or communicating, and so on. Rather than simply about facts, these disputes seemed to be about how research should be organized, or in general how science should be carried on. Studies of controversies provided a key to unlock the forms of life that constitute scientific practice. Thus, the response by sociologists and historians to the incommensurability between paradigms that Kuhn had highlighted was not to shun it, but to embrace it as an instrument that could lay bare the social dimension of science.

In order to do this, the analyst of scientific controversies had to maintain a stance of strict neutrality between the disputing parties. Here again, the Edinburgh school led the way methodologically. They did not shy away from relativism; in fact, they enshrined it as a principle of the field that became known as the Sociology of Scientific Knowledge. The stone that the philosophers had rejected became the keystone of a new approach. In Bloor's articulation of what he called the "Strong Programme," relativism was axiomatic; it was adopted on pragmatic or utilitarian grounds rather than established by philosophical argument.[25] Elsewhere, Barnes and Bloor ridiculed the notion that relativism opened the way to totalitarianism, claiming this was a prejudice of the political Right and what they scornfully labeled "the Cult of Rationalism."[26] Relativism was actually a strategic necessity for the social study of science, because all claims to knowledge had to be treated as equally in need of explanation in essentially the same sociological terms. Scientific belief was to be explained in the same manner as all other beliefs, with philosophical evaluations of its rationality or validity having no part to play. The Sociology of Scientific Knowledge, to which the Strong Programme gave rise, accepted relativism in this axiomatic role. To a large degree, this stopped the conversation with philosophers,

[23] Shapin (1995, esp. 294–98), Zammito (2004, 124–26).

[24] Collins's case-studies of scientific controversies were later collected in Collins (1985). See also Shapin and Schaffer (1985), Rudwick (1985), and Kuhn (1962/1970, quote on 94).

[25] Bloor (1991, 3–23, 157–61, 175–79).

[26] Barnes and Bloor (1982, 47).

who could not set aside evaluative questions so casually. But it did so in a way that was enabling for historians and sociologists, who could go about their business without being diverted by epistemological considerations.

This, then, was how Kuhn contributed to a split between historians and philosophers of science that has deepened in recent decades. It is not surprising that he reflected, toward the end of his life, that there seemed to be no single field of history and philosophy of science, but rather two fields with their own distinctive interests.[27] For Kuhn himself, a historical model that emphasized the insularity of scientific subcultures and the discontinuities in their experience over time raised a set of philosophical problems that he found profoundly engaging. Most historians, however, accepted certain features of the model without troubling about the philosophical implications. For them, Kuhn's significance lay in pointing the way to a more comprehensive historicization of scientific practice, the bringing under scrutiny of a range of topics that had not previously been subject to historical research. These included the social mechanics of disciplinary communities and institutions, the roles of instrumentation and laboratories, the functions of rhetoric in scientific texts, and so on. Underlying this program of study was a stance of pragmatic relativism. Historians simply accepted as scientific knowledge whatever appeared as such in the context under investigation; they declined to demarcate or prioritize according to predetermined criteria of what made for good or bad science. This move, made axiomatic in Bloor's "symmetry postulate," opened the way for a rich field of empirical investigation.[28]

For sociologists and historians, this seemed like a liberation, but for philosophers it left many crucial questions unanswered. Philosophers had interpreted Kuhn's work in an entirely different way and therefore had different ideas about what it led to. Looking back, they took quite another view of the developments that had followed Kuhn's appearance on the scene. John Zammito's *A Nice Derangement of Epistemes* (2004), reflects the outlook of those who see a great deal of unfinished philosophical business surrounding Kuhn's work and its reception. He notes that Bloor and Barnes continued to mount philosophical defenses of their relativist outlook, acting in effect as philosophers even as they claimed to have transcended philosophy in favor of a new kind of sociology of science.[29] Bloor proclaimed that sociological investigations were now "the heirs of the subject that used to be called philosophy"; but doing so did not free him from prolonged entanglement in disputes with philosophers (Bloor 1983, 182–84). Noting this, Michael Friedman charges the proponents of the Strong Programme with a cavalier attitude to philosophy, a pretense that deep philosophical issues can be tossed aside by appeal to empirical research and a superficial relativist stance. Friedman claims that Bloor misinterpreted Wittgenstein, as he misinterpreted Kuhn, to provide warrant for this approach. In Friedman's view, the Strong Programme's ambition to

[27] Kuhn (2000, 315–16).

[28] Bloor (1991, 7, 175–79).

[29] Zammito (2004, 137–50).

replace philosophy with a successor discipline can only be seen as a peculiar kind of philosophical hubris (Friedman 1998).

Zammito and Friedman signal the possibility of quite different interpretations of Kuhn's work than those that prevailed in the late 1960s and 1970s. Kuhn might with some justice assert that Bloor and Barnes had misunderstood him, as indeed Wittgenstein or W. V. Quine might claim that their fundamental philosophical arguments had been misinterpreted by the Edinburgh school. Such correctives can be salutary if they point to overlooked resources in texts that have generally been read in just one way. Perhaps a different reading of Kuhn could suggest possible grounds on which historians and philosophers of science might re-engage with one another? Philosophers are entitled to point out that the historians' stance of pragmatic relativism constitutes an implicit philosophical position that has often been embraced without sufficient justification. On the other hand, however, whatever scope there is for re-engagement between historians and philosophers of science will have to acknowledge the fundamental movement toward historicization that Kuhn introduced into science studies. This movement has now acquired its own historical inertia. There is no way back to the time before Kuhn's intervention. His sonorous words at the opening of *Structure* still hold good: taking history seriously challenges all preconceived ideas of the nature of science. It calls into question any assertions that presuppose the unity of science across time and space or its singularity as a cultural phenomenon. The move cannot be undone, even if the arguments for it do not have quite the force of logical compulsion. Future conversations between historians and philosophers will have to continue to be built on the realization that science is fundamentally a historical phenomenon.

That being so, it seems to me that judgments that Bloor and Barnes misinterpreted Kuhn or Wittgenstein are largely beside the point. The whole history of Kuhn's reception could be said to be a history of misinterpretations, albeit highly consequential ones. And this itself indicates some of the limitations of the historiographical concepts he bequeathed us. *Structure* did not become a paradigmatic text, or at least not in the way Kuhn wished. It did provoke a crisis among some philosophers, who were obliged to confront the implications of historicism for their models of scientific rationality, as Lakatos for example tried to do. Other philosophers kept their distance from historical studies and continue to do so, paying little attention to Kuhn or anyone who followed in his wake. Among historians, the text provided a welcome degree of legitimacy for their enterprise but no acceptable scheme of periodization. The book was too schematic, too oriented toward making a general philosophical argument, to provide a model of historical practice. Historians were in any case tending toward smaller-scale case-studies, away from the sweeping historical panoramas that had previously been demanded for pedagogical purposes. Moving in this direction, they learned some significant lessons from the Edinburgh reading of Kuhn. The sociologists did not accept *Structure* as a paradigmatic text any more than philosophers or historians did, but they creatively appropriated it in line with their own preoccupations and against the author's wishes. The situation can be understood in the terms that the Edinburgh school itself introduced in their somewhat strained interpretation of Kuhn. A text does not establish a paradigm by its own self-conferred authority; rather it is ascribed authority insofar as it can be read (or misread) to accord with the interests of a community. Thus, factors that Kuhn himself would have been inclined to label "external" impinge upon the interpretation of a text. Even an intellectual community that has acquired its own paradigm cannot be insulated from broader social forces.

For Kuhn personally, the lesson of the limits of his own historiographical paradigm was a hard one to learn. In his last recorded reflections on his career, a series of interviews conducted in Athens in October 1995, a few months before his death, he was still trying to control the interpretation of his work.[30] I think of these interviews as "Kuhn's Last Tape," in which, like Krapp in Samuel Beckett's play, he continually edits the tape of his own life, periodically rewinding to insert a clarification he should have made earlier and fast-forwarding to put down a marker for something he wants to say later. The aim throughout is to get everything in order and say it all clearly, so that nothing can be misunderstood. It is a Sisyphean task, of course, and one heavily laden with pathos, since the reader knows that this was Kuhn's last public pronouncement. It compounds the sense of tragedy attached to Kuhn's career, notwithstanding the brilliance of his intellectual accomplishment, that his last work saw him trying in vain to assert his authority over a text that had long since escaped his attempts to stipulate its meaning.

References

Andresen, Jensine. 1999. Crisis and Kuhn. *Isis* 90: S43–67.
Anderson, Perry. 1992. *English Questions*. London: Verso.
Barnes, Barry. 1982. *T. S. Kuhn and Social Science*. London: Macmillan.
Barnes, Barry. 1985. "Thomas Kuhn." In *The Return of Grand Theory in the Human Sciences*, edited by Quentin Skinner, 83–100. Cambridge: Cambridge University Press.
Barnes, Barry, and David Bloor. 1982. "Relativism, Rationalism and the Sociology of Knowledge." In *Rationality and Relativism*, edited by Martin Hollis and Steven Lukes, 21–47. Oxford: Basil Blackwell.
Bird, Alexander. 2000. *Thomas Kuhn*. Princeton, NJ: Princeton University Press.
Bloor, David. 1983. *Wittgenstein: A Social Theory of Knowledge*. London: Macmillan.
Bloor, David. 1991. *Knowledge and Social Imagery*. 2nd edition. Chicago, IL: University of Chicago Press.
Brown, Andrew. 2005. *J. D. Bernal: The Sage of Science*. Oxford: Oxford University Press.
Chimisso, Cristina. 2001. "Hélène Metzger: The History of Science Between the Study of Mentalities and Total History." *Studies in History and Philosophy of Science* 32: 203–41.
Collins, H.M. 1985. *Changing Order: Replication and Induction in Scientific Practice*. Beverly Hills and London: Sage Publications.
Dennis, Michael Aaron. 2003. "Historiography of Science: An American Perspective." In *Companion to Science in the Twentieth Century*, edited by John Krige and Dominique Pestre, 1–26. London: Routledge.
Friedman, Michael. 1998. "On the Sociology of Scientific Knowledge and Its Philosophical Agenda." *Studies in History and Philosophy of Science* 29: 239–71.

[30] Kuhn (2000, 255–323).

Fuller, Steve. 2000. *Thomas Kuhn: A Philosophical History for Our Times*. Chicago, IL: University of Chicago Press.

Gillispie, Charles C. 1957. "[Review of Joseph Needham, *Science and Civilization in China*, vol. 2]." *American Scientist* 45: 169–76.

Gillispie, Charles C. 1990. *The Edge of Objectivity: An Essay in the History of Scientific Ideas*. 2nd edition. Princeton, NJ: Princeton University Press.

Hacohen, Malachi Hai. 2000. *Karl Popper: The Formative Years, 1902–1945*. Cambridge: Cambridge University Press.

Herbert, Christopher. 2001. *Victorian Relativity: Radical Thought and Scientific Discovery*. Chicago, IL: University of Chicago Press.

Hollinger, David A. 1983. "The Defense of Democracy and Robert K. Merton's Formulation of the Scientific Ethos." *Knowledge and Society: Studies in the Sociology of Culture Past and Present* 4: 1–15.

Hollinger, David A. 1990. "Free Enterprise and Free Inquiry: The Emergence of Laissez-Faire Communitarianism in the Ideology of Science in the United States." *New Literary History* 21: 897–919.

Hollinger, David A. 1995. "Science as a Weapon in *Kulturkämpfe* in the United States During and After World War II." *Isis* 86: 440–54.

Hoyningen-Huene, Paul. 1993. *Reconstructing Scientific Revolutions: Thomas S. Kuhn's Philosophy of Science*. Chicago, IL: University of Chicago Press.

Kadvany, John. 2001. *Imre Lakatos and the Guises of Reason*. Durham, NC: Duke University Press.

Kaiser, David. 1998. "A Mannheim for all Seasons: Bloor, Merton, and the Roots of the Sociology of Scientific Knowledge." *Science in Context* 11: 51–87.

King, M.D. 1980. "Reason, Tradition, and the Progressiveness of Science." In *Paradigms and Revolutions: Appraisals and Applications of Thomas Kuhn's Philosophy of Science*, edited by Gary Gutting, 97–116. Notre Dame, IN: University of Notre Dame Press.

Kuhn, Thomas S. 1962/1970. *The Structure of Scientific Revolutions*. 2nd edition. Chicago, IL: University of Chicago Press.

Kuhn, Thomas S. 2000. *The Road Since Structure: Philosophical Essays, 1970–1993, with an Autobiographical Interview*, edited by James Conant and John Haugeland. Chicago, IL: University of Chicago Press.

Lakatos, Imre 1970. "Falsification and the Methodology of Scientific Research Programmes." In *Criticism and the Growth of Knowledge*, edited by Imre Lakatos and Alan Musgrave, 91–196. Cambridge: Cambridge University Press.

Mayer, Anna K. 2000. "Setting up a Discipline: Conflicting Agendas of the Cambridge History of Science Committee, 1936 –1950." *Studies in History and Philosophy of Science* 31A:665–89.

Mayer, Anna K. 2004. "Setting up a Discipline II: British History of Science and 'the End of Ideology', 1931–1948." *Studies in History and Philosophy of Science* 35A: 41–72.

Mendelsohn, Everett. 1989. "Robert K. Merton: The Celebration and Defense of Science." *Science in Context* 3: 269–89.

Merton, Robert K. 1973. "Science and the Social Order." In *The Sociology of Science: Theoretical and Empirical Investigations,* edited by Norman W. Storer. Chicago, IL: University of Chicago Press.

Moore, James R. 1979. *The Post-Darwinian Controversies: A Study of the Protestant Struggle to Come to Terms with Darwin in Great Britain and America, 1870–1900*. Cambridge: Cambridge University Press.

Novick, Peter. 1988. *That Noble Dream: The "Objectivity Question" and the American Historical Profession*. Cambridge: Cambridge University Press.

Orwell, George. 1970. *The Collected Essays, Journalism and Letters of George Orwell, e*dited by Sonia Orwell and Ian Angus. 4 vols. Harmondsworth, Middlesex: Penguin Books.

Popper, Karl R. 1945. *The Open Society and Its Enemies*. 2 vols. London: George Routledge and Sons.

Popper, Karl R. 1970. "Normal Science and Its Dangers." In *Criticism and the Growth of Knowledge*, edited by Imre Lakatos and Alan Musgrave, 51–58. Cambridge: Cambridge University Press.
Porter, Roy. 1986. "The Scientific Revolution: A Spoke in the Wheel?" In *Revolution in History*, edited by Roy Porter and Mikuláš Teich, 290–316. Cambridge: Cambridge University Press.
Rudwick, Martin. 1985. *The Great Devonian Controversy: The Shaping of Scientific Knowledge Among Gentlemanly Specialists*. Chicago, IL: University of Chicago Press.
Scott, William Taussig, and Martin X. Moleski, S.J. 2005. *Michael Polanyi: Scientist and Philosopher*. Oxford: Oxford University Press, 2005.
Shapin, Steven. 1992. "Discipline and Bounding: The History and Sociology of Science as Seen Through the Externalism-Internalism Debate." *History of Science* 30: 333–69.
Shapin, Steven. 1995. "Here and Everywhere: Sociology of Scientific Knowledge." *Annual Review of Sociology* 21: 289–321.
Shapin, Steven, and Simon Schaffer. 1985. *Leviathan and the Air-Pump: Hobbes, Boyle, and the Experimental Life*. Princeton, NJ: Princeton University Press.
Smith, Barbara Herrnstein. 2006. *Scandalous Knowledge: Science, Truth and the Human*. Durham, NC: Duke University Press.
Snobelen, S.D. 2001. "'God of Gods, and Lord of Lords': The Theology of Isaac Newton's General Scholium to the *Principia*." *Osiris* 16: 169–208.
Stewart, Larry. 1996. "Seeing Through the Scholium: Religion and Reading Newton in the Eighteenth Century." *History of Science* 34: 123–65
Suppe, Frederick, ed. 1977. *The Structure of Scientific Theories*. 2nd edition. Urbana, IL: University of Illinois Press.
Warwick, Andrew C. 1992 and 1993. "Cambridge Mathematics and Cavendish Physics: Cunningham, Campbell and Einstein's Relativity 1905–1911." *Studies in History and Philosophy of Science* 23: 625–56; 24: 1–25.
Werskey, Gary. 1988. *The Visible College: A Collective Biography of British Scientists and Socialists of the 1930s*. 2nd edition. London: Free Association Books.
Young, Robert M. 1985. "Darwin's Metaphor: Does Nature Select?" In *Darwin's Metaphor: Nature's Place in Victorian Culture*, 79–125. Cambridge: Cambridge University Press.
Zammito, John H. 2004. *A Nice Derangement of Epistemes: Post-Positivism in the Study of Science from Quine to Latour*. Chicago, IL: University of Chicago Press.

Chapter 3
The History and Philosophy of Science History

David Marshall Miller

Past science is studied from two perspectives. The intellectual history of science, which focuses on the development of ideas and arguments, and the social history of science, which focuses on the development of science as a social undertaking within its broader contexts, are both alive in the academy. Nevertheless, these two approaches do not interact very well, and the field of science history is bifurcated along these lines. Indeed, intellectual and cultural historians of science tend, basically, to ignore one another. They have different training, different aims, different audiences, and often different institutional homes. Intellectual historians of science tend to be conversant with philosophers, social historians of science associate with mainstream historians, but they do not often discourse with each other. In turn, this has led to remarkable naïveté on each side regarding the work of science historians across the disciplinary fence.

This disciplinary divide is signaled by the two dominant "brands" of science history. On the one hand, scholars focusing on social history constitute the majority of "History of Science" (HOS) graduate programs, which are often housed within History departments, where socio-cultural approaches likewise predominate.[1] One can include scholars of "STS" ("Science and Technology in Society" or "Science and Technology Studies") in this group, though STS comprises sociological studies of science more broadly. On the other hand, scholars of "History and Philosophy of Science" focus on intellectual history of science and are usually aligned with Philosophy departments, at least in practice if not explicitly. There are also many scholars in philosophy departments who examine the intellectual history of philosophy, including natural philosophy. The study of science history thus breaks down into an HOS approach on one side and an HPS approach on the other.

[1] It is important to note that "History of Science" is *not* identical to the academic study of past science. To avoid confusion, I use 'science history' for the latter.

D.M. Miller (✉)
Duke University, Durham, NC, USA
e-mail: david.m.miller@duke.edu

S. Mauskopf, T. Schmaltz (eds.), *Integrating History and Philosophy of Science*, Boston Studies in the Philosophy of Science 263, DOI 10.1007/978-94-007-1745-9_3,

I should note that I am describing, in overly general terms, *methodological* approaches to science history. Caveats abound. For one thing, the brands I describe here often, but not always, line up with the *institutional* names of departments and programs. HPS-style scholars can be found in "History of Science" departments, and vice versa. For instance, the Science and Technology Studies department at University College London maintains a heavy HPS cast, while the History and Philosophy of Science department at Indiana includes both HOS- and HPS-type scholars. Moreover, my overly general characterization does not capture all scholars working on science history. There are "straight" intellectual historians, for example, who are as skeptical of philosophy as they are of social history. I will try to locate their work in the intellectual landscape below.

The division of science history into two dominant brands has never been comfortable, and what started off as problematic has only gotten worse. There have been several attempts to redress the situation over the years, and a recent spate of conferences has revisited the issue. These meetings have lamented the failure of integration between intellectual and cultural history of science in particular and the history and philosophy of science more generally, though I will argue below that the conflation of these questions is part of the problem. Nevertheless, the conferences and their participants have reflected the strict disciplinary divides they sought to overcome.[2] To speak only of meetings I have attended, the program at the March 2007 conference at Duke, titled "Do Historians and Philosophers of Science Have Anything to Say to Each Other?" and the basis of the present volume, consisted primarily of professional historians of science—HOSers. Meanwhile, the "Conference in Integrated History and Philosophy of Science" at the University of Pittsburgh in October 2007 featured mainly historically-inclined professional philosophers—HPSers. In both cases, there was a remarkable amount of confusion and even disdain regarding the aims and activities of the other camp, which was not sufficiently represented. These conferences illustrate in microcosm the general recognition that the disciplinary boundaries between historically- and philosophically-inclined studies of past science should be broken down. But they also show the level of miscomprehension about where those boundaries lie, and thus the total lack of consensus as to how one should reconcile the two sides of the discipline. There is almost universal agreement that there is a problem, and nearly universal disagreement about what the problem is. HOS and HPS, it seems, have completely lost track of one another, to the point of not knowing where each other are.

I share the opinion that the entrenched distinction between intellectual and social history of science is deleterious to science history. A brief perusal of the history

[2]Symposia at meetings of the Philosophy of Science Association in 1970 and 1992 are representative. See Ruse (1992), Steinle and Burian (2002). More recent events also include "Do the history of science and the philosophy of science have a future together?", University College London, June 2006, and "Do the History of Science and the Philosophy of Science Have Anything to Say to Each Other?", Florida State, March 2008 (which shared no speakers with the similarly titled Duke conference). There have also been two subsequent conferences on "Integrated History and Philosophy of Science" (colloquially known as &HPS) in 2009 (at Notre Dame) and 2010 (at Indiana). Another is scheduled for 2012 in Athens.

and philosophy of the discipline reveals that the distinction is not native to science history. It was imposed from without as science historians sought institutional and intellectual refuge in Philosophy and in mainstream History. These allied disciplines then co-opted science history to their own ends, rending the field. The distinction, moreover, is pernicious since it does not derive from past science. By leading scholars to focus on one aspect of the object of study, the distinction artificially closes off legitimate routes of inquiry. Disciplinary prejudices blind the scholar to important interactions between the intellectual and the social that are clearly and unproblematically present in actual past science. Nevertheless, the distinction persists. It is continually reinforced by the training and practice of science historians.

The situation calls for critical self-reflection. Science historians need to identify the disciplinary boundaries that separate them as a first step toward overcoming them. It will be helpful, in particular, to situate the historical study of science in its historical and philosophical context. By recognizing the intellectual and social bases of their practices, scholars will be able to recognize where they stand in the disciplinary landscape and where others stand in relation to them—how they are separated and how they are continuous. This mapping of the field as a whole should at least engender a discussion that spans the discipline and will perhaps lead to the kind of interaction and cooperation that has been lacking heretofore. Both HOS and HPS approaches are necessary for a proper understanding of past science, and better dialogue between them would make scholarship more effective and more productive. Science historians will need to renegotiate the boundaries of their own discipline in order to integrate the segregated approaches and counteract the divisive, extrinsic demands of Philosophy and History. The divisions between HOS and HPS may be ultimately irreconcilable, but we should at least understand why this is so.

In the interest of the self-criticism I advocate, I admit that I am trained as an intellectual historian of science in the HPS mode, and I bear the prejudices and biases pertaining thereto. What follows will surely be a product of those presuppositions and may therefore strike my audience as odd or ill-founded, especially since I am likely representative of a minority view.[3] If so, I welcome criticism, since it will illuminate and evince defenses of the otherwise tacit prejudices of both sides. The resulting discussion would be precisely the kind of reflective criticism and renegotiation that I aim to promote. From my limited, individual perspective, I cannot foresee the ultimate consensus, if any, that may be reached by the discipline as a whole, and I am not advocating one brand of science history over another. If I am against anything, it is the unreflective adherence to the norms of the HPS and HOS brands that I think is all too common.

[3]HOS is by far the more common approach. To give one crude measure, of the fifty-nine American graduate programs listed on the History of Science Society's website, only two are separate HPS *departments* (Indiana and Pittsburgh) and only three more are listed as HPS "programs" (Notre Dame, Texas-Austin, and Montana State). Indiana, Pitt, and Notre Dame are excellent, well-respected programs, but Rachel Laudan's assertion that HPS-style historians of science are an "endangered species" still rings true (Laudan 1992).

3.1 A History of Science History

Scholarly disciplines are bounded by disciplinary prejudices. One must accept the presuppositions of a field in order to be counted as a member of it. Graduate training is intended to inculcate such assumptions, and the sublimation of foundational questions is the mark of a mature scholar. Of course, disciplinary prejudices are necessary, since the phenomena under scrutiny are simply too hoary to make sense of without them. The rules and methods of the discipline allow the scholar to filter out the subject of their interest from everything else. Disciplines differ because they focus on different aspects of phenomena—they use different biases. By the same token, though, scholars do not often scrutinize the foundational assumptions of their work. This makes it difficult to adjust methods and aims once a discipline has matured and foundational questions have been settled. Scholarly interest is understandably directed outward, not back on the scholar's own practice.

In most cases, disciplinary biases are not problematic, since disciplines tend to either spontaneously grow up around their subjects, cleave themselves off of parent disciplines, or combine the methods of existing disciplines. These developments are organic, occurring in the course of the dialectic between scholars and the phenomena they study. The disciplinary prejudices that emerge are natural, in the sense that they are motivated by and appropriate to the objects of study,[4] generating coherent and productive scholarly programs. However, when the disciplinary prejudices are awkward and uncomfortable, as in science history, one must question them to diagnose and resolve the problem. For instance, we can examine the history of the history of science in order to figure out how science history came to be divided into two distinct parts.

The history of science history has been told before,[5] but it bears repeating, at least in *very* broad outline, since it begins the process of self-reflection that is ultimately necessary to reconcile and reconstitute the field. Making their traditions explicit helps scholars of both HOS and HPS to recognize their activity in relation to that of the other brand. In particular, the history of the discipline and its separate brands partly explains the accidental and artificial nature of the prejudices afflicting its current practice. One finds that what began as an organic discipline came to be co-opted by Philosophy and mainstream History, and therefore became beholden to their extrinsic concerns. The disciplinary prejudices dividing HOS from HPS are, in this sense, artificial, since they did not arise from the history of science itself. The discomfort felt by scholars in the field is, in part, a recognition of this artificiality.

In the Enlightenment, natural science was held as the epitome of human accomplishment. Newton's achievements demonstrated the highest measure of illumination, and all the other disciplines sought to emulate the example. Thus, science

[4]For a more substantial biological metaphor for science historiography, see Machamer (1994).

[5]For example, Christie (1990), Cohen (1994), Kragh (1987), Nickles (1995).

history of a sort was practiced within various fields by those aiming to establish their disciplines as "sciences" in the style of mechanics. The histories produced were whiggish, since the point was to emphasize the inexorable progress made toward certainty, not the actual development of a discipline, with all its sidetracks and red herrings. Hence the "historical" chapters of science textbooks and the biographies of "great men" typical of eighteenth and nineteenth century science history. Joseph Priestley's histories of electricity and optics and Charles Darwin's historical introduction to the *Origin of Species* are tokens of this type.[6] Hence also the work of William Whewell, whose study of the "inductive sciences" was meant to show their progressive consolidation of knowledge.

In its original form, science history was also naturally associated with Philosophy, since the main interest was epistemological: the eventual establishment of sure knowledge. Whewell, for instance, thought of himself as primarily a philosopher, and his *History of the Inductive Sciences* was part of the philosophical project expressed by his *Philosophy of the Inductive Sciences*. The naturalness of this association can also be seen in the strong effects philosophy, especially positivism, had on the natural sciences themselves around the turn of the twentieth century in the work of scientist-philosophers like Mach, Poincaré, Einstein, and Reichenbach.[7] Out of this whiggish, positivist, intrascientific tradition emerged the first professional historians of science, most notably George Sarton and Alexandre Koyré. Sarton, originally trained as a mathematician, lionized Poincaré, whose portrait he put on the frontispiece of the first issue of *Isis* in 1913. Koyré's graduate work was in philosophy and mathematics, some of which he pursued at Göttingen under Husserl and Hilbert.

Meanwhile, mainstream History had been divided by the effects of nineteenth-century work by Hegel and Marx. Like their Enlightenment predecessors, Hegel and Marx believed in the progress of human history, but they evaluated progress differently. Hegel held it was an essentially intellectual phenomenon; Marx thought it was essentially material. So, for Hegel, the benefit of science was its ability to produce knowledge. For Marx, it was the ability to produce things.[8] The earliest incarnation of science history was naturally linked to the Hegelian approach, since it located progress in the approach toward certainty—i.e., in the intellectual realm. As we shall see, though, it is this initial difference of perspective, endemic to mainstream History, not science history, that laid the groundwork for the divisions that afflict science history today.

In the early twentieth century, the Hegelian and Marxist accounts of human progress became associated with broad political movements. Hegelianism was perverted into nationalism, which stressed allegiance to and the progress of a national *idea*, and from there into fascism. Marxism was folded into socialism and thence communism. The resulting ideological tensions soon came to be reflected in the way

[6]Christie (1990), Kragh (1987).

[7]Feigl (1970) also makes this argument.

[8]Christie (1990, 12).

science history was studied.[9] The Second International Congress for the History of Science, held in London in 1931, was a pivotal moment. The conference consisted mainly of intellectual historians with nationalist tendencies, but the seven delegates from the Soviet Union made a deep impression arguing the Marxist point of view. For them, science was a socially conditioned human practice, responsive to "external," non-intellectual factors. Famously, Boris Hessen gave a talk reducing Newton's science to the material and economic problems of his time, class, and so on. Though this was serious and striking historiography, it was also Soviet propaganda, and Moscow's embassy in London furiously translated and published the delegation's papers as *Science at the Cross Roads* within ten days.[10] Sympathetic scholars, like J. D. Bernal, Joseph Needham, and Edgar Zilsel, were converted to the Soviets' approach, and the event marked the beginning of the widely recognized "externalist" study of science history. The traditional Hegelian "internalism" continued to prevail, however, and externalism remained relatively marginal. Nevertheless, the Hegelian-Marxist split had injected itself into science history, partly as a result of global politics.

In mainstream History, Marxism eventually penetrated much deeper, almost to the complete exclusion of intellectual approaches, which were driven off into other fields, such as political science. This was especially true after the rise and ultimate self-immolation of fascism up through World War II and the holocaust, which tainted Hegelian-style intellectual history with vapors of totalitarianism and moral turpitude. Even today, intellectual historians in general struggle to find a place in History departments, where they are seen as conservative, old-fashioned, and vaguely sinister. Science history, however, was a special case. The field proved resistant to the general historiographical trend, precisely because science itself, insofar as it makes claims to rationality and truth, resists complete reduction to material or social considerations such as economy, race, class, and gender.[11]

Such resistance was a mixed blessing. On the one hand, science history was left outside the practical and institutional pale of mainstream historical studies and thus relegated to minority, outsider status. On the other hand, intellectual historians of science found a welcoming reception by the Western conservative establishment after the war, which sought to use science history to explicitly counter the Marxism infiltrating the humanities. Science history was also seen as a way of encouraging student interest in science, and thus formed an essential part of a Cold War curriculum designed to promote the American scientific excellence that would overcome the Soviets. The notion of a Scientific Revolution, coined and popularized by Koyré, was also useful in demonstrating a kind of revolution that did not involve violence or, for that matter, the overthrow of a capitalist system. Moreover, science

---

[9]As one might expect from their intellectual predilections, Sarton and Koyré had right-wing political inclinations. They were nationalists and anti-Marxist. Koyré, a white Russian who emigrated during the revolution, was particularly fervent in this respect.

[10]Cohen (1994), Young (1990, 80–84).

[11]Laudan (1990, esp. 51).

history fit with the broader cultural excitement about the power and profits of science and technology.[12] Altogether, then, these intellectual and cultural factors led to the first substantial institutionalization of science history. History of science programs (sometimes entitled "History and Philosophy of Science," reflecting the involvement of philosophers, though not yet representative of the HPS mode) were set up, usually by science faculties, at Harvard, Princeton, Columbia, Oxford, Cambridge, Leeds, Sydney, Melbourne, and elsewhere.[13] Thus, science history found an institutional home, but it was a home apart from and in opposition to mainstream History. This Hegelian, intellectual tradition of science history remained dominant throughout the 1950s and early 1960s. The "internalist" stalwarts successfully defended the discipline from a few "externalist" critiques, which were dismissed as "a bit Marxist."[14]

Everything changed with the publication of Thomas Kuhn's *Structure of Scientific Revolutions* in 1962. Kuhn himself was a typical, scientifically-trained, philosophically-inclined, intellectual historian of science. In *Structure*, Kuhn tried to give an intellectual account of scientific change, but his argument located scientific knowledge in the "paradigm"—a communal entity consisting of shared concepts, practices, problems, and specialized languages. In a move Kuhn neither foresaw nor approved of, *Structure* seemed to warrant a reduction of science to a socio-culturally constituted paradigm.[15] Kuhn's theory also raised the specter of radical incommensurability, which threatened the very notion of scientific objectivity, especially in the hands of Kuhn's Berkeley colleague, the philosopher Paul Feyerabend. Hence, *Structure* weakened science history's traditional defense against socio-cultural materialist reductions, the appeal to the fundamental rationality and objectivity of science itself. Soon enough, Marxist-style historians had taken over and assimilated science history into History departments proper. In some ways, this was quite welcome, since science history could now call on the resources of mainstream History. "Conservative" intellectual historians, however, were once again driven out of the "History of Science" as practiced in the universities.

Besides its Kuhnian justification, the turn toward a more sociological approach to science history found political motiovation in the New Left.[16] Older Marxists

---

[12]Mayer (1999).

[13]There were precursors. At University College London, a Department of History and Method of Science was founded in 1921–1922 and renamed History and Philosophy of Science in 1938. George Sarton had helped establish the Harvard Committee on Higher Degrees in the History of Science and Learning in the 1920s, and Harvard began granting PhDs in History of Science in 1936 (Bennett 1997; Cohen 1984; Hall 1984; Kuhn 1984; Smeaton 1997).

[14]Henry Guerlac reported this as a colleague's response to his own work (Guerlac 1977, 36). Cohen speculates that the colleague was Koyré (Cohen 1994, 561n168). More damaging to "internalism" were critiques from other intellectual historians, like Frances Yates, who challenged the presumption of a well-defined rational "science" apart from other cognitive, but "irrational," activities, such as magic (Hesse 1970; Turner 1990).

[15]Whether Kuhn's work actually warrants this move is still a point of vigorous debate.

[16]Porter (1990, 41).

had thought—in the Enlightenment style—that science was on their side. Science would provide the material culture needed for the post-revolutionary utopia. By the 1950s and 1960s, though, science had become part of the conservative, capitalistic establishment. In the view of the New Left, science was used to impose social norms and was therefore anti-democratic. Of course, this view was not entirely mistaken. The early to mid-twentieth century enthusiasm for race hygiene and eugenics is only one case in point. The social historians also had more recent examples of science's sometimes objectionable role in the military-industrial complex, such as nuclear weapons, MIT's Draper Laboratory, the Tuskegee Experiments, DDT, and so on. In order to undermine the establishment, they had to disparage science and reduce its social role. This made social historians, to some degree, anti-science. They sought to emphasize the social mechanisms by which scientists come to accept beliefs, and they downplayed the significance of the "objective" features of science, including its intellectual content and its predictive and explanatory power. Here, the social historians found allied interests amongst sociologists of science, descended from the work of Robert K. Merton. Merton's study of scientific values and institutions fit well with the study of paradigms, since the former help determine the latter.[17]

The conquest was as complete as it was sudden. In the 1950s, the internalists held the upper hand. By 1968, the so-called "internalism-externalism debate" had been declared over. The end of the debate did not represent a compromise so much as a sound defeat of the internalists, who ceded the field and the brand name to the externalist approach. Programs in "History of Science" were converted to the externalist approach, or even into STS programs outright. That the "internalism-externalism debate" is now outdated (and that it is gauche to revisit the issue) is evidence of total victory—there are simply too few internalists left in "History of Science" to make a stand. In fact, the very meaning of "internalism" itself has changed. In current usage, it often signifies the socio-cultural interactions *within* science, as opposed to science's interactions with the wider socio-cultural sphere. But even this would have been considered "external" on the earlier meaning of 'internalism,' which referred to a historiographical approach focused on the internal logic of scientific progress.[18] Internalism in this older sense has been practically effaced from "History of Science" as it is presently conducted. We should not confuse the practical victory of the Marxist-style externalists for rational propriety, though. The state of play does not mean the distinction is or ever was dissolved. The internalists have simply moved on.

Though no longer considered part of "History of Science" proper, intellectual science history found a ready reception amongst philosophers. Since the 1920s, the

[17]The alliance was somewhat ironic, though, since Merton himself was fiercely anti-Marxist and intended his work to show, like his predecessor Weber's, the essential importance of Western, capitalistic, and individualistic values for the development of science. Merton claimed that "external" factors could affect the rate and not the course of scientific development, which was determined only by its internal logic, but later Marxist-style historians claimed him for their purposes (Shapin 1992, 336–37; Young 1990, 83).

[18]Compare, for example, Cohen (1994) or Hesse (1970) to Shapin (1992).

logical positivists and, later, the analytic philosophers descended from them, had strived to protect their discipline from the same sort of political impositions that had injected themselves into History. The early logical positivists of the Vienna Circle of the 1930s, for instance, were especially concerned to show that Philosophy could achieve transcendent validity and value. Recalling Enlightenment attitudes and inspired by recent mathematical and scientific achievements (e.g., Einstein's relativity theory and the formalization of mathematics by Russell and Whitehead), these philosophers held science to be a model of the transcendent objectivity they sought for themselves. They tried to turn philosophy into science, eschewing non-empirical "metaphysics," on the one hand and seeking out the methods that make a discipline scientific, on the other.

At first, this program was pursued "logically"—through a priori reasoning. By the middle of the century, though, it was accepted that the so-called linguistic turn had failed. Famously, philosophers could not even establish criteria by which "science" could be distinguished from metaphysics or pseudoscience. Kuhn and Feyerabend, along with Imre Lakatos, Norwood Russell Hanson, and Stephen Toulmin, were part of a generation of philosophers who sought to base their analyses of science on its actual past and present practice. In their work, Hegelian-style intellectual history of science offered both raw data and proving ground for philosophical models of rationality. Their calls for empirical studies of science led to a cooperation with the holdouts of the "right wing" of science history, which led to the establishment or reinvention of a handful of HPS departments and programs across the globe around 1970, thereby initiating the modern HPS brand. It must be noted, however, that these historically-interested philosophers primarily sought to use history *for* philosophical insights, but denied that history per se contained anything of philosophical significance. Even Lakatos, one of the most vocal supporters of the cooperation of Philosophy and science history, advocated "rational reconstructions" of science—science as it "should have happened"—and relegated the reporting of historical facts to footnotes.[19]

After Kuhn, science history has developed along the two distinct paths that separate the discipline today. In HOS/STS, the first sustained movement was Social Constructivism (or "Social Structure of Knowledge"), which first flourished in the 1970s with work by David Bloor and Barry Barnes. The social constructivists took the reduction of science to social mechanisms to a radical extreme (following a similar development in mainstream History), arguing that scientific knowledge is entirely constituted by social interactions.[20] Social constructivism itself has lost much of its impetus, but similar trends remain powerful amongst science historians of the HOS variety. Foucault, whose views are widely influential, emphasized the relations between power and "knowledge"—including scientific knowledge. Another dominant tendency, partially inspired by a move towards narrative in

[19]Hanson (1962), Lakatos (1971). See also Kuhn (1977).

[20]In fact, Barry Barnes has argued that the historian is best served if she is completely ignorant of the beliefs held by the scientists she studies (Barnes 1990, 71).

mainstream History and exemplified by Shapin and Schaffer, is to study scientific practice in very particular geo-temporal contexts, to the exclusion of transcendent considerations.

The point of this brief history is that science history resisted the general trend of History during the twentieth century, since it was traditionally and constitutionally aligned with the Hegelian-style intellectual history falling out of favor in light of world events. The work of Kuhn and its surrounding counter-culture atmosphere then undermined the very features of science that science historians had used to organically develop and defend their own scholarship. Never very large and still relatively young, the field was cut intellectually and institutionally adrift, open to the imprecations of more established disciplines, particularly History and Philosophy, that co-opted science history to their own purposes. This has led to a bifurcated field, resting uncomfortably between History and Philosophy, where different practitioners identify more with their institutional peers than with each other.

3.2 Philosophy of Science History

The situation would be unproblematic if science history's cognate disciplines could cooperate. Science history would then be a natural overlap, in the way that, say, biochemistry bridges biology and chemistry. Painting with a broad brush, the trouble is that mainstream History and Philosophy are inherently at odds with one another. While both disciplines have legitimate affinities and interests in science history, they end up exerting centrifugal forces that artificially pull the discipline further apart.

Especially in the English-speaking world, Philosophy sees itself as a normative discipline in that it, broadly speaking, studies epistemic norms and examines arguments in order to figure out how beliefs should lead to each other. The aim is to understand the nature of human reasoning (of all kinds).[21] Science has a remarkable power to produce convincing claims about the world. Philosophers are intrigued by this argumentative power, and they seek to isolate the forms of argument—the rational method—by which science achieves its epistemic efficacy. Science history, therefore, provides a store of arguments with which philosophers can construct and evaluate their models of rational behavior. They look to science history as a way to determine which epistemologies "work" and which do not.

History, on the other hand, sees itself as a descriptive discipline. The aim is to get as close as possible to past events and figure out the conditions and motors of human activity. Science is interesting because of its importance amongst human practices, especially in the Western Tradition, where scientific activities, products,

[21] A simple description of Philosophy is difficult, and exceptions to my two-sentence definition are common. Still, I think this view of the general slant of the field holds. For instance, this characterization captures the many philosophers who see themselves as merely describing (as opposed to prescribing) norms, as in naturalized epistemology.

and values are central. Science has a remarkable power to affect the conditions of human existence, and historians seek the sources and effects of this power.

These disciplinary self-conceptions rest, in turn, on antithetical presuppositions. Philosophers (for the most part) presume that there is an absolute measure of rational conviction against which different epistemological methods can be measured, regardless of socio-cultural context. A good argument, most philosophers would contend, justifies its conclusion independently of the particular material and social conditions under which it is formulated. In philosophy, therefore, past arguments, including scientific ones, are shorn of any non-cognitive or even non-rational context. It makes perfect philosophical sense, for instance, to compare the views of two authors widely separated in time or to ask for an earlier author's possible response to a later author (e.g., how would Descartes respond to Newton?).[22]

Historians dismiss all of this as a "positivist" delusion. The assumption of a transcendental measure of argumentative "success" is unfounded. The failure of the demarcation project itself shows that one cannot decide a priori what counts as "rational" or "scientific." Arguments are always conditioned by their social and material context. Their power to engender belief depends, in part, on who makes them and to whom they are made.[23] Even the basic determinations "science" and "rational" must be contextualized in a particular time and place; the more detailed the contextualization, the better. In particular, there is no reason whatsoever to exclude non-cognitive factors as irrelevant. (Indeed, given the old Marxist prejudice against intellectualism, historians are likely to emphasize non-cognitive factors over cognitive ones.) The philosophers' preoccupation with the intellectual realm threatens to anachronistically distort the historical account. The philosophers' pursuit of rationality, says the historian, fails to say anything definitive about the essentially contextualized human condition. Hence, the philosophical study of science is ill-founded and uninteresting.[24]

To philosophers, meanwhile, historians commit a logical fallacy of their own by presuming the failure to describe transcendental rational norms entails that such norms do not exist. Historians, therefore, adopt a philosophical position for which no sustained argument is offered. It is impossible to study anything, history included, without some a priori framework by which phenomena are made meaningful. All

---

[22] A partial exception can be made for feminist philosophy and related studies of "science and values." These areas do acknowledge the effects of social values on reasoning and science, though the morals drawn are often still normative: how reasoning and science *should* respond to external values.

[23] See Laudan (1990), Shea (1983), Thackray (1970a).

[24] To be fair, philosophers of science themselves have recognized the failure of the universalizing project. As a result, they have turned to more specific studies of particular disciplines (viz. the philosophy of physics, the philosophy of biology, and so on). While this sort of philosophy of science remains preoccupied with the normative and transcendent features of science, it is more sensitive to specifics of practice and argument. On the other hand, historians are generally unaware of these philosophical developments, so their conception of "philosophy of science" is often an outdated caricature—which is all the more reason to encourage dialogue.

observations are theory-laden, and it is the historians who are overly "positivist"[25] in their insistence that anachronism can and must be avoided.[26] Philosophers also see no profit in complete contextualization and they are not interested in antiquarian description for description's sake. They want the payoff for their own, always present-day, essentially rational selves. Hence, the historical study of science is ill-founded and uninteresting.

Of course, both sides are, basically, right.[27] Within their respective disciplines, though, historians and philosophers are entitled (and expected) to ignore such criticisms and hold up their basic presuppositions as regulative ideals around which to organize their inquiry. After all, these prejudices grew up organically within those disciplines, and they effectively shape scholarly discourse into something manageable. History is meant to study the conditions of human experience, Philosophy is meant to study human reason, and both perform their functions well. However, the science historians now operating in the midst of these disciplines are forced to make an impossible choice. To be an accepted scholar within one discipline or the other, they must either place an inordinate focus on the cognitive and universal aspects of science and minimize the contingent and contextual, or they must contextualize away the very universal claims that give science its special socio-cultural status and mark it out as a distinct human activity. In other words, science historians face the "barren antithesis"[28] of studying science without history or history without science. Both approaches are risible. They completely vitiate the intensions of the scholarship. HPSers are left with a gross misunderstanding of scientific reasoning, and HOSers fail to recognize the primary motor of scientific activity. Still, far too many authors capitulate. One is frequently frustrated by attempts to make sense of a scientific episode without any reference to its historical context or to describe a historical context without making any sense of the science it surrounds. As Larry Laudan has put the point: "Many have evidently concluded that the only alternative to the disembodied history of scientific ideas is a lobotomized history of scientific institutions."[29]

The other distinctive features of HOS- and HPS-style scholarship derive from the prejudices and practices of History and Philosophy. In *very* broad generalization, science historians of the HOS stripe today tend to be trained as historians, not as scientists, which is to say they tend to be socio-cultural historians like their mainstream History peers.[30] They lean toward a deflationary view of science as just another human activity without any universal pretensions. The heartland of HOS/STS is

---

[25]The overlapping but different connotations of this pejorative further signify the incommensurability between the philosophical and historical worldviews.

[26]Baltas (1994), Burian (2003), Nickles (1995, esp. 151–55).

[27]See, for further discussion Burian (1977), McMullin (1970).

[28]Thackray (1970b).

[29]Laudan (1990).

[30]HOSers, in my experience, sometimes excuse their focus on socio-cultural factors by pleading ignorance about the actual workings of science, which forms a sort of "black box" at the center of the institutions and activities they study. However, this does not excuse the general lack of interest

post-war twentieth-century science, where funding structures, collaborations, and technological outputs exemplify the socio-cultural entanglements of the scientific enterprise.[31] Socio-cultural historians also favor "thick" explanations that include all relevant causes, including the "little losers" of science—those whose contribution was minor or even completely forgotten. Hence, HOSers, like their mainstream historian peers, favor books as the measure of scholarly contributions. These tend to be written for a non-scientific audience, and HOSers often find themselves pitted against scientists in the culture wars, since scientists see the relativism embodied in their contextual approach as a threat.[32]

On the other side, HPS has been largely co-opted by the disciplinary concerns of Philosophy more generally. Hence, intellectual historians of science today tend to be trained as philosophers, and to share the anti-historical prejudices of that discipline. They seek to construct transcendent models of scientific reasoning and anachronistically ignore the contingent, "irrational" factors in the development of science. HPSers focus on periods in which intellectual progress is most on display and the intellectual contributions of individuals is clearest. Thus, they concentrate on the scientific revolution and turn-of-the-twentieth-century physics. For the same reason, they also tend to focus on the "big winners" of science whose work was the most influential or "successful." Also, HPSers often write only for the benefit of their philosophical peers or interested scientists, without trying to reach a general audience. They focus on particular epistemological issues and write papers. And so on.[33]

To me, all these divergent tendencies in a discipline that is ostensibly about one thing—past science—are evidence that the disciplinary biases by which History and Philosophy distinguish themselves do not answer to any clear distinctions in past science itself.[34] The thing we call "science" lies at the intersection of ideas and society. It is a complex set of human practices that occur in a social,

---

in *trying* to understand, even to a small degree, scientific reasoning. Such interest, by contrast, is expected amongst intellectual historians and philosophers of science.

[31] I should note that HOS-type historians, following Shapin and Schaffer, have recently applied socio-cultural historiography to the scientific revolution. Westman and Biagioli, for instance, apply institutional analysis to the work of Galileo. The burgeoning interest in the history of renaissance and early-modern alchemy, astrology, and magic represents a similar trend. Though these were originally treated quite intellectually in the hands of Yates, Dobbs, Newman, Principe, and others, the *pseudo*-scientific status of these activities has allowed more recent authors to downplay the intellectual content of historical episodes. Also, the field of readership studies in mainstream History has allowed HOSers to focus on the conveyance of ideas, rather than ideas themselves. These trends might be considered an offensive maneuver, since they threaten to expel intellectual science historians from areas where they have hitherto enjoyed preeminence. On the other hand, these trends might be an opportunity for increased dialogue and cooperation. I fear the former and hope for the latter.

[32] Turner (1990).

[33] In a keynote paper at the &HPS1 conference, Peter Machamer listed over twenty such distinctions separating philosophy and history of science. See also Richards (1992).

[34] I do not mean to prejudge the issue as to whether there is any well-defined thing called "science." For the time being, let the term signify an actors' category—"science" is whatever is called

cultural, and material context. It has its institutions, its resources, its products, its traditions, it communities, its power relations, its values, and so on. But almost uniquely amongst human activities, science is also an argumentative discipline that makes claims to transcendent, objective truth. The scientific enterprise is a wonderfully efficient producer of accepted truths about the world. Science, in itself, has always chugged along quite happily both as intellectual endeavor with pretensions to universality and as socially conditioned cooperative enterprise. Scientists respond to a wide range of influences, some obviously cognitive, some obviously socio-cultural or material, and many in between. Moreover, they are unconcerned with Cartesian distinctions between their mental and physico-socio-cultural states or Reichenbachian distinctions between the "context of discovery" and the "context of justification." It might be impossible for them or anyone else to say *how* cognitive and social factors interact (they may even deny such interaction), but it is clear *that* they do, just as it is clear that scientists operate without separating epistemic contexts. It is therefore possible and, indeed, necessary to study all these aspects of scientific activity in conjunction, without privileging one or the other, according to the peculiar demands of the scientific enterprise itself. Importing the prejudices of History and Philosophy into science history has caused science historians to ignore fruitful paths of inquiry and artificially constrained their narratives. Science history is pulled to the extremes when it should naturally seek the middle.

For all of the forgoing reasons, I am deeply pessimistic about attempts to "marry" or "integrate" the history and philosophy of science.[35] I am also, with Steven Shapin, suspicious of eclectic historiography of science that aims to be part internalist and part externalist.[36] The broader disciplines are simply too much at odds with one another. Though they may occasionally have something to say to one another, they will mainly turn their backs on one another. For the most part, they already have.[37] I am more optimistic about the possibility of a unified science history, though the two brands cannot be reconciled as long as they remain beholden to the conflicting prejudices and practices of the cognate disciplines in which they are ensconced. Philosophy is too anti-historical and History is too anti-intellectual for HOS and HPS as presently constituted to meet on common ground. Science history must reconcile its distinctions on its own terms, as an integrated unity with

'science'. I suspect that there is more to it than that, but this is one of the many questions I seek to open for discussion.

35 See Feyerabend (1970), Kuhn (1977), McMullin (1970). I am more optimistic about the program recently expounded by Mary Domski and Michael Dickson (following Michael Friedman), which calls for a "reinvigoration" of a pre-Kuhnian (actually, Cassirer-ian) "synthetic approach" (Domski and Dickson 2010).

36 Shapin (1992). See Nickles (1995), Steinle and Burian (2002) for such suggestions.

37 Historians almost never reach out to philosophers. Philosophers are a little more circumspect, but proposals for dialogue with History usually amount to calls for the culling of History for philosophical ends, not a sincere interest in *doing* responsible history. See, e.g., Burian (1977), Hull (1992).

its own disciplinary bounds. Science historians should loosen the bonds of History and Philosophy and make their discipline anew, in dialogue with past science itself.[38]

3.3 Mapping the Field

One way to begin the unification of science history is to attempt to map the field as it now stands. This will allow us to formalize many of the distinctions between different kinds of science history, as well as encourage reflection, discussion, and discourse by allowing scholars to locate and defend their positions in relation to one another. If we seek to set up a big tent, a map will help us figure out where to pitch it, or at least have a coherent argument about where to pitch it.

As a start, I propose a representation of the historiography of science along two axes suggested by the historical and philosophical discussion above. There are many other ways of representing the field, but I think these axes offer a tidy way of constraining the discussion, since they seem to be both *orthogonal* and *comprehensive*. That is, a scholar's position along one axis does not determine his position along the other, and every historiographical approach in the field can be uniquely located somewhere in the space defined by the axes. I put forward these axes tentatively, however. They are meant merely as a starting point for discussion, not the last word. I welcome disagreement, since it forces us all to reflect on the nature of our own scholarly project, and it is precisely this reflexivity that will lead to compromise. Also, by plotting the field of science history on these axes, it becomes clear that intellectual and socio-cultural histories are continuous with one another. The disciplinary prejudices that separate them are more or less arbitrary boundaries on a homogeneous landscape. There may be reasons to accept them, but such reasons need to be clearly articulated and defended, since they do not follow apparent "natural" distinctions.

The first axis has to do with the causal role of intellectual and non-intellectual factors in the production of scientific knowledge. One can think of this axis as expressing a pseudo-numerical ratio between socio-cultural, external factors and intellectual, internal factors. The extreme internalist (at "0") denies any causal efficacy to socio-cultural context in the development of science. On this view, science develops entirely according to its own internal logic, via purely intellectual exchanges amongst its practitioners, as if they were all part of a single mental process. Moving away from this extreme, one allows more and more significance to socio-cultural, material, and institutional conditions, until one reaches (at "infinity") the extreme externalist, who believes that scientific reason is purely epiphenomenal, floating on top of the non-intellectual context surrounding scientific activity that is solely responsible for any change.

[38]On the other hand, I would reject calls to let the two brands go their separate ways. See Pinnick and Gale (2000), Strasser (2005).

The second axis concerns the temporal *telos* or aim of scholarship. At one extreme is the view that history is meant to reconstruct the past "as it really was." Extreme temporalists thus seek to immerse themselves in the past, with all of its twisting complexity, frighted by the specter of anachronism at every turn and hermeneutically circling ever closer to historical fact. At the other end of the scale are the extreme presentists, who study history in order to instruct the present or even the transcendent. These authors write from an anachronistic point of view and comfortably relate past events to their own concerns and interests, to the point of "rationally reconstructing" the past in their own image.

It seems that most, if not all, distinctive features of various historiographies can be read off from their position on this map. That is, one's position on the map determines the kind of scholarship one produces. For instance, internalists will base their work more on primary sources than externalists; presentists will be more interested in generalization than temporalists; and so on. As we have seen, HOS tends toward temporalist externalism, while HPS leans toward presentist internalism. There are those toward the off poles, however. For instance, most scientists who write history, many pre-Kuhn historians of science, and "straight" intellectual historians exemplify temporalist internalism.[39] Meanwhile, many authors write externalist histories with presentist punchlines (viz. "The story of ... and *how it changed the world*"), which can also be said of many popular science writers outside academe.[40] There are also any number of positions between the extremes. The axes, after all, are spectral, not binary, and since they are orthogonal, there is no position on the map that can be ruled out on a priori grounds. It remains for science historians to work out what part of the map they want to stake out for themselves.

3.4 A Hopeful Conclusion

Given the paradigmatic differences and institutional pressures acting on the HOS and HPS brands of science history, it is not altogether surprising that science historians have widely divergent views about what kinds of scholarship their discipline includes. Nor is it surprising that each brand of historian is remarkably ignorant regarding the methodology and aims of the other brand. Entrenched disciplinary prejudices have created insularity, lack of communication, an absence of cooperation, and even disdain. Meanwhile, those trying to mediate between the camps or operate in the middle ground gain recognition from neither side. Nevertheless, there is a feeling on both sides, I think, that something is wrong, both in their separation from one another and in their uneasy allegiances with other disciplines.

The prejudices dividing the discipline are actually accidental impositions, born in mainstream History and then extrinsically enforced when science history sought

[39] As exemplars of this type, I have in mind Jed Buchwald, William Newman, Mordechai Feingold, Peter Dear, Nicholas Jardine, and (in more general history) Quentin Skinner and Jonathan Israel.

[40] Peter Galison's *How Experiments End* and Stephen Johnson's popular *The Ghost Map* are examples, respectively (Galison 1987; Johnson 2006).

institutional and intellectual relationships with the more established disciplines of History and Philosophy. This led to the artificial segregation of the field into its HOS and HPS brands. Yet science history is neither Philosophy nor mainstream History. It has disciplinary demands of its own, stemming from the peculiar nature of science itself. If we are to begin respecting those unique demands, we need to seriously evaluate our own various approaches in our own specialized discipline. To reiterate, my aim here is not to advocate one historiographical position over another. There may be good reasons for emphasizing cultural factors over intellectual ones and vice versa. My point, rather, is to encourage science historians to explicate the reasons why they choose certain historiographical approaches over others. One should not adhere to disciplinary biases that are not appropriate to the object of study, even if they are the methodological dictates of a "home" discipline. Science history should be defined as a separate discipline in its own right, something *both* HPS- and HOS-like, from which morals for History and Philosophy can be drawn, but which does not serve the sole purpose of producing such morals.

I have argued that science historians should question the historical and philosophical prejudices they have used to define the divergent strands of their discipline. Admittedly, this is a somewhat dangerous proposal, since it leaves science historians without a clear understanding of what their discipline *is*, at least for the time being. It also threatens to undermine the institutional support HOS and HPS have received from History and Philosophy. On the other hand, the proposal is not *very* radical. It only concerns the emphasis placed on certain factors and aims in the production of scholarship, with the suggestion that the degree of emphasis be left indeterminate. There are features of science history that remain uncontroversial. One would expect universal agreement that the discipline is concerned with the description and explanation of past science, which is itself the human enterprise of describing and explaining natural phenomena.[41]

Above all, we can fall back on a basic aim of science history: the descriptions and explanations we produce should be plausible. They should at least seem worthy of refutation, if not convincing outright, to fellow practitioners. If this is our aim, we will seek intellectual causes where they seem important and necessary and socio-cultural factors where they seem important and necessary, without a priori expectations favoring one kind or the other. In an adequate account, each step of historical development will seem plausible. The skilled reader will understand how and why things happened as they did, and the plausibility of the explanation will be determined by past events, not by its instantiation of disciplinary norms. There will be some latent anachronism, of course. Just as the strength of arguments depend on their time and place, the plausibility of an account partially depends on its historical context. That is to be expected, since the only way to make sense of the past is to "fuse our horizon" with that of historical events. Just where such fusion

[41]The particulars of this definition might be open to dispute. For instance, depending on one's point of view, it might or might not include human interventions in natural phenomena through technology and medicine. Ultimately, the scope of the discipline should be another point open for discussion.

should take place—close to the past or close to the present—is another matter for negotiation. The communal evaluation of plausibility, meanwhile, implies the need for venues and outlets open to the questioning of disciplinary bounds. The fact that cross-disciplinary scholarship is usually refereed by a scholar in one camp or the other, and therefore according to one set of entrenched biases, only redoubles the present difficulties.

The looseness of the plausibility constraint forces a science historian sincerely interested in constituting a coherent, cooperative discipline to be reflexive. Scholars need to be continually conscious of what it is that they are doing, even as they do it. One way to do this is to apply the methods of the study to the study itself, so that one is always calling oneself to account to oneself. One should defend not only the plausibility of the histories one produces, but the plausibility of the production of those histories, and let this defense be made consciously and publicly, subject to the observation and criticism of one's peers. The account of the account is just as important as the account itself. Of course, this point is nothing new. Reflexivity is a common denominator in arguments for most scholarly methodologies. Moreover, the best and most interesting scholarship is *always* the best and most interesting precisely because it is consciously reflexive in the sense I suggest. Indeed, conscious and explicit critical attention to the presumptions underlying responsible scholarship is what accounts for such work's ability to shape disciplines.[42]

I have tried to practice what I preach to some extent in the present essay—producing an historio-philosophical account of the way science historians do and perhaps should behave. In particular, I have tried to pay heed to both intellectual and socio-cultural factors in the development of our discipline, without privileging one over the other. Luckily, there are many other models to follow. Among recent work, I can draw particular attention to Hasok Chang's *Inventing Temperature*, which conscientiously and explicitly combines HOS and HPS approaches into something he calls "Complementary Science." Chang also shows how self-critical reflection leads to progress in science through a process of "epistemic iteration." To my mind, this is reflexive evidence that a reflexive approach is natural in the study of science.[43]

Ultimately, science historians will need to renegotiate the boundaries of their discipline so that their disciplinary prejudices are more organically suited to the historical study of science and not so beholden to the demands of mainstream History and Philosophy. It will require a considerable period of consensus building, and it is impossible to predict how the discipline will be constituted beforehand. I hope the present paper forms a starting point for negotiation, even by eliciting dissent.[44]

---

[42]See Laudan (1990).

[43]Chang (2004). This estimation is not just mine. *Inventing Temperature* is a work of science history that shared the prestigious Lakatos Award for books in Philosophy of Science.

[44]This paper grew out of discussions with Bruno Strasser and was greatly improved by comments from Jutta Schickore, Andrew Janiak, and the editors of this volume. All error, overstatement, and ignorance is my own.

References

Baltas, Aristides. 1994. "On the Harmful Effects of Excessive Anti-Whiggism." In *Trends in the Historiography of Science*, edited by Kostas Gavroglou, Jean Christianidis and Efthymios Nicolaidis, 107–19. Dordrecht: Kluwer.

Barnes, Barry. 1990. "Sociological Theories of Scientific Knowledge." In *Companion to the History of Modern Science*, edited by R.C. Olby, G.N. Cantor, J.R.R. Christie and M.J.S. Hodge, 60–73. London: Routledge.

Bennett, J.A. 1997. "Museums and the Establishment of the History of Science at Oxford and Cambridge." *British Journal for the History of Science* 30: 29–46.

Burian, Richard M. 1977. "More than a Marriage of Convenience: On the Inextricability of History and Philosophy of Science." *Philosophy of Science* 44: 1–42.

Burian, Richard M. 2003. "Comments on the Precarious Relationship between History and Philosophy of Science." *Perspectives on Science* 10: 398–407.

Chang, Hasok. 2004. *Inventing Temperature: Measurement and Scientific Progress*. Oxford: Oxford University Press.

Christie, John R.R. 1990. "The Development of the Historiography of Science." In *Companion to the History of Modern Science*, edited by R.C. Olby, G.N. Cantor, J.R.R. Christie and M.J.S. Hodge, 5–22. London: Routledge.

Cohen, H. Floris. 1994. *The Scientific Revolution: A Historiographical Inquiry*. Chicago, IL: University of Chicago Press.

Cohen, I. Bernard. 1984. "A Harvard Education." *Isis* 75: 13–21.

Domski, Mary, and Michael Dickson. 2010. "Discourse on a New Method, or a Manifesto for a Synthetic Approach to History and Philosophy of Science." In *Discourse on a New Method: Reinvigorating the Marriage of History and Philosophy of Science*, edited by Mary Domski and Michael Dickson, 1–20. Chicago-LaSalle, IL: Open Court.

Feigl, Herbert. 1970. "Beyond Peaceful Coexistence." In *Historical and Philosophical Perspectives of Science*, edited by Roger H. Stuewer, 3–11. Minneapolis, MN: University of Minnesota Press.

Feyerabend, Paul K. 1970. "Philosophy of Science: A Subject with a Great Past." In *Historical and Philosophical Perspectives of Science*, edited by Roger H. Stuewer, 172–83. Minneapolis, MN: University of Minnesota Press.

Galison, Peter. 1987. *How Experiments End*. Chicago, IL: University of Chicago Press.

Guerlac, Henry. 1977. *Essays and Papers in the History of Modern Science*. Baltimore, MD: Johns Hopkins University Press.

Hall, Rupert A. 1984. "Beginnings in Cambridge." *Isis* 75: 22–25.

Hanson, Norwood Russell. 1962. "The Irrelevance of History of Science to Philosophy of Science." *Journal of Philosophy* 59: 574–86.

Hesse, Mary. 1970. "Hermeticism and Historiography: An Apology for the Internal History of Science." In *Historical and Philosophical Perspectives of Science*, edited by Roger H. Stuewer, 134–62. Minneapolis, MN: University of Minnesota Press.

Hull, David. 1992. "Testing Philosophical Claims about Science." *PSA: Proceedings of the Biennial Meeting of the Philosophy of Science Association 1992*, Volume Two: Symposia and Invited Papers: 468–75. East Lansing, MI: Philosophy of Science Association.

Johnson, Steven. 2006. *The Ghost Map: The Story of London's Most Terrifying Epidemic, and How It Changed Science, Cities, and the Modern World*. New York, NY: Riverhead Books.

Kragh, Helge. 1987. *An Introduction to the Historiography of Science*. Cambridge: Cambridge University Press.

Kuhn, Thomas S. 1977. "The Relations Between the History and the Philosophy of Science." In *The Essential Tension*, 3–20. Chicago, IL: University of Chicago Press.

Kuhn, Thomas S. 1984. "Professionalization Recollected in Tranquility." *Isis* 75: 29–32.

Lakatos, Imre. 1971. "History of Science and its Rational Reconstructions." In *PSA 1970*, edited by Roger C. Buck and Robert S. Cohen, 91–136. Dordrecht: D. Reidel.

Laudan, Larry. 1990. "The History of Science and the Philosophy of Science." In *Companion to the History of Modern Science*, edited by R.C. Olby, G.N. Cantor, J.R.R. Christie and M.J.S. Hodge, 47–59. London: Routledge.

Laudan, Rachel. 1992. "The 'New' History of Science: Implications for Philosophy of Science." *PSA: Proceedings of the Biennial Meeting of the Philosophy of Science Association 1992*, Volume Two: Symposia and Invited Papers: 476–81. East Lansing, MI: Philosophy of Science Association.

Machamer, Peter. 1994. "Selection, System and Historiography." In *Trends in the Historiography of Science*, edited by Kostas Gavroglou, Jean Christianidis and Efthymios Nicolaidis, 149–60. Dordrecht: Kluwer.

Mayer, Anna-K. 1999. "'I have been very fortunate...'. Brief Report on the BSHS Oral History Project: 'The history of science in Britain, 1945–65'." *British Journal for the History of Science* 32: 223–35.

McMullin, Ernan. 1970. "The History and Philosophy of Science: A Taxonomy." In *Historical and Philosophical Perspectives of Science*, edited by Roger H. Stuewer, 12–67. Minneapolis, MN: University of Minnesota Press.

Nickles, Thomas. 1995. "Philosophy of Science and History of Science." *Osiris* 10 (2nd series): 138–63.

Pinnick, Cassandra, and George Gale. 2000. "Philosophy of Science and History of Science: A Troubling Interaction." *Journal for the General Philosophy of Science* 31: 109–25.

Porter, Roy. 1990. "The History of Science and the History of Society." In *Companion to the History of Modern Science*, edited by R.C. Olby, G.N. Cantor, J.R.R. Christie and M.J.S. Hodge, 32–46. London: Routledge.

Richards, Robert J. 1992. "Arguments in a Sartorial Mode, or the Assymmetries of History and Philosophy of Science." *PSA: Proceedings of the Biennial Meeting of the Philosophy of Science Association 1992*, Volume Two: Symposia and Invited Papers: 482–89. East Lansing, MI: Philosophy of Science Association.

Ruse, Michael. 1992. "Do the History of Science and the Philosophy of Science Have Anything to Say to Each Other?" *PSA: Proceedings of the Biennial Meeting of the Philosophy of Science Association 1992*, Volume Two: Symposia and Invited Papers: 467. East Lansing, MI: Philosophy of Science Association.

Shapin, Steven. 1992. "Discipline and Bounding: The History and Sociology of Science as Seen Through the Externalism-Internalism Debate." *History of Science* 30: 333–69.

Shea, William R. 1983. "Do Historians and Philosophers of Science Share the Same Heritage." In *Nature Mathematized*, edited by William R. Shea, 3–20. Dordrecht: D. Reidel.

Smeaton, William A. 1997. "History of Science at University College London: 1919–47." *British Journal for the History of Science* 30: 25–28.

Steinle, Friedrich, and Richard M. Burian. 2002. "Introduction: History of Science and Philosophy of Science." *Perspectives on Science* 10: 391–97.

Strasser, Bruno J. 2005. "L'Histoire des Sciences, une Histoire à Part Entière?" *Revue Suisse d'Histoire* 55: 3–16.

Thackray, Arnold. 1970a. "Comment on 'Hermeticism and Historiography: An Apology for the Internal History of Science' by Mary Hesse." In *Historical and Philosophical Perspectives of Science*, edited by Roger H. Stuewer, 160–62. Minneapolis, MN: University of Minnesota Press.

Thackray, Arnold. 1970b. "Science: Has Its Present Past a Future?" In *Historical and Philosophical Perspectives of Science*, edited by Roger H. Stuewer, 112–33. Minneapolis, MN: University of Minnesota Press.

Turner, John R. G. 1990. "The History of Science and the Working Scientist." In *Companion to the History of Modern Science*, edited by R.C. Olby, G.N. Cantor, J.R.R. Christie and M.J.S. Hodge, 23–31. London: Routledge.

Young, Robert M. 1990. "Marxism and the History of Science." In *Companion to the History of Modern Science*, edited by R.C. Olby, G.N. Cantor, J.R.R. Christie and M.J.S. Hodge, 23–31. London: Routledge.

Chapter 4
What in Truth Divides Historians and Philosophers of Science?

Kenneth L. Caneva

Let me begin with a little personal history and two anecdotes.

I entered Princeton's graduate program in the History and Philosophy of Science in the fall of 1967. The program embraced students formally enrolled in either the history or the philosophy department. We on the history side were neither required nor especially encouraged to take philosophy courses, and very few of us did. Nor did we see many philosophers in our history of science seminars. The director of the program, Thomas Kuhn, rarely offered a philosophy of science seminar but was heavily involved on the history of science side. Kuhn's office and those of the other history of science professors were together in a separate building housing the program headquarters. That was also where our seminars were held. I neither knew nor cared where the philosophers were. Although nominally in the history department, and required to have a history minor, we historians of science were a group largely apart from the larger number of "straight" historians, as we called them. We had our own study room in Firestone Library where most of us gathered on a regular basis. My impression was that the philosophers of science were fully integrated into the larger philosophy department. In other words, as far as we history of science folks were concerned, the ostensible Program in History and Philosophy of Science was in effective reality a program in the history of science. There was almost no contact between the parts, let alone fruitful interaction. And no one seemed to care. When I think back on the situation, I suspect a tacit but strong attachment to a preoccupation with fostering a proper professional identity may have played a key role.

Let me make clear at the outset that, since the earliest years of my post-doctoral career, I've studied with great interest and great profit many of the leading historically oriented philosophers of science: Popper, Lakatos, Toulmin, Feyerabend, Laudan, Shapere, Nersessian, Nickles, Longino, and (of course) Kuhn. My intellectual development would have been very much impoverished if I hadn't. But I otherwise don't regularly read philosophers of science, and tend to find their work off-putting when I do. They're *other*.

K.L. Caneva (✉)
University of North Carolina at Greensboro, Greensboro, NC, USA
e-mail: klcaneva@uncg.edu

S. Mauskopf, T. Schmaltz (eds.), *Integrating History and Philosophy of Science*, Boston Studies in the Philosophy of Science 263, DOI 10.1007/978-94-007-1745-9_4,

I don't believe that this sense of otherness stems from the fact that historians of science have become increasingly "externalist" by leaving behind "an 'internalist' approach that focuses on the logic of scientific development."[1] For one thing, that's not what "internalist" means to many of us whose work might be so tagged. It means privileging the science, and bringing in any and all relevant considerations that contribute to an understanding of the development, meaning, and implications of that science. It's thoroughly contextualizing, but not "constructivist" in the sense of "focus[ing] on economic, political and cultural influences." Nor can it rightly be called "internalist" insofar as its practitioners don't spend much time trying to define boundaries of inside and outside: its province is everything that affects what a scientist does or thinks.

Which brings me to my anecdotes. At some point during my graduate student days it struck me that although we history of science students had devoted ourselves with full energy to the study of the history of science, we'd never given much attention to the question of what science *is*. The very question seemed irrelevant to what we were doing. Who cared—except maybe Charles Gillispie—whether Lamarck's work was bona-fide "science" or not? But of course demarcationist philosophers of science from Popper on down have cared very much what counts as "science."

On another vividly remembered occasion, probably sometime during my second year at Princeton, I registered a feeling of inferiority vis-à-vis the philosophers—never mind I scarcely knew them. After all, *they* were interested in *truth*, while we historians only told stories on the basis of arcane information drawn from old books. Years later I had a corrective epiphany, when I realized that it was in fact we historians who more directly deal with truth, while philosophers just want to be *right*, or at least unassailable.

Which brings me to my announced theme: What in truth divides historians and philosophers of science?

Is truth about the world—the world of science, if you will—found in the particulars of science as it has actually been done and as revealed by historians of science, or in a philosophically vetted abstraction from that historical reality? The history of science—especially in many of its postmodernist STS incarnations—is often seen as hostile both to the authority of science and to any notion of truth. Nevertheless, I would urge that the authority of science—and if not the absolute truth, then at least the substantial groundedness of its claims—is better grounded in the history of science than in the philosophy of science. Tom Nickles is one of the few philosophers of science to reject the notion that historicism entails epistemic relativism: "While the vagaries of history may seem to favor relativism, history can also provide a laboratory for establishing consilience and robustness."[2] *Pace* Gillispie

[1] The words quoted in this paragraph all come from the statement announcing the theme of the conference at which this paper was read, entitled "Do Historians and Philosophers of Science Have Anything to Say to Each Other?" held at Duke University, 23–24 March 2007.

[2] Nickles (1992, 118, n. 2) discussed in Caneva (2005, 259). Another is Theodore Arabatzis, whose work succeeds uncommonly well in integrating historical and philosophical perspectives; see, in the present connection, Arabatzis (1996, 407, 2002, 114, 123, 2006, 8, 23, and chapter 9, passim) (the first two of which are discussed in Caneva (2005, 247, n. 312, 249, n. 322)).

and Plato, truth lies rather in the historicity of becoming than in the rationality of being.[3]

Kuhn opened the first section of *The Structure of Scientific Revolutions*, headed "Introduction: A Role for History," with a ringing manifesto: "History, if viewed as a repository for more than anecdote or chronology, could produce a decisive transformation in the image of science by which we are now possessed."[4] As everyone knows, Kuhn quickly lost control over the image of science that people took from reading *Structure*, an image that was decidedly relativist even as Kuhn himself steadfastly, if illogically, tried to distance himself from the charge of relativism.[5]

As Kuhn moved away from the framework of *Structure*, he reconceptualized his chief problems so that their solution lay not in the details of the history of science but in logical analysis and analogical modeling. "The trouble with the historical philosophy of science" was, in Kuhn's 1991 critique by that name, "that by basing itself upon observations of the historical record it has undermined the pillars on which the authority of scientific knowledge was formerly thought to rest without supplying anything to replace them."[6] In a curious way he seems to have adopted the viewpoint that in *Structure* he invoked to explain scientists' inclination to suppress the historical context of science: "More historical detail, whether of science's present or of its past, or more responsibility to the historical details that are presented, could only give artificial status to human idiosyncrasy, error, and confusion."[7] Tellingly, his response was to turn (again) from history to a properly recast general philosophical account of science in order to defend its epistemological authority.[8] He seemingly did not see how to ground the authority of science on its actual history.[9] It's not the devil but the savior that lies in the details.

We historians deal with time-bound particulars, and our truths lie in those particulars. In contrast, philosophers seek timeless truths from which the historical particulars have been distilled off. What I would urge is that the proper context for the grounding of scientific knowledge claims be seen precisely in its history, in the route by which scientists have come to accept what they do. It's usually been for very good reasons, reasons necessarily at many points brought into contact with what can only sanely be regarded as constraints from the physical world. Philosophical analysis may clarify, but it cannot authorize.

Let me be clear. I don't mean to suggest *at all* that defending the authority of science be the universal or necessary goal of the history of science. That's certainly

[3]See Caneva (1991) for an autobiographical elaboration of this issue.

[4]Kuhn (1962, 1) = Kuhn (1970, 1).

[5]Kuhn seems quickly to have forgotten his assertion in the second edition of Structure that "I cannot see how the relativist loses anything needed to account for the nature and development of the sciences" (Kuhn, 1970, 207). See Caneva (1998) for a discussion of ways to reinterpret relativism in a way that serves to ground the established findings of science.

[6]Kuhn ([1991]1992, 18) = Kuhn (2000, 118).

[7]Kuhn (1962, 137) = Kuhn (1970, 138).

[8]Kuhn ([1991]1992, 10, 14–15, 18) = Kuhn (2000, 111–12, 115–16, 118–19).

[9]This paragraph is taken from Caneva (2000, 116–17).

not what *I* do. My point *here* pertains to the relationship between the history and the philosophy of science insofar as they address epistemological issues.

Of course the knowledge that counts as scientific has properly been shorn of all particularities of person, place, and time. That's what its touted "objectivity," its universal validity, principally consists in. The loss of historical particularity as one shifts one's attention from the process of production to the finished product nicely parallels the loss of "career line" Kuhn identified as accompanying the shift of attention from particular named individuals to generalized and abstract "natural kinds" in dealing with the problem of reference: "When one makes the transition from proper names to the names of natural kinds, one loses access to the career line or lifeline which, in the case of proper names, enables one to check the correctness of different applications of the same term. The individuals which constitute natural families do have lifelines, but the natural family itself does not."[10]

Insofar as philosophy deals with natural kinds, it cuts itself off from any grounding in the actual historical course of events. The product of historical work is a narrative of events, either descriptive or explanatory; the product of philosophical analysis is a schema, either descriptive or normative. The historical unconnectedness of that schema is supposed to be a mark of its universality, of its objectivity. But unconnected objectivity is a mirage. Popper's objective "third world" is perhaps the best example of philosophers' mystification of the question of knowledge.

Nor have historians ever been enthusiastic about the historical applicability of Popperian demarcationism and falsificationism. Indeed, in their recent book, *Making Modern Science*, Peter Bowler and Iwan Morus assign special significance to historians' rejection of falsificationism: "Here was the point at which the history and philosophy of science began to part company. It seemed to many historians that the more they studied the actual behavior of scientists, the less it fit the idealized picture of the scientific method that the philosophers were devising."[11] More generally, historians can't live with overly precise concepts, whereas the name of the game in philosophy is precisely to make sharp distinctions, to establish precise criteria. At the same time, we historians know full well the extent to which scientists great and small have had to make do with less than precise concepts, less than precise criteria, and they're the ones, after all, who have produced the putative knowledge we lesser beings are scrambling to make sense of.

A closely related issue concerns the proper language, the proper conceptual armamentarium with which to render the past. We're rightly enjoined to use actors' categories lest we distort *their* context of understanding, their context of meaning. But *our* understanding must in the end be rendered in terms of categories meaningful to *us*. Has anyone ever attempted to produce an intelligible account of (say) the replacement of the phlogiston theory by the oxygen theory without a retrospective clarification in terms of what was *really* going on in chemists' laboratories? There

---

[10]Kuhn ([1977]1979, 411) = Kuhn (2000, 199). This paragraph is adapted from Caneva (2000, 105–106, n. 69).

[11]Bowler and Morus (2005, 9).

is a Bohrian complementarity between the two interpretive strategies: both are necessary for an adequate understanding of the past, but they can't be fully deployed simultaneously. This necessary complementarity opens up a space for the fruitful interaction of historians and philosophers of science by dehistoricizing some aspects of historians' understanding.

I can't take it upon myself here to pass in review the many scholars whose work embodies de facto such fruitful interaction. We historians and philosophers of science don't in fact work within non-communicating paradigms, even if an element of incommensurability hangs over our respective notions of where truth lies. In some ways the difficulty with establishing a fruitful dialogue is less an intellectual or conceptual problem than a matter of audience and focus. In surveying the literature on scientific discovery, I was struck not only by the relatively small role historians of science have played, but also by the extent of the non-intersection of the philosophical, sociological, and historical literature even when that work was clearly relevant to researchers in other disciplines.[12] To be sure, most philosophical work on discoveries has dealt with concepts and theories, whereas what scientists are usually famous for discovering are phenomena, laws, and entities, which are more in line with what historians concerned with discovery look at. Yet there nevertheless seems to me to have developed a much more sophisticated understanding of the erstwhile dichotomous enterprises of discovery and justification precisely on the basis of fruitful interaction between historians and philosophers of science.[13] To put it somewhat simply, philosophers have clarified these concepts in conjunction with historians' increasingly sophisticated stories. Perhaps most importantly, few people now see the matter in terms of a presentation-then-evaluation model, but rather recognize that what eventually gets accepted is ongoingly transformed in an extended process of vetting.

Insofar as the history of science is dominated by "externalist" and "constructivist" approaches, as characterized earlier, I see little prospect, and indeed little reason, for any meeting of the minds. However, insofar as the history of science places the science at the center of its attention, then there is common territory on which to pitch our respective tents. I personally have always seen the history of science as an especially promising way of getting at fundamental epistemological and ontological questions: What do we know and how have we come to know it? What's real in the world? What kinds of entities do we suppose—do we allow—to exist there? These are, of course, also deeply philosophical questions. What divides us is where we seek answers: in the particulars of the history of science, or in a sanitized abstraction from all particulars. Historians deal with real people in real situations. For us, there *is* no generic knower; there are only particular concrete people who would see and judge things otherwise if their experiences had been otherwise.[14]

[12]This was especially striking with regard to the abundant—and good—work of Thomas Nickles; see Caneva (2005, 253–54).

[13]Exemplary of this work is that of Theodore Arabatzis (1996, 2002, 2006).

[14]This sentence is adapted from Caneva (1998, 340).

Philosophers typically imagine scientists as decision-making algorithms. In my own work I've been repeatedly struck by how much can change without scientists making a lot of conscious decisions, let alone Kuhnian commitments. Our images of what the typical scientist actually *does* are often very far apart. In which case, maybe we don't have anything to say to each other.

But not so fast. Let me throw out one more topic of potential joint interest where fruitful interaction might be possible: the meaning of terms and the meaning of meaning. Although I don't have any philosophically sophisticated understanding of these issues, a lot of what I've done as an historian has been to tease out meanings and track their changes.[15] And for me as an historian, the meaning of a term comes from the connections and associations supplied by the ensemble of its occurrences, whereby it is of the utmost importance not to assume that a clear *concept* must lie behind a *term*. While important developments are taking place, the opposite is rather the case. I must leave to the judgment of philosophers whether they see anything of use to them in this approach to a common problem.

In this context I would remark that Kuhn's never instantiated attempt to identify a "lexicon" which precisely defines the terms of an erstwhile paradigm is exceedingly unlikely to bring clarity to the history of science, whatever service it might do to Kuhn's increasingly dehistoricized notion of incommensurability.[16] The inappropriateness of trying to identify such a lexicon, especially during creative periods of scientific change, stems from the fact that important terms *don't* necessarily have precise meanings, and that that imprecision facilitates the introduction and eventual acceptance of new categories. I've illustrated this elsewhere with regard to Mayer's and Colding's understanding of "force" and with regard to the collective transformation of Seebeckian thermoelectricity and Ørstedian electromagnetism.[17]

The complex contextuality of meaning—relating on the one hand to the meanings assigned by an individual scientist, on the other to ostensibly community-sanctioned meanings—suggests another place where historians and philosophers might fruitfully engage each other. Insofar as it is generally accepted that scientific knowledge claims are advanced by individuals—however much the statements expressing those claims become modified à la Ludwik Fleck in their transit through a collectivity of interacting individuals—it is also generally accepted that scientific knowledge claims qua scientific knowledge must acquire a depersonalized, community-sanctioned status. It seems to me that this essential epistemological transition is little understood by either historians or philosophers of science, and that its elucidation properly concerns both camps.[18]

---

[15] Most thematically, with regard to "force," in (Caneva 1993 and Caneva 1997[1998]).

[16] See Kuhn ([1986]1989).

[17] Caneva (1993, 1997[1998], 2005). Among many annoying misprints in the last-named, one is serious: in the last line of the first paragraph on p. 258, the words "Ørstedian electrodynamics" should be "Ørstedian electromagnetism."

[18] This issue was discussed in Caneva (1998).

Part of its solution will entail a reconceptualization of what one might usefully mean by "objectivity." As I've argued elsewhere, the notion of objectivity typically connotes something abstract and universal, truth independent of particular people or circumstances; in a word, unconnectedness. But a knowledge claim is quite literally unintelligible outside of communally constituted fields of meaning, nor can it be tested or judged independently of communally sanctioned tests and standards of judgement. It is properly the history of science that exhibits the epistemic warrant of a scientific knowledge claim, precisely by returning to awareness the processes by which it was and continues to be created.[19] Such epistemic warrant has an essential and constitutive diachronic element, the route by which we got to where we believe what we do.[20] In this view, neither our personal views nor the grounding of scientific knowledge can be honestly or adequately rendered in purely synchronic terms.

The question here is whether philosophers of science will be happy with such historical relativizing of a concept ostensibly not bound by personal and apparently contingent circumstances. Yet this kind of relativism should be seen as less threatening when one recognizes that the things to which scientific knowledge is relative are broad and deep intellectual, social, and instrumental connections, repeatedly scrutinized, challenged, defended, and modified.[21]

Historians for the most part accept as true the world validated by modern science, but don't usually make an issue of it. Phlogiston isn't real, miracles can't happen, and disembodied spirits don't exist. In *this* regard we historians might profit from a dose of philosophical skepticism, were it not for the fact that self-styled skeptics are typically among the most uncritical defenders of orthodox ontologies.[22]

On the other hand, the professional skepticism of many contemporary historians of science, a standoffishness with regard to the truth claims of contemporary science, is often a professional sham. Capturing the spirit of our times, Bowler and Morus observed that "[n]owadays, any history that treats the past as a series of steppingstones toward the present—and assumes that the present is superior to the past—is called Whig history."[23] But of course we *all* believe that our cosmology is better than Aristotle's precisely by being closer to the truth of things, and insofar as history succeeds in being an explanatory narrative of events, it must be guided by some endpoint in its story even as it need not assume any inevitability of the path historically traveled.

Let me close with a few further remarks about Kuhn. Although Kuhn was probably the most influential historically minded philosopher of science—or philosophically minded historian of science—of the past 40 years, neither his work nor his pronouncements shed much light on our topic. Indeed, to the frustration and annoyance of his critics, he always insisted that he tried to keep his philosophical

[19]This and the prior two sentences are taken from Caneva (1998, 329).

[20]This point is taken from Caneva (1998, 341).

[21]This sentence is taken from Caneva (1998, 331).

[22]See Caneva (2001[2007]) for an analysis of a range of related issues.

[23]Bowler and Morus (2005, 2).

and historical work separate. Indeed, in the last few decades of his life he showed almost no concern that his philosophy of science be controlled by any history of science.

One might have expected some kind of enlightenment with regard to the coherence of the Kuhnian enterprise from the paper he delivered in 1968 on "The Relations between the History and the Philosophy of Science"—revised in 1976 for publication in *The Essential Tension*—but in fact what we're mostly presented with is how different those separate enterprises are in general. The history of science and the philosophy of science have, we are told, different goals: for the former, a narrative that seeks to understand particular past events; for the latter, the discovery of timeless general truths about science.[24] Moreover, "in philosophy of science, there is no role for the multitude of particulars, the idiosyncratic details, which seem to be the stuff of history," while "there is an autonomy (and integrity) of historical understanding" that resists reduction to general formulas.[25] In the end, Kuhn saw little of value for the historian of science in the literature of the philosophy of science proper, but urged philosophers of science to exploit historical accounts as sources of problems and data grounded in the details of actual scientific practice. In the event, Kuhn the philosopher of science wished to free the understanding of scientific knowledge from dependence on the multifarious particulars that ground an historical accounting.[26] I maintain that that's precisely the wrong way to go, a way that impoverishes both the history and the philosophy of science.

References

Arabatzis, Theodore. 1996. "Rethinking the 'Discovery' of the Electron." *Studies in History and Philosophy of Modern Physics* 27: 405–35.

Arabatzis, Theodore. 2002. "On the Inextricability of the Context of Discovery and the Context of Justification." In *Revisiting Discovery and Justification*, edited by Jutta Schickore and Friedrich Steinle, 111–23. Workshop at the Max Planck Institute, 28 February–2 March 2002 ("Max-Planck-Institut für Wissenschaftsgeschichte," Preprint 211). Berlin: Max-Planck-Institut für Wissenschaftsgeschichte.

Arabatzis, Theodore. 2006. *Representing Electrons: A Biographical Approach to Theoretical Entities*. Chicago, IL and London: University of Chicago Press.

Bowler, Peter J., and Iwan Rhys Morus. 2005. *Making Modern Science: A Historical Survey*. Chicago, IL and London: University of Chicago Press.

Caneva, Kenneth L. 1991. "'Why Are You a Vegetarian?' On the Historicity of Becoming vs. the Rationality of Being and Other Practical Matters." *Mad River: A Journal of Essays* 2: 5–10.

[24] Kuhn ([1968]1977, 5).

[25] Kuhn ([1968]1977, 14, 18).

[26] This paragraph is adapted from Caneva (2000, 96–97). Note that while Kuhn the philosopher emphasized discontinuities (e.g., incommensurability and gestalt switches), Kuhn the historian emphasized continuities (e.g., the historical structure of scientific discoveries, Planck's conceptual trajectory). He never reconciled his conception of discovery as an extended process with his conception of scientific revolutions as sudden, holistic events.

Caneva, Kenneth L. 1993. *Robert Mayer and the Conservation of Energy*. Princeton, NJ: Princeton University Press.

Caneva, Kenneth L. 1997[1998]. "Colding, Ørsted, and the Meanings of Force." *Historical Studies in the Physical and Biological Sciences*. 28: 1–138 [published May 1998].

Caneva, Kenneth L. 1998. "Objectivity, Relativism, and the Individual: A Role for a Post-Kuhnian History of Science." *Studies in History and Philosophy of Science* 29: 327–44.

Caneva, Kenneth L. 2000. "Possible Kuhns in the History of Science: Anomalies of Incommensurable Paradigms." *Studies in History and Philosophy of Science* 31: 87–124.

Caneva, Kenneth L. 2001[2007]. "Are Spirits the Untouchables of Academia? Situated *No Legitimacy* in Feminism." *International Journal of Parapsychology* 12: 41–67 [published February 2007].

Caneva, Kenneth L. 2005. "'Discovery' as a Site for the Collective Construction of Scientific Knowledge." *Historical Studies in the Physical and Biological Sciences*, 35: 175–291.

Kuhn, Thomas S. 1962. *The Structure of Scientific Revolutions*. Chicago, IL: University of Chicago Press.

Kuhn, Thomas S. [1968]1977. "The Relations Between the History and the Philosophy of Science." In *The Essential Tension: Selected Studies in Scientific Tradition and Change*, 3–20. Chicago, IL and London: University of Chicago Press. Revised from a paper delivered at Michigan State University, 1 March 1968.

Kuhn, Thomas S. 1970. *The Structure of Scientific Revolutions*. 2nd edition enlarged. Chicago, IL: University of Chicago Press. "Postscript–1969" on 174–210.

Kuhn, Thomas S. [1977]1979. "Metaphor in Science." In *Metaphor and Thought*, edited by Andrew Ortony, 409–19. Cambridge: Cambridge University Press. Originally presented at a conference at the University of Illinois at Urbana-Champaign, September 1977. Reprinted in Kuhn 2000, 196–207.

Kuhn, Thomas S. [1986]1989. "Possible Worlds in History of Science." In *Possible Worlds in Humanities, Arts and Sciences: Proceedings of Nobel Symposium 65*, edited by Sture Allén, 49–51. Berlin and New York: Walter de Gruyter. "[C]onsiderably revised" since the symposium on 11–15 August 1986. Reprinted in Kuhn 2000, 58–89.

Kuhn, Thomas S. [1991]1992. *The Trouble with the Historical Philosophy of Science*. Robert and Maurine Rothschild Distinguished Lecture, 19 November 1991. Cambridge: Department of the History of Science, Harvard University. Reprinted with slight editing in Kuhn 2000, 105–20.

Kuhn, Thomas S. 2000. *The Road since Structure: Philosophical Essays, 1970–1993, with an Autobiographical Interview*, edited by James Conant and John Haugeland. Chicago, IL and London: University of Chicago Press.

Nickles, Thomas, 1992. "Good Science as Bad History: From Order of Knowing to Order of Being." In *The Social Dimensions of Science*, edited by Ernan McMullin, 85–129. Notre Dame, IN: University of Notre Dame Press.

Chapter 5
History and Philosophy of Science: Thirty-Five Years Later

Ronald N. Giere

5.1 Introduction

I must begin by thanking the editors of this volume, Tad Schmaltz and Sy Mauskopf, for inviting me to reflect on my views of the changing relationship between the history of science and the philosophy of science since publication of my review article, "History and Philosophy of Science: Intimate Relationship or Marriage of Convenience" (Giere 1973).[1] Of course, not only has the relationship between these two fields changed; my views of the relationship have also changed.

5.2 The Problem of Normativity

I was hired ABD in 1966 in the Department of History and Philosophy of Science at Indiana University. Founded in 1960, it was the first department of its kind in the United States. So, from the very beginning of my career, I was daily confronted with the fact that I was not in a philosophy department but an HPS department. I could not avoid reflecting on relationships between the two disciplines. By the time I arrived, the founder of the department, and a staunch advocate of a close relationship between the two disciplines, Russ Hanson, had departed for Yale. There was no such advocate left among the remaining faculty members. The "and" in HPS was just a conjunction. This separation was physically marked by the fact that all the historians' offices were on one side of the hall and the philosophers' offices on the other. The only attempt at combining the two disciplines was in the curriculum. All first year graduate students were required to take a two-semester survey in both

[1] As a sidelight, I should mention that my choice of the marriage metaphor might have been influenced by the fact that I was at that time in the midst of a divorce.

R.N. Giere (✉)
University of Minnesota, Minnesota, MN, USA
e-mail: giere@umn.edu

S. Mauskopf, T. Schmaltz (eds.), *Integrating History and Philosophy of Science*, Boston Studies in the Philosophy of Science 263, DOI 10.1007/978-94-007-1745-9_5,

the history of science and the philosophy of science.[2] But there was no connection between the two surveys. The history survey was a chronological presentation of the whole history of science from the fourth century BCE to the end of the nineteenth century CE. In these early years, the philosophy of science survey consisted of one semester on scientific explanation and one semester on confirmation. All students ended up as either historians of science or philosophers of science, but all ended up knowing more about the other discipline than the faculty.

It was in this context that I approached a review of a volume of *Minnesota Studies in the Philosophy of Science* entitled: *Historical and Philosophical Perspectives of Science* (Stuewer 1970).[3] The volume had papers by both historians and philosophers of science, but no one seriously faced what seemed to me the crucial issue. How can a normative philosophy of science be reconciled with a descriptive history of science? If, as Kuhn had suggested, history of science can serve as evidence for philosophical claims about science, we have a circle. Before one can use the history of science as evidence one needs a philosophical account of what constitutes evidence. So the epistemology of science must precede any use of the history of science as evidence for philosophical claims.[4] And this problem follows after the obvious prior question of how factual claims could ever be evidence for normative claims. I returned to my philosophical work on the foundations of probability and statistics.

5.3 Naturalism

It was a decade before my views on this subject underwent any radical revision. Among the influences on my thinking in the early to mid 1980s were interactions with the social psychologist and philosopher, Don Campbell, and encounters with the then new sociology of science as well as the emerging cognitive sciences. In any case, the result was that I came to the conclusion that the philosophy of science

[2]HPS at Indiana is solely a graduate department. Although some undergraduate courses are taught, there has never been an undergraduate major. That was part of the agreement with the Philosophy Department when HPS was founded.

[3]This review was personally commissioned by the then editor of BJPS, Imre Lakatos, over drinks in London. He had some very definite instructions on what the review should say. I don't remember what they were, but, in any case, I did not follow them.

[4]It is worth recalling that Hanson's view of HPS prominently featured the idea that philosophers should help historians assess the validity of arguments offered by historical figures for various hypotheses, given the evidence available at the time (Hanson 1962). Standards of validity, however, are arrived at independently of any historical facts. While acknowledging that judgments of the weight of evidence are not necessarily deductive, he did not much concerned himself with debates over the best inductive methods then going on among such major figures as Carnap, Reichenbach, and Popper, who (initially) advocated a pure deductivism. I owe these notes on Hanson's views to Matthew Lund (2010).

should be transformed into something like the theory of science. That is, philosophers should be in the business of constructing a theoretical account of how science works.[5] Philosophical claims about science would then have the status of empirical theories. In short, the philosophy of science should be naturalized.[6] This means, among other things, giving up pretensions to finding autonomous standards for the practice of science.

But what about the problem of circularity that had so bothered me earlier? That charge assumes that there is an a priori way of determining what a good scientific argument should be. If one gives up that assumption, then one can investigate the reliability of various methods for judging scientific claims using other methods which, in turn, depend on some prior empirical claims. To take an especially clear and simple example, we can explain theoretically why, when testing the effectiveness of some treatment, randomized designs are better than prospective designs. The latter require that the experimental and control groups be matched for a set of known variables. A randomized trial in effect controls, probabilistically, for all variables, known and unknown. It presumes, however, that one already has very good evidence that the method for randomizing is sound.

Here one is more or less forced to embrace aspects of Pragmatism, particularly the idea that there is no foundational method. Rather, inquiry always beings with the beliefs one has. In that context, anything can be questioned, but not everything at once. And there has to be a reason to question claims hitherto accepted, such as conflict with new data. Science is to be valued not only for the theories it has given us, but also for the methods it uses to modify old theories and establish new ones.

It is also possible to recover a form of normativity. Our theories of how science works are normative in the same way other scientific theories are normative, namely in a conditional rather than categorical form. If one wants to get a rocket to the moon, then one should rely on classical mechanics. We know it works well for that purpose.

It must be stressed that a naturalized philosophy of science is not descriptive in a simple-minded way, just describing the gross behavior of scientists in everyday categories. Rather, it is theoretical in the way most sciences are theoretical. It seeks to uncover underlying processes in the practice of science. Thus, for example, during the last several decades, many philosophers of science have investigated the

[5]This applies only to what is called the "general" philosophy of science. The philosophy of the special sciences should be treated separately. This work is sometimes done within an autonomous philosophical framework, such as logic. It is also sometimes done in the scientists' own frameworks, in which case it is automatically naturalized.

[6]I announced this program in my 1985 paper, "Philosophy of Science Naturalized" (Giere 1985). The title was obviously modeled on Quine's "Epistemology Naturalized" (1969), although I was little influenced by Quine. In fact, I came up with the title first and then hurried to write a paper to go with it since others were sure to come up with the same idea. Indeed, shortly before the paper came out, Don Campbell telephoned me to suggest we write a paper together with just that title. I sheepishly informed him that my paper with that title was forthcoming.

nature and role of models in science. The concept of a model provides a theoretically richer way of understanding scientific practice than just blanket notions of "theory" and "observation." Most recently, various notions of modeling have proven fruitful for understanding the growing role of computer simulations in science (Winsberg 2010).

Finally, of course, a naturalized philosophy of science is fully compatible with the history of science which, by its nature, is a naturalistic study of past science and scientists. Nevertheless, in spite of a historical turn among some philosophers of science, few, if any, thought that the philosophy of science could be wholly assimilated into the history of science.

5.4 Cognitive Science and the Sociology of Science

If one is going to naturalize the philosophy of science, to what is it to be naturalized? At the time there were two obvious candidates: the cognitive sciences and the sociology of science. I leaned toward the cognitive sciences, and the subtitle of *Explaining Science* (Giere 1988) was "A Cognitive Approach". But I never regarded these options as exclusive. On the contrary, I assumed that a comprehensive theory of science would have to include elements of both. Much of the book was organized around the two categories of representation and judgment, both major topics in the cognitive sciences. On the representation side, I emphasized the use of models. On the judgment side, I urged understanding scientific inference not as a kind of logic but as decision making, with individuals and groups of scientists deciding which hypotheses provisionally to accept or reject.

My reason for emphasizing the cognitive over the social was the latter's rejection of any form of scientific realism. Realistic talk among scientists was not to be taken at all literally. And there are no standards for scientific reasoning. In rejecting these views I was in agreement with most philosophers of science. But I was also more sympathetic to the new sociology of science than most philosophers because it emphasized the role of scientists, that is, agents, in the production of scientific knowledge. I had already concluded that a naturalized philosophy of science, like most history of science, should focus on the activities of scientists (Giere 1989). I thus explicitly rejected he positivist attempt to objectify scientific representation in terms of semantic relations between symbols and the world, and scientific judgment in terms of logical relations among statements.

Meanwhile, historians of science were busy assimilating the new sociology of science. They were prepared for such assimilation since the field of history as a whole had for some time been moving in the direction of social history. At that time, this was thus a much more natural match for historians of science than the philosophy of science. The change in the history of science was quite rapid. By the end of the century, a stalwart supporter of history of science as intellectual history sadly admitted to me that "they have won," the "they" being either social historians of science or sociologists of science.

5.5 History of Science, Philosophy of Science, and Sociology of Science

In many ways, Science and Technology Studies (STS) is the natural successor to History and Philosophy of Science (HPS). Its major components are the History of Science and Technology, the Philosophy of Science and Technology and the Sociology (also Anthropology) of Science and Technology. Both the History of Technology and the Sociology of Technology are well developed; the Philosophy of Technology less so. The explicit inclusion of technology is significant because the connections between technology and the rest of society are much tighter than those between science and the rest of society. From the standpoint of the culture at large, it is probably more important that we understand the workings of technology than the workings of science. Nevertheless, I will focus on science, mainly because I have had little engagement with studies of technology.

In many ways, my current views of the relationship between the history of science and the philosophy of science are similar to what they were at the beginning. In particular, I still think these are different disciplines with different goals and methods. The goal of a naturalized philosophy of science is to construct a theory of how science works. It need not assume, however, that science has worked the same way throughout its history. But, as a matter of fact, the philosophy of science is quite present oriented. Historians of science seek to tell us something about how particular historical episodes large and small, unfolded. They rarely try to construct general theories of science, even ones focused on a particular historical period. The different methods of naturalized philosophy of science and history of science are suited to their respective goals.

On the other hand, and already noted, relations between the history of science and the sociology of science are much closer. It may be difficult to tell the difference between a work in the social history of science and one in the historical sociology of science. But some sociologists of science also construct theories of science, sometimes ones intended to be all inclusive. The many works of Bruno Latour provide a prominent example. And the sociologist of science, Andy Pickering (1995), has tried to subsume all of STS under the umbrella of cultural studies. I have always viewed sociological theories of science as complementary to a generally naturalized philosophy of science, and the cognitive study of science in particular.[7]

The main difference in the relationship between the history of science and the philosophy of science since I began is that, whereas there used to be almost no interaction between the two fields, there is now considerable interaction. No longer do philosophers of science write about black ravens and the shadows of flagpoles. Their examples are taken from real science, both contemporary and historical. Sometimes philosophers of science do their own history of science.

---

[7] The foremost advocate of the cognitive study of science (Nersessian 2005), has recently attempted to cross the divide between cognitive and social studies of science, coming from the cognitive side.

For a naturalized philosophy of science, Kuhn's suggestion that the history of science can provided evidence useful to the philosophy of science can be realized. But the evidential relationship here is not especially crisp. There are no crucial experiments to be had. It is more a matter of developing a general interpretive framework to understand scientific activities, past or present. Not all interpretations are equally good, of course, but the criteria for judging differing frameworks are multiple.

Unfortunately, the interaction tends not to be symmetrical. Many philosophers would like to think that the concepts they are developing can be useful to historians in their approach to understanding historical episodes in science. But the presumption that what philosophers say is irrelevant to the work of historians of science seems still strong. And, indeed, a few decades ago even historically minded philosophers of science were busy constructing theories of "methodology" with the aim of showing that the development of science is "rational" or "progressive."[8] Historians were wise to reject these ideas.

My hope for the future is that, as the philosophy of science becomes more thoroughly naturalized, this side of the relationship between history of science and philosophy of science will improve. Again, the distinctions between theories and models, and the general importance of models in science, would seem to provide good candidates for deployment by historians of science. Whether they will take up such ideas remains to be seen.

For me, the big question remains: How are we humans capable of knowing such things as that we ourselves are the product of millions of years of organic evolution and billions of years of the physical evolution of the universe? That, of course, is a philosophical question, but it requires a scientific answer, to which all of us, historians, philosophers and sociologists of science, can contribute.

References

Giere, R.N. 1973. "History and Philosophy of Science: Intimate Relationship or Marriage of Convenience?" *British Journal for the Philosophy of Science* 24: 282–97.

Giere, R.N. 1985. "Philosophy of Science Naturalized." *Philosophy of Science* 52: 331–56.

Giere, R.N. 1988. *Explaining Science: A Cognitive Approach*. Chicago, IL: University of Chicago Press.

Giere, R.N. 1989. "The Units of Analysis in Science Studies." In *The Cognitive Turn: Sociological and Psychological Perspectives on Science*, Sociology of the Sciences Yearbook, Vol. XIII, edited by S. Fuller, M. DeMey, T. Shinn, and S. Woolgar, 3–11. Dordrecht: D. Reidel.

Hanson, N.R. 1962. "The Irrelevance of History of Science to Philosophy of Science." *Journal of Philosophy* 59: 570–86.

Kuhn, T.S. 1962. *The Structure of Scientific Revolutions*. Chicago, IL: University of Chicago Press (2nd edition, 1970).

Laudan, L. 1977. *Progress and Its Problems*. Berkeley, CA: University of California Press.

[8] I am referring, of course, to works such as those of Lakatos (1970) and Laudan (1977), which were responses to the perceived "irrationality" in Kuhn's (1962) picture of science.

Lakatos, I. 1970. "Falsification and the Methodology of Scientific Research Programmes." In *Criticism and the Growth of Knowledge*, edited by I. Lakatos and A. Musgrave, 91–195. Cambridge: Cambridge University Press.
Lund, M. 2010. *N. R. Hanson: Observation, Discovery, and Scientific Change*. Amherst, NY: Humanity Books.
Nersessian, N.J. 2005. "Interpreting Scientific and Engineering Practices: Integrating the Cognitive, Social, and Cultural Dimensions." In *Scientific and Technological Thinking*, edited by M.E. Gorman, R. Tweney, D. Gooding, and A. Kincannon, 17–56. Mahwah, NJ: Lawrence Erlbaum.
Pickering, A. 1995. *The Mangle of Practice: Time, Agency, and Science*. Chicago, IL: University of Chicago Press.
Quine, W.V.O. 1969. "Epistemology Naturalized." In *Ontological Relativity and Other Essays*, 69–90. New York, NY: Columbia University Press.
Stuewer, Roger H. 1970. *Historical and Philosophical Perspectives of Science*. Minnesota Studies in the Philosophy of Science, Vol. 5. Minneapolis, MN: University of Minnesota Press.
Winsberg, E. 2010. *Science in the Age of Computer Simulation*. Chicago, IL: University of Chicago Press.

Chapter 6
Philosophy of Science and Its Historical Reconstructions

Peter Dear

Philosophy of science without history of science is empty;
history of science without philosophy of science is blind.
– Imre Lakatos[1]

6.1 History and Philosophy of Science

The title of this chapter alludes, of course, to a famous article by Imre Lakatos, "History of Science and Its Rational Reconstructions." It is also meant to invoke the title of another celebrated article, itself teasingly resonating with Lakatos's, by Steven Shapin, "History of Science and Its Sociological Reconstructions" (Shapin 1982). Both Lakatos and Shapin wanted to talk about how best to do the history of science, but my own intention is not to tell people how to do the philosophy of science. However, I do value the disciplinary intersection of "history and philosophy of science" (HPS); consequently, my subject is what I think HPS used to be about, and what it could still offer us.

Lakatos's basic point, rather like Kuhn's in *Structure of Scientific Revolutions*,[2] was that the history of science cannot be studied without some theory of what constitutes science itself. For him, a theory of science was necessary to enable the historian to say anything meaningful.[3] There is clearly a basic logical truth here: if an historian of science cannot say what science is, then how can he identify his own subject-matter so as to study its history? (Whether it follows that an elaborated *theory* of science is therefore required is another matter.) One of the implicit assumptions of Lakatos's view is that science is essentially a cognitive enterprise,

[1] Lakatos (1971, 102) (the tag echoes Kant, of course). Cf. versions of the same aphoristic statement in Hanson (1962, e.g., 580).

[2] Kuhn (1970, 207).

[3] Hanson's rather different point, in his (1962), was that the philosopher should use solid historical scientific cases as starting points for analysis, for both heuristic and pragmatic reasons.

P. Dear (✉)
Cornell University, Ithaca, NY, USA
e-mail: prd3@cornell.edu

S. Mauskopf, T. Schmaltz (eds.), *Integrating History and Philosophy of Science*, Boston Studies in the Philosophy of Science 263, DOI 10.1007/978-94-007-1745-9_6,

a way of making knowledge of the sort that can be represented adequately by truth-statements about the world. For most anglophone philosophers of science, this assumption is probably still generally accepted. But historians of science, whether at the time or now, were reluctant to embrace Lakatos's vision of HPS, with philosophy leading history by the hand. Quite what were the characteristic questions for historians of science, in contrast to those of the philosophers, is somewhat unclear, since their work still focused at that time on the history of ideas; perhaps it would not be incorrect to say that, at some level, they wanted to trace how modern scientific ideas had come into being, where the philosopher of science wanted to know why those modern ideas were justified (the assumption usually being that they *were* justified).

Conflicts of interest were there, incipiently, from the outset, and were basically a problem of academic disciplines. Disciplines not only have their own subject matter, but they also have their own problems and ways of addressing them; their own ways of doing things.[4] If HPS seemed a good idea in the 1950s, it was because it apparently had a single subject matter. But it was surely not clear that history and philosophy had the same way of doing things; that they saw their common subject matter, science, in the same way, or raised the same questions about it. William Whewell is sometimes seen as a founder of HPS,[5] but he wrote well before the establishment of a professionally entrenched specialty called "philosophy of science." Philosophy there was, but not philosophy of science, and by the 1950s the latter was much more mature than was a professional "history of science." So the advent of HPS was spearheaded by philosophers of science; the "Harvard Case Studies in Experimental Science,"[6] the *International Encyclopedia of Unified Science*,[7] and the work of Norwood Russell Hanson and Thomas Kuhn, were what HPS meant by the 1960s, rather than the important historical studies of such as Richard Westfall or Charles Gillispie.[8]

Nowadays, a case might be made that professional roles have reversed: where once philosophers were taken to lead historians in setting the agenda and questions for HPS, contemporary philosophers of science who look to history do so by following the lead of historians; many philosophers of science are now *historians* of the philosophy of science. The online discussion board HOPOS, "History of Philosophy of Science," has been well established for many years, and there is a considerable associated coterie of philosophers who specialize in the history of early-modern philosophy, from Descartes to Kant, and who hold regular regional

---

[4]Cf. on related themes Shapin (1992).

[5]Yeo (1994), Fisch and Schaffer (1991).

[6]The individual pamphlets began appearing in 1948, collected in Conant (1957).

[7]An abortive project of logical empiricism promoted by Otto Neurath that began publishing in 1938 and ended unfinished in 1969; its sections (bibliographically awkward but published by Chicago University Press) included Thomas Kuhn's *Structure of Scientific Revolutions* (1962) as well as other sections that fell far from the ambitions of the Vienna Circle in the 1930s.

[8]British parallels could also be cited: Mary Hesse, Gerd Buchdahl, and Cambridge HPS rather than A. Rupert Hall, for example.

conferences. Perhaps the chief movers in this endeavor are Daniel Garber and Roger Ariew, both specialists in Cartesianism. A leading Kant scholar, Michael Friedman, contributed an essay to a recent number of *Isis* as part of a set on the topic "Changing Directions in the History and Philosophy of Science."[9] Of the four pieces, two were by philosophers of science, Friedman and Alan Richardson, and the solution offered by the philosophers to the conundrum of HPS was, in effect, to turn their share of the acronym into the History of the Philosophy of Science; the HOPOS formula. Richardson's essay is called "Scientific Philosophy as a Topic for History of Science,"[10] while Friedman's "History and Philosophy of Science in a New Key," refers at the end to "the relationship between the history of science and (the history of) the philosophy of science."[11] In effect, both authors concede an intellectual priority to historical enterprises. Changing directions indeed.

My own intellectual roots lie in HPS of the Cambridge sort in the 1970s. It can be difficult to break away from those fundamental perceptions of what the *point* of studying science from a scholarly perspective actually is. When the intellectual dynamic of the "philosophy" part of the complex gave way for some of us in the 1980s to that of the "sociology of scientific knowledge" (SSK), it provided a new, but by no means unrelated, meaning to studying the history of science. This new meaning relied fundamentally on many of the assumptions from the philosophy of science that had formed integral parts of HPS, since those were the targets that SSK attacked (many would say in sometimes strawman form).[12] SSK was therefore exciting; it was transgressive and idolatrous. It was also the context for Shapin's explicitly normative "History of Science and Its Sociological Reconstructions" in 1982, which made no bones about expelling Lakatos (and, by implication, philosophers of science generally) from the history of science. Not all historians of science applauded, especially in the United States; the idea of sociology of scientific knowledge evidently felt to them rather too ideological and doctrinaire. HPS had seemed to promise that studying the history of science would enable greater insight into what this special kind of knowledge called science was all about—a glimpse of the transcendental, in a way; and aspirations to a philosophical profundity implied by HPS flattered that idea. SSK, by contrast, seemed rather too mundane and reductionist for any such project, and even, in some cases, a little Marxist. Nonetheless, some of us found SSK liberating. It seemed to turn the historian's ways of doing things into effective means of answering versions of longstanding philosophical questions without having to defer to the philosophers: you didn't have to explain why one theory was better than its predecessors, you gave a local, socially-rooted account

[9]"Focus: Changing Directions in the History and Philosophy of Science," *Isis* 99 (2007), 88–134.

[10]Richardson (2007).

[11]Friedman (2007, quote on 134).

[12]The work of David Bloor in the 1970s was central to the philosophical assaults of SSK, together with that of Barry Barnes and Harry Collins; see for a neat summation of some central arguments (Barnes and Bloor 1982). Historians of science in Britain began to engage with it from about 1980 onwards, while Anglophone philosophers of science reacted soon thereafter.

of why particular groups of people at particular times preferred it.[13] Normativity turned into a philosopher's conceit that historians of science, following the new sociologists, could sneer at. So much for HPS, it appeared.

But of course philosophy of science still haunted the SSK-inspired historians, hidden by the guise of its opposition. Among other things, relativism was proclaimed as a central badge of SSK, which gave people lots of harmless entertainment for several years, and which on occasion involved a good deal of moralizing by some philosophers.[14] That moral outrage seems a bit more moderated now, particularly in the wake of the more recent "science wars," in part because of a strong emphasis on making a distinction between methodological and epistemic relativism.[15] Methodological relativism has been pushed for quite a while now by Harry Collins and is subscribed to by several other contributors to a 2001 essay-collection that Collins edited with Jay Labinger, called *The One Culture?* (Labinger and Collins 2001). The idea is that the social researcher or historian simply proceeds, when studying some aspect or passage of scientific activity, as if the knowledge-claims made by the participants were neither true nor false in an absolute sense; truth or falsity of the claims made cannot therefore be used as elements in an understanding or explanation of what happened. However, no inference is to be drawn from this methodological stance as to any commitment to *epistemic* relativism, wherein the knowledge-claims would be seen as necessarily having no absolute truth or falsity independent of the individuals or groups who adhere to them.

Methodological relativism seems to have been a fairly effective means of deflating the moral outrage of those scientists or science-enthusiasts who bristled at the idea that scientific truth-claims might not be absolutely true in a context-independent sense (although a clear-cut demarcation between methodological and epistemic relativism might not be so easy to establish "absolutely"...). But something of the sort, whether one calls it methodological or pragmatic relativism—devoted to the science-studies scholar's avoidance of taking sides over truth-claims about the natural world—seems to be in practice a fairly stable solution for the time being. An objection to this trick, however, has been made by the sociologist Andy Pickering, in a review of the Labinger/Collins collection (Pickering 2001). Pickering made the point that the representatives of science studies in the book were not themselves representative of the entire sweep of approaches that make up contemporary science studies. However, this seems to leave unaffected the idea that the distinction between methodological and epistemic relativism is useful to pretty much everyone in the field—even if some people might not always want to make it (ideological critiques of science sometimes eschew relativism entirely, for instance).

---

[13]Thus Collins (1985, 145–48), even claimed to offer a generally applicable "sociological resolution of the problem of induction."

[14]A good exemplar is Brown (1984). Later serious engagements with SSK by philosophers include Friedman (1998).

[15]A good deal of science-wars intemperance may be found in Gross and Levitt (1994), Koertge (1998).

6.2 Epistemography and Historical Explanation

Do such considerations elevate SSK, or some other form of science studies, to the status of a kind of successor discipline to HPS? Or, alternatively, is HPS itself flexible enough to take on board some of the central techniques and ambitions of science studies? There are various points of possible rapprochement between them, and between the history and the philosophy of science themselves, but the issue of relativism is surely a part of the mix that needs to be placed under arrest.

My own contribution to the Labinger/Collins collection was an attempt to argue that pretty much everyone doing "science studies" takes a certain core aspect of their endeavor for granted, one that relates directly to the methodological/epistemic distinction in relativism (Dear 2001). Insofar as all science-studies scholars, of whatever kind, necessarily rely at various times on providing accounts of what science or scientists are, or do, in some circumstance or other, then this unavoidable element of purported naturalistic description of science should be seen as the irreducible core of all science-studies work. To emphasize that dimension, and to underline the difference between it and assertions about the validity or invalidity of particular scientific knowledge-claims, of the kind that might offend scientists, I suggested that the core of science studies should be seen as "epistemography"—the attempt to give an empirical account of knowledge-practices. This seems to be a useful pragmatic stance to take, even despite the obvious objection that one person's empirical account might be somebody else's distorted misrepresentation; or some objection based on the "theory-ladenness of observation." The idea is that all science-studies scholars are, as at least part of their work, attempting to present a true account of some aspect of science. That *attempt*, rather than the possibility of ultimate success, is a crucial moral component of all work in science studies.

"Epistemography" thus adopts a moral stance that can position itself quite effectively against the impression that science studies sometimes appears to undermine the trustworthiness and truth-production capacity of the sciences themselves. Much like methodological relativism, the notion of epistemography confronts dismissals of science studies that embody a kind of moral outrage deriving from a view of science that sees it as a transcendental sort of human activity. After all, most people think twice these days before openly basing their position on transcendence.

Furthermore, an epistemographical approach still allows particular pieces of work in science studies—such as studies in the history of science—to be vigorously challenged on grounds that the epistemography is wrong or fallacious, or that the arguments built with it are invalid in some way; science-studies scholars do that to one another all the time, "relativism" notwithstanding.

This issue recalls a striking point once made by David Bloor concerning negative reactions to SSK: he drew parallels between criticisms of his "Strong Programme in the Sociology of Knowledge" and those made in the mid-nineteenth century of Higher Criticism of the Bible and research on the history of church doctrines (Bloor 1988). Turning religious doctrines into historically explicable positions, as Higher Criticism did, was seen by some as supplanting the sanctity of those doctrines with something less exalted: they no longer seemed to be genuine religious doctrines, because they no longer owed their existence to the sanction of God. Similarly, said Bloor, giving sociological explanations for scientific beliefs is often seen as undermining them, because those beliefs will no longer owe their existence to the sanction of a transcendent Nature. Bloor's observations underline the degree to which the enterprise of epistemography has no interest in issues of transcendence.

One of the features of an epistemographical enterprise is that it possesses, if only implicitly, a sense of temporality. The abandonment of Mertonian-style structural-functionalist explanations in understanding the actual enterprises of science surely implied an equal abandonment of the timelessness implied by Mertonian assumptions about the essential nature of science as a social activity—sociologists of science such as Merton himself, or Joseph Ben-David,[16] never worried very much about what they meant by the word "science"; they knew it when they saw it. So the epistemographical core of science studies really implies taking on board fully temporalized understandings of history. Studies of the past should do more than just treat historical episodes as mere "case-studies" located in some kind of "virtual present, which is what I think a lot of sociological studies in the SSK tradition have often done (with honorable exceptions such as Shapin and Schaffer's *Leviathan and the Air-Pump* (Shapin and Schaffer 1985)). What virtual-present, case-study approaches tend to *lose* by discarding a true temporal dimension in their accounts is a certain form of meaning itself: meaning made by time. A commonplace of cultural-historical analysis is that the meanings of ideas or actions in any particular time and place are dependent on what counts as "normal" in that time and place; but the "normal" contains such a plethora of inter-relations, connotations, and ineffable complexities that they cannot really be understood in broadly structural terms (the typical form of SSK accounts), but only temporally or narratively.[17]

Other ways of talking about time within science studies include Bruno Latour's notion of "making time," used most effectively in *The Pasteurization of France*, and Pickering's attempt at a new metaphysics of time, in his talk of "temporal emergence" in passages of scientific work.[18] Each approach expresses, in very different ways, the notion that time is a constitutive element of events, not just a backdrop against which things occur. But oddly, neither of these ways of talking about time has really caught on in science studies. Epistemography itself can be seen as necessarily historicist, in the sense that its accounts must be heavily contextualist, but even more in the sense that it makes passage through time a constitutive element of understanding knowledge and its making. However the enterprise is carried out, and whether or not done by people with an academic training in historical scholarship, the making of knowledge through time is an inseparable part of an epistemographical enterprise.

[16]In particular, Ben-David (1984).

[17]See, for more on cultural history of science, Dear (1995). The narrative understanding of historiographical temporality is classically discussed in White (1978), explicitly applied to works in the history of science in Clark (1995). One (non-invidious) example of an SSK-inspired historical case study in a "virtual present" would be Ashmore (1993).

[18]Latour (1988, esp. 49–52), also Latour (2004, 188–94), Pickering (1995).

None of this is intended to say that, somehow, historians have always "got it right"; one of the glories of historical scholarship is the difficulty of laying down precise rules about how to do it. In fact, precisely what is and is not illegitimate in talking about the history of science is, I think, a major issue for science studies in general, despite the current lack of attention to it. The principal challenge, once more, concerns the category "science" itself: since philosophers gave up on finding any straightforward demarcation criteria by which to give it a decontextualized, universal definition, historians of science, particularly of the pre-modern period, have found it increasingly difficult to establish that the things they like to investigate necessarily have much to do with the modern category "science" at all—because that modern category is itself unclear. HPS with an ambiguous "S," just like science studies that is unsure of what science is, seems to reorient its central concerns around precisely this question of what the field's subject matter is; and there, even epistemography will be inadequate to provide answers.

Stephen Gaukroger has long been an exemplar of the HPS scholar, particularly on the old Cambridge model. His most recent book, *The Emergence of a Scientific Culture* (Gaukroger 2006), with its claims to providing a sweeping historical view and even invoking, in its title, the presumptive and ambiguous word "culture," provides a valuable opportunity to consider what kind of scholarship that HPS model produces. One of the book's most striking aspects is that it shows Gaukroger as a philosopher; in effect, he writes disciplinarily as a philosopher. The very ways in which he sets up his topics and problematics are very much along the lines of his classic *Explanatory Structures*: explication and conceptual analysis of historical texts (Gaukroger 1978). The virtues of Gaukroger's work resemble those of Alexandre Koyré's: although both scholars, with all their differences, are each enormously rewarding to read, neither represents what *historians* of science do nowadays. Koyré's and now Gaukroger's enterprises recall Kuhnian precepts on reconstructing the intellectual coherence of alien scientific enterprises; a hermeneutic endeavor that stops when everything seems to click. That's an enormously stimulating experience, to do for oneself or to read a masterly account devised by somebody else, but it leaves too much hanging. To the historian of science nowadays, it often seems like a propaedeutic enterprise, one that may clarify part of what the historian wants to understand or explain, but which doesn't itself do that job. This is where SSK got its foot in the historian's door 30 years ago.

Gaukroger's work does a lot more than reconstruct "explanatory structures"; he's a philosopher in the sense in which historians of philosophy such as Dan Garber, Roger Ariew, or Tad Schmaltz are philosophers, producing historically situated, intellectual-contextual history of philosophy, or history of science. This enterprise itself, however, can still be seen as something subtly different from the work of many historians of science. One of the great maxims of SSK held that one should not account for people's ideas by reference to other ideas; on this view, ideas don't causally generate new ideas. Instead, one was to seek socio-cultural explanations or understandings of why *these* sorts of people would have preferred *those* sorts of ideas. In practice, of course, this was an unapproachable goal; any account, whether concerning contemporary or past science, will always smuggle in assumptions about how a particular idea unproblematically implied another, and so on. Nonetheless, if an explanatory question concerning beliefs comes front-and-centre, many historians of science, in keeping with that general perspective, will now look for something from an ontological realm distinct from that of ideas: institutional considerations or cognate cultural practices, for example. That recent tendency has perhaps moved the history of science farther away from much philosophy of science.

6.3 Time and Philosophy of Science

In understanding science as an historical enterprise, philosophy has more to offer than simply "philosophy of science" in the sense of epistemology. The philosophy of history is itself important, in making sense of how to engage with the making of an enterprise, an endeavour, or a body of knowledge *over time*, as with the work of Latour and Pickering, or aspects of Hans-Jörg Rheinberger's work (Rheinberger 1997). But the most obvious and immediate intersections are with the work of Thomas Kuhn and Michel Foucault. Each, with differing presuppositions and audiences, attempted to stabilize his objects of study while also allowing temporal movement in historical accounts: in Kuhn's case, the stability of paradigms, or "disciplinary matrices," punctuated by "scientific revolutions"; in Foucault's, stable, all-encompassing "epistemes" separated chronologically by mysterious "ruptures."[19] Each theorist was in some sense a structuralist, and that anthropological approach directly or indirectly helped to render their ideas approachable and attractive to many. But those who did not find such approaches attractive balked precisely at the stop-motion view of science as a knowledge-making enterprise.

Latour's views, to the extent that they can be fixed and characterized, themselves drew explicitly on anthropological models, but rather than emphasizing the "cognitive content" of successive quasi-stable structures in looking at the temporal paths of science, he has looked at the stability of the *processes* of science. In proposing to place a moratorium on talk of "theories" when interpreting science, in his *Science in Action* (Latour 1987), he implicitly criticized the construct "theory" in science-studies itself: "As soon as a divide is made between theories and what they are theories *of*, the tip of technoscience is immediately shrouded in fog. Theories, now made abstract and autonomous objects, float like flying saucers above the rest of science, which by contrast becomes 'experimental' or 'empirical'.... Doing a history of scientific 'theories' would be as meaningless as doing a history of hammers without considering the nails, the planks, the houses, the carpenter and the people who are housed, or a history of cheques without the bank system."[20] Much less tangibly than hammers, theories are presented by Latour as abstractions with

[19] Kuhn (1970), Foucault (1973).

[20] Latour (1987, 241–43, quote 242).

no real ontological status (especially given that Latour is here denying the ontology/epistemology dichotomy itself). Conversely, his position also eschewed the very category of "cognitive content," since the "cognitive" has no role in this play; there are, instead, knowledge-claims, statements that can be made stronger or weaker as they are bandied about "in action."

Latour's provocations and insights were salutary, and rapidly gave rise within science studies, including the history of science, to an emphasis on "science as practice."[21] Although the term "practice" quickly became overused and somewhat evacuated of content, its original purpose was to refocus the study of science away from "ideas" and onto "what scientists do" (or "did"). It thereby took the focus away from the epistemic character of science; this was no longer its interesting feature, the reason for paying attention to science in the first place. Instead, science was interesting as a social and cultural institution.

But it would be hard to be a philosopher of science if one had no interest in science-as-epistemology; indeed, an alternative interest in science as a social and cultural institution would mean that one would not be a philosopher at all, but a sociologist or anthropologist. Perhaps the most relevant shift in the philosophy of science since the rise of SSK was a much increased focus on so-called "naturalistic" approaches in the philosophy of science, involving the examination of "science as she is practiced" instead of more prescriptive or normative reconstructions of scientific work. While Lakatos's "rational reconstructions" have sometimes given way to applications of Bayesian logic, in general philosophers of science are much more concerned with understanding what scientific practitioners do and have done, and how best to make sense of it. That surely brings work in the philosophy of science into the arena of the history of science, but the question then remains as to what the relationship might be between the questions asked by the philosopher and those asked by the historian.

Sometimes philosophy of science of the more "naturalistic" kind seems to be done on an implicit assumption that scientists clearly know what they're about, so a close study of them will reveal the immanent wisdom behind their work—otherwise, why bother studying them? To some extent this is a long-standing tradition: figures like William Whewell were engaged in that sort of enterprise, fitting historical accounts to a broad normative methodological picture of science so as to have each support the other (again, perhaps echoes of Lakatos, too). Another dimension of "naturalistic" philosophical study of science is interventionist work on contemporary science, where the prime example would be evolutionary biology, where philosophers of science such as Michael Ruse and Elizabeth Lloyd involve themselves in ongoing scientific debates, applying philosophical acumen to what one might call natural-philosophical arguments.[22] Nancy Cartwright and recently Ron Giere are among those who have focused especially on this "natural-philosophical" dimension of the philosophy of science, in a tradition reminiscent

[21] Notably, Pickering (1992).

[22] Ruse (2006), Lloyd (2005).

(and, in Cartwright's case, more than reminiscent) of Aristotle.[23] This second enterprise is the least disturbing to historians of science, perhaps; whereas the first, to the extent that it involves on at least some implicit level notions of proper procedure or legitimate "method" in science as the quarry to be chased down, goes against the grain of most historians, who have long since abandoned the idea that there is any such thing as a real, timeless "scientific method": Paolo Rossi attributes to Alexandre Koyré a paraphrase of Napoleon on strategy, that scientific method is easy; it's the application that's hard.[24]

6.4 Epistemic Themes as a Commonality in HPS

These issues surrounding the possibility of a revivified HPS are especially difficult for historians of the pre-modern period, because of the aforementioned problems of the very word "science." A radically different way of posing the question may help to resolve the difficulties, however, and serve to integrate with a newer HPS sensibility. We have always tended to consider the history of science as being about the development and establishment of particular ideas about nature, and the ways in which people have gone about doing those things. But there are other ways of addressing the past of science, which I would like to consider briefly from my own recent work.

The first approach I really owe to my Cornell colleague Mike Lynch. In his *Scientific Knowledge and Ordinary Action*, Lynch discusses ways in which ethnomethodology can contribute things of value to the social studies of science. In particular, he argues that much of what we study when we look at scientific practice is the elaboration and deployment of a host of specific epistemic themes that are reiterated constantly in scientific discourse; themes such as observation, description, replication, testing, measurement, explanation, proof (Lynch 1993, esp. 280, 300). A focus on epistemic themes, reminiscent of Raymond Williams's focus on keywords (Williams 1976), is an attractive way to describe much of what the domain of science studies tries to do. For the history of science, the approach is a way of relating together the diverse array of subjects and time-periods that field tries to address. Since hardly anyone tends to claim nowadays that there really is a single, essential kind of thing called "Science," a focus on recurrent epistemic themes indicates how it might be possible to examine comparable problematics over a considerable sweep of history—even despite the likely absence of a transhistorical constancy underlying them. The history of science can readily be approached by combining Lynch's epistemic themes with a Foucauldian notion of *genealogy*[25]; this amounts to a conception of history that eschews the treatment of the past

[23]Cartwright (1989), Cartwright (1992), Giere (1999), Giere (2006), these latter representing a move away from the author's earlier cognitivism.
[24]Rossi (1982, 5).
[25]Classically exemplified in Foucault (1997).

from a perspective of anthropological strangeness in favor of historical investigation oriented towards understanding better the kinds of cultural and intellectual practices that we're possessed with in the present. And in that enterprise philosophical concerns and analysis must play an essential role.

Reason and rationality have been major epistemic themes in the sciences, mathematics, natural history and various other knowledge enterprises for many centuries; but at the same time, it would obviously be ahistorical to imagine that "reason" has always been the same thing. But it is clearly an appropriate subject for investigation as an epistemic theme of the kind discussed by Lynch.[26] For the longest time (about three centuries), it's been usual to speak of philosophical approaches in the seventeenth century as falling into two main camps, the "empiricist" and the "rationalist." Attempts in the seventeenth century to draw sharp battle-lines between the two were often vigorous, especially among English "empiricists" such as Locke and Boyle. However, self-styled empiricists or not, seventeenth-century philosophers all spoke of "reason" as a central component of knowledge-generation. "Experience" meant using one's senses to learn new things, but it also involved thinking about those new things; "reason" had to do with that appropriate way of thinking.

Most practical uses of the term "reason", or else "rational", and their cognates in this period did not subject it to much interrogation, any more than is usually done nowadays; philosophers in most circumstances talked as though everyone knew perfectly well what such terms meant, and as though these obvious terms always represented something *good*. Lexicographically, the Latin term "*ratio*", like its vernacular equivalents, had many senses in different disciplinary contexts, such as mathematics, logic, metaphysics, or physics. In the context of logic, reason was explicated in terms of deductive, inductive, analogical, probable, or demonstrative kinds, in various combinations.[27] At the same time, actual uses of the term or its cognates seldom came with explicit indications of which relevant sense of the term "reason" they meant to invoke; it was presumably supposed to be evident from context, or else functionally ambiguous—like language in general.

One celebrated locus in which the word "*ratio*" appears is a comparatively unpublicized one in modern scholarship, because it's usually rendered into modern languages as something other than "reason": this is Francis Bacon's talk about "method." Rather notoriously in Baconian scholarship, Bacon does not use the perfectly contemporary word *methodus*, or "method," to label what's usually translated that way; instead, his term, usually translated as "method," in the "Great Instauration" or the *New Organon* is *via et ratio*. Bacon's famous method is a *via*, in other words, a *way* of going about things, and in this phrase "*via et ratio*" we can see that same sense incorporated into Bacon's sonorous periods, using the word

[26] I have also considered the epistemic theme of "intelligibility" from this same perspective in Dear (2003).

[27] See, for examples, Goclenius (1613, 954–59), Micraelius (1662, 1200–202), cf. Chauvin (1713, 555).

"*ratio*" itself. And this might suggest that "reason" is here for Bacon a way of *going along*.[28]

One way of "going along" that was highly prized in the seventeenth century, as well as since, concerns inferential procedures in mathematics. Deductive proofs were the gold standard in seventeenth-century mathematics, and they require, in good Euclidean style, the step-wise movement from one set of statements to a new statement that the previous ones were held to imply. But there were those who failed to see those sorts of movements as unproblematic. There was no difficulty in actually performing them (except perhaps for sceptics), but understanding what justified them was another matter.

Descartes's *Rules for the Direction of the Mind*, from the 1620s, analyzes deductive inference, much as had Aristotle, down to its fundamentals. Rule 10 characterizes the procedures of syllogistic logic as a model for disciplined reasoning, but then criticizes them for being too rote. His point is effectively psychological: syllogistic logic makes reasoning so automatic that the mind fails to pay attention to what it's doing. Logicians prescribe precepts that supposedly govern human reason, "in which the conclusions follow with such irresistible necessity that if our reason relies on them, even though it takes, as it were, a rest from considering a particular inference clearly and attentively, it can nevertheless draw a conclusion which is certain simply in virtue of the form." That way lies the risk of being taken in by sophisms, whereas by contrast "the cleverest sophisms hardly ever deceive anyone who makes use of his untrammeled reason."[29] Reason appears here as something that precedes formal argumentation, rather than being captured by the precepts of formal argumentation. Instead, it involves what Descartes calls "intuition."

"The self-evidence and certainty of intuition is required," says Rule 3, "not only for apprehending single propositions, but also for any discourse whatever.... Take, for example, the inference that 2 plus 2 equals 3 plus one: not only must we intuitively perceive that 2 plus two make four, and that 3 plus one make four, but also that the original proposition follows necessarily from the other two."[30] This kind of intuition, moving from one set of statements to the next, is a kind of seeing: you just *see* that the conclusion follows.[31]

Deductive reasoning is grounded, for Descartes, in primitive intuition: irreducible leaps from one statement to the next that you just have to *see*, and in a more *capacious* sense, a trained capacity to encompass an entire deductive *sequence* as a single intuitive perception. In that sense, the core of deductive reasoning was anything *but* mechanical. Even deduction was not *simple* deduction.

Blaise Pascal, like Descartes, spoke of formal reasoning with reference to the "natural light," which enables us to obtain certainty in something like geometry

[28]Cf. Silverthorne's translation "way and method" in Bacon (2000, 22).

[29]Descartes (1996, 10: 405–406); translated in Descartes (1985–1991, 1: 36).

[30]Descartes (1996, 10: 369); modified from Descartes (1985–1991, 1: 14–15).

[31]The analogy is explicit in Rule 9, Descartes (1996, 10: 400–401).

even despite the impossibility of defining and proving every element of the requisite starting principles. In the *Pensées*, Pascal describes the matter compendiously:

> The knowledge of first principles—space, time, motion, and numbers—is as firm as any of those that our reasoning gives us, and it is on this knowledge of the heart and instinct that reason must rely and must base all its discourse. The heart feels that there are three spatial dimensions and that numbers are infinite, and reason then demonstrates that there are no two square numbers of which one is double the other. Principles are felt, propositions are proved, and both with certitude, although by different ways—and it is as useless and ridiculous for reason to demand of the heart proof of its first principles before willingly accepting them as it would be ridiculous for the heart to demand of reason a sentiment of all the propositions that it demonstrates in order to be willing to accept those. This powerlessness ought only to serve to humiliate reason... but not to undermine our certitude.
> (Pascal 1963, 512 col.II (L110; my translation))

One might say that, for Pascal, "reason" is more of a human practice than a transcendent route towards truth.

"Reason" as an epistemic theme is historically a rather flexible resource, a means of coercion and persuasion. One of the advantages of examining early-modern discussions of such modern categories and philosophical shibboleths is that those discussions often tell us a great deal about what we've unwittingly accepted as moderns—such as disciplined inner conviction as a mark of truth.

I want to finish by outlining another way of addressing science historically. This also comes from particular understandings of the early-modern period, and comes closest to a genealogical approach, in Foucault's sense of a non-teleological historical emergence of the present. As historians of science increasingly confront the question of what, for any given period, should count as "science" in the first place, it behooves them to consider the kinds of categories and practices that went into the creation of the cultural category "science" as it emerges, according to general consensus, by the nineteenth century.

Early-modern engagements with formal knowledge of nature, and especially the employment of new uses of the term "natural philosophy" in the seventeenth century, seem to show a movement towards an uneasy accommodation between, on the one hand, natural philosophy in its classical sense of a contemplative branch of philosophy, and on the other hand an endeavour aimed at practical utility and instrumental application. That accommodation was facilitated by the development from the seventeenth century onwards of an explicit, theorized kind of experimental practice that could link claims about the nature of the world to instrumental techniques for exploiting it. But the accommodation never truly clicked; it never became truly "natural"—even though it was routinely represented as if it was.

A crucial paradox seems to be at stake here: to the extent that people ever since the eighteenth century have tended to understand "science" as being fundamentally "natural philosophy"—contemplative understanding of the world—then the nineteenth-century's category of "applied science" can be regarded as something that emerges unproblematically from natural philosophy taken as "pure" science. But complementarily, to the extent that people since the eighteenth century have understood "science" to mean, fundamentally, "instrumentality"—the capability of utilitarian, operational control of nature—then to that extent the status of natural philosophy is simply that of a set of accounts of nature, belief in the truth of which is justified by the very fact of their discursive implication in the instrumental work itself. Those alternatives have been two distinct ways of representing what "science" is, but they have not generally been clearly distinguished from one another. Instead, there has been an implicit relationship of bootstrapping between them, each supporting the other by virtue of only one of them being attended to at a time. If both the natural-philosophy view and the instrumentality view are interrogated side-by-side, however, the circularity and non-necessity of their mutual support becomes evident. And according to my argument, this situation, one that modern science has directly inherited from early-modern traditions, incorporates what one can see as the basic ideology of modern science.[32]

The Indiana University Department of History and Philosophy of Science asks the question on its website "What is History and Philosophy of Science?" Its answer begins with this statement: "Studies take many different forms, all with the common aim of understanding how science works."[33] The careful ambiguity in that description is crucially important in light of the ideology just outlined: what is meant by the phrase "how science works"? What are its implications—that science "works" because of its instrumental efficacy? Does this mean that science is just a set of techniques? Or does it mean that science successfully discovers truths about the natural world? Genealogically, historically, perhaps we can say that it should mean both at once. Such an analysis, and a tracing-out of its meaning and implications, would indeed be a fitting endeavour for HPS.

References

Ashmore, Malcolm. 1993. "The Theatre of the Blind: Starring a Promethean Prankster, a Phoney Phenomenon, a Prism, a Pocket and a Piece of Wood." *Social Studies of Science* 23: 67–106.

Bacon, Francis. 2000. *The New Organon,* edited and translated by Lisa Jardine and Michael Silverthorne. Cambridge: Cambridge University Press.

Barnes, Barry, and David Bloor. 1982. "Relativism, Rationalism and the Sociology of Knowledge." In *Rationality and Relativism*, edited by Martin Hollis and Steven Lukes, 21–47. Cambridge, MA: MIT Press,.

Ben-David, Joseph. 1984. *The Scientist's Role in Society: A Comparative Study*. Chicago, IL: University of Chicago Press.

Bloor, David. 1988. "Rationalism, Supernaturalism, and the Sociology of Knowledge." In *Scientific Knowledge Socialized*, edited by Imre Hronsky, Márta Fehér and Balázs Dajka, 59–74. Dordrecht: Kluwer.

Brown, James R., ed. 1984. *Scientific Rationality: The Sociological Turn*. Dordrecht: D. Reidel.

Cartwright, Nancy. 1989. *Nature's Capacities and Their Measurement*. Oxford: Clarendon Press.

[32] This argument is pursued at greater length in Dear (2005); these themes also run through Dear (2006).

[33] http://www.indiana.edu/~hpscdept/ (accessed May 15, 2008).

Cartwright, Nancy. 1992. "Aristotelian Natures and the Modern Experimental Method." In *Inference, Explanation and Other Frustrations*, edited by John Earman, 44–71. Berkeley, CA: University of California Press.
Chauvin, Étienne. 1713. *Lexicon philosophicum*, Facs. rpt. 1967. Dusseldorf: Janssen.
Clark, William. 1995. "Narratology and the History of Science." *Studies in the History and Philosophy of Science* 26: 1–71.
Collins, H.M. 1985. *Changing Order: Replication and Induction in Scientific Practice*. London: Sage.
Conant, James Bryant, ed. 1957. *Harvard Case Histories in Experimental Science*, 2 vols. Cambridge, MA: Harvard University Press.
Dear, Peter. 1995. "Cultural History of Science: An Overview, with Reflections." *Science, Technology, and Human Values* 20: 150–70.
Dear, Peter. 2001. "Science Studies as Epistemography." In *The One Culture? A Conversation About Science*, edited by Jay A. Labinger and Harry Collins, 128–41. Chicago, IL: University of Chicago Press.
Dear, Peter. 2003. "Intelligibility in Science." *Configurations* 11: 145–61.
Dear, Peter. 2005. "What Is the History of Science the History *Of*? Early Modern Roots of the Ideology of Modern Science," *Isis* 96: 390–406.
Dear, Peter. 2006. *The Intelligibility of Nature: How Science Makes Sense of the World*. Chicago, IL: University of Chicago Press.
Descartes, René. 1985–1991. *The Philosophical Writings of Descartes*. Translated by John Cottingham, Robert Stoothoff, and Dugald Murdoch, 3 vols. Cambridge: Cambridge University Press.
Descartes, René. 1996. *Oeuvres de Descartes*, edited by Charles Adam and Paul Tannery, 11 vols. Paris: J. Vrin.
Fisch, Menachem, and Simon Schaffer, eds. 1991. *William Whewell: A Composite Portrait*. Oxford: Clarendon Press.
Foucault, Michel. 1973. *The Order of Things: An Archaeology of the Human Sciences*. New York, NY: Vintage Books.
Foucault, Michel. 1977. *Discipline and Punish: The Birth of the Prison*. Translated by Alan Sheridan. New York, NY: Pantheon.
Friedman, Michael. 1998. "On the Sociology of Scientific Knowledge and Its Philosophical Agenda." *Studies in History and Philosophy of Science* 29: 239–71.
Friedman, Michael. 2007. "History and Philosophy of Science in a New Key." *Isis* 99: 125–134.
Gaukroger, Stephen. 1978. *Explanatory Structures: A Study of Concepts of Explanation in Early Physics and Philosophy*. Atlantic Highlands, NJ: Humanities Press.
Gaukroger, Stephen. 2006. *The Emergence of a Scientific Culture: Science and the Shaping of Modernity 1210–1685*. Oxford: Oxford University Press.
Giere, Ronald N. 1999. *Science Without Laws*. Chicago, IL: University of Chicago Press.
Giere, Ronald N. 2006. *Scientific Perspectivism*. Chicago, IL: University of Chicago Press.
Goclenius, Rudolph. 1613. *Lexicon Philosophicum*, Facs. rpt. 1964. Hildesheim: Olms.
Gross, Paul R., and Norman Levitt. 1994. *Higher Superstition: The Academic Left and Its Quarrels with Science*. Baltimore, MD: The Johns Hopkins University Press.
Hanson, Norwood Russell. 1962. "The Irrelevance of History of Science to Philosophy of Science." *The Journal of Philosophy* 59: 574–86.
Koertge, Noretta, ed. 1998. *A House Built on Sand: Exposing Postmodern Myths About Science*. Oxford: Oxford University Press.
Kuhn, Thomas S. 1970. *The Structure of Scientific Revolutions*, 2nd edition. Chicago, IL: University of Chicago Press.
Labinger, Jay A., and Harry Collins, eds. 2001. *The One Culture? A Conversation About Science*. Chicago, IL: University of Chicago Press.

Lakatos, Imre. 1971. "History of Science and Its Rational Reconstructions." In *PSA 1970*. Boston Studies in the Philosophy of Science, vol. viii, edited by Roger C. Buck and Robert S. Cohen, 91–108. Dordrecht: D. Reidel.

Latour, Bruno. 1987. *Science in Action: How to Follow Scientists and Engineers through Society*. Cambridge, MA: Harvard University Press.

Latour, Bruno. 1988. *The Pasteurization of France*. Translated by Alan Sheridan and John Law. Cambridge, MA: Harvard University Press.

Latour, Bruno. 2004. *Politics of Nature: How to Bring the Sciences into Democracy*. Translated by Catherine Porter. Cambridge, MA: Harvard University Press.

Lloyd, Elisabeth A. 2005. *The Case of the Female Orgasm: Bias in the Science of Evolution*. Cambridge, MA: Harvard University Press.

Lynch, Michael. 1993. *Scientific Practice and Ordinary Action: Ethnomethodological and Social Studies of Science*. Cambridge: Cambridge University Press.

Micraelius, Johannes. 1662. *Lexicon philosophicum*, 2nd edition Stettin, Facs. rpt. 1966. Dusseldorf: Janssen.

Pascal, Blaise. 1963. *Oeuvres complètes*, edited by Louis Lafuma. Paris: Éditions du Seuil.

Pickering, Andrew, ed. 1992. *Science as Practice and Culture*. Chicago, IL: University of Chicago Press.

Pickering, Andrew. 1995. *The Mangle of Practice: Time, Agency and Science*. Chicago, IL: University of Chicago Press.

Pickering, Andrew. 2001. "Sociology vs. Science: The War Drags On." Review of Labinger and Collins, ed., *The One Culture*? *Physics World* 14:45–46.

Rheinberger, Hans-Jörg. 1997. *Toward a History of Epistemic Things: Synthesizing Proteins in the Test Tube*. Stanford, CA: Stanford University Press.

Richardson, Alan. 2007. "Scientific Philosophy as a Topic for History of Science." *Isis* 99: 88–96.

Rossi, Paolo. 1982. "The Aristotelians and the Moderns: Hypothesis and Nature." *Annali dell'Istituto e Museo di storia della scienza di Firenze* 7: 3–28.

Ruse, Michael. 2006. *Darwinism and Its Discontents*. Cambridge: Cambridge University Press.

Shapin, Steven. 1982. "History of Science and Its Sociological Reconstructions." *History of Science* 20: 157–211.

Shapin, Steven. 1992. "Discipline and Bounding: The History and Sociology of Science as Seen Through the Externalism-Internalism Debate." *History of Science* 30: 333–69.

Shapin, Steven, and Simon Schaffer. 1985. *Leviathan and the Air-Pump: Hobbes, Boyle, and the Experimental Life*. Princeton, NJ: Princeton University Press.

White, Hayden V. 1978. *Tropics of Discourse: Essays in Cultural Criticism*. Baltimore, MD: Johns Hopkins University Press.

Williams, Raymond. 1976. *Keywords: A Vocabulary of Culture and Society*. New York: Oxford University Press.

Yeo, Richard. 1994. *Defining Science: William Whewell, Natural Knowledge and Public Debate in Early Victorian Britain*. New York, NY: Cambridge University Press.

Chapter 7
The Underdetermination Debate: How Lack of History Leads to Bad Philosophy

Wolfgang Pietsch

7.1 Introduction

Over the course of a century, the debate on underdetermination has produced an abundance of versions of the thesis that evidence does not uniquely determine scientific theories. Almost everybody agrees that some weak transitory underdetermination is a historical reality while several strong renderings are clearly implausible. Thus, the real challenge of the debate consists in formulating the underdetermination thesis in a way that strikes the right balance between the extremes. Such a formulation reaches beyond the trivial observation that theories are underdetermined if relevant evidence is missing. It should be methodologically useful both for the working scientist and for the historian of science while evading the common objections.

We will show in this essay that as a guideline to a philosophically viable conception of underdetermination the historical perspective proves to be essential. Conversely, attempts at an ahistorical formulation of the thesis will lead to a distorted notion of underdetermination that is open to all kinds of objections. The story will rely on a simplified construal of the debate on underdetermination in the twentieth century. It will contrast the views of Pierre Duhem and W. V. O. Quine, respectively the most prominent proponents of a historical and an ahistorical rendering of underdetermination. We will show that the common objections against underdetermination are fatal only to Quine's version. Thus, in this case historical ignorance indeed led to bad philosophy.

Ultimately, the underdetermination thesis should be able to make sense of those episodes in the history of science, where a situation of underdetermination was diagnosed by at least some of the leading scientists involved. Examples are plenty, of which I will mention three. One that is much discussed concerns the transition from Euclidean to non-Euclidean physics at the turn from the nineteenth to the twentieth century. There was a fairly short period, when several of the leading geometers and physicists of the time, including Hermann von Helmholtz, Henri Poincaré, and

W. Pietsch (✉)
Carl von Linde-Akademie, Technische Universität München, München, Germany
e-mail: pietsch@cvl-a.tum.de

S. Mauskopf, T. Schmaltz (eds.), *Integrating History and Philosophy of Science*, Boston Studies in the Philosophy of Science 263, DOI 10.1007/978-94-007-1745-9_7,

Albert Einstein, held the choice between the different axiomatizations of geometry to be of conventional nature, each axiomatization corresponding to a different formulation of the laws of physics (e.g. Helmholtz 1870; Poincaré 1902; Einstein 1921). The demand for simplicity of those laws would somewhat narrow down sensible choices of the axiomatization of geometry.

Another often-cited example is the equivalence between wave and matrix mechanics in the early years of quantum theory as pointed out in particular by Erwin Schrödinger: "Considering the extraordinary differences between the starting-points and the concepts of Heisenberg's quantum mechanics and of the theory which has been designated 'undulatory' or 'physical' [...] it is very strange that these two new theories agree with one another with regard to the known facts, where they differ from the old quantum theory. [...] That is really very remarkable, because starting-points, presentations, methods, and in fact the whole mathematical apparatus, seem fundamentally different" (Schrödinger 1926, 45). As Schrödinger stresses, matrix mechanics emphasizes the discontinuous nature of matter, while wave mechanics emphasizes the continuous aspects. Consequently, wave mechanics is naturally formulated in terms of differential equations, while matrix mechanics introduced its own peculiar algebraic language into physics. As in most historical cases of underdetermination, the exact interpretation of this situation has been quite controversial both among the scientists directly involved and later on among historians and philosophers of quantum theory (Muller 1997; Perovic 2008).

While a considerable philosophical literature exists on the equivalence of matrix and wave mechanics and even more on the conventionality of geometry, another pertinent example has evaded almost completely the attention of modern philosophers of science.[1] Underdetermination occurs in the transition from Newtonian action-at-a-distance theories to field theoretic formulations in classical physics, most notably in the case of electrodynamics (Pietsch 2010). Reading Maxwell's *Treatise on Electricity and Magnetism*, nothing short of being the *Principia* for classical electrodynamics, most modern readers will be surprised that a large part of the preface is devoted to an analysis of the relation between Faraday's field electrodynamics and the action-at-a-distance electrodynamics of Coulomb or Ampère. Maxwell emphasizes the equivalence of both theories, not only in terms of empirical adequacy, but also of non-empirical epistemic criteria: "In a philosophical point of view, moreover it is exceedingly important that two methods should be compared, both of which have succeeded in explaining the principal electromagnetic phenomena, and both of which have attempted to explain the propagation of light as an electromagnetic phenomenon, and have actually calculated its velocity, while at the same time the fundamental conceptions of what actually takes place, as well as most of the secondary conceptions of the quantities concerned, are radically different" (Maxwell 1873, xii).

---

[1]Complaints about an alleged lack of examples are quite frequent in the literature (e.g. Norton 2008, 25). However, the fact that the underdetermination in classical electrodynamics has barely been discussed suggests that examples were searched for in the wrong place and that there are many more to be unearthed from the history of science (see also Section 7.3, item iv).

My criteria for formulating the "correct" version of the underdetermination thesis are of pragmatic nature: that it is of methodological importance for actual scientific theorizing and that it proves a useful tool for reconstructing certain episodes in the history of science. Of course, it should also evade all objections that have been raised against the thesis. With an eye on actual scientific practice both in its historical and methodological dimensions, the underdetermination thesis that we will embrace will be largely in Duhemian spirit. By contrast, the logical and linguistic remarks in Quine's rendering will turn out largely irrelevant to our formulation of the thesis. Basically, the historical version of underdetermination that we are eventually going to defend is the following: *In the history of science, especially of physics, one repeatedly encounters situations where several theories are equally strong in terms of their empirical consequences and with regard to epistemic virtues but rely on different metaphysics that provide the scientists with different instructions what to do next and what to expect from nature. There are no convincing reasons to exclude the possibility of such situations in the future.*[2] None of the known objections is fatal to this formulation of the underdetermination thesis, while it nevertheless holds important implications for scientific method.

Section 7.2 will provide an overview of the most important versions of the underdetermination thesis. In Section 7.3, we will compare Duhem and Quine's versions of underdetermination and link the differences to their respective attitudes towards the history of science. The attitude towards history can also serve as a guideline for detecting further differences beyond those that are usually discussed in the literature. In the next step of the argument, it will be pointed out in Section 7.4 how the neglect of historical perspective has been detrimental to the underdetermination thesis since the common objections are fatal only to the ahistorical view. Section 7.5 will briefly summarize from the literature the two most important arguments in favour of the underdetermination thesis and point out their weaknesses. Concluding from Sections 7.4 and 7.5 we are confronted with a stalemate, where neither the arguments for nor those against underdetermination turn out ultimately convincing. Section 7.6 will try to ameliorate this situation by searching for further arguments in support of underdetermination that are informed by the historical perspective. We will show that denying underdetermination can seriously hinder progress in certain instances of the evolution of science. Also, denying underdetermination leads to an implausible conception of scientific theories, in which conventional elements can be unequivocally separated from non-conventional elements. The relevance of a historical perspective for thinking about scientific method will be briefly addressed in Section 7.7.

---

[2]We leave open the question if future evidence can decide between competing approaches. The issue is somewhat overrated since in the historical context one often finds disagreement about which assumptions are central to a framework and which are merely auxiliary. Still, given that the different approaches have all proved fruitful in the past, it is plausible that some core assumptions can be upheld no matter what evidence comes up. Otherwise, the previous successes of the various frameworks would seem a miracle.

7.2 Underdeterminations

More than a century after the publication of the *Aim and Structure of Physical Theories*, Duhem's *opus magnum* on the methodology of physics, the often-heated debate concerning underdetermination is difficult to disentangle. Various renderings of the underdetermination thesis are around and more than once critics have pointed out that some fairly uncontroversial version is being defended while far-reaching conclusions are then drawn from another much more dubious version (e.g. Laudan 1990, 324). Conversely, defendants of underdetermination might argue that all objections concern only some distorted rendering of the underdetermination thesis, while the real thing stays undefeated. In this vein, we will argue in this paper.

In any case, it is crucial to distinguish the different versions of the underdetermination thesis that have been proposed in the literature. The common denominator in all of them is that *in some manner evidence underdetermines theory*. Various claims result from specifying the three central concepts, i.e. underdetermination, evidence, and theory. The most fundamental distinction concerns the methodological toolbox by which underdetermination is established—resulting in two generic types of underdetermination (Laudan 1990): *deductive underdetermination*, sometimes also referred to as Humean underdetermination, and *ampliative underdetermination*. In case of the former, underdetermination is established by means of a pure hypothetico-deductive method, where theories are evaluated solely on the basis of their observable, deductive consequences. By contrast, ampliative underdetermination takes into account further criteria. Thus, ampliative underdetermination is deductive underdetermination plus underdetermination with respect to X, where X might be some non-empirical epistemic virtues like simplicity, fruitfulness, scope or consistency.[3] For genuine inductivists, X would incorporate sophisticated inductive methods, e.g. eliminative induction or Bayesian probabilism.

Another distinction arises when specifying the extent of evidence that is taken into account. Most importantly, do we refer to the actual evidence in a specific historical context, or to all possible evidence, i.e. all observation statements implied by a theory. Both approaches have been proposed in the literature: the former goes by the name *transient underdetermination*, the latter is often called *permanent underdetermination* (Stanford 2009; see also Sklar 1975, 380–81). In the first case, the underdetermination predicament usually goes away when further evidence accumulates; in the second, no evidence can decide between rival theories. Sometimes, transient underdetermination is taken to imply that there necessarily exists a piece of evidence which will decide between competing approaches. We will not require that. In our usage of the terms, a case of transient underdetermination, i.e. with respect to actual evidence, can also constitute a case of permanent underdetermination, i.e. with respect to possible evidence.

---

[3] A comment on notation: To adequately mirror the distinction between deductive and ampliative underdetermination, I distinguish between empirical adequacy and non-empirical epistemic virtues.

Transient underdetermination has recently been examined extensively in the work of Kyle Stanford (2001, 2006, 2009). Stanford has formulated a novel epistemic challenge to scientific method, his so-called *problem of unconceived alternatives*, which might also be termed the problem of *recurrent transient underdetermination*. His claim is that underdetermination can be a challenge to realism not only if it is permanent but also if it is transient, while recurrent. If there continue to be currently unthought-of alternatives even to our best-established theories then underdetermination clearly constitutes a challenge to scientific realism. According to Stanford, the history of science provides ample evidence that the threat of unconceived alternatives is real.

Finally, one can distinguish versions of underdetermination by specifying what exactly one understands by theory. A conservative proponent of underdetermination will insist on some, however fuzzy dividing line between observation statements, theoretical hypotheses, and analytic statements. Given this assumption, underdetermination concerns only some esoteric "isles" of interconnected theoretical hypotheses in abstract scientific theories. By contrast, a more radical proponent might deny any meaningful distinction between observation statements, theoretical hypotheses, and analytic statements. From this perspective, one will quickly conclude that not only some restricted areas, but *all* our knowledge is underdetermined. While familiar from the literature, these brands of underdetermination have not yet been given concrete names. We will call them *isolated* and *ubiquitous* underdetermination. Further distinctions have been proposed, but the mentioned three are the most important ones or at least the most relevant for our purposes.

Following this systematic exposition let us now locate Duhem and Quine on this matrix. For this purpose we rely on Duhem's *The Aim and Structure of Physical Theory*, especially chapter IV of part one,[4] and on *Two Dogmas of Empiricism*, the *locus classicus* for Quine's version of underdetermination. Duhem turns out a careful advocate of ampliative underdetermination taking into account both non-empirical epistemic virtues, which he summarizes as economy of thought, and inductive methods, whose effectiveness he criticizes. To account for actual theory choice in cases where rational criteria fail, Duhem proposes his theory of good sense. For him, theories are evaluated according to empirical adequacy, non-empirical epistemic virtues and inductive support. He clearly leans toward transient underdetermination in that he generally considers the historical context with a specific situation of evidence. This can be most clearly perceived from the fact that all his examples are drawn from the history of science, especially from physics. Finally, he clearly endorses isolated underdetermination by denying that phenomenological sciences like physiology are affected by underdetermination at all. According to Duhem, underdetermination concerns only theoretical hypotheses in advanced ("symbolic") scientific theories.

---

[4]The better-known chapter VI of part two is much less telling about Duhem's stance concerning underdetermination.

Quine is in many ways more radical than Duhem. He is also less clear about what his views really are, at least in the influential (Quine 1951). For example, he wavers between deductive and ampliative underdetermination. Still, most of his examples indicate that he has the deductive rendering in mind, e.g. when he writes: "Even a statement very close to the periphery [of the web of knowledge] can be held true in the face of recalcitrant experience by pleading hallucinations or by amending certain statements of the kind called logical laws" (Quine 1951, 43). Clearly, underdetermination is established here in terms of observable deductive consequences only. This impression is also supported by his radical denial of the analytic-synthetic distinction, rendering both epistemic virtues and inductive methods revisable and thus ultimately incapable of deciding between competing approaches. In principle, for Quine even deductive logic is open to revision, leading to an extreme relativism where everyone can claim whatever pleases him. Detailed discussions of scientific method are remarkably absent from Quine's main writings on underdetermination (i.e. 1951 and 1975). There is no critique of inductive methods and nothing comparable to Duhem's theory of good sense.[5] At least from his most explicit formulations, the reader gains the impression that Quine (1951) leans toward the more radical options of permanent underdetermination ("Any statement can be held true come what may"; 43) and ubiquitous underdetermination ("The totality of our so-called knowledge or beliefs, from the most casual matters of geography and history to the profoundest laws of atomic physics or even of pure mathematics and logic, is a man-made fabric which impinges on experience only along the edges"; 42).

7.3 Chasing Duhem, Fleeing Quine

In this part of the argument we will show how the main differences between Duhem and Quine's versions of underdetermination derive from their respective stances toward history. We will dub Duhem's viewpoint the historical and Quine's the logical rendering of underdetermination. Even a superficial comparison between Duhem's *The Aim and Structure* and Quine's *Two Dogmas of Empiricism* reveals the fundamentally different perspectives. Duhem's text is rich with examples from the history of physics, while Quine—"indifferent as Americans often are concerning history" (Vuillemin 1979, 598)—indulges in fundamental considerations of logic and language that remain detached from any historical reality of science. Of course, this corresponds to their philosophical heritage: Duhem was a working physicist and an acclaimed historian of physics, while Quine was a logician and philosopher of language.

It is quite straightforward why someone with the historical interests of Duhem should impose careful and sensible limitations on the underdetermination thesis.

---

[5] Quine elsewhere discusses the epistemic virtues simplicity, familiarity, scope, and fecundity (1955). However, his elaboration remains rather detached from scientific practice mixing examples from everyday life with some superficial expositions of examples from physics.

Summarized from the last section, the main issues that distinguish Duhem's account from Quine's are[6]: (i) the limited extent of holism; (ii) that only theories involving symbolic representations are affected; (iii) the invocation of *good sense* to account for actual theory change; (iv) that the thesis is illustrated by means of historical examples; (v) that the thesis is framed in terms of actual evidence pertaining to a specific historical situation; (vi) that complete empirical equivalence is not necessarily required, but rather equal strength in terms of empirical adequacy and non-empirical epistemic virtues like simplicity, fruitfulness or coherence; (vii) that underdetermination has immediate implications for the methodology of physics.

Let us now point out in a somewhat tedious exercise, how each of these differences can be traced back to Duhem's historical and Quine's logical outlook. (i) A crucial distinction concerns the extent of holism that underpins the underdetermination thesis. Confirmational holism means the idea that scientific hypotheses are not vulnerable to experiments in isolation, but only as a group (Duhem 1991, 183–88). Whenever a prediction turns out wrong, there is considerable ambiguity which hypotheses to abandon. In the words of Duhem, "the validity of [the scientist's] conclusion is as great as the validity of his confidence [in the accuracy of all other propositions he has used in addition to the examined hypothesis]" (185). While Duhem advances a holism that is restricted to limited groups of hypotheses within physics, Quine's holism comprises everything from mathematics and logic to the purely phenomenological sciences: "The unit of empirical significance is the whole of science" (Quine 1951, 42).

How is this difference connected with the presence or absence of historical perspective? While it may in principle be true that all knowledge is connected via logic and mathematics, the implications of Quine's holism are far too unrealistic for any serious scientist or historian of science to accept. A physicist at CERN formulating a hypothesis about the Higgs boson will hopefully never be worried by the fact that you are reading my paper in this very moment, although these two statements are in principle connected via the laws of logic. An astronomer concerned with the mechanics of the solar system will hopefully never dare to question the principles of mathematics, if he finds Mercury to behave unexpectedly. Quine's unlimited holism bears no insight as to why, in the past, certain hypotheses were rejected and others kept.

By contrast, Duhem's holism is historically accurate. It is a historical truism that unexpected experimental results in physics never lead to prompt and unanimous conclusions by the scientists involved. Rather, "[n]o absolute principle directs this inquiry, which different physicists may conduct in very different ways without having the right to accuse one another of illogicality" (Duhem 1991, 216). Often dissenters remain who do not accept the consensus reached by the majority. While Quine's holism can account for the existence of dissenters in principle, it is inaccurate about where, when, and to what extent dissent surfaces. For example,

[6]The differences between a Quinean and a Duhemian rendering of underdetermination are addressed in among others (Vuillemin 1979; Ariew 1984; Quine 1986; and Gillies 1993).

unexpected trajectories of Mercury will never produce dissenters concerning the principles of deductive logic. Duhem's holism can account for that, Quine's cannot.

(ii) Somewhat relatedly, Duhem restricts the range of underdetermination to theories that involve symbolic representations and do not reason directly on facts. Accordingly, only mature theories involving an abstract mathematical layer that is not accessible to direct observation are affected by underdetermination. Only when observation requires sophisticated scientific instruments, we encounter the kind of theory-ladenness that eventually leads to holism and underdetermination. Thus, underdetermination will occur in physics but not in physiology (Duhem 1990, 180). By contrast, Quine (1951) famously frames his discussion in an outright denial of the analytic/synthetic distinction per se, which for him constitutes one of the dogmas of empiricism. He thus renders the distinction between theoretical and observational statements, between symbolic representation and statements of facts largely meaningless. To be fair, Quine does introduce a distance-measure indicating how far away a statement is from the sensory periphery (1951, 43). But his wording suggests that for him the distinction between observation statements and theoretical hypotheses is only a matter of degree and not of qualitative nature. It therefore cannot yield the conceptual basis for restricting underdetermination to specific areas in the web of experience.

It is not difficult to connect this difference with Duhem's interest and Quine's disinterest in the history of science. Someone interested in a historically adequate account of physical methodology must allow for a distinction between observational and theoretical statements that is sufficiently robust to ground qualitative differences between them. From the pragmatic, historical point of view it would just be outrageous to treat in the same manner highly abstract theoretical statements, e.g. concerning the properties of quarks and strings, and pure observation statements like "The needle of my measuring device points to 5". Contrary to Quine, this is not just a matter of degree.

(iii) The historical fact that in spite of underdetermination most physicists eventually agree on the implications of experiments is accounted for by Duhem's theory of "good sense". It has sometimes been suggested that good sense should be understood entirely in terms of inductive methods and of non-empirical epistemic virtues. Such an interpretation must be rejected since Duhem insists that good sense is not rationally reconstructable; rather, it refers to "reasons which reason does not know" (1991, 217), "that confused collection of tendencies, aspirations, and intuitions" which cannot be further analysed and which cannot be rigorously formulated (104). Presumably, Duhem is aware of the large gap between actual practice in physics and toy models of scientific inference as discussed in philosophy of science. While Duhem acknowledges the utility of inductive frameworks, his holism points to the ambiguities of these frameworks and to subjective elements involved.

Duhem's account of good sense, as unsatisfying as it may be, is historically accurate, since it accounts for the eventual convergence of opinions, while nevertheless acknowledging that there is no universally accepted inductive method in physics and that leading scientists like Einstein have repeatedly stressed the power of intuitions and creativity in the development of scientific theories. On the other side, it is not surprising that Quine with his lack of interest in history does not come up with a concept comparable to good sense.

Issues (i)–(iii) are frequently cited in comparisons between Duhem and Quine (e.g. in Gillies 1993). However, once it is understood that the historical perspective is crucial, further differences can be detected which are otherwise easily overlooked. (iv) One rather obvious aspect concerns the choice of examples that are used to illustrate the underdetermination thesis. Duhem's examples are all taken from the history of physics, while (Quine 1951) starts an unfortunate chain of examples that are construed from existing theories involving redefinition of terms, abandonment of logic, hallucinations and the like: Brutus may not have killed Caesar, if "killed" happened to have the sense of "begat" (36); an allegedly failed prediction can be held true if pleading hallucination or changing the laws of logic (43).[7] None of these "examples" will convince working scientists that underdetermination actually constitutes an interesting epistemological problem with relevance for scientific practice. Admittedly, Quine's examples have an advantage as well, in that they can be constructed starting from our currently best theories. By contrast, Duhem's examples are today largely outdated—in fact most of them already were in Duhem's days. Working with Duhem's examples, one could argue that underdetermination is only a problem for science in its immature stages—irrelevant to our modern theories. So, why didn't Duhem worry? Presumably, he was illustrating a methodological point about certain epistemic situations that can arise in the evolution of physics. Implicit in his choice of examples is the claim that such situations may always recur in the future. This shows Duhem to be one of the ancestors of Stanford's recurrent transient underdetermination leading to the problem of unconceived alternatives (Stanford 2001, 2006, 2009).

(v) Another obvious consequence from the historical perspective is that underdetermination must be considered with respect to the actual evidence in a specific historical context. For the historian the most interesting situations are those, where theories are equally confirmed by past evidence but through different metaphysics provide different research programs for the future. By contrast, Quine insisted on rendering underdetermination in terms of possible evidence, abstracting from a specific historical situation: "natural science is empirically under-determined [. . .] not just by past observation but by all observable events" (Quine 1975, 313). Duhem's historical viewpoint is largely indifferent concerning the question if some future evidence might decide between competing approaches. Plausibly, given the commitment and the ingenuity of the scientists involved, the approaches are often potentially equivalent with regards to future evidence.

(vi) In actual historical contexts, one should not expect exact empirical equivalence between rival theories, not even with respect to past evidence. For example,

[7]Quine (1975) later corrects his stance and denies that these are genuine examples of underdetermination (cp. also Laudan 1990, 332–35). He demands that underdetermination "needs to be read as a thesis about the world" (1975, 324) concerning theories that are "equally sustained by all experience, equally simple, and irreconcilable by reconstrual of predicates" (328). By that time, however, Pandora's box was already open.

different approaches will deviate from each other in the periphery, the domain of application may be somewhat different, or they will differ with regard to the accuracy with which they can describe certain phenomena. In actual historical contexts, competing theories will be largely on a par in terms of epistemic virtues and inductive support, but they won't necessarily be empirically equivalent. By contrast, Quine prefers the logically clean formulation of empirically equivalent theories with respect to possible evidence. In general, actual history fails to provide such rivals.

(vii) Finally, there is a difference in aim: Duhem's historical perspective mirrors an interest in scientific method that can already be deduced from the title *Aim and Structure of Physical Theory* and that is lacking in Quine's account. If Duhem succeeds in making sense of certain historical episodes by employing the underdetermination thesis, this will also reveal the function of underdetermination for the methodology of physics. In this way, historical case studies of underdetermination can provide methodological insights that are relevant for contemporary physics as well.

In summary, we have seen in this section that all differences between Duhem's and Quine's renderings of underdetermination can be traced back to an interest or lack of interest in the history of science, respectively.

7.4 All Objections Refuted

In the next step of the argument we will now show that the familiar objections against underdetermination from the literature are fatal only to Quine's logical rendering of the thesis and not to Duhem's historical formulation. The historical perspective thus turns out essential for formulating a defensible version of the underdetermination thesis. Conversely, the lack of historical perspective has opened up the underdetermination thesis to easy criticism. Two objections are usually given particular weight: the argument from an impoverished account of scientific method, which has most prominently been voiced in Larry Laudan's influential *Demystifying Underdetermination* (1990, 346), and the identical rivals objection (Norton 2008; cf. Quine 1975; Magnus 2003; Frost-Arnold and Magnus 2009). We will briefly discuss some further objections against the underdetermination thesis that play a more minor role in the literature: Grünbaum's point that there is no general argument proving the existence of alternative theories, Laudan and Leplin's claim that empirical equivalence is contextual, and finally the objection from scientific import.

According to the identical rivals objection, in many alleged cases of underdetermination we are actually dealing with different formulations of one and the same theory. John Norton puts it in the following way: "The very fact that observational equivalence can be demonstrated by arguments brief enough to be included in a journal article means that we cannot preclude the possibility that the theories are merely variant formulations of the same theory" (2008, 17). The proponent of the historical rendering need not be worried since a closer look reveals that this attacks only the logical version of underdetermination. First, Norton evokes equivalence with respect to possible evidence rather than actual evidence. Also, he requires only that both theories are observationally equivalent and not that they are on a par regarding non-empirical epistemic virtues. Finally, good examples of historical underdetermination are decidedly those that can*not* be demonstrated—in full detail—within a journal article. Rather, such rival approaches differ in all important aspects, in particular ontology, mathematical framework as well as experimental practices (cp. the quotes by Schrödinger and Maxwell in the introduction). As Stanford rightly points out genuine examples of underdetermination require "the sort of difficult conceptual achievement that demands the sustained efforts of real scientists over years, decades, and even careers" (2006, 15)—an achievement that cannot be laid out in a few paragraphs.

The crux of evaluating the identical rivals objection lies in the quest for good criteria, in which situations we are dealing with variant formulations of theories and in which we are not (Magnus 2003). Norton's somewhat ingenious suggestion—referring to the possibility of formulating the equivalence within a journal article—is quite helpful as a first guess, but clearly fails as a reliable criterion. Quine proposes as a criterion that the theories can be translated by means of a reconstrual of predicates, i.e. essentially by a redefinition of terms (1975, 320). However, this is neither necessary nor sufficient. It is not sufficient, since it ensures only empirical equivalence but not equivalence with respect to epistemic virtues. It is not necessary, since even in reformulations sometimes interesting shifts in meaning can occur. For example, a remarkable shift in the fundamental ontology occurs when Euclidean geometry is reformulated in non-Euclidean terms. It is not altogether clear why we should speak of equivalent theories in this case (Magnus 2003, 1258).[8]

While there may not be a universal criterion for the identity of theories, the historical perspective can provide some insights. Genuine cases of underdetermination are those in which the theories are potentially equivalent in terms of their empirical consequences but also with regard to epistemic virtues, but where they differ enough to provide the scientists with different outlooks on the world suggesting different research programs. In summary, the identical rivals objection does not expose serious problems of the underdetermination thesis, but disqualifies those ahistorical examples where the competing accounts are easily mapped onto each other, in particular Quine's suggestions referring to redefinition of terms, hallucination and the like.

Taking examples from the history of science also immunizes the underdetermination thesis against the second main objection, namely that the underdetermination thesis allegedly presupposes an impoverished hypothetico-deductive account of scientific method. This objection belongs to the standard repertoire of arguments against underdetermination and has been most prominently voiced in (Laudan 1990, 346). Laudan correctly emphasizes that deductive underdetermination does not warrant ampliative underdetermination. The underlying claim is that

[8]Magnus (2003, 1263) suggests that there is a sufficient condition for non-identity of theories, namely empirical inequivalence. However, this criterion falls prey to Laudan and Leplin's point about the contextuality of empirical equivalence to be discussed below.

underdetermination is usually established on the grounds of observable deductive consequences only, leaving aside all the wonderful inductive tools belonging to more sophisticated accounts of scientific method. Surely, deduction cannot tell us which of the several assumptions that enter into the derivation of a false prediction is wrong. However, sophisticated inductive accounts like eliminative induction or Bayesian probabilism are supposedly able to provide such advice, or so the objection goes.

It is certainly correct to criticize the lack of detailed discussions of inductive methods in most arguments for underdetermination. Still, the objection does not succeed against the historical rendering of underdetermination that we are defending here. Most importantly, if the historical examples would really fall prey to this objection, it would imply that scientists like James Clerk Maxwell, William Thomson, or Erwin Schrödinger—all of whom acknowledged underdetermination at some point—worked with an impoverished scientific method. This would clearly be an absurd consequence given that the work of such leading physicists should provide role models for scientific method. We can therefore conclude that the objection from an impoverished account of scientific method does not affect the historical rendering of underdetermination while strongly discrediting the strict logical rendering which indeed relies on equivalence of observable consequences alone.

Of course, several accounts of induction exclude underdetermination to some degree, as for example (Laudan 1990, 332) and (Norton 2008, 29–32) show in some detail. But both seem to be fighting windmills here, at least if one assumes Duhem to be one of the targets. Nowhere in *The Aim and Structure* does Duhem claim that when a prediction turns out wrong, we are given free choice, which of the assumptions to abandon that entered the derivation of the prediction. Rather, Duhem admits that often scientists will readily agree on this issue. He also credits induction with a fruitful role for scientific inference: “induction may indicate to some extent the path leading to certain hypotheses” (1991, 259). The essential disagreement between Duhem and critics of underdetermination like Norton consists not in the fact that one allows for inductive methods and the other does not. Rather, Duhem is much more sceptical than Norton concerning the reach of induction: “[n]o system of hypotheses can be obtained by experimental induction alone” (ibid.). Duhem offers a general critique of inductive methods, examining Newton and Ampère’s claims that they deduced their theories uniquely from experience (190–200). While critics of underdetermination often seem to imply that scientists’ choices are always rationally reconstructable in terms of inductive methods, Duhem insists that there are elements involved which defy a full rational reconstruction. Such elements Duhem subsumes under his notion of good sense.

We are now on much more elusive grounds. Eventually, an opponent of underdetermination would have to show that (i) induction as used in physics is indeed *fully* formalizable, and also (ii) that the various inductive methods always lead to the same results in the same situations, or alternatively that only one of these inductive methods is correct. Furthermore, he would have to prove (iii) that there are no subjective elements involved in the inductive process which could easily destroy any consensus between scientists, e.g. he would have to exclude such accounts of induction as subjective Bayesianism.

Neither Norton nor Laudan's survey of inductive methods establishes any of this. In the end, there is a large gap between showing that some toy models of scientific inference do not exhibit underdetermination and establishing that underdetermination is not an issue for real science. That scientific inference in the real world is a much trickier business can be deduced from the fact that no working scientist has ever been able to come up with a full-fledged account of scientific method. And those that have tried to develop a universal inductive method like Francis Bacon or John Stuart Mill have eventually been ridiculed for their alleged naiveté. Also, some of our most esteemed scientists like Albert Einstein have repeatedly stressed the role of intuitions or even of a decisively creative element in the development of physical theories.

Still, the objection from an impoverished account of scientific method exposes a weakness in the holist argument for underdetermination which is embraced both by Quine and by Duhem and which will be discussed in more detail in the next section. It constitutes therefore an objection against one of the main arguments in support of underdetermination rather than against the underdetermination thesis itself. Certainly, undermining an argument for a thesis does not necessarily refute the thesis. In the same vein, Laudan acknowledges that there may be ampliative underdetermination even though "[t]he fact that a theory is deductively underdetermined [. . .] does not warrant the claim that it is ampliatively underdetermined" (1990, 346).

This point turns out closely related to an objection against underdetermination that has been voiced prominently by Adolf Grünbaum. He stresses that from the impossibility of an *experimentum crucis* one cannot deduce the necessary existence of alternative formulations: "Duhem *cannot* guarantee on any general logical grounds the deducibility of O [empirical findings] from an *explanans* constituted by the conjunction of H [empirical hypothesis] and some revised *non*-trivial version R of A [auxiliary assumptions]" (Grünbaum 1960, 75). Granted, but neither can one exclude the existence of alternative formulations. Grünbaum's argument seems to imply a stalemate between proponents and opponents of underdetermination.

Let us briefly address some further worries that have sometimes been voiced against underdetermination but which once again turn only on the logical rendering. The notion of empirical equivalence, which is central to the logical view, has rightly been criticized by Laudan and Leplin, who argue in an influential paper that empirical equivalence is "both contextual and defeasible" (1991, 454). Indeed, judgments of empirical equivalence have sometimes been revoked, when the context changed. For example, while in the end of the nineteenth century Euclidean and non-Euclidean geometry were deemed empirically equivalent, many hold today that physical geometry has been empirically proven to be non-Euclidean. As pointed out in the last section, underdetermination requires the commitment of the scientists to develop competing frameworks in a way that they remain empirically equivalent. Thus, the contextuality of empirical equivalence is just a further argument for rendering underdetermination with an eye to the historical context.

Another objection that is frequently voiced by working scientists could be termed objection from scientific import. If fully equivalent theories existed, then science would not lose much if we just chose an arbitrary one and discarded all others. Maybe underdetermination with full empirical equivalence would still be interesting for the metaphysician, who could learn from underdetermination that different metaphysics are able to account for the same observations. According to this view, underdetermination could serve as a mine for arguments in favour of or against metaphysical realism while the working scientist would just shrug his shoulders and reply that the debate is of little import for the practice of science. However, this picture is mistaken about the notion of empirical equivalence that we have just seen to depend on the continuous commitment of the scientists. It also misconstrues the role of metaphysics for science. Metaphysics is not irrelevant for scientific research; rather it serves as a necessary and indispensable guideline for the inductive business of science. It leads the scientist to consider what theoretical problems to tackle next or which experiments to do.

In the example of underdetermination between field theory and action at a distance in classical electrodynamics, each perspective had its merits linked to the specific metaphysics of the approaches. The field ontology led to the discovery of many phenomena that concern the "medium" between charges and currents, for example the theories of dielectrics and diamagnetism (Faraday) or the unification of optics with electrodynamics (Maxwell). On the other hand, the action-at-a-distance ontology proved fruitful for finding the Newtonian force laws governing electrostatics and magnetostatics (Coulomb) or the unification of electrostatic and electrodynamic interaction (Weber). Progress in electrodynamics would have been seriously hampered by an exaggerated dogmatism concerning ontology as well as by an outright denial of underdetermination in scientific method.

In summary, underdetermination is really about equally strong theories with different metaphysics that provide the scientist in a specific historical context with different instructions what to do next and what to expect from nature. If this historical perspective on underdetermination is presupposed then none of the known objections from the literature is fatal to the underdetermination thesis.

7.5 "Arguments" for Underdetermination

The plausibility of the underdetermination thesis suffers less from the objections leveled against it, which were shown not to be very persuasive, but from the weakness of the arguments for it (cf. Norton 2008, 21–26). In the following we will outline both Duhem and Quine's defense of underdetermination. Both authors rely heavily on confirmational holism. A second strain of arguments for underdetermination is of (meta-)inductive nature enumerating examples. Finally, a third strain will be developed in the next section. There, we will argue that a denial of underdetermination leads to a crippled account of scientific method. We will show that in essential periods of scientific evolution at least some scientists must acknowledge underdetermination (as Maxwell and Schrödinger in the quotes from the introduction) while a universal denial of underdetermination would seriously block progress. Also, we will show that a denial of underdetermination implies an unrealistic conception of scientific theories, misconstruing the roles of conventions and research ideals.

The most widely accepted argument for underdetermination relies on confirmational holism. It can be found clearly articulated in the argument against *crucial experiments* of Duhem's *Aim and Structure* (1991, 183–90). According to Duhem, in abstract sciences like physics it is impossible to test hypotheses without the help of auxiliary assumptions, which usually include the basic tenets of several physical theories. For instance, a hypothesis about the nature of light can be verified or falsified only by presupposing some core assumptions of optics, thermodynamics, and mechanics that are used in the construction of the respective scientific instruments and in the design of the experimental set-up. Consequently, the test of a hypothesis is only as good as the confidence in those other tenets. In principle, one can always hold on to any hypothesis and in the case of recalcitrant evidence blame some of the auxiliary assumptions. As Duhem states provocatively physicists could for example have saved the particle nature of light in spite of Foucault's experiment, if they had only attached some value to this task (187).

The holist argument also builds the backbone of Quine's defense of underdetermination. The *locus classicus* is *Two Dogmas of Empiricism*, in which he famously attacks the analytic/synthetic distinction as well as the reductionist claim that statements can be confirmed or disconfirmed in isolation. This leads him directly to confirmational holism, from which underdetermination is then derived. According to Quine, changing a statement in the interior of the conceptual net of a theory can always be compensated by other adjustments in the interior, leaving the edge of the net, representing the empirical results of the theory, unchanged. Of course, this involves the assumption that such compensatory adjustments are always possible for which a rigorous argument is missing. In a later essay, Quine points out that mature scientific theories always encompass extra "stuffing" which is not determined by the infinite conjunction of relevant observation statements. Even by the best standards of rationality, a certain freedom of choice remains implying underdetermination (1975, 324).

While the holist argument is widely accepted, it clearly relies on a purely hypothetico-deductive method and therefore only establishes deductive underdetermination, i.e. identity of the observational consequences (cf. the argument from an impoverished account of scientific method in the last section). Regarding the more interesting ampliative underdetermination, the argument can provide good reasons neither for nor against it (Laudan 1990, 346). Unfortunately then, there seems to be no shortcut to establishing ampliative underdetermination.

Do we have to resort to the (meta-)inductive justification of underdetermination on a case-by-case basis? Proponents of underdetermination should point out a sufficient number of examples in order to conclude that underdetermination is something everyone should worry about. By contrast, opponents of the thesis would have to show that these examples are few and that the few are trivial or belong to immature theories. In any case, the inductive argument for underdetermination (if it is to establish ampliative underdetermination) involves detailed historical studies, a profound knowledge of the respective sciences and is in general beyond the reach of a single philosophy paper. Another weakness of inductive justification is that it can establish underdetermination in general only for the narrow realm of the special science from which the examples are taken. Given the vast differences in outlook and methodology, it is not clear why an inductive justification in, say, biology should immediately carry over to physical theories.

Ultimately, the inductive argument results once again in a stalemate between opponents and proponents of underdetermination, where one side will claim that there are a large number of examples for underdetermination and the other side will deny these examples on a piecemeal basis. Proponents will produce historical evidence that theories will always be troubled by underdetermination. Opponents will try to establish that mature theories are not implicated. Given the difficulties in determining what exactly a mature theory is, the prospects of both an inductive justification or a piecemeal confutation of underdetermination are rather dim. Godfrey-Smith (2008) has pointed out a related symmetry of arguments for and against underdetermination, where supporters will claim that there are always several theories that can account for a given body of evidence, while opponents will respond that for every two rival theories there always exists some piece of evidence that can discriminate between them.

Taking stock, the holist and the inductive arguments do not provide sufficient grounds for establishing the more interesting ampliative underdetermination.

7.6 Arguments from History and Scientific Method

In the previous two sections we have detected a lack of good arguments for and against underdetermination. The common objections turned out either to be directed against a misconstrued version of the thesis or to concern only arguments for the thesis. On the other side, neither the holist nor the inductive argument proved particularly strong for establishing underdetermination. Thus, we are confronted with a situation, where the arguments are neither strong enough to establish nor to discount the underdetermination thesis.

In view of this situation, the present section will be devoted to developing additional arguments in order to make a more convincing case for underdetermination. The main trick will consist in a change of perspective away from treating underdetermination as a crucial argument in the realism-antirealism debate towards considering the implications of underdetermination for scientific method. I will present arguments to the effect that denying underdetermination leads to a crippled scientific method. First, a methodology excluding underdetermination would hinder progress in crucial episodes during the evolution of science. Second, a denial of underdetermination would imply an implausible role for conventions and research ideals in scientific theories. These methodological arguments for underdetermination are all informed by the historical perspective.

7.6.1 Underdetermination and Scientific Progress

In this section, we argue that progress in science would be seriously hindered if some scientists would not allow for underdetermination. Throughout the history of science we encounter repeatedly episodes where leading scientists have acknowledged underdetermination—as in the quotes by Maxwell and Schrödinger from the introduction. We will see that their attitude contributed considerably to the progress in their respective scientific discipline.

First of all, proponents of underdetermination are in a good position for integrating virtues of competing programs. Their creative work in building coherent theories is much facilitated by the fact that they have at their disposal a plethora of possibly useful analogies between rival approaches. By contrast, an opponent of underdetermination will insist on there being only one true theory, rendering pointless any elaboration of analogies between different programs. If only one of the approaches eventually tells the true story and if by consequence all other programs are wrong, why should one expect scientific progress from developing competing programs in parallel and adapting them to each other. Even if a combination of several programs will turn out the correct theory, the elaboration of analogies should not be helpful since where one theory is right the other(s) must necessarily be wrong. Thus, accepting underdetermination is a necessary premise for fruitfully exploiting analogies between competing programs.

Furthermore, establishing underdetermination is a helpful manoeuvre to facilitate the transition between different paradigms during a scientific revolution. It enables a comparably smooth and undogmatic paradigm change that allows proponents of unsuccessful approaches to save face to a certain extent. After all, given underdetermination, the acceptance of a new paradigm does not render wrong what someone once believed in and preached to students.

Let us illustrate these considerations by means of the rivalry between action-at-a-distance and field electrodynamics (Pietsch 2010). The acknowledgment of underdetermination allowed William Thomson and James Clerk Maxwell to develop analogies between both frameworks and thereby contribute to progress in electrodynamics. As an example, the electric potential ϕ and the vector potential $\boldsymbol{A}$ were developed to be effectively employed in both programs, while however designating different things. In field theory they describe the state of a continuous entity, while in action at a distance they describe the relation between particles. As Olivier Darrigol writes: "Thomson forged multi-purpose concepts that transcended cultural barriers and individual theoretical preferences [...] Physicists conversant with French [action-at-a-distance] electrostatics could easily express the potential in terms of electric fluid densities. The followers of Faraday's views, if any, could draw the lines perpendicular to the equipotential surfaces and call them force lines" (2000, 136). Also, the eventual transition from the dominant action-at-a-distance paradigm to the outsider view of Faraday would have been much more difficult without a previous demonstration of some equivalence between the approaches. Similar arguments can be given for almost any case of historical underdetermination, e.g. Schrödinger's assessment of equivalence between matrix and wave mechanics.

If physicists like Maxwell or Schrödinger had not acknowledged underdetermination, they would have dogmatically insisted on their preferred formulation—as admittedly many scientists did in the same situation, for example Faraday on the field view or Heisenberg on matrix mechanics. Nobody would have invested in developing the links between the different approaches and thus a split in the scientific community would most probably have ensued and would have taken up much intellectual esprit in essentially pointless debates. In short, if noone would have been prepared to take the stance of underdetermination, progress in physics would have been seriously hindered.

Admittedly, these considerations do not establish that scientists should necessarily and always endorse underdetermination. Rather, a scientific method *excluding* underdetermination might have its heuristic merits as well, as for example Kuhn emphasizes: "[The] invention of alternates is just what scientists seldom undertake except during the pre-paradigm stage of their science's development and at very special occasions during its subsequent evolution. So long as the tools a paradigm supplies continue to prove capable of solving the problems it defines, science moves fastest and penetrates most deeply through confident employment of those tools. The reason is clear. As in manufacture so in science—retooling is an extravagance to be reserved for the occasion that demands it" (1996, 76). In a sense, we are confronted with two different modes of conducting scientific research, one allowing for and the other denying underdetermination. Each has its respective merits and therefore its *raison d'être*. However, if the possibility of underdetermination is excluded in total, progress in science will at some point be severely hindered.

7.6.2 Conventions and Underdetermination

An intimate relationship between conventional elements in scientific theories and the underdetermination thesis has been pointed out by several authors (e.g. Brown 1989, 50–56). Conventions are not determined empirically but rather by pragmatic considerations, they are in many ways relative to us who decide on them. Thus, there exists no unique, correct choice for conventions. Since, broadly speaking, different choices of conventions yield different theories, scientific theories are underdetermined with respect to these choices. An argument for underdetermination results, if we can show that scientific theories contain *non-trivial* conventional elements leading to non-trivial underdetermination.

For the opponent of underdetermination, two different responses to this argument are feasible. First, he could argue that scientific theories do not necessarily contain conventional elements.[9] Second, he could claim that the kind of underdetermination

[9]Brown's conventionalist twist has been criticized in Okasha (2000, 289–90), on the grounds that conventionalism is "just one possible response—and one which, with the exception of Poincaré,

that results from conventional choices is trivial in the sense that it can easily be detected and isolated. To counter these rejoinders, we will now show that every scientific theory necessarily contains conventions, e.g. in the units of measurable quantities, in fundamental constants and in symmetries and invariances, and that it is far from trivial to determine conventional elements within scientific theories.

Over the course of history, metrology as a scientific discipline (not mete*o*rology!) has involved some of the finest minds in science. Metrology is concerned with measuring in general; it is the science that defines the basic units of all those quantities necessary for the description of the physical world. In every major country there is at least one large research institution that is exclusively concerned with metrology, e.g. the National Institute of Standards and Technology in the United States or the Physikalisch-Technische Bundesanstalt in Germany. No doubt, metrology deals mainly with conventions, e.g. that we measure length with meters, time with seconds, mass with kilograms etc. In a way, the existence of metrology institutes with a large scientific staff and with enormous budgets is already proof that the fixing of conventions is far from trivial.

There is obviously no shortage of conventional elements in scientific theories, at least when they incorporate measurable continuous quantities. It is beyond question that scientific theories are underdetermined with respect to the choice of the fundamental units of those quantities. For example, physical geometry is underdetermined with respect to the choice of foot or meter as the fundamental unit of length. Of course, an opponent of underdetermination will be quick to point out that such underdetermination is trivial and therefore cannot have the profound methodological implications that we attributed to the underdetermination thesis.

Everything depends on the question if conventional elements can always be identified and isolated as easily as in the case of meter vs. foot. If so, then the opponent of underdetermination can rightly claim that the choice of units only leads to trivial examples. However, the prospects of such an endeavour seem rather dim. After all, the conventional choice of a measure for a continuous quantity does not always concern a trivial factor between different units as in: 1 meter = 3.28 feet. One can imagine much more complicated relations, for example a length measure being a complex function of the position. This possibility cannot be ruled out a priori for reasons of simplicity. After all, whichever measure one regards as fundamental, the other is complex in relation to it.

As first pointed out by Hermann von Helmholtz (1870), the physical equations have to be adjusted in order to compensate for different choices of spatial measure. Thus, when we consider more complicated choices, the conventional element cannot be easily isolated and separated from presumably non-conventional elements like the fundamental laws. Indeed, the very formulation of the fundamental laws depends on the exact choice of measure. Eventually, a spatial measure should be

has usually appealed more to philosophers than to scientists". This is historically incorrect. Geometric conventionalism was in various shades accepted by a considerable number of leading physicists at the turn from the nineteenth to the twentieth century—among them (von Helmholtz 1870; Einstein 1921).

chosen that renders the physical equations as *simple* as possible. Consequently, the underdetermination due to conventional elements is limited by considerations of simplicity. However, simplicity is a far too malleable concept to fully rule out underdetermination arising from the choice of spatial measure.

An illustration of the complex issues involved in the choices of measures for physical quantities can be found in Hasok Chang's admirable book *Inventing Temperature*. Chang terms epistemic iteration the complex process how a reliable measure for temperature is constructed in parallel with scientific progress in the theory of heat. Chang's historical case study provides extensive evidence that the choice of measure for fundamental quantities cannot be isolated from the theory itself, that conventional elements cannot be isolated from non-conventional elements. Certainly, Chang does not embrace a "simplistic type of conventionalism" which would allow for arbitrary choices of measures. Rather, considerations of simplicity narrow down the choices. However, it is highly unlikely that simplicity can single out a unique true theory. In any case, no convincing argument has ever been given in that respect. The sophisticated conventionalism of Poincaré is closely related with the coherentism advocated by Chang. While arbitrary choices of conventions are ruled out, there is no ultimate empirical or epistemic justification for the choice of measures for fundamental physical quantities (Chang 2004, 223).

Fundamental constants also involve conventional elements that are often tied to the choices of measures of fundamental quantities. Fundamental constants provide another example how difficult it is to determine the boundary between conventional and empirical elements further establishing that conventions cannot be easily isolated and that therefore underdetermination resulting from conventional choices is not trivial. Consider the velocity of light c. The exact status of this constant continues to be debated. Following the establishment of special relativity, some have held c to simply result from the erroneous assumption that space and time are distinct concepts. Rather, we supposedly live in a unified space-time where space and time should be measured with the same units and consequently the velocity of light should be one. Others have questioned this viewpoint insisting on the conceptual distinctness of space and time. Then, the velocity of light is of much more pronounced empirical nature, for instance it may even undergo change over time.[10]

There are further somewhat less tractable conventions in scientific theories. Consider for example the three quantities force $\boldsymbol{F}$, mass m, and acceleration $\boldsymbol{a}$ connected via Newton's relation $\boldsymbol{F} = m\boldsymbol{a}$. Several interpretations are feasible. First, we could interpret the equation as providing a definition for one of the quantities $\boldsymbol{F}$, m, or $\boldsymbol{a}$. In this case only two of the quantities are independently measurable.

[10] A variable speed of light has been proposed in various physical contexts, most notably in cosmology. For example, Arnot (1941) suggests it to account for the Hubble expansion, a possibility discussed also by Popper (1940). In the more modern literature, a varying speed of light has been proposed to solve various problems of big bang theory, for example the flatness, the horizon, and the cosmological constant problems (e.g. Barrow 1999). The literature on changing natural constants in physics is rich and controversial. Over the years it has involved some of the finest minds in science; cp. for example Dirac's widely-discussed suggestion of a varying gravitational constant (1937).

Whichever two we take to be fundamental depends on a conventional choice that is not necessitated by empirical facts. This being an ontological convention concerning attributes of fundamental entities, we are thus led to ontological underdetermination. Essentially, the underdetermination of action-at-a-distance and field electrodynamics is a sophisticated case of such an ontological underdetermination that results from a reformulation of the corresponding physical theory and a different choice about what is considered fundamental and what is not. Second, we could interpret the equation $\boldsymbol{F} = m\boldsymbol{a}$ as an empirical law connecting three quantities that are independently measurable. However, it is a tricky question how one can establish that there exist independent measurement procedures for quantities connected by a deterministic law. As of today, there seems to be no common agreement on the exact status of $\boldsymbol{F} = m\boldsymbol{a}$, whether it is an empirical law or a definition and, in case of the latter, which quantity is being defined. Once again, the boundary between conventional and empirical elements is quite blurred.

A further argument for the existence of non-trivial conventional elements in scientific theories can be based on symmetries and invariances. Every invariance of a scientific theory directly implies a conventional choice. Once again, trivial examples are readily available. Homogeneity of space implies the conventional choice of the origin of the coordinate system in physical geometry. Isotropy of space implies the conventional choice of the direction of the axes of the coordinate system. Gauge invariance in classical electrodynamics implies a certain conventional choice of the potentials ϕ and $\boldsymbol{A}$. Lorentz invariance in the theory of relativity implies the conventional choice of the reference system, i.e. the velocity of the observer. The historical fact that invariances and symmetries have often been debated, or that invariances that were long thought trivial have eventually been drawn into question, as left–right mirroring in particle physics, points to the fact that the conventions implied by symmetries and invariances are not trivial in the sense that they can be easily isolated.

In general, the structure of the bare theory itself does not allow us to determine which parts are conventional and which parts are empirical. Rather, we also have to take into account how propositions are treated by the people working with them. Often, a proposition becomes a convention only if it is treated as such. This is the reason why without detailed historical stories like Chang's exposition of the "invention" of temperature, the argument for underdetermination based on conventions cannot be explicated.

7.6.3 Research Ideals and Underdetermination

If one allows that the broader metaphysical world-view can influence scientific theorizing and if metaphysical pluralism is accepted in the sense that several world-views are possible, then underdetermination is a plausible consequence. Metaphysical propositions which exert influence on scientific theorizing will be called research ideals in the following.

Consider as an example the issue of determinism vs. indeterminism, i.e. the question whether every event is determined by events prior to it or not. Arguably, this is not an empirical question. After all, to prove determinism one would have to show that every event is determined by some other events prior to it. Given the finiteness of our experience, such an endeavour is surely impossible. To prove indeterminism one would have to show that there are events which are not fully determined by prior causes. Given that there are arbitrarily many candidates for such prior causes and again the finiteness of our experience, indeterminism cannot be established either. Only if further restrictions are admitted, e.g. locality, determinism and indeterminism become empirically distinguishable.

Different choices of research ideals can eventually lead to underdetermination. For example, the choice between determinism and indeterminism lies at the root of the underdetermination between orthodox non-relativistic quantum theory and Bohmian mechanics, the former based on an indeterministic and the latter on a deterministic metaphysics. Other cases of underdetermination originating in different choices of research ideals are readily available, e.g. concerning the choice between a fundamentally discrete or a continuous nature of matter. The underdetermination between field and action-at-a-distance electrodynamics falls broadly into this category. As (Schrödinger 1926, 45) points out, matrix and wave mechanics also originate in different conceptions of matter, the former stressing the discontinuous and the later the continuous aspects.

An opponent of underdetermination would have to deny the role sketched for research ideals in this section. He would either have to question that metaphysics exerts an influence on science or claim that there is only one correct metaphysics. Both standpoints are implausible if judged from the history of science.

7.7 Conclusion

The historical perspective provides a reliable guide to formulating the underdetermination thesis. It helps to avoid some common distortions of the thesis which have in the past opened it up to the usual objections. If history is taken as a guide then underdetermination will automatically be construed in a way that it can account for certain episodes in which underdetermination was explicitly acknowledged by leading scientists. The historical perspective will also allow to comprehend in which ways acknowledging underdetermination can enable scientific progress. In this way, underdetermination turns out a powerful tool both for the historian to make sense of certain episodes in the history of science and in the hands of the working scientist to enable progress in certain contexts. In a slight variation of a well-known quote by Duhem[11] we conclude that to give the history of underdetermination is at the same time to make a (methodo-)logical analysis of it.[12]

[11] "To give the history of a physical principle is at the same time to make a logical analysis of it" (Duhem 1991, 269).

[12] I thank the editors of this volume, Tad Schmaltz and Seymour Mauskopf, for helpful comments and criticism. I am also grateful to Mauricio Suárez for suggestions on an earlier version of this article as well as to audiences at EPSA09 in Amsterdam and at &HPS2 in Notre Dame.

References

Ariew, Roger. 1984. "The Duhem Thesis." *British Journal for the Philosophy of Science* 35: 313–25.

Arnot, Frederick L. 1941. *Time and the Universe*. Sydney: Australasian Medical Publishing.

Barrow, John D. 1999. "Cosmologies with Varying Speed of Light." *Physical Review D* 59, article 043515.

Brown, James R. 1989. *The Rational and the Social*. London: Routledge.

Chang, Hasok. 2004. *Inventing Temperature. Measurement and Scientific Progress*. Oxford: Oxford University Press.

Darrigol, Olivier. 2000. *Electrodynamics from Ampère to Einstein*. Oxford: Oxford University Press.

Dirac, Paul A.M. 1937. "The Cosmological Constants." *Nature* 139: 323.

Duhem, Pierre. 1990. "Logical Examination of Physical Theory." *Synthese* 83: 183–88.

Duhem, Pierre. 1991. *The Aim and Structure of Physical Theory*. Princeton, NJ: Princeton University Press. Translation of *La théorie physique. Son objet et sa structure*. Paris: Chevalier & Rivière, 1906.

Einstein, Albert. 1921. "Geometry and Experience." In *Sidelights on Relativity*, 25–56. Mineola, NY: Dover, 2010.

Frost-Arnold, Greg and P.D. Magnus. 2009. "The Identical Rivals Response to Underdetermination." *PhilSci archive*. http://philsci-archive.pitt.edu/archive/00003390/. Accessed 15 February 2010.

Gillies, Donald. 1993. *Philosophy of Science in the Twentieth Century*. Oxford: Blackwell. Excerpt reprinted in *Philosophy of Science. The Central Issues*, edited by M. Curd and J.A. Cover, 302–19. New York, NY: Norton, 1998.

Godfrey-Smith, Peter. 2008. "Recurrent Transient Underdetermination and the Glass Half Full." *Philosophical Studies* 137: 141–48.

Grünbaum, Adolf. 1960. "The Duhemian Argument." *Philosophy of Science* 27: 75–87.

Helmholtz, Hermann von. 1870. "Über den Ursprung und die Bedeutung der geometrischen Axiome." In *Populäre Wissenschaftliche Vorträge von H. Helmholtz. Drittes Heft,* 21–54. Braunschweig: Vieweg und Sohn. 1876.

Kuhn, Thomas. 1996. *The Structure of Scientific Revolutions*. Chicago, IL: University of Chicago Press.

Laudan, Larry. 1990. "Demystifying Underdetermination." In *Scientific Theories*, edited by C. Wade Savage, 267–97. Minneapolis, MN: University of Minnesota Press. Reprinted in *Philosophy of Science. The Central Issues*, edited by M. Curd and J.A. Cover, 320–53. New York, NY: Norton, 1998.

Laudan, Larry and Jarrett Leplin. 1991. "Empirical Equivalence and Unterdetermination." *The Journal of Philosophy* 88: 449–72.

Magnus, P.D. 2003. "Underdetermination and the Problem of Identical Rivals." *Philosophy of Science* 70: 1256–64.

Maxwell, James C. 1873. *A Treatise on Electricity and Magnetism*. Oxford: Clarendon Press.

Muller, F.A. 1997. "The Equivalence Myth of Quantum Mechanics." *Studies in the History and Philosophy of Modern Physics* 28: 35–61, 219–47.

Norton, John D. 2008. "Must Evidence Underdetermine Theory?" In *The Challenge of the Social and the Pressure of Practice: Science and Values Revisited*, edited by M. Carrier, D. Howard, and J. Kourany, 17–44. Pittsburgh, PA: University of Pittsburgh Press.

Okasha, Samir. 2000. "The Underdetermination of Theory by Data and the 'Strong Programme' in the Sociology of Knowledge." *International Studies in the Philosophy of Science* 14: 283–97.

Perovic, Slobodan. 2008. "Why Were Matrix Mechanics and Wave Mechanics Considered Equivalent?" *Studies in History and Philosophy of Modern Physics* 39: 444–61.

Pietsch, Wolfgang. 2010. "On Conceptual Issues in Classical Electrodynamics: Prospects and Problems of an Action-at-a-distance Interpretation." *Studies in History and Philosophy of Modern Physics* 41: 67–77.

Poincaré, Henri. 1902. *La Science et l'Hypothèse*. Paris: Flammarion.
Popper, Karl R. 1940. "Interpretation of Nebular Red-Shifts." *Nature* 145: 69–70.
Quine, Willard V.O. 1951. "Two Dogmas of Empiricism." Reprinted in *From a Logical Point of View: 9 Logico-Philosophical Essays*, 2nd edition, 20–46. Cambridge, MA: Harvard University Press, 1980.
Quine, Willard V.O. 1955. "Posits and Reality." Reprinted in *The Ways of Paradox and Other Essays*. 2nd edition, 246–54. Cambridge, MA: Harvard University Press.
Quine, Willard V.O. 1975. "On Empirically Equivalent Systems of the World." *Erkenntnis* 9: 313–28.
Quine, Willard V.O. 1986. "Reply to Jules Vuillemin." In *The Philosophy of W. V. Quine*, edited by Lewis E. Hahn and Paul A. Schilpp, 619–22. La Salle, IL: Open Court.
Schrödinger, Erwin. 1926. "On the Relation between the Quantum Mechanics of Heisenberg, Born, and Jordan, and that of Schrödinger." In *Collected Papers on Wave Mechanics*, 45–61. New York, NY: Chelsea, 1982.
Sklar, Lawrence. 1975. "Methodological Conservatism." *Philosophical Review* 84: 384–400.
Stanford, P. Kyle. 2001. "Refusing the Devil's Bargain: What Kind of Underdetermination Should We Take Seriously?" *Philosophy of Science* 68: S1–12.
Stanford, P. Kyle. 2006. *Exceeding Our Grasp*. Oxford: Oxford University Press.
Stanford, P. Kyle. 2009. "Underdetermination of Scientific Theory." In *Stanford Encyclopedia of Philosophy (Winter 2009 Edition)*, editd by Edward N. Zalta. http://plato.stanford.edu/archives/win2009/entries/scientific-underdetermination/. Accessed 15 February 2010.
Vuillemin, Jules. 1979. "On Duhem's and Quine's Theses." *Grazer philosophische Studien* 9: 69–96. Reprinted in *The Philosophy of W. V. Quine*, edited by Lewis E. Hahn and Paul A. Schilpp, 595–618. La Salle, IL: Open Court.

Part II
Case Studies

Chapter 8
Beyond Case-Studies: History as Philosophy

Hasok Chang

8.1 The Trouble with Case-Studies, and the Active Philosophical Function of History

What can we conclude from a mere handful of case-studies? This has been a vexing question for integrated history and philosophy of science (HPS). The field of HPS has witnessed too many hasty philosophical generalizations based on a small number of conveniently chosen case-studies. This was seen as detrimental to philosophy and history both. On the philosophical side, case-studies may end up as empty gestures parading as evidence confirming one's pre-existing biases about the nature of science and its methods. At best what we get is "grand conclusions by induction from absurdly small samples", in Richard Burian's words (2001, 388). The deeper problem, as Joseph Pitt (2001, 374) put it, is that "if philosophers wish to use historical cases to bolster their positions, then ... we will have to figure out how to relate the history to the philosophical point without begging the question." On the historical side, even philosophically sympathetic historians despaired of the oversimplifications that philosophers were apt to make of complex historical material through the case-study approach. John Hedley Brooke's complaint is typical and apt (1981, 257): "When the circumstances and the problems were so complex, the isolation of a single philosophical or methodological point as the key to an adequate explanation must lead to a distortion of emphasis." I believe that the neglect to clarify the nature of the history–philosophy relationship in case-studies has contributed decisively to a widespread disillusionment with the whole HPS enterprise.

Emblematic of this disillusionment is Thomas Kuhn. He had a particularly sharp reaction against Imre Lakatos's explicit treatment of history as testing-ground for philosophical theories of scientific method, with his predilection for rational reconstructions of history. Against Lakatos's dictum (1971, 91) that "history of science without philosophy of science is blind", Kuhn retorted (1971, 143): "What Lakatos

H. Chang (✉)
University of Cambridge, Cambridge, UK
e-mail: hc372@cam.ac.uk

S. Mauskopf, T. Schmaltz (eds.), *Integrating History and Philosophy of Science*, Boston Studies in the Philosophy of Science 263, DOI 10.1007/978-94-007-1745-9_8,

conceives as history is not history at all but philosophy fabricating examples." But how did Kuhn, who thought that even science did not have paradigm-independent facts, assume that historical facts could be discerned without a philosophical framework? Kuhn's answer, given five years later: "The historian's problem is not simply that the facts do not speak for themselves but that, unlike the scientist's data, they speak exceedingly softly. Quiet is required if they are to be heard at all." (Kuhn 1980, 183) Kuhn argued that history of science and philosophy of science had fundamentally different goals, and therefore "no one can practice them both at the same time" (1977a, 5). All this is quite ironic coming from the man who seemed to cross the history–philosophy boundary so freely, and whose work was probably more responsible than anyone else's for encouraging the drawing of philosophical conclusions from historical cases. Wasn't it Kuhn who had uttered these famous words as the very first sentence of *The Structure of Scientific Revolutions* (1962, 1): "History ... could produce a decisive transformation in the image of science by which we are now possessed"? Perhaps, as pioneers often do, he felt that he had unleashed a Frankenstein, and the tone of his latter-day condemnation of "historical philosophy of science" is consistent with this interpretation (Kuhn [1991] 2000). Kuhn did agree, of course, that philosophy of science should ultimately be informed by some knowledge of scientific practice (historical or contemporary). But he never specified a clear method for the history–philosophy interaction, and without such a method we are condemned to the dilemma between making unwarranted generalizations from historical cases and doing entirely "local" histories with no bearing on an overall understanding of the scientific process.

In attempting to transcend this dilemma, I believe that the first thing we need to do is to see if we can get beyond an inductive view of the history–philosophy relation, which takes history as *particular* and philosophy as *general*. Of course we cannot get away from inductive thinking entirely, but it is instructive to try seeing the history–philosophy relation as one between the *concrete* and the *abstract*, instead of one between the particular and the general. Abstract ideas are needed for the understanding of *any* concrete episode, so we could not avoid them even if we only ever had one episode to deal with. We cannot understand scientists' actions, not to mention judge them, without considering them in abstract terms (such as "confirmed", "coherent", "observation", "measurement", "explanation", "simple", "novel", "self-interested", "curiosity-driven", "collective", "honest", "open-minded", etc.). Any concrete account requires abstract notions in the characterization of the relevant events, characters, circumstances and decisions. If we extract abstract insights from the account of a specific concrete episode that we have produced ourselves, that is not so much a process of *generalization*, as an *articulation* of what was already put into it.[1] To highlight this change of perspective, I prefer to speak of historical "episodes" rather than "cases". When we have an episode of *The Simpsons*, or *Buffy*

[1] It may even be an act of self-analysis, in case the episode was initially narrated without a good awareness of the abstractions that guided its construction.

*the Vampire Slaye*r, or what have you, the episode is not really a case or an example of whatever the general idea of the show might be. Rather, the episode is a concrete instantiation of the general concepts (the characters, the setting, the type of events to be expected, etc.), and each episode also contributes to the articulation of the general concepts. To be sure, this analogy is very imperfect, but it does express something relevant about the relation between concrete historical episodes and abstract philosophical conceptions.

Philosophers will be pleased but not surprised to hear that abstract conceptions are necessary for the telling of history. The more novel idea I would like to propose here concerns the opposite direction of dependency—doing history can help our philosophizing, even as we refrain from making crude inductions. When there are no ready-made philosophical concepts through which a given historical episode can be properly understood, the historian needs to craft new abstract philosophical concepts. *This necessity should not be resisted or avoided, but actively embraced as a great intellectual opportunity*. Instead of sitting back and complaining that existing philosophical frameworks are inadequate for understanding real history, historians can actively engage in the creation of new philosophical ideas through their concrete investigations. If the historians won't do it, philosophers can jump in. Regardless of the official affiliation of the scholars involved, we will have history and philosophy being done at the same time, in spite of Kuhn's verdict on the impossibility of this enterprise. History-writing can be a very effective method of philosophical discovery.

Once an abstract idea has been generated, it needs to show its worth in two different ways. First of all, its *cogency* needs to be demonstrated through further abstract considerations and arguments. This is where philosophy takes up the reins again, to examine carefully what has been generated through historiographical necessity. If the abstract idea is deemed to be cogent in itself, then its range of applicability needs to be checked. This can only be done by trying to the framing of various other concrete episodes, with history in the driving seat again. An abstraction becomes *general* only when it has been *applied* widely. Successful application functions as confirmation, but without the presumption of universality in what is confirmed. This is consistent with Burian's view on how a bottom-up approach to case-studies can contribute to the establishment of "regional standards" rather than "universal methodologies or epistemologies" (Burian 2001, 400).

I will illustrate these claims through two investigations in HPS from my own recent work. In both cases, I am not going to present significant new content that is not already in my other publications, but I do hope to present some significant new reflections on the existing material. This also raises a question of reflexivity, already noted by Brooke and Burian: how can we use case-studies to show how to go beyond case-studies?! I hope my answer will be clear as I go on: it is through wrestling with these concrete studies that I crafted my abstract insights about the history–philosophy relation, as the existing ideas concerning that relation did not seem to be helpful.

8.2 Temperature Measurement and Epistemic Iteration

8.2.1 Circularity and Reliability in Measurement

The first episode comes from my work on the history and philosophy of temperature measurement (Chang 2004, esp. chapter 2). When I started this project, I was thinking about the theory-ladenness of observation, and the question of how observations can be justified despite theory-ladenness without falling into a vicious circularity. Having got tired of trying to explain quantum physics to people who didn't really want to know,[2] I wanted to deal with simpler scientific material. I thought, temperature would be nice, since it is a fundamental physical concept that everyone has some sense of. At least everyone knows how to use a basic thermometer.

Does something as simple as an ordinary mercury thermometer rely on theory? Yes: The key assumption is that mercury expands uniformly (or, linearly) with increasing temperature. Think about how simple thermometers are made. We take a glass tube partly filled with mercury. We put it into freezing water, see how far the mercury comes up, and mark that point 0; we put it into boiling water, and mark the position of the mercury column 100; then we divide up the interval evenly to make a scale, calling the halfway point between zero and hundred "50", and so on. So what we assume is that when the temperature is exactly 50°C, the mercury comes up exactly to the halfway point. But is that true? Table 8.1 shows some simple data from the mid-nineteenth century, which illustrates why practicing scientists would have worried about this philosophical-sounding problem. What this table shows is that thermometers filled with different liquids will differ very seriously in the middle of the scale, even if they are graduated to agree exactly at the fixed points. How can we tell which one is correct?

A good physicist will think that assumptions about the behavior of thermometric fluids should of course be tested by experiment. So we take some data, and make a plot of the volume (height) of mercury against temperature, and see if the points lie on a straight line. But how would we get the temperature values, unless we already had a thermometer we can trust, which is exactly what we don't have yet? We could

Table 8.1 Comparison of thermometers filled with different liquids

Mercury (°C)	Alcohol	Water
0	0	0
25	22	5
50	44	26
75	70	57
100	100	100

Data from Lamé (1836, 1: 208)

[2]For instance, Chang (1995) addressed these concerns as they pertain to energy measurements in early quantum physics; almost exactly the same issues are played out in temperature measurement, as discussed in Chang (2004, chapter 3).

try using a different kind of thermometer to take the temperature values, but then we would have to ask how we know *that* thermometer to be correct. Alternatively, we could try to make a theoretical argument about whether mercury expands linearly or not, but that would require a detailed theory of thermal physics, which we would, in all likelihood, have to test by experiments that involve the measurement of temperature.

Like a good philosopher, I made an abstract and precise formulation of the problem:

(i) We want to measure quantity *X*;
(ii) If quantity *X* is not directly observable, we infer it from another quantity Y, which is directly observable.
(iii) For this inference we need a law that expresses *X* as a function of *Y*.
(iv) But the form of this function cannot be discovered or tested empirically, because that would involve knowing the values of both *Y* and *X*, but *X* is the unknown variable that we are trying to measure.

I called this "the problem of nomic measurement" (Chang 2004, 59, 89–90). The precise formulation of the problem made me realize two things: first, it must be a general problem pertaining to nearly every attempt to justify a method of measurement; second, there was no apparent solution to it!

So I was about to give up. Then it occurred to me: "But they have *done* this!" Scientists today do know exactly how mercury expands with temperature (and they tell us that the expansion is far from linear). Someone, sometime, must have figured out how to solve the problem, and the least I could do was to go and look at the actual *history* to learn how it was done. So I began to look up the history of thermometry, in which I became immersed for nearly a decade. Clearly, right from the start I approached this history with a philosophical problem in mind. But I did not go in with a particular philosophical view (hypothesis, even) to confirm, or refute. Rather, I went to the history with an unresolved philosophical question, with a necessarily open mind about what answers I might find. This is a mode of work consonant with Burian's response to Pitt: "case studies, properly deployed, illustrate styles of scientific work and modes of argumentation that are not well handled by currently standard philosophical analyses" (Burian 2001, abstract, 383).

Looking deeper into the history at first deepened my philosophical puzzle rather than resolve it. Every trail I followed into the past looking for the moment of rational foundation of thermometry seemed to lead into a thicket, not a clearing. I couldn't find any satisfactory foundationalist answers on how thermometers were validated, which is what I was after. Some scientists did propose foundationalist solutions, but they were all duly shot down by other scientists. For instance, Jean-André De Luc's "method of mixtures" seemed to offer a solution: mix up freezing water (at 0°C by definition) and boiling water (at 100°C by definition) in the ratio of *a:b* by weight; and the resulting mixture should be at [b/(a+b)]°C, and whichever thermometer indicates that temperature is correct (see Chang 2004, 60–68, 93–94 for details). With this method De Luc nearly generated a consensus on the foundations of thermometry, but then his critics eventually pointed out that his method assumed that the specific heat of water was constant, which was a groundless assumption. Worse than that, it was impossible to test that assumption of constant specific heat, without reliable temperature data. De Luc's critics destroyed his solution, without offering a better one.

What became clear through such examples was the failure of foundationalism, which was the basic philosophical framework with which I began this work, without even realizing that clearly. Now, with my limited philosophical imagination, I thought that if foundationalism did not work, then the fallback position could only be coherentism. Quine's web of belief and Neurath's boat beckoned, as I drowned in the historical mess. But I also found that the primary literature I was looking at clearly conveyed a sense of *progress* in thermometry, which could not easily be captured in straightforward coherentism in which the only validation of a proposition comes from its logical consistency with other propositions we believe. I had a great difficulty in making sense of this historical episode, as long as my understanding was limited by the philosophical framework defined by the opposition between foundationalism and coherentism. My limited philosophical imagination was stuck in a false dilemma: justification is by a self-evident foundation, or it is merely circular.

8.2.2 Epistemic Iteration and Progressive Coherentism

I managed to get out of that impasse by crafting the idea of "epistemic iteration". The concept of iteration allowed me to see that the circularity in question was more like a helix, in which scientists began by *presuming* the validity of an untested measurement method, starting a process of inquiry which eventually doubled back on itself to refine and correct its own starting assumptions. Epistemic iteration is a process in which we create successive stages of knowledge, each building on the preceding one, in order to enhance the achievement of certain epistemic goals, such as precision, consistency, scope, explanatory power, simplicity, and so on. We do not start with indubitable facts, or unrevisable axioms. Instead, we start with a system of knowledge that we recognize as imperfect or even faulty, which is used for its own improvement. No fixed algorithm tells us how to proceed. But we have the impetus and constraints provided by the epistemic values and aims that we adopt. (The idea of epistemic iteration is explained fully in (Chang 2004, chapter 5); I will only provide an introductory summary here.)

Inventing the concept of epistemic iteration helped me capture the historical development of thermometry more sensibly. First, people started with sensation, assuming that the feeling of hot and cold corresponded to high and low temperatures. Then they created "thermoscopes", by finding certain materials that seemed to expand and contract reliably as things felt hotter and colder. Thermoscopes extended the observable temperature range, and they were also more sensitive than human sensation. When thermoscopes were well enough established, they also began to *correct* sensations, as people ceded epistemic authority about temperature to these simple machines: so you might say "I feel colder but the mercury is not moving—I must be getting a fever." And so on. This kind of refinement and self-correction happened again when people made numerical thermometers starting from thermoscopes, by attaching scales to them.

The making of numerical thermometers led to further iterative improvements. Although there never was any indubitable reason for believing that any particular thermometric substance expanded uniformly, various numerical thermometers were made, each on the presumption of uniformity, and eventually those that did not return self-consistent numbers were rejected; Victor Regnault's work in the 1840s was crucial in that development. The most self-consistent thermometric fluid turned out to be air, and the readings of air-thermometers provided the basis for William Thomson's thermodynamic theory and the Joule–Thomson experiment, with the results of which the physicists corrected the air-thermometer used for that very experiment. In this whole process there was progress *without* a firm foundation, in terms of increasing precision, consistency, and scope of the measurements.

So, that was the origin of my ideas about epistemic iteration, born out of historiographical necessity. The next step was to examine, develop and apply these ideas more broadly. As I do not have the space to go through the details of the doctrine of "progressive coherentism" that I developed (see Chang 2007), I will just show you a metaphorical picture of it (Fig. 8.1). I mentioned my disillusionment with foundationalism. The traditional foundationalist picture of building knowledge on a firm ground does not work, since there is no such thing in empirical science

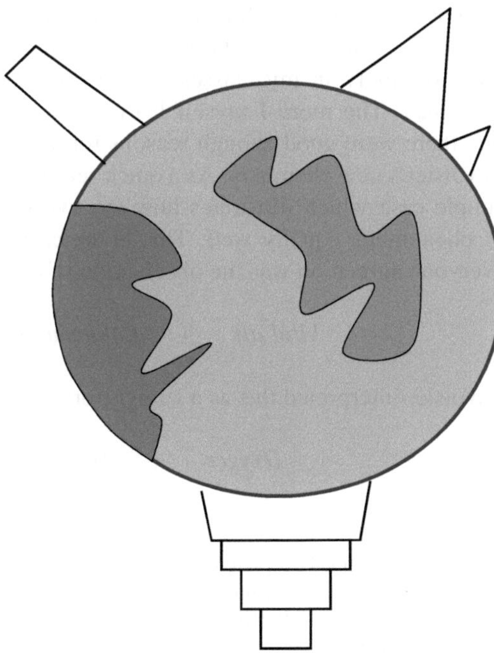

Fig. 8.1 The progressive-coherentist metaphor of building on a round earth

corresponding to that firm ground. But foundationalists have actually been sitting on the perfect metaphor for the kind of ideas I want to express—we only have to remember that the earth is not flat. In real construction, we don't build upward on a flat earth, but outward on a round earth; that is how the metaphor has to work, too. There are no fixed points and no up-and-down in the universe. We build on the earth not because it is firmly fixed anywhere, but because it is a large and dense body that attracts other things, and we happen to live on it.

I have found this framework of progressive coherentism quite useful in thinking about other historical episodes, so at least to my own mind it has proven its broad applicability. The clearest applications are to other cases of nomic measurements. A number of other system in which numerical precision is improved also seem to rely on a process of epistemic iteration. I have also briefly indicated how progressive coherentism applies to the development of chemical analysis (Chang 2007, section 4). Even instances of scientific revolutions could be illuminated by seeing them in progressive-coherentist light, since new paradigms do build, partly and at least initially, on the empirical basis of the older paradigms that they replace.

8.3 The Chemical Revolution: Pluralism and Systems of Practice

8.3.1 The Chemical Revolution as a Puzzle

The second episode I want to discuss is the Chemical Revolution. This comes from current research that will be incorporated into my forthcoming book, titled *Is Water H_2O?* This was research born out of teaching. I covered the Chemical Revolution quite routinely in my classes, but always had to leave it as a puzzle yet to be explained. The more I myself learned about it, the less I could convince myself that there were good enough reasons for a clear majority of chemists to go over to Lavoisier's new chemistry. As a quick reminder for the incredulous, here is one very simple case which illustrates how the two theories could explain the same range of phenomena equally well. This is the case of the composition of water.[3] What everyone agreed on was the observation of the following reaction:

$$\textit{Vital air} \quad + \quad \textit{Inflammable air} \quad \rightarrow \quad \textit{Water}$$

Lavoisier interpreted this as a straightforward proof that water was a compound:

$$\textit{Oxygen} \quad + \quad \textit{Hydrogen} \quad \rightarrow \quad \textit{Water}$$

[3] Priestley maintained his preference for the phlogiston theory until his death, and it is quite telling that his last stance was a book titled *The Doctrine of Phlogiston Established and That of the Composition of Water Refuted*, published in 1803. By that time, most others had converted to Lavoisier's theory.

Isn't this enough to refute the phlogiston theory, which held that water was an element? Not so, as there is a coherent alternative interpretation based on phlogistonist ideas (advanced by Cavendish and Priestly):

$$\textit{Dephlogisticated water} + \textit{Phlogisticated water} \rightarrow \textit{Water}$$

And there are many other similar instances of empirical equivalence between the two theories.

So how do we explain the mass-conversion of late eighteenth-century chemists to the Lavoisierian doctrine? It is possible that scientific change happens due to rhetoric, interests, ideology or fashion, but I think in this case those factors give only a partial explanation. Yes, it was Joseph Priestley in Britain that led the resistance to the new "French chemistry", but this was the same Priestley who was hounded out of his native land for his support of the French Revolution, the same revolution that would two years later guillotine Lavoisier. And so on. The contextual picture is very complex, and will not give us a simple explanation of anything. (In this day and age, we might have to say: when *external* explanations are insufficient, we must invoke *internal* ones.) In any case, as philosophers we may well want to retain the question of whether the decisions that the past scientists took were scientifically justified or not.

Elsewhere (Chang 2009, 2010) I have argued in some detail that none of the standard philosophical criteria show sufficiently that the phlogiston theory was clearly inferior to the oxygen theory. Just to summarize the outcome here: it is not the case that the phlogiston theory was in the end refuted by empirical evidence while Lavoisier's theory was not. It is, for one thing, not the case that the phlogiston theorists had to assign negative weight to phlogiston (only a very few did). The idea that the phlogiston theory lacked empirical adequacy compared to Lavoisier's theory does not stand up to scrutiny, even in Philip Kitcher's (1993) very subtle version of it. Perhaps the most common and fatal failure in the standard discourse is the ignoring of the many empirical problems that Lavoisier's theory had. It is also not the case that Lavoisier's theory was inherently simpler than the phlogiston theory, despite Andrew Pyle's (2000) sophisticated defence of this idea. Even Pyle's version of the argument is ultimately founded in the sense that the phlogiston theory unnecessarily complicated things by postulating the existence of an unobservable and redundant substance, namely phlogiston; this argument ignores the fact that Lavoisier also had to postulate just such a substance, caloric, which he proudly put at the head of his table of simple substances.

I think the best integrated-HPS work we have on this subject is still Alan Musgrave's paper of 1976. Musgrave starts by rejecting some old misconceptions (1976, 182–86), anticipating the difficulties of the later works just cited, too. But I don't believe Musgrave's own story works, either. His Lakatosian answer proposes that the crucial factor was the comparative progressiveness of the two competing research programmes. It was rational for chemists to abandon the phlogiston programme because it stopped making successful new predictions and only resorted to ad hoc hypotheses. Musgrave maintains (1976, 205): "Between 1770 and 1785 the oxygen programme. . .developed coherently and each new version was theoretically and empirically progressive, whereas after 1770 the phlogiston programme did neither." But Musgrave himself states (1976, 199): "While Lavoisier was failing, Priestley was having great success with the 1766 version of phlogistonism. . .. the most impressive experiment of all came in early 1783." This was the confirmation of the phlogistonist prediction that a calx would be reduced to metal by heating in inflammable air. To support the Lakatosian argument, then, we need to find successful novel predictions that Lavoisier made around 1783 or later. Where are these predictions? Musgrave doesn't tell us. Was he thinking of the prediction that the oxidation of inflammable air (hydrogen) would produce an acid?[4] Or the prediction that muriatic (hydrochloric) acid would be decomposed into oxygen and the "muriatic radical", which even appears in Lavoisier's list of elements?

8.3.2 Epistemic Pluralism

Faced with these philosophical failures, I started looking for a different framework for understanding the Chemical Revolution. The most promising one seemed to be theory-choice on the basis of epistemic values, as implicit in Kuhn's *Structure of Scientific Revolutions* but articulated more clearly in his later paper "Objectivity, Value Judgment, and Theory Choice" (1977b). My preliminary conclusion in those terms was that the opposing camps of chemists did uphold different epistemic values. For example, Priestley had an ideal of completeness in recording and explaining every little thing happening in his lab (which he achieved by twisting his theory if necessary), while Lavoisier pursued clean and elegant theoretical uniformity. There was a clear trade-off between these two values, as Kuhn said we should often expect. So it makes sense that many people followed Lavoisier in pursuing uniformity, while some remained with Priestley in maintaining a stronger respect for all the messiness and particularity of phenomena. There was no compelling objective argument that uniformity was more important (or truth-conducive) than completeness. And so on.

If I had stopped there, the payoff from this case-study would have been a small vindication of the Kuhnian view on scientific change against the more traditional view that theory-change in science occurs (or should occur) only for reasons that all good scientists can commonly accept. That would have been a reasonable enough outcome, but I was left with a strong sense of unfinished business. Having studied the Chemical Revolution in depth and detail, my feeling was that if dissenters like Priestley dissented because they had different yet respectable values, that dissent should have been allowed, even fostered. If the phlogiston theory had its own distinct merits, it should have been kept. But it's not that I thought Lavoisier's theory should not have been developed and adopted. *They should have both lived.* That was the new insight that the case of the Chemical Revolution had finally forced upon me a full-fledged pluralism that says multiple theories should have been kept, not a

[4]Again, Musgrave himself points out this failed Lavoisierian prediction, and how long he had struggled with it (1976, 199–200).

petulant relativism that just says whichever theory would have been just as good or bad.

Thus I found that an immediate implication of the Kuhnian conclusion on the Chemical Revolution went against a central assumption in Kuhn's philosophy of science, which has been shared by most of his enemies, and had been accepted implicitly by myself. This assumption is monism. Kuhn maintained (as a matter of both fact and ideal) that in the normal state of science there was one paradigm dominating any given discipline. When normal science goes into a crisis this dominance is challenged, and when there is a revolution the dominant paradigm is overthrown and replaced by a new paradigm, which then enjoys its own era of monopolistic dominance. One paradigm or the other must win in the revolutionary struggle, even if there aren't unequivocally good reasons for the winner to win; that is the way it is, and that is the way it should be.

Rejecting this monist assumption and adopting a pluralist framework of analysis has improved and enriched my account of the Chemical Revolution, presented in full in Chang (forthcoming, chapter 1). Pluralism is of course not an entirely new idea (and epistemic iteration, too, has been a common enough idea, in various guises), but I believe that it has never seriously been articulated in the context of the historiography of the Chemical Revolution. Freed from the monist straightjacket, my puzzle about the dominance of Lavoisier now receives a new framing as a question about the failure of pluralism. Both epistemic and social factors are smoothly incorporated into the story of the triumph of Lavoisierian dogmatism, with myself as the historian reserving the right to retrospective dissent. Within a pluralist framework it is also easier to make sense of the wide variety of reactions to the rise of Lavoisier, of the various forms that each theory could and did take, and of the interactions between the two theories and their variations. Moving beyond that stage of work, I have also made a full articulation of pluralism in science and its merits (ibid., chapter 5), and I look forward to the task of applying a pluralist framework to the study of other historical episodes.

8.3.3 Systems of Practice

Thinking about the Chemical Revolution has led me to another major philosophical innovation, which is to take *systems of practice* instead of theories as units of epistemic appraisal. Thoughts related to this had bubbled up in my mind on various occasions, but it was the Chemical Revolution episode that made them crystallize. I have already mentioned how I departed from the traditional model of theory-choice in which different theories vie for the title of one exclusive truth, in favor of one in which theories are appraised in terms of various epistemic values. Setting truth aside, as an inoperable aim, I wanted to learn how to make sense of how scientists actually make theory-choice. Therefore considering various kinds of success rather than the ultimate truth, I also came to see that it was not theories in themselves, but what one did with theories that was successful or unsuccessful—and, of course, that success came not only in the realm of theory but also in the realm of experiment, instrumentation, modeling, simulation, and many other activities. Each of these epistemic activities has a distinct kind of aim, and can be assessed in relation to how successful it is in meeting its own aim. So it becomes necessary to consider how various epistemic activities fit coherently together with each other, to form a scientific system of practice (or, a system of knowledge). This is similar to a Kuhnian paradigm (in the sense of "disciplinary matrix" rather than "exemplar"), but I refrain from using the term "paradigm", both because Kuhn did not specify how different elements of it came together, and also because there are strong monist connotations to the Kuhnian term due to his assumption of the monopoly that a paradigm does and should enjoy in normal science.

To be more precise: an *epistemic activity* is a coherent set of mental or physical actions (or operations) that are intended to contribute to the production or improvement of knowledge in a particular way, in accordance with some discernible rules (though the rules may be unarticulated). It is important to keep in our view the aims that scientists are trying to achieve in each and every situation. The presence and operation of an identifiable aim (even if not articulated explicitly by the actors themselves) is what distinguishes actions and activities from mere physical happenings involving human bodies. Common types of epistemic activities include measurement, detection, prediction, hypothesis testing, etc. When we start thinking of scientific work as a collection of activities, an immediately obvious thing is the sheer variety in the types of epistemic activities that scientists engage in. Here is a partial list of types of epistemic activities: describing, predicting, explaining, hypothesizing, testing, observing, detecting, measuring, classifying, representing, modelling, simulating, synthesizing, analyzing, abstracting, idealizing.

Epistemic activities normally do not, and should not, occur in isolation. Rather, each one tends to be practiced in relation to others, constituting a whole system. A scientific *system of practice* is formed by a coherent and interacting set of epistemic activities performed with a view to achieve certain aims. It is the overall aims of a system of practice that define what it means for the system to be coherent. For instance, Lavoisier created a system of chemistry whose main activities included making various chemical reactions involving gases, tracking chemical substances through weight-measurement, classifying compounds according to their compositions, and analyzing organic substances by combustion. The overall aims of this system included determining the composition of various substances, and explaining chemical reactions in terms of the composition of the substances. The coherence of a system goes beyond mere consistency between the propositions involved in its activities; rather, coherence consists in various activities coming together in an effective way toward the achievement of the aims of the system. Coherence comes in degrees and different shapes, and it is necessarily a less precise concept than consistency, which comes well defined through logical axioms.

It may seem difficult to make a sharp distinction between epistemic activities and systems of practice, and this is intentional. When I distinguish higher and lower levels of description, that is only relative and context-dependent. Each epistemic activity is in itself a system of activities. For example, the combustion-analysis of a chemical substance consists of various simpler activities, such as burning, the absorption of combustion-products using other chemicals, weighing, and percentage-calculations. And even those component activities in themselves consist of other activities (the act of weight consisting in the placing of samples and weights on balance-pans, reading the number off the scale, etc.). In each situation in which we study a body of scientific practice, I am proposing to call the overall object a *system*; when it is desired that we should study more closely different aspects of that system, we can analyze the system into different *subordinate activities*, without implying a reductionist metaphysics in which a system is made by a simple addition of various activities which do not really have any connection with each other.

At least in the Anglophone traditions, philosophical analyses of science have been unduly limited by the common habit of viewing science as a body of propositions, focusing on the truth-value of those propositions and the logical relationships between them. This has led to the neglect of experimentation and other non-verbal and non-propositional dimensions of science in philosophical analyses. Many historians, sociologists and philosophers have pointed out this problem, but so far no clear alternative philosophical framework has been agreed upon to provide a language for fuller analyses of scientific practice. In an attempt to make improvements on this situation, the first step to take is to go beyond talking about theories and theory-choice, while not neglecting the theoretical dimension of science.

I have only just begun to articulate this way of analyzing science (see Chang, forthcoming, chapter 1, section B1.1). There is still much philosophical work to do in examining the cogency of the concepts of epistemic activity and system of practice. I expect to be able to draw much useful inspiration from pragmatist philosophy, among other sources. However, I have already begun to apply these abstract ideas to frame further historical episodes, starting with chapters 2 and 3 of the forthcoming book. There seems much promise in this direction.

8.4 Concluding Remarks

I started by problematizing the common use of historical case-studies in relation to philosophy of science. And then I tried to illustrate one particular mode of history–philosophy interaction, which can be schematized as follows:

Existing philosophical framework
→ Historiographical puzzle: an episode that is difficult to understand
→ Search for a new philosophical framework
→ Better understanding of the episode, in the new philosophical framework
→ Further development of the new philosophical framework
→ Application of the new framework to other episodes

We start with an existing philosophical framework, and find historiographical puzzles, namely episodes that are difficult to describe and understand. In attempts to find an apposite description of these episodes, historians can *generate* new concepts and ways of thinking that philosophers may not come up with from their entirely abstract work. And then philosophers can take up the new abstract ideas generated through the historical work, and develop and apply them further. In this business it is difficult to say whether a given investigator is working as a historian or a philosopher, but that is just the point. Contrary to Kuhn, I think this process works out best if the historian and the philosopher *is* the same person doing both at the same time. Now, it may be objected that *good* philosophers should be able to come up with good philosophical ideas without the help of historiographical stimulus, but in reality I think abstract philosophical thinking has not been very effective in generating ideas that are useful for the analysis of scientific practice. And at any rate, I do not see anything wrong with getting help where we can, rather than insisting on restricting the source of our ideas.

There is also a mirror-image of this process, in which philosophy helps history:
Existing historiography

→ Philosophical puzzle: a set of putative actions/decisions by past scientists that does not make sense
→ Search for better historiography
→ Removal of philosophical puzzle
→ Empirical work to complete new historical account
→ Reflections on other related history

This is a reminder that when there is a tension between history and philosophy, it is not always philosophy that is to blame, or at least not entirely. (This is similar to how a conflict between theory and observation does not necessarily mean that the theory is incorrect.) In fact, I believe that such correction of history through philosophical puzzlement is also necessary in the case of the Chemical Revolution (see Chang 2010).

I believe that both of these processes have already been at work in much of the best research in integrated HPS, and I hope that my articulation will help them continue in more conscious, thorough and methodologically sound ways. Finally, I would like to close by placing these processes into a broader framework of history–philosophy interaction. Table 8.2 summarizes what I have worked out so far.

Table 8.2 Modes of history–philosophy engagement

Mode of engagement	History gives philosophy	Philosophy gives history
Necessary	Subject matter ("philosophy of science without history of science is empty")	Conceptual framework ("history of science without philosophy of science is blind")
Critical	Counter-examples	Detection of implausibility
Heuristic	New concepts for better understanding of puzzling episodes	Discovery of new historical facts to remove philosophical puzzles
??? ("other things")		

There are at least three modes of history–philosophy engagement. First, they are indeed necessary for each other, history giving philosophy its very subject matter, and philosophy providing the necessary conceptual framework. Then there is the critical function that history and philosophy can serve for each other. And what I have addressed in this paper is mainly the heuristic function. My table, just like Francis Bacon's list of "instances meeting in the nature of heat", ends with a bunch of question-marks, or, "other things" (Bacon 2000, 111). That is my way of indicating that we are at the start, not at the end, of our business.[5]

References

Bacon, Francis. 2000. *Novum Organum*, edited by by Lisa Jardine and Micheal Silverthorne. Cambridge: Cambridge University Press.

Brooke, John Hedley. 1981. "Avogadro's Hypothesis and Its Fate: A Case-Study in the Failure of Case-Studies." *History of Science* 19: 235–73.

Burian, Richard M. 2001. "The Dilemma of Case Studies Resolved: The Virtues of Using Case Studies in the History and Philosophy of Science." *Perspectives on Science* 9: 383–404.

Chang, Hasok. 1995. "Circularity and Reliability in Measurement." *Perspectives on Science* 3: 153–72.

Chang, Hasok. 2004. *Inventing Temperature: Measurement and Scientific Progress*. New York, NY: Oxford University Press.

Chang, Hasok. 2007. "Scientific Progress: Beyond Foundationalism and Coherentism." In *Philosophy of Science*, edited by Anthony O'Hear. Royal Institute of Philosophy Supplement 61, 1–20. Cambridge: Cambridge University Press.

Chang, Hasok. 2009. "We Have Never Been Whiggish (About Phlogiston)." *Centaurus* 51: 239–64.

Chang, Hasok. 2010. "The Hidden History of Phlogiston: How Philosophical Failure Can Generate Historiographical Refinement." *HYLE – International Journal for Philosophy of Chemistry* 16(2): 47–79.

Chang, Hasok. Forthcoming. Is Water H_2O? Evidence, Realism and Pluralism. Dordrecht: Springer.

Kitcher, Philip. 1993. *The Advancement of Science: Science Without Legend, Objectivity Without Illusions*. New York, NY, and Oxford: Oxford University Press.

Kuhn, Thomas S. 1962. *The Structure of Scientific Revolutions*. Chicago, IL: The University of Chicago Press.

Kuhn, Thomas S. 1971. "Notes on Lakatos." In *PSA 1970: In Memory of Rudolf Carnap*, edited by Roger C. Buck and Robert S. Cohen. Vol. 8 of Boston Studies in the Philosophy of Science, 137–46. Dordrecht: Reidel.

Kuhn, Thomas S. 1977a. "The Relations Between the History and the Philosophy of Science." In *The Essential Tension*, 3–20. Chicago, IL: University of Chicago Press.

Kuhn, Thomas S. 1977b. "Objectivity, Value Judgment, and Theory Choice." In *The Essential Tension*, 320–39. Chicago, IL: University of Chicago Press.

Kuhn, Thomas S. 1980. "The Halt and the Blind." *British Journal for the Philosophy of Science* 31: 181–92.

[5]This paper originated as a presentation at the first Integrated History and Philosophy of Science conference ("&HPS1"), at the University of Pittsburgh, 13 October 2007. I would like to take this opportunity to thank John Norton and colleagues for hosting that meeting, and various other participants for their helpful comments.

Kuhn, Thomas S. [1991] 2000. "The Trouble with the Historical Philosophy of Science." In *The Road Since Structure*, edited by James Conant and John Haugeland, 105–20. Chicago, IL: University of Chicago Press. From a lecture delivered in 1991 at Harvard University.

Lakatos, Imre. 1971. "History of Science and its Rational Reconstructions." In *PSA 1970: In Memory of Rudolf Carnap*, edited by Roger C. Buck and Robert S. Cohen. Vol. 8 of Boston Studies in the Philosophy of Science, 91–136. Dordrecht: Reidel. Also reprinted in Colin Howson, ed., *Method and Appraisal in the Physical Sciences* (Cambridge: Cambridge University Press, 1976).

Lamé, Gabriel. 1836. *Cours de physique de l'Ecole Polytechnique*. Paris: Bachelier.

Musgrave, Alan. 1976. "Why Did Oxygen Supplant Phlogiston? Research Programmes in the Chemical Revolution." In *Method and Appraisal in the Physical Sciences*, edited by C. Howson, 181–209. Cambridge: Cambridge University Press.

Pitt, Joseph C. 2001. "The Dilemma of Case Studies: Toward a Heraclitian Philosophy of Science." *Perspectives on Science* 9: 373–82.

Pyle, Andrew. 2000. "The Rationality of the Chemical Revolution." In *After Popper, Kuhn and Feyerabend*, edited by Robert Nola and Howard Sankey, 99–124. Dordrecht: Kluwer.

Chapter 9
Hidden Entities and Experimental Practice: Renewing the Dialogue Between History and Philosophy of Science

Theodore Arabatzis

9.1 Introduction

The voluminous literature on the relationship between history of science and philosophy of science has been one-sided—occupied for the most part with the significance of the former for the latter. Historically oriented philosophers of science have viewed the history of science as a repository of empirical material for testing philosophical theories of scientific rationality or scientific change. Historians of science, on the other hand, have often doubted the "pragmatic value" of the philosophy of science (Buchwald 1992, 39). Even philosophically inclined historians, such as Thomas Kuhn, have denied the relevance of "current philosophy of science ... for the historian of science" (Kuhn 1977, 12).[1] The widespread skepticism, among historians, about the historiographical utility of philosophy of science may have been reinforced by some philosophers' forays into history of science, which were blatantly insensitive to the categories of historical actors (see, for instance, Lakatos 1970). Be that as it may, philosophy of science, as I have argued elsewhere, may enrich historiography by scrutinizing the philosophical underpinnings of historiographical categories and choices (Arabatzis 2006a). When I advocate a philosophical historiography of science I do not, thereby, recommend the importation of ready-made philosophical positions into historiography. Rather my point is that an engagement with certain philosophical issues and debates may deepen historical analysis. If none of the available philosophical positions can do justice to the complexities of the historical record, then philosophically inclined historians of science should develop their own historiographically-driven philosophy of science.

[1] I should note that this asymmetry is primarily a feature of Anglo-American history and philosophy of science. In France, on the other hand, history of science has had a much stronger connection to philosophy of science, as testified to by the work of Georges Canguilhem and Michel Foucault. For this point, I would like to thank Bernadette Bensaude-Vincent and Henning Schmidgen.

T. Arabatzis (✉)
Department of Philosophy and History of Science, University of Athens, University Campus, Ano Ilisia, 157 71 Athens, Greece
e-mail: tarabatz@phs.uoa.gr

S. Mauskopf, T. Schmaltz (eds.), *Integrating History and Philosophy of Science*, Boston Studies in the Philosophy of Science 263, DOI 10.1007/978-94-007-1745-9_9,

In the process, they may come up with novel philosophical insights (cf. Chapter 8 by Chang, this volume).

Let me present very briefly two examples from my previous work that illustrate in a concrete manner what I have in mind. The first concerns scientific discovery (see Arabatzis 2006b). The apparently descriptive statement "X discovered Y" involves an epistemic judgment, namely that the evidence mustered by X was sufficient to establish Y's existence. Furthermore, the concept of scientific discovery has a realist flavor: if something is discovered then it is ipso facto real. Thus, by employing scientific discovery as a historiographical category, one runs into the issue of scientific realism. In order to narrate a discovery-episode historians would profit from taking into account the complexities of that issue. To identify the object of a discovery and those who were responsible for it calls for conceptual analysis, on top of empirical research. The point of such an analysis should be, in my view, to chart a neutral ground that is shared by realists and anti-realists alike and, thus, to enable the narration of discovery episodes that would be equally acceptable to both groups.

My second example concerns the philosophical issue of conceptual change and its implications for choosing the subject of a historical narrative. If concepts evolve and cease to refer to the same entities, as Kuhn and Feyerabend have famously argued, then, prima facie, they are not good candidates for historical subjects. The fluidity of scientific concepts seems to preclude the possibility of framing coherent historical narratives around them. Quentin Skinner has made this point in no uncertain terms:

> as soon as we see there *is* no determinate idea to which various writers contributed, but only a variety of statements made with the words by a variety of different agents with a variety of intentions, then what we are seeing is equally that there *is* no history of the idea to be written. (Skinner 1969, 38)

In my work on the history of the electron I tried to address Skinner's challenge, as regards the history of scientific concepts. To that effect I have drawn upon the considerable philosophical literature on conceptual change in science. In the process I hope I have shed new light on some of the philosophical issues involved. It would take me too far astray to present, even in outline, this literature and my own take on it.[2] For our purposes here, the important point is the relevance of a philosophical issue to a historiographical problem.

In this paper I want to investigate further the prospects of integrated history and philosophy of science, by examining how philosophical issues concerning experimental practice and scientific realism can enrich the historical investigation of the careers of "hidden entities", entities that are not accessible to unmediated observation. Conversely, I will suggest that the history of those entities has important lessons to teach to the philosophy of science. Thus, my aim is to indicate some ways in which the dialogue between history and philosophy of science could be renewed.

---

[2] I refer the interested reader to (Arabatzis 2006a).

9.2 Why Use the Term "Hidden Entities"?[3]

Let me start with a comment on my choice of terms. I have chosen the term "hidden entities" instead of other more familiar terms, such as "unobservable entities" or "theoretical entities", for the following reasons. First, I wanted to avoid the thorny issues surrounding the observable-unobservable distinction. This distinction immediately invites questions about the boundary between the observable and the unobservable and about its epistemic significance. Forty five years ago Grover Maxwell argued that it is not possible to draw a sharp dividing line between the observable and the unobservable realms and, therefore, the distinction in question lacked any epistemological and ontological significance (Maxwell 1962). This issue has been debated by philosophers of science ever since, especially after Van Fraassen reinstated the distinction and placed it at the centre of his constructive empiricist epistemology. The advantage of using the term "hidden", in this respect, is that we leave open the possibility of the hidden becoming disclosed.

Second, I have also avoided the term "theoretical entities", even though I used it elsewhere, because it conveys the misleading impression that hidden entities do not transcend the theoretical framework in which they are embedded. In fact, these entities are trans-theoretical objects, which cut across different theories or even entire disciplines. Several philosophers of science have stressed their trans-theoretical character. On the one hand, philosophers such as Nancy Cartrwright and Ian Hacking have emphasized the synchronic dimension of the trans-theoretical character of hidden entities. Witness Cartwright's remark concerning "the electron, about which we have a large number of incomplete and sometimes conflicting theories" (Cartwright 1983, 92). On the other hand, philosophers such as Dudley Shapere and Hillary Putnam have pointed out the diachronic dimension of the trans-theoretical character of hidden entities, that is, the fact that these entities are usually the objects of consecutive scientific theories. Furthermore, the term "theoretical entities" undervalues completely the fact that many of the entities in question become experimental objects that are investigated in the laboratory, often without any guidance from a systematic theory about their nature.

Of course, I could have used other terms, such as "inferred entities" or "hypothetical entities". For the period in which my work has focused so far (the late nineteenth and early-twentieth centuries) the terms "hidden" or "invisible" entities have the additional advantage that they denote a category of historical actors, atomists and anti-atomists alike. Heinrich Hertz, for instance, pointed out in his posthumously published *Principles of Mechanics* (1894) that "the form of the atoms, their connection, their motion in most cases—all these are entirely hidden from us" (Hertz 1956, 18). The lack of direct epistemic access to those characteristics of atoms, however, did not diminish Hertz's long-time conviction in the atomic constitution of matter

[3]I have borrowed this term from the title of an international laboratory for the history of science organized by the Dibner Institute in June 1998.

and in the possibility of determining some of the hidden properties of atoms (e.g., their size) experimentally (cf. Lützen 2005, 46, 55–56).[4]

Another well-known advocate of the atomic theory, the French experimental physicist Jean Perrin, described the aim of science in these colourful terms:

> In studying a machine, we do not confine ourselves only to the consideration of its visible parts . . . We certainly observe these visible pieces as closely as we can, but at the same time we seek to divine the *hidden* gears and parts that explain its apparent motions.
>
> To divine in this way the existence and properties of objects that still lie outside our ken, *to explain the complications of the visible in terms of invisible simplicity*, is the function of the intuitive intelligence which, thanks to men such as Dalton and Boltzmann, has given us the doctrine of Atoms. (Perrin 1916, vii)

Furthermore, according to Perrin, the line between the visible and the invisible may shift as a result of technological developments (ibid.). As a matter of fact, he strived throughout his career to lift the veil that hided molecular reality and to render molecular motions visible (see Nye 1972, 54; Bigg 2008).

Anti-atomists also employed a similar terminology. In an impassioned advocacy of "energetics" Pierre Duhem claimed that its principles "do not aspire at all to resolve the bodies we perceive or the motions we report into imperceptible bodies or hidden motions" (Duhem 1913/1996, 233). In stark contrast to the "neo-atomists", he thought that the hidden realm behind the phenomena is not epistemically accessible (ibid., 238; cf. Nye 1972, 166).

As a final example, consider Henri Poincaré's response to the popular fin-de-siècle view that the history of scientific theories resembles a heap of "ruins piled upon ruins" (Poincaré 1905, 160; cf. Nye 1972, 35–38). He argued that the prima facie plausibility of that view derived from neglecting to attend to the proper aim of scientific theories. The aim in question, according to Poincaré, is not to reveal the hidden objects that give rise to physical phenomena, because "Nature will hide for ever [those objects] from our eyes" (Poincaré 1905, 161). Rather, the aim of theorizing is to discover "The true relations between these real objects". These "are the only reality we can attain" (ibid.). Furthermore, the discovery of these relations is an enduring achievement that will not be undermined by the subsequent development of science. Thus, Poincaré's relational realism makes possible to salvage the continuity and permanent value of scientific knowledge.

Notwithstanding the popularity of the term "hidden" among historical actors, in our constructivist age this term may have some objectionable overtones, suggesting a pre-existing reality waiting to be disclosed. I think, however, that one may adopt a distinction between a hidden and a manifest realm, while remaining neutral in metaphysical disputes concerning the nature of reality.

[4]It should be noted that in the *Principles of Mechanics* Hertz used the term "hidden" mainly in connection with mass. The introduction of "hidden masses" served a theoretical purpose, namely to dispense with the notion of force. I would like to thank Giora Hon for pointing out to me the nuances of the term "hidden" in Hertz's text.

9.3 A Glance at the Role of Hidden Entities in the History of the Physical Sciences: The Historical Roots of a Philosophical Problem

The explanation of phenomena by postulating hidden entities has been a significant aspect of the sciences, at least since the seventeenth century. Think, for instance, of the central tenet of the mechanical philosophy, namely that the fundamental constituents of the world are imperceptible material particles in constant motion. Those particles were introduced for explanatory purposes, to accommodate various phenomena within a mechanical framework. Descartes, for instance, attempted to account for magnetic attraction by postulating screw-shaped particles, "which in passing through the pores in magnets and iron, drive the air from between the two and cause them to move together" (Westfall 1977, 37). In the following centuries we witness a multiplication of hidden entities, many of which were introduced for a similar reason, that is, to accommodate, within a mechanical framework, phenomena that were not easily susceptible to mechanical explanation. In the eighteenth century, for example, subtle fluids were posited to make mechanical sense of phenomena, such as electricity and magnetism, which seemed to involve action at a distance. Those imponderable fluids were supposed to be self-repelled and attracted to matter, which they permeated. By the end of the eighteenth century they had proved their fertility and promised to offer a unified quantitative framework for investigating electricity, magnetism, light, heat, and combustion (Heilbron 1993). Similarly, in the nineteenth century the "luminiferous" ether was put forward to incorporate light within a mechanical framework. The subsequent development of field theory led to a unification of light with electromagnetic processes and an identification of the optical and the electromagnetic ether. By the last quarter of the nineteenth century, the prospects of understanding a dazzling variety of disparate phenomena as manifestations of a hidden mechanical medium seemed bright indeed (Cantor and Hodge 1981).

Furthermore, the mechanical tradition was reinforced by the postulation of another hidden entity, the atom, which was originally invoked by John Dalton in response to problems in meteorology and chemistry. In the latter its main functions were to simplify, systematize and explain empirical regularities, such as the laws of definite and multiple proportions. It was soon appropriated by physicists, who employed it to develop a successful mechanical account of heat as a form of motion. Throughout the century, however, many scientists thought of atoms as dispensable fictions and the question of their ontological status remained open (Gardner 1979; Chalmers 2009; Rocke 2010). In the early-twentieth century the atomic debates were finally resolved, mainly as a result of Perrin's experimental investigations of Brownian motion which provided striking evidence in favor of the existence of atoms. The subsequent development of microphysics led to a real explosion in the number of the hidden constituents of matter, ranging from electrons to quarks.

This brief and impressionistic historical sketch indicates that hidden entities have often (always?) been introduced for explanatory purposes. Some of them (e.g., the subtle fluids) were subjected to experimental investigation, whereas others (e.g., the ether) were resistant to experimental detection. Thus, entire domains of theoretical and experimental practice have been structured around hidden entities. This fact alone would suffice to render these entities historiographically significant. Furthermore, they are puzzling from a philosophical point of view. Several of them, notwithstanding their explanatory fertility, turned out to be fictitious. Phlogiston, caloric, and the ether, to mention the most salient cases, are no longer recognized as real entities. For this reason, perhaps, the philosophical literature concerning hidden ("unobservable"/"theoretical") entities has focused on the problem of scientific realism, that is, on the grounds that we have for believing in their existence.

Among the origins of this problem is the so-called underdetermination of theory by evidence, namely the fact that there can be more than one hypotheses or theories that are compatible with the phenomena. This problem had been discussed since antiquity. The introduction and proliferation of hidden entities, however, made it more intractable. Any inductive generalization faces "horizontal" underdetermination, but with the hypothetical postulation of entities "underneath" the phenomena one has to worry also about "vertical" underdetermination.[5]

9.4 Bypassing Underdetermination: Cartwright and Hacking on Entity Realism

There have been various attempts to come to terms with the problem of underdetermination. The one I will discuss here was put forward by Ian Hacking, who tried to bypass this problem by focusing on experimental practice and the specific mode of causal reasoning that is employed in that practice. A similar view has been adopted and further developed by Nancy Cartwright. Instrumentation and experimentation, in Hacking's and Cartwright's view, can provide, under certain circumstances, unmediated (largely theory-free) access to the hidden reality behind the appearances. Hacking has argued that the manipulation of hidden entities in the laboratory compels us to be realists about them. The uses of hidden entities as investigative probes and as engineering tools leave little room for doubting their existence. Hidden entities cease to be hypothetical when we succeed in manipulating them. For instance, the reality of electrons is beyond reasonable doubt, since we have devices with which we can spray them. In Hacking's seductive words, "if you can spray them, then they are real" (see Hacking 1983, 22ff.). Of course, it may turn out that our theoretical representations of electrons and their properties are mistaken, but it is highly unlikely that electrons will turn out to be fictitious. Cartwright concurs:

> I agree with Hacking that when we can manipulate our theoretical entities in fine and detailed ways to intervene in other processes, then we have the best evidence possible for our claims about what they can and cannot do; and theoretical entities that have been

[5]I borrow these terms from (Worrall 2000).

> warranted by well-tested causal claims like that are seldom discarded in the progress of science. (Cartwright 1983, 98)

This version of realism, as many commentators have pointed out, faces several difficulties.[6]

9.5 Problems of Entity Realism: A Role for History of Science

Perhaps the main difficulty is that Hacking begs the question by assuming "what is under dispute", namely that we can spray electrons (cf. van Fraassen 1985, 298). The identification of an act of laboratory manipulation with the spraying of electrons cannot be the premise of an argument purporting to demonstrate the existence of electrons.[7] To put it another way, our confidence in the existence of electrons must precede our claim that in a certain laboratory setting we manipulate electrons (cf. Seager 1995, 467–68). Of course, "manipulation" is a success term—we cannot manipulate something that does not exist (cf. Nola 2002, 5). Perhaps that is why Hacking calls his "conclusion ... obvious, even trifling" (Hacking 1983, 146). The real question, though, concerns the identity of the objects we manipulate.

I will call this difficulty "the manipulation of what?" problem: before we invoke manipulability as a demonstrative principle, we need to identify the entity that we manipulate. There are experimental situations, however, where we manipulate *something* without knowing *what kind of thing* we manipulate. For instance, in the last quarter of the nineteenth century several physicists manipulated cathode rays, experimental objects that were produced in the discharge of electricity through gases at very low pressure.[8] The identification of cathode rays with electrons at the end of the nineteenth century revealed that the earlier manipulations of cathode rays had been, in fact, manipulations of electrons. Prior to that identification, however, the physicists who manipulated cathode rays did not know what kind of thing they manipulated. Hacking has claimed that "from the very beginning people were less testing the existence of electrons than interacting with them" (Hacking 1983, 262). Actually, people were interacting with electrons well before they even suspected their existence. Thus, manipulability, by itself, cannot establish the existence of, say, electrons, as opposed to cathode rays or an "I know-not-what" something (cf. Achinstein 2001a, 412; Boon 2004, 229).

To put it another way, the "material realization"[9] of an experiment can be compatible with a plurality of descriptions (and theoretical interpretations) of what is going on in the experiment. Since the material realization of an experiment underdetermines its theoretical interpretation, the question "What entity is being manipulated

[6]See, for instance (Arabatzis 2001; Elsamahi 1994; Gross 1990; Morrison 1990; Reiner and Pierson 1995; Resnik 1994).

[7]See the illuminating discussion in (Suárez 2008, 154).

[8]For a concise history of those objects see (Arabatzis 2009a).

[9]The term is from (Radder 1995, 69).

in the experiment in question?" cannot be answered merely on the basis of the experimental operations performed by the experimenter. The epistemic gap from our manipulations of "apparent" entities to the existence of hidden entities can only be bridged by our representations of the hidden world.

And this brings me back to the problem of underdetermination. One would expect that theoretical explanations as well as entity-based explanations of phenomena face equally this problem. Nancy Cartwright, however, has argued that there is an asymmetry in these two kinds of explanation. Only entity-based explanations are exempt from underdetermination:

> We can infer the truth of an explanation only if there are no alternatives that account in an equally satisfactory way for the phenomena. In physics nowadays, I shall argue, an acceptable causal story is supposed to satisfy this requirement. But exactly the opposite is the case with the specific equations and models that make up our theoretical explanations. There is redundancy of theoretical treatment, but not of causal account. (Cartwright 1983, p. 76)

The problem here, as I see it, is that Cartwright assumes that the current absence of alternatives implies the absence of alternatives period. One could very well conceive of the existence of two or more causal accounts of the same phenomena, based on the existence of altogether different entities. After all, in the history of the sciences there have been such cases—for instance, a phlogiston-based and an oxygen-based account of combustion (Arabatzis 2001, S534; Carrier 1993, 401–03). I don't see how this possibility could be excluded (cf. Clarke 2001, 719; Gelfert 2003, 248). Actually, a proponent of "experimental realism", Mauricio Suárez, has admitted this possibility. According to Suárez, "We arguably once had causal warrant for phlogiston but no longer do." And Priestley "was led by his prior belief in phlogiston to interpret all his experimental manipulations as providing grounds for the nonredundant role of phlogiston in the explanation of combustion" (Suárez 2008, 156, 157). This is exactly right, but, *pace* Suárez, I think that the phlogiston case undermines entity realism, by showing that the non-redundancy of entity-based causal explanations may be just a temporary feature of our knowledge. Even if, at a given stage of scientific development, we lack more than one causal explanation of certain phenomena, the future development of knowledge may bring to light "unconceived alternatives".

I have argued, so far, that the putative manipulation of a hidden entity is not a sufficient criterion for establishing its existence. Is it a necessary one? In response to his critics, Hacking has recognized the variety of standards of proof, in addition to manipulability, that are brought to bear, *within* scientific practice, on the existence of hidden entities.

> My experimental argument for entity realism may imply a sufficient (epistemological) condition for holding that an entity exists. But it does not imply a necessary condition. There may be many kinds of evidence that an entity exists. I hold only that manipulationability is the best evidence. (Hacking 1995/1996, 540)

Thus, manipulability should not be interpreted as a necessary condition for belief in the existence of a hidden entity. A difficulty remains, however: within scientific practice manipulability is sometimes (often?) not considered the "best proof" or the "best evidence" in favour of an entity (Gelfert 2003; Massimi 2004; Morrison 1990). So if we applied Hacking's criterion we would, sometimes, end up accepting entities that are contentious among the relevant experts or even admitted to be fictitious. In other words, the criterion may recommend ontological commitment even in cases where the scientific community has not unambiguously decided in favour of the existence of an entity.

Cartwright's exclusive emphasis on causal inference faces the same problem. Consider her account of

> the radiometer, invented by William Crookes in 1853. It is a little windmill whose vanes, black on one side, white on the other, are enclosed in an evacuated glass bowl. When light falls on the radiometer, the vanes rotate. It was ... agreed that the rotation is due to the action of the gas molecules left inside the evacuated bowl. ... in 1879 James Clerk Maxwell, using the kinetic theory of gases, argued that ... differential heating in the gas produces tangential stresses, which cause slippage of the gas over the surface. As the gas flows around the edge, it pulls the vanes with it.
>
> ...
>
> The molecules in Crookes's radiometer are invisible, and the tangential stresses are not the kinds of things one would have expected to see in the first place. Yet, ... I believe in both. I believe in them because I accept Maxwell's causal account of why the vanes move around. (Cartwright 1983, 5–6)

As with Hacking's manipulability criterion, the problem here is the anticipation of the verdict of the scientific community. Molecules remained controversial entities till the beginning of the twentieth century. Apparently, many physicists and chemists were not (and, I think, should not have been) swayed by Maxwell's causal account of the radiometer's function to believe in molecules. The moral of this case is that philosophers of science should not anticipate (or even supplant) the judgements of the scientific community by oversimplifying the issues at stake. Rather they should attend to the multitude of theoretical and experimental practices that are brought to bear, over extended periods, on the existence of hidden entities. Philosophy of science has to accommodate the complexity of its subject matter. To that effect, history of science has an indispensable role to play.

9.6 Towards a Historiographically Adequate Philosophical Attitude

It is clear, to my mind at least, that manipulability cannot get around the hypothetical status of hidden entities. Is there a philosophical attitude towards those entities that can do justice to their history? Among other things, we have to do justice to the historical fact that important scientists believed passionately (and, I think, for good reasons) in entities that turned out to be fictitious. We have to understand, *in epistemic terms*, how it was possible, or even reasonable, for a physicist of J. J. Thomson's caliber to claim in 1909 that "The ether is not a fantastic creation of the speculative philosopher; it is as essential to us as the air we breathe" (Thomson 1909, 267). In the same vein, we should be able to fathom Lord Kelvin's belief that "We know the luminiferous ether better than we know any other kind of matter in some particulars. ... we know more about it than we do about air or water, glass or iron" (Kelvin 1904, 10–11). By immersing ourselves in the theoretical, instrumental, and experimental practices of past scientists, in their "virtual reality" as it were (Seager 1995), it becomes possible to understand the plausibility, coherence, and success (relative to the then current epistemic standards) of their beliefs. Thus, it will occasion no surprise that the scientists in question developed an, often strong, conviction in the reality of their objects of study. At the same time, however, the fact that some of those objects have perished motivates us to distance ourselves from the ontological commitments of the historical actors. Thus, the attitude I am recommending drives a wedge between immersion in a worldview (and a set of practices) and belief in the hidden entities associated with it. It has some parallels with Husserl's *epoché*, an attitude of abstention from ontological questions. I will call it "attitude of ontological bracketing".[10]

9.7 Sidestepping the Problem of Realism

The attitude of ontological bracketing does not amount to antirealism. The realism issue concerns the proper epistemic attitude towards contemporary science, whereas the attitude I'm recommending is directed towards the scientific past. To extend the scope of this detached attitude to present-day science, one would have to show that contemporary science is epistemically on a par with past science. Furthermore, the aim of ontological bracketing is to sidestep the normative aspects of the problem of realism and focus on issues which, though related to it, have a predominantly descriptive and interpretative character. I will touch upon three of those issues:

First, there is a descriptive counterpart to the normative philosophical problem. How do the scientists themselves become convinced that a hidden entity is real? Although I hesitate to give a simple answer to such a complex question, I would stress two factors that are important in this respect: The first factor has to do with theory. The empirical adequacy, the explanatory power, and the fertility of the theory positing a hidden entity are usually considered among the most important reasons for believing in its existence.[11] The second factor is related to experiment. The over-determination of a hidden entity's properties in different experimental settings is often an important reason in favour of its existence. For example, in the late nineteenth century the charge to mass ratio of the electron was determined by different methods and in different kinds of experiments: on cathode rays, on ß-rays, on thermionic emission, and in spectroscopy. The approximate agreement of the results obtained convinced many physicists that electrons were real entities (see Arabatzis 2006a). Another prominent example concerns the resolution

[10] I would like to thank Mitchell Ash for pointing out the similarities between Husserl's ontological attitude and the historiographical-cum-philosophical stance I am trying to articulate. Cf. (van Fraassen 1980a, 81).

[11] The importance of these values of theory appraisal for the realism debate has been stressed by Ernan McMullin. See, for instance (McMullin 1984).

of the atomic debates in the early twentieth century. Perrin's convergent multiple determinations of Avogadro's number, on the basis of very different experimental procedures, tipped the scales in favor of the existence of atoms.[12]

The second issue concerns the role of experimentation on hidden entities in the construction of their representations. How do scientists infer the characteristics of such entities by experimenting on them? Here I will draw on two philosophers: Pierre Duhem and Norwood Russell Hanson. As Duhem argued, a hidden entity is associated with a constellation of effects: an electric current, for instance, "may manifest itself not only in mechanical effects but in effects that are chemical, thermal, luminous, etc" (Duhem 1954, 151). What we need to understand in specific cases is how these different effects are held together as manifestations of a single entity.[13]

Furthermore, we need to understand how specific characteristics are attributed to those entities. Hanson's remark that "The idea of... atomic particles is a conceptual construction 'backwards' from what we observe in the large" is particularly helpful in this respect (Hanson 1963, 47). When an experimentally produced phenomenon is attributed to a hidden entity, the characteristics of the phenomenon that are of interest to the scientist(s) must be linked with the putative properties and behaviour of the entity in question. As Cartwright has put it, echoing Hanson's idea,

> Given our general knowledge about what kinds of conditions and happenings are possible in the circumstances, we reason backwards from the detailed structure of the effects to exactly what characteristics the causes must have in order to bring them about. (Cartwright 1983, 6)

For instance, in late nineteenth-century spectroscopy the phenomena observed in the laboratory had three salient characteristics: the frequency, intensity, and polarization of spectral lines. Once spectral lines were attributed to a hidden entity, the electron within the atom, their characteristics had to be linked with the properties and behaviour of that entity. The frequency, intensity, and polarization of spectral lines were correlated with the frequency, amplitude, and direction of vibration of the electron within the atom. In that way, experimentally obtained information guided the articulation of the representation of the electron.

A related question concerns the *measurement* of hidden entities. Since the late nineteenth century various properties of hidden entities have been measured, the mass and charge of elementary particles being among the most prominent. How is it possible to measure something that is hidden? The process of measurement in this case is very similar to Newton's "deduction from the phenomena". Given the hypothesis that an entity exists and that it is subject to certain laws, it is possible to use experimental results to fill in the blanks in the description of the entity. Thus, the measurement of hidden entities can be represented as "the continuation of theory construction by other means" (van Fraassen 1980b, 673). Again, one sees the potential significance of philosophy of science to history of science. Philosophical

---

[12]See (Nye 1972, 160ff). This episode has been the subject of divergent philosophical analyses. See, for instance (Cartwright 1983; Salmon 1984; Achinstein 2001b; van Fraassen 2009).

[13]For a preliminary attempt to answer this question, see (Arabatzis 2006a).

views about the character and function of hidden entities may stimulate and enrich historical analysis.

We should grant, I think, that theory is crucial for the experimental investigation of hidden entities. We should still ask, however, whether these entities qua experimental objects have any independence from their theoretical representations. In other words, do they have a life of their own? I think that they do, and this is an insight of lasting value in Hacking's and Cartwright's "experimentalism" that is borne out by the history of hidden entities. A substantial part of our knowledge of them derives from experiment and is, in an important sense, independent from theory. First, it is often the case that scientists are involved in exploratory experimentation on hidden entities, without being guided by a full-fledged theoretical account of their nature (Clarke 2001, 711; Steinle 1997, 2002). That was the case, for example, in experimentation on cathode rays during the last quarter of the nineteenth century (Hiebert 1995). Furthermore, experimentally determined properties of hidden entities are often incorporated into very different theoretical representations of them. Scientists who may disagree about the ultimate nature of those entities may come to agree about their experimentally determined properties. Those properties may, in turn, become essential for identifying their carriers in different experimental settings. For instance, J. J. Thomson in England, Walter Kaufmann in Germany, and Paul Villard in France had very different ideas about the ultimate nature of cathode rays. Thomson identified them with subatomic particles; Kaufmann represented them as ether waves; and Villard believed that they were charged hydrogen particles. All of them, however, agreed on the value of their mass to charge ratio.[14] Finally, the existence of conflicting theoretical representations of a hidden entity does not necessarily call into question its identity in experimental contexts. For example, in the early twentieth century several incompatible accounts of the shape and structure of the electron were put on the table. Those accounts led to different predictions about the velocity dependence of the mass of the electron. Walter Kaufmann's experiments on β-rays (high speed electrons) were set up to resolve that issue. What is significant for my purposes is that the entities experimentally investigated by Kaufmann were taken, by all parties in the dispute, to be the common referent of the divergent theoretical representations of the electron (see Arabatzis 2009b; cf. Galison 1997, 812–13).

9.8 Concluding Remarks

To conclude, I hope I have showed that our understanding of hidden entities and their role in experimental practice can be enhanced by adopting an integrated historical-cum-philosophical approach. On the one hand, philosophical reflection on the problem of entity realism has a lot to gain by examining historically how those entities were introduced and investigated. On the other hand, the historical

[14] See (Arabatzis 2004; Lelong 2001).

analysis of the careers of those entities may profit from philosophical reflection on their existence and their role in scientific practice.

Acknowledgments Early versions of this paper were presented at the 1st Conference in Integrated History and Philosophy of Science at the University of Pittsburgh, the Initiativkolleg "Naturwissenschaft im historischen Kontext" of the University of Vienna, the staff seminar of the Department of Philosophy and History of Science at the University of Athens, the Equipe REHSEIS—Université Paris 7, and the Max Planck Institute for the History of Science in Berlin. I am indebted to the audiences for perceptive comments and suggestions. I am particularly grateful to Hasok Chang, Melinda Fagan, Gérard Jorland, and Vasso Kindi for encouraging me to bring out the implications of my discussion of hidden entities for the integration between history and philosophy of science. I am deeply thankful to the École des hautes études en sciences sociales and the Max Planck Institute for the History of Science, where part of the work for this paper was carried out. Finally, I would like to thank the University of Athens for supporting my work with a research grant and the State Scholarships Foundation for funding my research through the IKYDA program.

References

Achinstein, P. 2001a. "Who Really Discovered the Electron?" In *Histories of the Electron: The Birth of Microphysics*, edited by. J.Z. Buchwald and A. Warwick, 403–24. Cambridge, MA: MIT Press.

Achinstein, P. 2001b. *The Book of Evidence*. Oxford: Oxford University Press.

Arabatzis, T. 2001. "Can a Historian of Science be a Scientific Realist?" *Philosophy of Science* 68, suppl.: S531–541.

Arabatzis, T. 2004. Unpublished. "Misinterpreting (correct) experimental results: Kaufmann's rejection of the particulate interpretation of cathode rays". Paper presented at the History of Science Society Meeting in Austin, Texas.

Arabatzis, T. 2006a. *Representing Electrons: A Biographical Approach to Theoretical Entities*. Chicago, IL: The University of Chicago Press.

Arabatzis, T. 2006b. "On the Inextricability of the Context of Discovery and the Context of Justification". In *Revisiting Discovery and Justification: Historical and Philosophical Perspectives on the Context Distinction*, Archimedes 14, edited by J. Schickore and F. Steinle, 215–30. Dordrecht: Springer.

Arabatzis, T. 2009a. "Cathode Rays". In *Compendium of Quantum Physics: Concepts, Experiments, History and Philosophy*, edited by F. Weinert, K. Hentschel, and D. Greenberger, 89–92. Dordrecht: Springer.

Arabatzis, T. 2009b. "Electrons". In *Compendium of Quantum Physics: Concepts, Experiments, History and Philosophy*, edited by F. Weinert, K. Hentschel, and D. Greenberger, 195–99. Dordrecht: Springer.

Bigg, C. 2008. "Evident Atoms: Visuality in Jean Perrin's Brownian Motion Research". *Studies in History and Philosophy of Science* 39: 312–22.

Boon, M. 2004. "Technological Instruments in Scientific Experimentation". *International Studies in the Philosophy of Science* 18: 221–30.

Buchwald, J.Z. 1992. "Kinds and the Wave Theory of Light". *Studies in History and Philosophy of Science* 23: 39–74.

Cantor, G.N., and M.J.S. Hodge, eds. 1981. *Conceptions of Ether: Studies in the History of Ether Theories 1740–1900*. Cambridge: Cambridge University Press.

Carrier, M. 1993. "What Is Right with the Miracle Argument: Establishing a Taxonomy of Natural Kinds". *Studies in History and Philosophy of Science* 24: 391–409.

Cartwright, N. 1983. *How the Laws of Physics Lie*. Oxford: Oxford University Press.

Chalmers, A. 2009. *The Scientist's Atom and the Philosopher's Stone: How Science Succeeded and Philosophy Failed to Gain Knowledge of Atoms*. Dordrecht: Springer.

Clarke, S. 2001. "Defensible Territory for Entity Realism". *British Journal for the Philosophy of Science* 52: 701–22.
Duhem, P. 1913. "Logical Examination of Physical Theory". In *Pierre Duhem: Essays in the History and Philosophy of Science*, edited by R. Ariew and P. Barker, 232–38. Indianapolis: Hackett, 1996.
Duhem, P. 1954. *The Aim and Structure of Physical Theory*; trans. from the 1914 edn by P.P. Wiener. Princeton, NJ: Princeton University Press.
Elsamahi, M. 1994. "Could Theoretical Entities Save Realism?" *PSA 1994* 1: 173–80.
Galison, P. 1997. *Image & Logic: A Material Culture of Microphysics*. Chicago, IL: University of Chicago Press.
Gardner, M.R. 1979. "Realism and Instrumentalism in 19th-Century Atomism". *Philosophy of Science* 46: 1–34.
Gelfert, A. 2003. "Manipulative Success and the Unreal". *International Studies in the Philosophy of Science* 17: 245–63.
Gross, A. 1990. "Reinventing Certainty: The Significance of Ian Hacking's Realism". *PSA 1990* 1: 421–31.
Hacking, I. 1983. *Representing and Intervening*. Cambridge: Cambridge University Press.
Hacking, I. 1995/1996. "Comments on Zeidler & Sobczynska's Paper". *Foundations of Science* 4: 537–42.
Hanson, N.R. 1963. *The Concept of the Positron: A Philosophical Analysis*. Cambridge: Cambridge University Press.
Heilbron, J.L. 1993. *Weighing Imponderables and Other Quantitative Science Around 1800,* A Supplement of *Historical Studies in the Physical and Biological Sciences*, XXIV. Berkeley, CA: University of California Press.
Hertz, H. 1956. *The Principles of Mechanics: Presented in a New Form.* New York, NY: Dover.
Hiebert, E. 1995. "Electric Discharge in Rarefied Gases: The Dominion of Experiment. Faraday. Plücker. Hittorf". In *No Truth Except in the Details. Essays in Honor of Martin Klein*, edited by A.J. Kox and D.M. Siegel, 95–134. Dordrecht: Kluwer.
Kelvin, L. 1904. *Baltimore Lectures on Molecular Dynamics and the Wave Theory of Light.* London: C. J. Clay and Sons.
Kuhn, T.S. 1977. *The Essential Tension*. Chicago, IL: University of Chicago Press.
Lakatos, I. 1970. "History of Science and its Rational Reconstructions". In *PSA 1970, Boston Studies in the Philosophy of Science*, vol. 8, edited by R.C. Buck and R.S. Cohen, 91–136. Dordrecht: Reidel.
Lelong, B. 2001. "Paul Villard, J.J. Thomson and the Composition of Cathode Rays". In *Histories of the Electron: The Birth of Microphysics*, edited by J.Z. Buchwald and A. Warwick, 135–67. Cambridge, MA: MIT Press.
Lützen, J. 2005. *Mechanistic Images in Geometric Form: Heinrich Hertz's Principles of Mechanics*. Oxford: Oxford University Press.
Massimi, M. 2004. "Non-defensible Middle Ground for Experimental Realism: Why We Are Justified to Believe in Colored Quarks". *Philosophy of Science* 71: 36–60.
Maxwell, G. 1962. "The Ontological Status of Theoretical Entities". In *Scientific Explanation, Space and Time, Minnesota Studies in the Philosophy of Science*, vol. 3, edited by H. Feigl and G. Maxwell, 3–27. Minneapolis, MN: University of Minnesota Press.
McMullin, E. 1984. "A Case for Scientific Realism". In *Scientific Realism*, edited by J. Leplin, 8–40. Berkeley, CA and Los Angeles, CA: University of California Press.
Morrison, M. 1990. "Theory, Intervention and Realism". *Synthese* 82: 1–22.
Nola, R. 2002. "Realism Through Manipulation, and by Hypothesis". In *Recent Themes in the Philosophy of Science*, edited by S.P. Clarke and T.D. Lyons, 1–23. Dordrecht: Kluwer.
Nye, M.J. 1972. *Molecular Reality: A Perspective on the Scientific Work of Jean Perrin*. London: MacDonald & New York, NY: Elsevier.
Perrin, J. 1916. *Atoms*. London: Constable & Company.

Poincaré, H. 1905. *Science and Hypothesis*. New York, NY: Walter Scott.
Radder, H. 1995. "Experimenting in the Natural Sciences: A Philosophical Approach". In *Scientific Practice*, edited by J.Z. Buchwald, 56–86. Chicago, IL: The University of Chicago Press.
Reiner, R., and R. Pierson. 1995. "Hacking's Experimental Realism: An Untenable Middle Ground". *Philosophy of Science* 62: 60–9.
Resnik, D. 1994. "Hacking's Experimental Realism". *Canadian Journal of Philosophy* 24: 395–412.
Rocke, A.J. 2010. *Image and Reality: Kekulé, Kopp, and the Scientific Imagination*. Chicago, IL: The University of Chicago Press.
Salmon, W.C. 1984. *Scientific Explanation and the Causal Structure of the World*. Princeton, NJ: Princeton University Press.
Seager, W. 1995. "Ground Truth and Virtual Reality: Hacking vs. Van Fraassen". *Philosophy of Science* 62: 451–78.
Skinner, Q. 1969. "Meaning and Understanding in the History of Ideas". *History and Theory* 8: 3–53.
Steinle, F. 1997. "Entering New Fields: Exploratory Uses of Experimentation". *Philosophy of Science* 64, suppl.: S65–74.
Steinle, F. 2002. "Experiments in History and Philosophy of Science". *Perspectives on Science* 10: 408–32.
Suárez, M. 2008. "Experimental Realism Reconsidered: How Inference to the Most Likely Cause Might Be Sound". In *Nancy Cartwright's Philosophy of Science*, edited by S. Hartmann, C. Hoefer and L. Bovens, 137–63. New York, NY: Routledge.
Thomson, J.J. 1909. "Address of the President of the British Association for the Advancement of Science". *Science* 30: 257–79.
van Fraassen, B.C. 1980a. *The Scientific Image*. New York, NY: Oxford University Press.
van Fraassen, B.C. 1980b. "Theory Construction and Experiment: An Empiricist View". *PSA 1980* 2: 663–78.
van Fraassen, B.C. 1985. "Empiricism in the Philosophy of Science". In *Images of Science*, edited by P. Churchland and C. Hooker, 245–308. Chicago, IL: The University of Chicago Press.
van Fraassen, B.C. 2009. "The Perils of Perrin, in the Hands of Philosophers". *Philosophical Studies* 143: 5–24.
Westfall, R.S. 1977. *The Construction of Modern Science: Mechanisms and Mechanics*. Cambridge: Cambridge University Press.
Worrall, J. 2000. "The Scope, Limits, and Distinctiveness of the Method of 'Deduction from the Phenomena': Some Lessons from Newton's 'Demonstrations' in Optics". *British Journal for the Philosophy of Science* 51: 45–80.

Chapter 10
Scientists' Methods Accounts: S. Weir Mitchell's Research on the Venom of Poisonous Snakes

Jutta Schickore

10.1 Introduction

In this essay I pursue two related goals. I draw attention to a key yet neglected element of scientific writing about experiments: methods accounts. By "methods accounts" I mean scientists' accounts of the rules one should apply in experimental practice, the problems one may encounter in doing so, and the extent to which the investigators believed they had followed these rules. I then utilize the study of methods accounts to consider if and how historical and philosophical analyses might be brought together to elucidate past scientific episodes.

At first glance, scientists' conceptions of good experimental methods and their development seem to be an important focal point for joint philosophical and historical analysis and thus a theme that fits squarely into the overall scope of the volume. I contend, however, that the question of how two scholarly fields, the history and the philosophy of science, should be combined, is ill conceived because it is based on a misconception of the practice of philosophical analysis. Metascientific analysis – the analysis of science – can best be understood as a dynamic process, whereby preliminary interpretations of the historical record are brought together with provisional conceptual frameworks; and the initial outcome of this exercise is being reworked until one has reached a cogent interpretation in which descriptions of episodes deemed exemplary and metascientific concepts coincide. Metascientific analysis is interpretive, and insofar as it involves historical reflection, it does so from the outset.

This essay deals with a series of reports of experiments with snake venom. Their author, S. Weir Mitchell, is perhaps best known as a neurologist and literary writer,[1] but the investigation of snake venom was an important part of his scientific work that occupied him for decades.[2] Between 1860 and 1890, he published a number

---

[1] See, e.g. Canale 2002; Otis 1999; Goetz 1997.

[2] See Cervetti (2007), who focuses on the cultural context of Mitchell's work on snakes. A detailed historical study of Mitchell's experiments is lacking.

J. Schickore (✉)
Indiana University, Bloomington, IN, USA
e-mail: jschicko@indiana.edu

S. Mauskopf, T. Schmaltz (eds.), *Integrating History and Philosophy of Science*,
Boston Studies in the Philosophy of Science 263, DOI 10.1007/978-94-007-1745-9_10,

of works on this topic, which present his research on the chemical nature of snake venom and its effects on organs, body fluids, and tissues. I am interested in the structure of these reports, particularly in the way in which Mitchell addressed issues of experimental methods. In other words, I am not addressing the question of how Mitchell actually proceeded. Rather, I am concerned with what Peter Galison has recently called "technologies of argumentation," the concepts, tools, and procedures needed at a given time to construct an acceptable scientific argument (Galison 2008, 116). I focus on one aspect of these technologies: the part that is concerned with experimental methods.

Mitchell's work on venom and his methods accounts in particular are of interest because Mitchell is situated at a turning point. He followed in the footsteps of previous experimenters and explicitly acknowledged their insights and methodological principles for generating and assessing experimental evidence. But he also took the investigation to a new level, cellular tissue, and used new techniques of analysis; and his experimental results disagreed with several established findings. One may thus expect that his methods accounts reflect these changes. Moreover, he was active at a time when American medical men debated the merits and demerits of the experimental method in medicine. In the following, I seek to identify and characterize the accounts and arguments related to experimental techniques and procedures in Mitchell's work, consider how innovative they are, and examine how Mitchell deployed methodological terms to make his case. I utilize this analysis to reflect on the problem of the relation between history of science and philosophy of science more generally.

10.2 A Topic for Multi-Perspectival Metascientific Analysis: Scientists' Methods Accounts

As a whole, the present volume seeks to test the mutual relevance of history and philosophy of science through concrete cases. I should thus say at the outset that while I do think that historical reflection has an important role to play for the analysis of science, I have come to find notions of "integrated HPS" and "history and philosophy of science" problematic and misleading.[3] The terms suggest that two scholarly fields are somehow combined or confronted with one another. It is commonly assumed that philosophy provides the general framework and history the empirical data against which philosophical claims are tested or from which general claims are derived (see, e.g., Burian 2002; Laudan 1989).

This combination or confrontation model of HPS misrepresents the very nature of metascientific analysis. Actually, metascientific analysis is neither a bottom-up generalization from historical data nor a top-down "test" of preconceived philosophical frameworks. Rather, it is interpretive and hermeneutic in the sense that we approach a portion of science that we deem interesting with a preliminary set of

[3] I explain this in more detail in Schickore (forthcoming-a).

tools we deem appropriate, drawn from our background knowledge, and see how far it takes us. Often, those conceptual tools that at first appear useful for our concrete analytic goals will turn out to have certain shortcomings vis-à-vis the portion of science we are interested in. The preliminary concepts will be specified, modified, and revised in light of our findings. More often than not, we will arrive far from the starting points. The research question, the evidence, the analytic framework, and the interpretation are, in a sense, outcomes of the analysis.

I do think that historical reflection is essential for metascientific analysis. I contend that our understanding of scientific concepts, practices, and rules is comprehensive only if we take seriously that the concepts, practices, and rules in use in science today have developed over time. Historical—or, as I prefer to call it, *historicist*—reflection can aid the analysis on two levels[4]: as the history of the methodological, the epistemological, or scientific concepts and practices used by the people whose work we are studying or as reflection on the history of the very concepts we are utilizing in our metascientific analysis. Historicist reflection is an integral part of metascientific analysis; it does not make sense to think of it as a separate pursuit that has to be "combined" or "confronted" with philosophical reflection on science.

Methods accounts—information and arguments concerning experimental techniques and procedures—are an important element of the argument in the report. But this aspect of scientific activity has not received much attention. A few studies have focused on how scientists' methodological pronouncements are utilized in public speeches to promote programs such as Newtonianism or Baconianism. Other studies have unearthed scientists' metaphysical and epistemological stances, such as their position in the realism-antirealism debates. But analyses of scientists' views about methodological issues such as reliability, reproducibility, robustness, and the like are rare, and the existing conceptual tools for such analyses are rather diverse. And the scientists' views are difficult to grasp. One needs to scrutinize past scientific writings to expose and reconstruct how scientists present their findings, support their arguments, and utilize statements and reflections about methods to confer epistemic force on the results presented—a task for conceptual analysis. To carry out such analysis, one needs a framework of methodological tenets and rules. In addition, one needs empirical information, e.g. about the institutional context that may impinge

[4]Historicist thinking is the project of understanding the present through tracing the past. Note that in the context of the present discussion, "historicism" does not mean "radical context-dependence" as an imperative of historical analysis, nor does it mean an acknowledgement of "laws of historical development" (the kind of historicism Popper criticized). Rather, "historicism" refers to the historicist-hermeneutic maxim that "understanding something" means "understanding how it came into being". Gustav Droysen, Wilhelm Dilthey, and other philosophers of history and knowledge advocated this maxim, according to which historicist philosophy is ultimately concerned with the present. We need to historicize our knowledge in order fully to understand it (for instructive accounts of different versions of historicism and the development of the historicist maxim, see Beiser 2007; Schnädelbach 1987).

on methods accounts. Methods accounts may also be informed by traditions of writing about experiment (Galison's technologies of argumentation), and by disputes about the boundaries of disciplines.

This essay presents an analysis of Mitchell's methods accounts. Obviously, I cannot provide a full description of how my analysis came about, but I will illustrate some salient features of the process. I indicate two provisional starting points: a brief outline of Mitchell's works on snake venom and a review of recent discussions about the methodology of experiment. To expose Mitchell's methods accounts I work from both points inwards, bringing in additional information about the intellectual and institutional context of physiology in the second half of the nineteenth century, the relevance of organic chemistry for medicine, the methodological concerns of Mitchell's predecessors, as well as the development of methodological thought about experiments in recent philosophy of science.[5]

10.3 Starting Point I: Mitchell's Experimental Reports

Mitchell's work on snake venom spanned several decades. He began experimenting with rattle snakes, but in later years he also experimented on cobras and vipers. He published a number of articles in medical and popular journals as well as three book-length treatises, which will be the main focus of my analysis. His first book-length report on snake venom experiments, entitled *Researches on the Venom of the Rattlesnake: With an Investigation of the Anatomy and Physiology of the Organs Concerned*, appeared in 1860 (Mitchell 1860). Most of the themes Mitchell covered and the questions he asked were not new: Similar investigations had been carried out since the seventeenth century; and indeed, Mitchell presented his work as a continuation of earlier studies on snake venom. One particularly important reference point for him was the work of the late eighteenth-century Italian experimental philosopher Felice Fontana who, in turn, owed much to Francesco Redi, court physician, naturalist to the Duke of Tuscany, and member of the Accademia del Cimento.[6]

In the early modern period, the meat of vipers was an essential ingredient of theriac, a remedy for snake bites and various other health troubles. For their experiments, investigators often made use of the numerous vipers that had been collected for the production of theriac. Mitchell, by contrast, kept the snakes he used in his experiments in order to have venom readily available at all times. He was thus also able to provide general information about the habits of rattlesnakes in captivity.

The main experimental part of Mitchell's book deals with the physical and chemical character of the venom, with the action of venom on plants and animals, and more specifically with its impact on specific organs, organic systems, and body fluids. Mitchell advanced two overall points. First, venom was a composite of several

[5] This essay is part of a larger study of the development of methodological thought. See, in addition Schickore (2010, forthcoming-b).

[6] On earlier contributions to snake venom research, see the references in the previous footnote.

components, and not all of these were toxic. Secondly, venom could produce acute or chronic effects. In acute cases, death occurred rapidly, within minutes, and in these cases respiration and the heart became enfeebled. In secondary or chronic poisoning, death occurred only after several hours, and post-mortem dissection showed changes in the blood.

The year after the publication of the first report of experiments, a much shorter study on the treatment of rattlesnake bites appeared that discussed the efficacy of certain antidotes, in particular, "Bibron's antidote," a remedy for rattle snake bites the French zoologist Bibron had recently developed (Mitchell 1861). In the book, Mitchell presented a summary of the relevant findings from the 1860 report, along with descriptions of a few experiments he considered particularly informative for the purpose taken. The descriptions were almost verbatim from the earlier work.

The last book-length report on snake venom experiments, *Researches Upon the Venoms of Poisonous Serpents*, co-authored by Edward Reichert, was published in 1886 (Mitchell and Reichert 1886). Its scope is considerably wider, covering the venoms of different kinds of poisonous snakes. Also, its structure is far more complex, because at that time Mitchell had modified his earlier ideas about the chemical nature and composition of venom. The book reports experiments showing that snake venom was of proteid nature; that it had two main toxic components, peptones and three kinds of globulins; that each component had specific toxic effects; and that venoms from different kinds of snakes differed in their chemical composition. So the argument becomes multilayered: Mitchell reported especially designed experiments that examine the effects on body systems, tissues, and fluids of pure venoms from different kinds of snakes as well as of each of the venom components.

Mitchell's father had been a chemist, and Mitchell's biographers note that Mitchell had been interested in chemistry from an early age (e.g. Burr 1929, 28). At the beginning of his career, Mitchell went to Paris and studied among others with Claude Bernard, who had a pronounced interest in physiological applications of chemistry. So it is not surprising to find that the chemical analysis of snake venom and its specific physiological effects on particular tissues are at the heart of Mitchell's work. The chemical investigation of venom as such was not a new thing. For instance, already in the early modern period, researchers had speculated that the working of venom could be explained as a process of fermentation. Since the seventeenth century it had been common to try to determine the acidity and alkalinity of venom and of mixtures of blood and venom using color tests. But the results had been inconclusive. Mitchell repeated the experiments, finding the venom uniformly acid whether moist, dry, fresh, and old. He found the venom acid both when drops of venom were applied to litmus paper and when the snake bit on the paper so that the paper received the venom directly from the fang.

Mitchell's experiments resonate with early nineteenth-century plant and animal chemistry in both theme and approach. The discovery of alkaloids in plant chemistry transformed medical science in the second third of the nineteenth century (Lesch 1984). The interest in alkaloids clearly motivated Mitchell's work in the late 1850s. At the time Mitchell embarked on venom research, he was also studying plant alkaloids, such as a new alkaloid contained in two varieties of woorara (curare), as well as the alkaloids of sassy-bark and cichoria (Hammond and Mitchell 1860; Mitchell and Hammond 1859). The chemical part of the 1886 study still begins with the question of alkaloids and whether or not they are present in venom (Mitchell and Reichert 1886, 9).[7]

In his researches, Mitchell made use of investigative tools that had emerged in early nineteenth-century plant and animal chemistry, namely the isolation of the "active principle", the physiologically effective substances of plants.[8] An early example of such an endeavor is Pierre Joseph Pelletier and François Magendie's work on ipecacuanha. To isolate the active principle of the substance—the "emetine" or vomitive matter—the researchers studied its physical and chemical properties and its physiological effects and compared these effects with those of the substance in its original form (Lesch 1984, 137). Similar investigations were carried out on *strychnos* plants and other plant groups. All these investigations had a common basic structure: isolation and chemical study of the active principle, animal experiments to study its effects, as well as the expectation that the active principle would be a salifiable organic base, and clinical trials (Lesch 1984, 138–42).

The same investigative steps are described in Mitchell's early works on alkaloids, for instance in the essay co-authored by Mitchell and Hammond on corroval and vao, two varieties of woorara (curare). The authors explicitly refer to Pelletier's earlier analyses of curare. They describe the physical characteristics of the two substances and the extraction of the "poisonous principle" through water and alcohol. They identify the principle obtained from corroval as an alkaloid and proposed to name it "corrovalia". Animal experiments could show that the substance was highly toxic (Hammond and Mitchell 1860, 8).

Mitchell's reports on snake venom experiments follows the same steps. Mitchell describes the physical character of the venom, the analysis of its chemical properties, and the effect on the character of the fluid of various chemicals. He lists trials with about a dozen chemicals and described their visible effects on the venom; in particular, whether or not they yielded precipitates. Included in the list are common tests for albuminous matter and nitrogenous nature (Mitchell and Reichert 1886, 34, 36).

The central section of the chemical part refers to a cluster of experiments devoted to the question of whether the venom was a composite and whether it was possible to isolate the physiologically active substance. The section describes how, through the boiling of venom, filtering, and washing with cold water Mitchell had obtained a coagulum and a fluid. Injections of each into the breasts of pigeons showed that the fluid was deadly, while the coagulum was innocent. He had also found that treating

[7] At that time, it was still under discussion whether plant and animal alkali differed in their nature.

[8] Lesch shows that the interest in physiologically active plant principles was part of an effort to rationalize drug therapy. It was this motivation that eventually led to the formation of the class of alkaloids (Lesch 1984, 81)

the fluid with alcohol yielded a poisonous precipitate. Thus he had identified the active toxicological element or "essential principle" (Mitchell 1860, 36) *Crotaline*, an albuminoid body.

The final sections of the chemical portion of Mitchell's work are devoted to investigations of the influence on the activity of venom of temperature and various chemicals, all of which showed that the venom (i.e. the efficacy of the active principle) was toxically unaltered by boiling, freezing, and chemical agents. Later chapters deal with the physiological effects of venom; they are thus again in line with common practices of animal and plant chemistry.

Mitchell and Reichert's 1886 study is an extension and modification of the earlier monograph. It reports experiments with venom from different kinds of snakes, and it offers a much more complex chemical analysis of the composite venom itself. But the overall structure of the report resembles the first one. The earlier work distinguishes between a precipitate and the fluid venom proper. It claims that the precipitate is innocent and that the fluid is the true physiological principle; and the physiological effects of the principle is described. The later work shows that the fluid is in fact a composite; it is composed of a liquid part (the peptones) and a precipitate (the globulins). The globulins could be further resolved in three different "principles", which Mitchell named "water-venom-globulin", "copper-venom-globulin", and "dialysis-venom-globulin" according to the three chemical processes by which they could be isolated (Mitchell and Reichert 1886, 10). All these elements were of proteid nature, but no evidence of the presence of alkaloids could be found. Different kinds of venoms differed both in their proportion of globulins and in the proportions of globulins and peptones. This new analysis makes the investigation extremely complex because the effects of each of the elements—the peptones and the three globulins—had to be studied separately in each kind of venom. The 1886 report is not comprehensive in this respect, in part because Mitchell and Reichert did not have enough venom available to carry out all the necessary.

Each physiological chapter in the 1886 book follows the same scheme. First, the action of pure venom on a particular body function is described in its action on normal animals, animals with cut pneumogastric nerves (nerves that supply the lungs and stomach), and animals with sections of the upper spinal cord. The sections isolate the heart from the nerve centers and thus permit separate investigations of the effects of venom on body systems. This description is followed by reports of two repetitions of the three series of experiments, once with venom globulins, and once with venom peptones.[9]

[9]For instance, to elucidate the action of venom on the pulse rate, Mitchell presented schematic descriptions of experiments (68 overall), in which a drop of pure venom, globulin, or peptone was injected intravenously into a rabbit. In each of the three series of experiments, the first set was made on a healthy animal, followed by one set of experiments on animals with cut pneumogastric nerves and another on animals with cut pneumogastric nerves and a section of the upper part of the spinal cord. The venom stemmed from different kinds of snakes.

10.4 Starting Point II: Concept to Analyze Methods Accounts

To identify the methods-related portions in Mitchell's research reports, I use as a preliminary guide recent work on scientists' criteria for successful experimentation. I already noted that not many scholars have analyzed strategies for securing experimental evidence and that studies of how these strategies have emerged are rare. Philosophers' historical surveys of scientific method mostly focus on past instances of rules for scientific—inductive, deductive, abductive—reasoning (e.g. Achinstein 2004) or on scientists' stances in realism-antirealism debates. Harry Collins, Jim Bogen, and Allan Franklin are notable exceptions. All three deal with experimental methods and strategies for securing experimental evidence.

In his 1985 book *Changing Order, Replication and Induction in Scientific* Practice—a synopsis of previously published work—Collins offers a critical analysis of the idea (advocated, among others, by Popper) that replication is the central condition of experimental confirmation. He suggests that the most powerful kind of replication is located on a continuum between the re-run of an experiment that is identical in all respects (except time) with the first one and one that shares nothing more than the subject matter with the original experiment.[10] But he introduces this analysis of experimental confirmation by replication only to dismiss it as irrelevant. His point is that it is impossible to generate an algorithm that can single out the kind of replication that is most powerful. Successful replication of an experiment is ultimately a social accomplishment. Collins's analytic theory of replications is subsidiary to his rejection of "formal" philosophy of science and his general defense of social constructivism. For Collins, the study of Mitchell's methods account would be part and parcel of a study of how Mitchell's experimental results fared in the negotiations concerning the validity of claims about the composite nature and toxicity of snake venom, whereby attempts to replicate experimental results would take center stage.

Collins's critique of "formal sets of criteria for replication" is leveled against the received view of philosophy of science in the 1970s. Allan Franklin's set of "epistemological strategies" is leveled against Collins. Franklin draws together a long list of procedures that have been used to validate and secure experimental results. The set includes (1) Observing the predicted effect of an intervention. (2) Independent confirmation through a different technique or apparatus.[11] (3) Experimental checks and calibration, whereby the experimental apparatus reproduces known phenomena. (4) The reproduction of known artifacts. (5) The elimination of plausible

[10] Collins offers an example for this (tongue firmly in cheek), noting that if a gypsy "confirmed" an experimental effect by reading the entrails of a goat, we wouldn't be too impressed. But if the differences between the experimental techniques are decreased—if the gypsy used old-fashioned scientific instruments instead of entrails, say—the confirmatory power of the evidence thus produced will increase.

[11] The first two strategies are drawn from Ian Hacking's famous article on microscopes (Hacking 1981).

sources of error and alternative explanations of the result. (6) Using the results themselves to argue for their validity. (7) Using an independently well-corroborated theory of the phenomena to explain the results. (8) Using an apparatus based on a well-corroborated theory. (9) Statistical validation of the evidence.[12]

The set is, as Franklin explicitly points out, neither exclusive nor exhaustive. Reproduction of data and phenomena is on the list, but nowhere does he claim that replication is a necessary condition for experimental confirmation.[13] Introducing the various strategies, Franklin illustrates them with different episodes from the history of science. But he does not mean to suggest that in any given case, all strategies apply. None of these procedures is necessary for validation, nor are they meant to be together sufficient to guarantee a valid result (Franklin 1989, 459). Past scientific episodes across different scientific fields and periods are a quarry to unearth a number of strategies that have been applied to confirm experimental evidence. Franklin does not trace the emergence of these criteria. He is interested in the question of strategies for the validation of experimental results only insofar as they demonstrate that scientific practice is rational. One may of course ask whether particular investigators applied any of these strategies to secure their experimental results. But to Franklin, this is not the point of his approach.

James Bogen, by contrast, does study the strategies applied by a particular individual, and he does assess the significance of replication as a strategy for validating experimental results. He analyzes the works of the nineteenth-century neuroscientist John Hughlings Jackson. Bogen would happily agree with Collins that there are no preconceived, formal transhistorical criteria for replication and that actual scientific practice should be the measure of successful confirmation. But for him, this does not mean that analyses and explications of the concept of replication and other methodological concepts are superfluous or useless. On the contrary, analytic distinctions of, for instance, different forms and purposes of replications are an important part of philosophical studies of scientific methodologies.

In this spirit, Bogen presents a threefold distinction between different sorts of re-doing experiments. He distinguishes (a) procedural replication (rerunning experiments), (b) the replication of data, and (c) the replication of effects. Procedural replications—replications of experimental processes—are performed for purposes of calibration, identifying systematic sources of error, and estimating ranges of random error. These reruns are not expected to produce exactly the same results. The replication of data also requires rerunning experiments, but here the purpose is to make sure that the experimental effect is not merely due to coincidence—and again,

[12] He develops the set in Franklin (1986, chapter 8), and subsequently it has appeared with minor variations in many publications, including (Franklin and Howson 1988; Franklin 1989, 1990; Franklin and Howson 1998) as well as the entry "Experiment in Physics" in the *Stanford Encyclopedia of Philosophy* (Franklin 2010). I am referring to the latest version here because I find that presentation of the strategies the most concise.

[13] Franklin does not discuss the concept of replication. He frequently uses the term in the section on biological experiments, but only to refer to mechanisms for the replication of DNA.

the reruns are not expected to yield exactly the same results. Finally, the replication of effects involves reasoning from different kinds of data, which are obtained by different types of experiments. In recent philosophical literature, this is regarded as the epistemically most valuable type of replication with the strongest confirmatory power (cf. Bogen 2001, 512–14).

Based on a reconstruction of Jackson's theory of the organization of the brain, Bogen argues that none of the three forms of replication is necessary for the establishment of empirical evidence. Bogen ends his analysis of nineteenth-century neurological work with a puzzle: He shows that in some cases, the argument that was made in the text could have been supported with a reference to a single instance of positive evidence. Nevertheless, for nineteenth-century neurologists, "a few" instances seem to have been preferable to a single instance.

None of these approaches to replication is entirely satisfactory, but this need not concern us here. Together they draw attention to a number of things to look for in Mitchell's reports, such as references to replications, references to independent confirmations, as well as discussions of checks, calibrations, and sources of error. In particular, three of Bogen's insights initially appear useful and potentially relevant for the study of Mitchell's methods account: the insight that replication comes in different forms; that a small number of replications is deemed to have strongest confirmatory power; and that replication is important but not necessary for confirmation. These insights are all the more significant for my purposes because Jackson is a neuroscientist and a contemporary of Mitchell, who happened to be engaged in neurological research, too.

10.5 Methods in Mitchell's Argument

In the context of an analysis of chemical papers published at the Paris Academy around 1700, Larry Holmes proposed a distinction between two components of experimental reports: "argument" and "narrative" (Holmes 1991). The narrative element comprises (detailed) descriptions of what experiments were performed, the experimental setups and procedures, the specific circumstances, and the results obtained. The argument deploys evidence from observations and experiments to advance and support a claim to knowledge. In terms of Holmes's distinction, one might say that in Mitchell's treatises argument outweighs the narrative element. But the reports are not as rigidly and schematically divided into "experiments", "methods", "data", and "discussion" as contemporary articles in science journals. The experiments are briefly introduced. Most of them are presented as synopses (apparently quoted from lab notebook entries (Mitchell 1860, 31)); only the information that is deemed salient is included (e.g. the exact amount of venom and chemical reagents, the temperature of the agents, the length of time of the experiment). The experiments that are included in the reports were selected from a larger number, "being the most illustrative cases in my possession" (Mitchell 1860, 72). Especially in the 1886 report, many results are simply given as lists or in tabular form with only a minimum of additional information or interpretation.[14] Interspersed with these descriptions are passages in which Mitchell presents his inferences and conclusions drawn from the results he had obtained, and most chapters end with a short concluding summary.

Informed by the works of Collins, Bogen, and Franklin, one also expects to find in Mitchell's report references to successful or unsuccessful replications, independent confirmations, calibrations, and so on. Mitchell's reports do touch upon these issues, but two things are immediately obvious. First, several of Mitchell's remarks concerning methods are fairly brief. Secondly, he addressed methods-related issues in different ways. Some of his remarks concern the particulars of snake venom research, others experimentation in general. For instance, Mitchell framed his 1860 study with a declaration that his views were based solely in experimental evidence, that he had carried out his research without preconceived views and with a love of truth, and that, due to the difficulty and complexity of his subject, errors had to be anticipated (Mitchell 1860, iv).

Pronouncements like these are plentiful in experimental reports from the late seventeenth century; in fact Francesco Redi introduced the report of his experiments with snake venom in the same fashion. Redi's plea for experimentation resonates with the Proem of the *Saggi di naturali esperienze fatte nell' Accademia del Cimento* of 1667, the Academy's collaborative (and only) publication. Like the *Saggi*, Redi's letter is framed with a plea for experience or experiment[15] as the ultimate arbiter of truth. Redi begins with the affirmation that he found himself ever more "firm in my attention of not trusting the phenomena of nature if I do not see them with my own eyes and if they are not confirmed by iterated and reiterated experience".[16] This declaration echoes the motto of the Cimento, "provando e riprovando", the commitment to the empirical test of knowledge claims.[17]

Such an expression of the experimentalist's creed may at first appear largely gratuitous for a mid-nineteenth-century medical scientist, and one may think that the imperatives Mitchell invoked mainly underscored the continuation of the experimentalist program. Having completed his medical training in 1850, Mitchell had traveled to Paris and studied with Bernard. He shared Bernard's appreciation for experiment-based physiology. His biographers note that Mitchell fondly remembered a conversation with Bernard: "I [Mitchell] said 'I think so and so must be

[14]In the experiments on pulse rates, for example, lists of measurements suggest that the pulse was measured frequently (often several times for a couple of minutes, and then every minute or so). But Mitchell did not mention how the pulse rate was measured. He made only a few comments on the experiments themselves, pointing out that the venom was usually dissolved in 1 c.c. of distilled water and injected intravenously into the external jugular vein, and that the amounts of globulin and peptone used equaled the doses contained in the pure venom.

[15]As Jay Tribby notes, at the Tuscan court, there does not seem to have been a systematic distinction between experience, experiment, and observation (or between the equivalent terms in Latin, French, or Italian). See Tribby (1991, 425), see also Findlen (1994, 203–5).

[16]Knoefel (1988, 3).

[17]Baldwin (1995, 404), Findlen (1993, 38). For more details, see Schickore (2010).

the case.' 'Why think,' he replied, 'when you can experiment? Exhaust experiment and then think'" (Burr 1929, 64). This exchange suggests that Mitchell simply wanted to affirm the experimentalist creed. But given Mitchell's precarious position in American physiology and the medical community in the second half of the nineteenth century, another interpretation suggests itself. In the US at that time, laboratory-based physiology was still far less well established than across the Atlantic (Shortt 1983). Mitchell attempted twice to get a chair a Philadelphia, but on both occasions he lost the competition to a medical practitioner who was not an experimentalist and had better social connections to the medical and academic establishment in Philadelphia (Fye 1983). So it is reasonable to assume that Mitchell wanted to do more than just link his work to a long established tradition of experimentation; he wanted to stress that new insights in the medical sciences required a sound foundation in experimentation.

The architecture of the treatise on the treatment of rattlesnake bites of 1861 suggests the same attitude and line of reasoning. It contains some systematic methodological reflections on the experimental study of life as part of his discussion of common remedies. Mitchell included what he called "experimental criticisms", discussions about the conditions of successful experimentation with antidotes and potential sources of fallacies. In his book, the experimental approach to toxicology is justified indirectly yet forcefully. Mitchell discussed three points, all of which he considered crucial for successful experimentation with antidotes:

1. Fallacies in regard to the use of antidotes of all kinds, arising from want of exact knowledge as to the secretion of venom, and the mode in which the serpent uses its fangs and ejects the poison.
2. Fallacies as to antidotes, arising from the want of information on the natural history of the disease caused by the venom.
3. General considerations as to antidotes, and as to the mode of conducting researches in this direction so as to avoid errors. (Mitchell 1861, 6)

Mitchell presented these points as key motivation for the study he had published in 1860. He explicitly noted that he had originally planned to assess the value of Bibron's antidote, but his attempts had not yielded any reliable results; he was "working in the dark" (Mitchell 1861, 4). Only after sustained laboratory research in accordance with the above guidelines had he succeeded in achieving his goal.

From the list of potential fallacies in the assessment of the efficacy of antidotes, Mitchell derived a number of methodological imperatives pertaining specifically to experiments with antidotes. Investigators had to take into account that even if two snakes that are alike with respect to age, size, and vigor, and even if the amount of venom contained in their fangs is the same, it does not mean that their bites are alike, too. In fact, the mechanisms of the bite differed so much that "the danger of the bite is utterly unequal" (Mitchell 1861, 6), and therefore the effect of the antidote might differ greatly. Mitchell also took the occasion to argue for the importance of medical statistics, in particular, the statistical assessment of the mortality of snake bites. Contrary to popular opinion it was not true that rattlesnake bites were always fatal, so what seemed a successful treatment might as well be a case of recovery by natural causes. But since there were no comprehensive statistics about the fatality of the bites, it was not possible to make any concrete assessments. Thirdly, it was important to distinguish between antidotes that chemically altered the venomous fluid and antidotes that counteracted the effects of the venom. The latter offered a more promising way to proceed in the battle against poisoning. But to do so, it was absolutely crucial to know exactly what effects the venom caused.

These reflections are geared toward the specific therapeutic project that Mitchell had in mind. The methodological imperatives he derived might have some use in other clinical contexts, but Mitchell did not comment on this. He introduced and presented these considerations to justify both the aims and the experimental nature of his 1860s study, emphasizing that in his experiments, he had followed the guidelines laid out in the 1861 treatise.

Notably, the points Mitchell highlighted in the 1861 study are not the ones he brought up in the report of 1860. Here he drew attention to a number of issues that were directly tied to assessments of the validity of experimental outcomes, reflecting concrete problems arising from Mitchell's chemical and physiological experiments.

To bring out the differences between the different kinds of remarks concerning methods, it is helpful to distinguish between explicit *methodological imperatives* concerning rules that should be observed; *methodological reflections* discussing and defending particular ways to proceed; and *methods statements*, statements that characterize procedures and strategies actually applied. On the basis of this threefold distinction one can make a nuanced assessment of the roles of methods accounts in Mitchell's overall argument as well as a detailed comparison between Mitchell's text and earlier methods accounts.

Given Mitchell's professional situation, it is quite likely that the general methodological imperatives in the 1860 report were meant to emphasize that medicine and physiology should be experiment-based. But Mitchell also made more specific methods statements that were linked to reports of concrete experiments. These statements served to support the experimental results Mitchell presented in his texts. These statements are of particular interest to the analyst because they indicate what methodological concerns Mitchell had.

To expose and characterize Mitchell's statements concerning methods, the conceptual frameworks by Collins et al. are instructive both because they suggest items to look for and because there is a significant contrast between Mitchell's methods statements and recent methodological thought. As I noted above, in the recent philosophical literature, replications of effects are hallmarks of experimental confirmation. Mitchell, by contrast, was much more concerned with repetitions. He occasionally mentioned that he had redone experiments performed by his predecessors (such as the test for acidity) but he put the emphasis on repetitions he had carried out of his own experiments.

Given the recent emphasis on independent confirmation of experimental results (or "replication of effects") and the disdain for repetitions as confirmatory instances, two things are noteworthy: First, the replication of effects was not part of the repository of methodological notions that Mitchell brought to bear on his experiments. Secondly, Mitchell did explicitly mention repetitions, and he appears to have attached a great deal of significance to them. Mitchell's model, Fontana, too, highlighted the repetitions he had performed, as did his early modern predecessors. Mitchell's predecessors frequently stated that they had carried out "hundreds" of repetitions. Here, this practice was regarded as a means to deal with both the accidents and contingencies of nature and with the ineptness of the experimenter (Schickore 2010). References to re-runs of experimental trials indicated that the experimenters had succeeded in establishing the regular course of things despite the contingent factors that might impinge on experimental practice.

Mitchell often indicated that repetitions had been performed—sometimes "a few", occasionally "numerous". For Mitchell, however, the practice of re-running experiments created a methodological challenge of a new kind: the outcomes of similar experiments did not always produce similar results. In the late seventeenth century, the identity of the outcomes of the same experimental trials was the ideal that guided methodological reflections and the evaluation of experiments. The general expectation was that repetitions of the same experimental trial yielded the same results. The point of multiple repetitions was that they elucidated the regular course of things by making contingent factors inconspicuous. Investigators mentioned the problem that contingencies might impede, and had impeded, experiments, but they eventually picked the "right" answer from the outcomes of a number of repeated trials even in cases where they had encountered discrepancies.[18]

Mitchell approached discrepancies differently: Not only did he expect discrepancies of experimental outcomes to occur, but he also assumed that not all discrepancies could be circumvented. Mitchell did not draw up a taxonomy of discrepancies in experiments but it becomes clear from his experimental report that he thought discrepancies among experimental outcomes might be due to a number of causes. These causes included occasional flaws in the experimental procedures; the great complexity of the experimental situation that made it impossible to identify all relevant parameters that needed to be kept under control; and the variability of the features of the experimental objects (e.g. the individual features of the snakes, the mechanisms of their bite) which made it impossible to keep even the known parameters completely stable. So the task was twofold. Acceptable (and to an extent, unavoidable) discrepancies had to be distinguished from those that were due to faults and flaws and hence unacceptable; and the actual results had to be extracted from the domain of acceptable discrepancies.

Mitchell's report of his investigation of the action of venom on blood is a rich source for the study of how Mitchell handled the problem of discrepancies and how methods statements are integrated in the argument. Since the seventeenth century, investigators had surmised that the alteration of the blood was one of the main causes of death through snakebite. Mitchell's distinction between primary or acute and secondary or chronic poisoning is crucial here because he found that in acute

[18]On the history of managing discrepant measurements, see Buchwald (2006).

poisoning no alteration of blood takes place, whereas in secondary poisoning the changes are very pronounced. The occurrence of changes in blood is thus the demarcation criterion between primary and secondary poisoning. Mitchell described three sets of experimental trials that explored the effects of mixing venom and blood. Mitchell investigated this in vitro, by mixing drops of fresh blood drawn from different animals with drops of venom. The first set of experiments was concerned with the procedure of clotting, the second and third investigated which component was affected. Describing his investigation of blood clotting, Mitchell drew attention to the "liability to fallacies" of the experimental procedure (Mitchell 1860, 90). He cautioned that the blood might coagulate very quickly, that is, before it was mixed with venom, or that the mixing if it took too long, might break new clots. A table presents the results of seven experiments (a selection) with blood from a rattlesnake, frogs, small birds, a dog and man. It is safe to assume that the selection of experiments presented in the table comprises those that Mitchell deemed free from these influences. The results in the table support his main argument, viz. that the mixture of venom and blood at first does not produce alterations in the blood; the mixture coagulates, and softening and dissolution of the clots occurred only after a time. The table also displays discrepancies as the mixture, while coagulating in all cases, did not coagulate uniformly. Mitchell noted that the clots became "more or less" altered and dissolved "partly or entirely" (Mitchell 1860, 90), but he did not attempt to explain these differences.

The next set of experiments investigated whether or not the blood disks suffered alterations. Mitchell stated that he had made "very many careful examinations of the blood-disks of frogs, birds, dogs, etc., which had been killed by snake bites." Again, the results were not completely uniform—in rare cases he had found some alteration, but he noted that they had not been very remarkable (Mitchell 1860, 91). It was thus likely that venom affected the blood plasma, and the next set of experiments addressed this possibility. Mitchell reported a series of four experiments performed to elucidate how much time it took for the fibrin to disappear from the blood. In each case the blood lost the power to coagulate. But the times of death differed widely. Two dogs died rapidly, one only after several hours. Moreover, one dog recovered. (This last experiment was so remarkable that Mitchell cited it in the treatise on antidotes of 1861.) Again, Mitchell did not attempt to explain the differences among the results. He mentioned a number of contingent factors that might explain why the times of death were so different—for instance, the dog that survived longer was larger, the most rapid death occurred after simultaneous bites from several snakes. But there is no explanation of why the fourth dog survived.

Like the 1860 report, the 1886 treatise also describes sets of trials with obviously discrepant results. Measuring the action of venom on the pulse rate, Mitchell and Reichert saw themselves confronted with a range of measured values. They stated: "In all of our observations we find that the results produced in animals, under apparently the same conditions and by using the same doses, vary very greatly; sometimes the pulse is quickened from the first and remains beyond the normal until death ensures, sometimes there is a primary diminution followed by an increase, at others there is a diminution which continues until death. The pulse is generally found to vary much in frequency. These facts all suggest that the action of the pure venom is of a complex nature..." (Mitchell and Reichert 1886, 56). Indeed, of six experiments with pure rattlesnake venom on normal animals "in three the pulse-rate was diminished and remained below normal, in two there was a primary increase followed by a diminution, and in one of these the pulse-rate afterwards went above the normal, while in another there was a primary diminution followed by an increase" (Mitchell and Reichert 1886, 56). Only in the six experiments with cobra venom there was a general increase. Mitchell and Reichert explained the discrepancy among the measured values as a result of the complexity of the phenomenon under investigation and stated that it was impossible to make any predictions of the alterations in the pulse rate in any given case. The only conclusion that they drew was that it was the "tendency" of venoms to cause an increase of the pulse rate, that the section of the pneumogastric nerve increased that tendency, and that the conjoined section of the nerve and the spinal cord rarely caused an increase. Apparently they simply gleaned this tendency from the measurements. The chapter presents the series of measured values obtained in altogether 34 experiments with pure venom, but there is no evidence for statistical calculations.

But even if repetitions yielded sufficiently similar results or results that were so regular that "tendencies" could be inferred, there was of course the possibility that some permanent factor distorted the outcome. The reports of experiments on the influence of heat on the efficacy of venom show that Mitchell reckoned with this possibility. Mitchell took care to describe how the venom was heated and included a table that showed the results of a series of ten experiments. The venom was heated to successively higher temperatures, and then injected into the breasts of reed-birds; and the time of death was noted. The fact that death occurred after longer and longer time periods suggested that the venom was losing its power when heated, and Mitchell stated this (Mitchell 1860, 44). In the subsequent paragraph, however, he declared that he had been mistaken. In other words, the results shown in the table are actually misleading because they falsely support the view that venom was less efficacious when heated. Mitchell described the problem: Due to the scant supply of venom, he had been forced to work with very small quantities. He had realized that during the boiling process, most of the fluid would cling to the test tube and would thus not be injected. Repetitions of the experiment with larger quantities and on pigeons showed different results, all birds died "with the usual symptoms" (Mitchell 1860, 45), so in fact heat did not affect the efficacy of the venom.

This episode also shows how important it is to distinguish between methods statements as they are actually utilized as tools of argumentation and strategies of validation as they are routinely applied in practice. Franklin does not attend to this distinction. He lists "using an apparatus based on a well-corroborated theory" as one "epistemological strategy" among many for the validation of experimental results. Mitchell utilized a number of apparatus for his experiments such as manometers, microscopes, and—in the episode at stake—test tubes for heating samples. We may assume that the theory behind these instruments was considered to be "well corroborated" or even trivial. The apparatus are black-boxed in Mitchell's report, and if they were based on "well-corroborated theories" these theories were not explicitly utilized in the argument. In general, one would not expect that routine experimental checks are explicitly mentioned and utilized as tools in an experimental report. But of course, one can easily think of situations in which apparatuses become problematic, and *then* they become part of the argument, as the episode of the influence on the efficacy on venom of heat shows.

In contexts like these, Mitchell stressed the importance of systematic variations of the parameters of experimental settings as a means to identify sources of error and discrepancy. Several times he described flaws in the experimental set-up he had identified through parameter variations. Upon investigating whether gland tissue or infusions of gland tissue had toxic effects, for example, he realized that it was extremely difficult to clear the glands of poison. It might thus seem that delicate animals had died from an injection or infusion of gland tissue even though they had in fact died from remaining traces of venom. Mitchell described very carefully how he had carried out this particular experiment to demonstrate that in his case, the sample had not been contaminated (Mitchell 1860, 39). He went on to highlight the role of multiple variations as a tool to identify systematic flaws in the set-up, stating: "it is still desirable that these experiments should be repeated, with every possible modification; since, as I have endeavoured to show, this, like all other portions of our subject, is girt about with such difficulties as may well baffle the most careful" (Mitchell 1860, 42).

Mitchell described specific kinds of parameter variations which are conducted to compare a parameter of the experimental setting under investigation with a benchmark parameter. These he explicitly designated as "checks". For instance, when Mitchell determined the toxic effects of chemically altered venom, a test of the efficacy of the original pure venom served as the benchmark against which the toxicity of the altered venom was assessed (Mitchell 1860, 35). And when Mitchell investigated the effect of venom on muscles, he conducted an experiment whereby he punctured exposed muscles with dry clean fangs whose ducts had been stopped with wax and compared the time and intensity of the twitching with the effects of an injection of venom through the fangs (Mitchell 1860, 78).

The above instances show that for Mitchell, discrepancies among experimental results were expected and that they were produced by different kinds of causes: contingent impediments, variations intrinsic to the experimental objects, and the complexity of experimental parameters. There were discrepancies due to the intrinsic variability of experimental parameters, such as the variation among individual living organisms. Discrepancies might be due to the sheer complexity of experimental parameters, which could not be completely grasped. Discrepancies might also be due to contingent factors occasionally impinging on experimental settings. The differences among these kinds of discrepancies are clearly methodologically important. Discrepancies due to the variability of experimental objects could not be completely avoided.[19] Discrepancies due to occasional flaws, however, could be

[19] On the history of the standardization of experimental animals, see Logan (2002).

removed—at least in principle. Multiple variations appeared as a suitable tool to identify potential flaws in the experimental procedure. On the other hand, even if discrepancies were small and thus seemingly insignificant and negligible, there may be a persistent permanent flaw in the experimental setting, which, as in the case of the apparent sensitivity to heat of venom, produced regular, if flawed, results. Again, Mitchell claimed that multiple variations of experimental parameters were a good way to identify such flaws.

My brief survey of Mitchell's experimental reports suggests that Mitchell deployed methods statements in a number of ways. He highlighted that he had performed repetitions of experiments in cases when it was a possibility that contingencies and flaws might have yielded questionable results. He pointed out sources of experimental error that he had identified and stressed that multiple variations of experimental parameters had been instrumental to identify them; and he drew attention to unavoidable discrepancies, either due to the variation of experimental objects or to the complexity of the experimental setting.

Given the recent emphasis on the confirmatory power of replications of effects, it is notable that replications of effects did not play a role in Mitchell's methods accounts. On the other hand, his emphasis on repetitions with variations can be understood as a continuation of previous methodological concerns, as I have shown.

Repetitions (successive performances of one's own experiments) have been largely neglected in current philosophical discussions. They are dismissed as events with little confirmatory power (Collins) or relegated to the realm of preparatory manipulations (Bogen). Focusing largely on present science, Franklin and other philosophers highlight the reproduction of an experimentally generated effect through *different* experimental techniques and procedures (Bogen's "replication of effects") as a strategy to obtain independent confirmation. This is considered a particularly powerful way of confirming a phenomenon and of validating empirical results. As far as I can see, Mitchell did not explicitly require an array of independent procedures to secure empirical evidence. If it is correct that today experimenters are centrally concerned with multiple determinations of experimental outcomes, major transformations of the practical rules of experimentation must have occurred sometime after the mid-nineteenth century. When, why, and in what contexts did the concerns with multiple determinations by different means arise? Because multiple determinations are deemed crucial for modern experimental practice, tracing them is a key part of our understanding of methodological thought and its development.

10.6 Conclusion

I do not claim to have provided an exhaustive analysis of Mitchell's methods accounts, let alone of his toxicological work. But I hope to have shown that methods accounts are a significant part of the "technology of argumentation" experimenters deploy to produce sound scientific arguments. Such methods accounts can take different forms: they may be expressed as general methodological imperatives or as concrete methods statements in the context of particular experimental endeavors;

and they may or may not be accompanied by additional justificatory reflections and arguments.

Is the study of methods accounts a good example of "integrated HPS"? I think it has become clear that the analysis of methods accounts calls for a multitude of analytic perspectives. But in my view, the idea of a "confrontation" of history of science and philosophy of science does not capture the nature of this pursuit. The historical record, the interpretation, as well as the conceptual scaffolding I have presented in the previous sections are all *outcomes* of my analysis. My study is hermeneutic in the sense that the starting points for my investigation were provisional. New and different aspects of the historical record have become important once I began looking for certain features of experimental reports, such as instances of "replication". I have tried to capture some of the dynamic of the interpretive procedure in my presentation, although it is of course impossible (and undesirable) to make this procedure fully transparent. The initial conceptual tools have taken shape during the analysis for instance through distinguishing methods statements, imperatives, and reflections. On the basis of this distinction, one can make nuanced assessments of the roles of methods accounts in Mitchell's overall argument, as well as a detailed comparison between Mitchell's text and earlier methods accounts. The comparison highlights that Mitchell emphasized repetitions with variations just like some of his predecessors. Unlike his predecessors, Mitchell expected discrepant results from similar trials. And unlike his successors, Mitchell did not place emphasis on multiple determinations of experimental outcomes.

My analysis is metascientific, it is interpretive, and it involves historicist reflection from the outset. It does so in two ways. It identifies and distinguishes methodological concepts and traces these concepts backward to Mitchell's predecessors and (admittedly in a rather sketchy fashion) forward to the present. Moreover, my analysis indicates how metascientific concepts themselves have developed since the days of Popper. But my analysis cannot be characterized as "confronting a philosophical framework" or a "philosophical methodology" (whatever that may be) with a "historical case".

References

Achinstein, P., ed. 2004. *Science Rules. A Historical Introduction to Scientific Methods*. Baltimore, MD: Johns Hopkins University Press.

Baldwin, M. 1995. "The Snakestone Experiments: An Early Modern Medical Debate". *Isis* 86: 394–418.

Beiser, F. 2007. "Historicism". In *Oxford Handbook to Continental Philosophy*, edited by B. Leiter and M. Rosen, 155–79. Oxford: Oxford University Press.

Bogen, J. 2001. "'Two as Good as a Hundred': Poorly Replicated Evidence in Some Nineteenth-Century Neuroscientific Research". *Studies in History and Philosophy of Biology and Biomedical Sciences* 32: 491–533.

Buchwald, J.Z. 2006. "Discrepant Measurements and Experimental Knowledge in the Early Modern Era". *Archive for the History of the Exact Sciences* 60: 565–649.

Burian, R.M. 2002. "Comments on the Precarious Relation Between History of Science and Philosophy of Science". *Perspectives on Science* 10: 398–407.

Burr, A.R. 1929. *Weir Mitchell – His Life and Letters*. New York, NY: Duffield & Co.

Canale, D.J. 2002. "Civil War Medicine from the Perspective of S. Weit Mitchell's 'The Case of George Dedlow'". *Journal of the History of the Neurosciences* 11: 11–18.

Cervetti, N. 2007. "S. Weir Mitchell and His Snakes: Unraveling the 'United Web and Woof of Popular and Scientific Beliefs'". *Journal of Medical Humanities* 28: 119–33.

Findlen, P. 1993. "Controlling the Experiment: Rhetoric, Court Patronage and the Experimental Method of Francesco Redi". *History of Science* 31: 35–64.

Findlen, P. 1994. *Possessing Nature. Museums, Collecting, and Scientific Culture in Early Modern Italy*. Berkeley, CA: University of California Press.

Franklin, A. 1986. *The Neglect of Experiment*. Cambridge: Cambridge University Press.

Franklin, A. 1989. "The Epistemology of Experiment". In *The Uses of Experiment. Studies in the Natural Sciences*, edited by D. Gooding et al., 437–60. Cambridge: Cambridge University Press.

Franklin, A. 1990. *Experiment Right or Wrong*. Cambridge: Cambridge University Press.

Franklin, A. 2010. "Experiment in Physics". In *The Stanford Encyclopedia of Philosophy*, edited by E.N. Zalta (Spring 2010 Edition): URL = <http://plato.stanford.edu/archives/spr2010/entries/physics-experiment/>.

Franklin, A., and C. Howson. 1988. "It Probably Is a Valid Experimental Result: A Bayesian Approach to the Epistemology of Experiment". *Studies in History and Philosophy of Science* 19: 419–27.

Franklin, A., and C. Howson. 1998. "Comment on 'The Structure of a Scientific Paper' by Frederick Suppe". *Philosophy of Science* 65: 411–16.

Fye, W.B. 1983. "S. Weir Mitchell, Philadelphia's 'Lost' Physiologist". *Bulletin of the History of Medicine* 57: 188–202.

Galison, P. 2008. "Ten Problems in History and Philosophy of Science". *Isis* 99: 111–24.

Goetz, C. 1997. "Jean Martin Charcot and Silas Weir Mitchell". *Neurology* 48: 1128–132.

Hacking, I. 1981. "Do We See Through a Microscope?" *Pacific Philosophical Quarterly* 63: 305–22.

Hammond, W.A., and Mitchell, S.W. 1860. On the Physical and Chemical Characterization of Corroval and Vao, Two Recently Discovered Varieties of Woorara, and on a New Alkaloid Constituting their Active Principle. *Proceedings of the Academy of Natural Sciences of Philadelphia* 12: 4–9.

Holmes, F.L. 1991. "Argument and Narrative in Scientific Writing". In *The Literary Structure of Scientific Argument: Historical Studies*, edited by P. Dear, 164–81. Philadelphia, PA: University of Pennsylvania Press.

Knoefel, P.K. 1988. *Francesco Redi on Vipers*. Leiden: Brill.

Laudan, L. 1989. "Thoughts on HPS: 20 Years Later". *Studies in History and Philosophy of Science* 20: 9–13.

Lesch, J.E. 1984. *Science and Medicine in France. The Emergence of Experimental Physiology, 1790–1855*. Cambridge: Cambridge University press.

Logan, C.A. 2002. "Before There were Standards: The Role of Test Animals in the Production of Empirical Generality in Physiology". *Jounal of the History of Biology* 35: 329–63.

Mitchell, S.W. 1860. *Researches on the Venom of the Rattlesnake: With an Investigation of the Anatomy and Physiology of the Organs Concerned*. Philadelphia, PA: Smithsonian Institution.

Mitchell, S.W. 1861. *On the Treatment of Rattlesnake Bites, with Experimental Criticisms upon the Various Remedies Now in Use*. Philadelphia, PA: Lippincott & Co.

Mitchell, S.W. and W.A. Hammond 1859. An Experimental Examination of the Physiological Effect of Sassy-Bark, the Ordeal Poison of the Western Coast of Africa. *Proceedings of the Academy of Natural Sciences of Philadelphia* 11:13–16.

Mitchell, S.W. and E.T. Reichert 1886. *Researches Upon the Venoms of Poisonous Serpents*. Washington, DC: Smithsonian Institution.

Otis, L. 1999. *Membranes: Metaphors of Invasion in Nineteenth-Century Literature, Science and Politics*. Baltimore, MD: Johns Hopkins University Press.

Schickore, J. 2010. Trying Again and Again: Multiple Repetitions in Early Modern Reports of Experiments on Snake Bites. *Early Science and Medicine* 15.

Schickore, J. forthcoming-a. More Thoughts on HPS: Another 20 Years Later. *Perspectives on Science*.

Schickore, J. forthcoming-b. The Significance of Re-Doing Experiments: A Contribution to Historically Informed Methodology. *Erkenntnis*.

Schnädelbach, H. 1987. 'Etwas Verstehen heisst Verstehen, wie es geworden ist' – Variationen über eine hermeneutische Maxime. *Vernunft und Geschichte*. Frankfurt am Main: Suhrkamp.

Shortt, S.E.D. 1983. "Physicians, Science, and Status: Issues in the Professionalization of Anglo-American Medicine in the Nineteenth Century". *Medical History* 27: 51–68.

Tribby, J. 1991. "Cooking (with) Clio and Cleo: Eloquence and Experiment in Seventeenth-Century Florence". *Journal of the History of Ideas* 52: 417–39.

Chapter 11
Quantum Gravity Meets &HPS

Dean Rickles

The world of the very small is a quantum world, and that must be as true of space and time and gravity as of electrons and photons and quarks.
John Wheeler

Science is what scientists have done, not what a philosopher tells us the scientist meant to do, were really *doing, or should have done.*
James Cushing

11.1 Peeking Inside the Black Box

To paraphrase Otto von Bismarck, as far as most philosophers are concerned, scientific theories are like sausages: it is better not to see them being made! Standard practice amongst philosophers of science is to investigate the finished products of science: the theories that emerge from the scientific process. However, as Kuhn taught us, the finished product, as presented in textbooks for example (usually providing the "raw data' for philosophers" investigations), usually bears no trace of what is often a highly non-trivial path towards victory—though often for good pedagogical reasons.[1] What philosophers play with are so many black boxes. Until fairly recently they only focused on a handful of such black boxes, often from physics

[1]Note that I certainly don't mean to disparage the textbook genre. Textbooks are significant in very many ways, beyond the merely pedagogical; and *qua* historical objects, they are as interesting as any other such objects. The textbooks one learns from can forge social identity and define a community. In the context of quantum gravity research they are especially interesting because the arrival of a textbook signals a certain degree of "maturity" of the field. Only relatively recently have textbooks on quantum gravity begun to appear. See the chapters in Part III of Kaiser (2005) for more on this fascinating topic.

D. Rickles (✉)
University of Sydney, Sydney, NSW, Australia
e-mail: dean.rickles@sydney.edu.au

S. Mauskopf, T. Schmaltz (eds.), *Integrating History and Philosophy of Science*, Boston Studies in the Philosophy of Science 263, DOI 10.1007/978-94-007-1745-9_11,

alone.[2] However, they have expanded their horizons to include a broader range of physical theories, and even theories from the social and biomedical sciences. Still, most individual accounts of how science works focus on a small selection of scientific theories, and ignore the historical and sociological details behind their construction *and* their evaluation.

This situation clearly falls way short of an integration of history and philosophy. While it is true that for some philosophical purposes this is as fine a grain of detail as one needs,[3] if we are considering methodological issues, a black box approach cannot be sufficient: we need to probe inside to see what factors accounted for the success of some theory (or failure of another), and whether, in hindsight, they were good ones and/or the *only* ones. As James Cushing and others have so ably demonstrated, there is in fact often an enormous amount of contingency in theory-selection, and what appeared to be "the only theory for the job" was really only one amongst several (quite distinct, yet empirically adequate) possibilities. Given such contingency, a variety of non-epistemic factors can enter into the analysis, supposedly leading to the additional input of psychology and sociology. It is precisely this intrusion that so offends philosophers of science—or at least those who cling to the distinction between the contexts of discovery and of justification: yes, social/psychological factors can enter into science *weakly*, in the discovery phase, but they should never spill over into the justificatory phase.

Though, superficially, philosophers, historians, and sociologists of science share the same object of investigation, there are, of course, many subtle (and not so subtle) differences. Chief amongst these differences is the fact that philosophers of science, inasmuch as they think about it at all, usually wish to *use* historical and sociological data to inform a *general* theory of science, or at least some explanatory thesis about the way science works—they ever seek grist for their mills.[4] Hence,

---

[2]For example, Newtonian physics, general relativity (very minimally construed: i.e. tending to focus on the "famous" light-bending experiment), and astronomy. Or else, various classic "dead" theories, such as phlogiston and the geocentric view. (Note, this black box analogy was traced, by Trevor Pinch, back to his supervisor, Richard Whitley—see Pinch (1992, 488)

[3]Philosophers of physics, for example, often need only inspect the formal representation of a theory and consider its space of possible interpretations. For this, one usually does not need to know the intimate historical details of the theory's construction, though even here I would have grave doubts about the quality of such a *wholly* ahistorical approach. For example, it was his deep knowledge of the historical complexities of general relativity that led John Stachel to uncover the hole argument, surely one of the most important arguments in contemporary philosophy of spacetime physics.

[4]As Richard Burian puts it, "[w]hen philosophers turn to particular historical materials or case studies, they often begin with pre-established concepts and sometimes with expected conclusions in mind ... [and the] concepts employed often contain presuppositions about the nature of theory, evidence, and explanation, about the relation of experiment to theory, the objectivity and intellectual autonomy of scientific work, and the like" (Burian 2002, 398). Case studies can, of course, be useful for exploratory purposes (as they are in the social sciences for example). However, unless one performs an analysis of a sufficiently large sample of scientific theories (preferably chosen at random), the evidence they confer on some *general* theory of science is rather weak. (I might add that changing the terminology from "case studies" to "episodes" will not improve the quality vis-à-vis evidence for methodological theories.)

their interest in history and sociology tends to be indirect, concerned with their utility (in the form of degree of support) rather than their intrinsic worth. Historians and sociologists tend to favour a more descriptive (and they might say, objective) approach; wishing to describe as faithfully as possible, present and preconceptions to one side, some scientific episode.[5] They will trawl a wide variety of sources in order to piece together an image of what happened—albeit an imperfect image, as they will acknowledge, tainted with various biases. They like to see what goes in to the sausage machine, and observe how it works, rather than just focusing on what comes out at the other end.

These professional differences can lead to some animosity between the various disciplines, and any "integrative" approach to philosophy of science will have to try to balance these differences in outlook. Finally, it has to be said that when philosophers do employ history, it is often *bad* (or lazy) history (*cf.* Pearce Williams 1975), for example, failing to take proper account of the different modes of presentation of a theory (and its manner of construction and justification) depending on the intended audience—journals, notebooks, interviews, textbooks, and public lectures can reveal an enormous disparity despite sharing common subject matter. Whether philosophers buy into sociological elements deep within science or not, this feature of tailoring a description to an intended audience simply cannot be doubted. As I indicated above, philosophers usually refer to textbooks. But textbooks are just as bespoke as any other public account. For the historian, "primary sources" (especially original notebooks and correspondence) weigh especially heavily in terms of understanding theory construction and justification—*cf* (Hoddeson 2002). For sociologists, it is the *actual practice* of science (as it unfolds) that is most revealing.

How do we encourage and enable philosophers to look inside the black box of science? How do we persuade them to look beyond the slender "internal" (or rationally reconstructed) histories that they favour? The strong programme advocates a perfectly symmetrical treatment of "true" (or "good" or "selected") science and "false" (or "bad" or "rejected") science. In practice, this is rather difficult to achieve in an objective fashion because of the spectre of the way history *actually* unfolded![6] However, the most trouble-free way to achieve what this so-called symmetry postulate set out to achieve, is to probe what Bruno Latour labeled "science in action": situations in which the truth values of the theories aren't yet settled. In a sense, such cases render the symmetry postulate redundant, for there is no fact of the matter and so no broken symmetry in need of repair. Unfortunately, there aren't many philosophically interesting (that is, interesting to *philosophers*) situations of this sort, and

[5]As von Humboldt famously (though, from our present temporal location, somewhat naively) put it: "The historian's task is to present what actually happened. The more purely and completely he achieves this, the more perfectly has he solved his problem" (Humboldt, 1822, 57).

[6]The strong programme is also just as problematically generalist as the standard philosophers' accounts: we need evidence to convince us that some family of once competing theories were indeed *equally* viable before consensus was reached. This might well be true of some episodes but not in others.

the examples that have been conducted have tended to be of a somewhat mundane character, often involving the discovery of some substance (such as a particular vaccine) rather than the construction of a theory. Philosophers, for better or for worse (though this is changing somewhat), tend to be attracted by the bright lights of *revolutionary* episodes in science and especially by overarching, universal, fundamental theories.

However, quantum gravity research offers exactly a such a situation, where the symmetry between the various competing approaches appears not yet to have been broken *and* in which we have an example of a revolution, albeit a revolution *in process*.[7] Moreover, it directly involves theory construction, and fundamental theory construction at that. Indeed, it is widely believed to be the greatest unresolved problem in fundamental physics. Despite over 80 years of hard labour, by the finest physicists, all agreed on the importance of the problem posed by quantum gravity, there is still no finished product to speak of: no culminating theory packaged in a neat black box that philosophers of science can utilise without worrying about its complex historical trajectory. Or, to return to my earlier metaphor: there's no sausage to speak of; it's still in the machine!

Quantum gravity research is all the more enticing from the point of view of (integrated) history and philosophy of science [&HPS] since (for reasons to be discussed more fully below) it is not principally guided by the standard methodological devices of empirical testing via experiments, novel predictions, or observations. Yet one can still find all of the evaluative moves (selections and rejections) ordinarily seen in "run of the mill" scientific endeavours: theories of quantum gravity have come and gone despite being experimentally inaccessible. If not the standard methodological virtues, what is guiding theory construction and selection in this case?

In this chapter I aim to answer this question, but more generally I aim to to highlight the ways in which quantum gravity research provides an excellent example for &HPS. It enforces a "mixed methods" approach since it involves a situation with no definitive theory coupled with an awful lot of nontrivial history containing several important theoretical casualties, despite the absence of direct experimental

[7]Schweber defines Whiggish history as "the writing of history with the final, culminating event or set of events in focus, with all prior events selected and polarized so as to lead to that climax" (Ashtekar 2005, 41). While I am not strongly anti-Whiggish (I don't see that the whiff of the present can ever sensibly be eradicated from historical studies), evidently, since quantum gravity is still under construction, there is no definitive "endpoint" towards which Whiggish histories can retrospectively chart the progression of the theory—though one can envisage the possibility of "local" Whiggism, involving smaller historical steps. Moreover, the "justification/discovery" distinction (the central culprit behind the disconnect between history of science and philosophy of science) looks far more flimsy in the context of quantum gravity research since the circumstances surround the construction of the theory (such the desire to have universal theories that do not have limitations of scale) become the very mode of justification. That is, a successful theory (i.e. successful to the extent that it ought to be pursued) is simply one that meets this desire in a consistent way.

support.[8] This points quite naturally to a greater consideration of "external" factors (if we must persist with this notion) controlling theory evaluation. Hence, we have a natural convergence of history, philosophy, and sociology. I submit that a study of quantum gravity along any one of these lines (philosophical, historical, sociological) will inevitably soon find itself incorporating the others.

Let me nail my colours firmly to the mast regarding matters methodological: I advocate a view broadly similar to that espoused by James Cushing (himself borrowing crucial ideas from Arthur Fine), according to which history is of vital importance to philosophical theses, but if one looks sufficiently closely at a wide enough sample of historical episodes in science one very quickly sees that there is no one size fits all scheme: even methodology can change if the context so demands it. The process of constructing and evaluating scientific theories, much like an economic time-series, is distinctly non-stationary. Just when it seems to be acting according to some pattern, the pattern shifts. We find this to be especially true in the case of quantum gravity research.[9] The methodological lessons of quantum gravity do not stop there: quantum gravity research is important too in our primary theories; namely, the standard model of particle physics and classical general relativity (both of which inform the standard model of cosmology). These theories would look very different were it not for the impact of quantum gravity research and the concepts and tools it has generated—indeed, this external utility has been adopted at various times to support continued research on quantum gravity.

I begin with a brief description of some peculiarities of historical research on quantum gravity, introducing the basic idea of the problem of quantum gravity by

---

[8]The historical nature of quantum gravity will also please those historians who bemoan the trend towards specialization. It has almost a century of development with no closure. Quantum gravity is a distinctively international field of research; it incorporates elements from a very wide variety of theories, and many branches of mathematics. It has witnessed both military and industrial support, in addition to standard university-based support. More recently it has begun to utilise cosmology, computer simulation, condensed mater physics, and the new range of particle accelerators. A historian would be hard pressed to give a local account of quantum gravity.

[9]Quantum gravity might look unappealing for those philosophers steeped in "the new experimentalism", for, *prima facie*, there simply are no experiments to analyse! However, recent work in quantum gravity attempts to make contact with experiments (using astrophysical data and the LHC, for example), though so far without success. The reasons behind the lack of success is interesting in itself. But even the early work which lacked experiments *simpliciter* is interesting from the point of view of how scientists go about evaluating their theories when so important a resource as experiment is unavailable. Thought experiments play a more important role (I will discuss below a foundational one, associated with Bohr and Rosenfeld's analysis of measurability of quantum fields). But also, theoretical predictions of the ingredient theories of quantum gravity (i.e. general relativity and quantum field theory) are used as (proxy) experimental data points for quantum gravity research. Most notably, perhaps, is the computation of the black hole entropy formula, that any approach worth its salt must be able to derive—Eric Curiel (2001) has argued that this kind of usage of still-unconfirmed claims as *evidence* (if an approach is able to reproduce it) is illegitimate (here stemming from the semiclassical theory involving quantum fields on a classical, black hole background). However, the illegitimacy depends on what one views as the "laws of scientific development".

way of its beginnings. I then describe the problem of quantum gravity in more detail, focusing on the energy, length, and mass scales that characterise it, and consider the role that these scales played in early work. I then go on to introduce a variety of the main ways of proceeding with respect to the problem in these early days. My focus throughout is on the early history, pre-1960s, since beyond this the entanglement with cosmology significantly complicates matters (but see Kaiser (2007, 2006, 1998) for some interesting work on the entanglement of *classical* general relativity and elementary particle physics), as does the emergence of string theory (on which, see: Rickles (forthcoming-b)). Finally, I consider, rather more directly, the implications of the development of quantum gravity research for &HPS.[10]

11.2 One Revolution Too Few?

In 1940 Einstein wrote the following words:

> The development during the present century is characterized by two theoretical systems essentially independent of each other: the theory of relativity and the quantum theory. The two systems do not directly contradict each other; but they seem little adapted to fusion into one unified theory. [...] [T]his theory, like the earlier field theories, has not up till now supplied an explanation of the atomistic structure of matter. This failure has probably some connection with the fact that so far it has contributed nothing to the understanding of quantum phenomena. To take in these phenomena, physicists have been driven to the adoption of entirely new method. [...] [T]he quantum theory of to-day differs fundamentally from all previous theories of physics, mechanistic as well as field theories. Instead of a model description of actual space-time events, it gives the probability distributions for possible measurements as functions of time. [...] All attempts to represent the particle and wave features displayed in the phenomena of light and matter, by direct course to a space-time model, have so far ended in failure. [...] For the time being, we have to admit that we do not possess any general theoretical basis for physics, which can be regarded as its logical foundation. [.] Some physicists, among them myself, can not believe that we must abandon, actually and forever, the idea of direct representation of physical reality in space and time; or that we must accept the view that events in nature are analogous to a game of chance. (Einstein 1940, 489–92)

It is a little curious that so many great revolutionary episodes happened almost simultaneously at the beginning of the twentieth century. Perhaps one revolution made it easier for others to follow, via some kind of snowball effect? Whatever the reason, the revolution that resulted in general relativity and the revolution that resulted in quantum theory were close neighbours in time. Einstein was profoundly involved in the creation of both theoretical frameworks, though the former more so than the latter. At the time of the construction of the general theory of relativity he firmly believed in the existence of quanta of radiation. But this only involved a belief in the property of discreteness (with no real sense of ontological substrate beyond this), rather than belief in what would become quantum mechanics (or quantum field theory—though here too his contributions on emission and absorption of

[10] I am indebted to the brief review of the early history of quantum gravity by John Stachel: (1998).

radiation proved crucial). Most physicists believe *another* revolution is required to bring quantum theory and general relativity—*cf.* (Rovelli 2002).

Since such quanta, with their discrete energies and other properties, would inevitably couple to the gravitational field (in however small a way, the gravitational interaction being *universal*), Einstein couldn't ignore the fact that *something* would need to be said about the nature of this interaction.[11] Indeed, almost as soon as general relativity was completed, Einstein became aware of a possible conflict between it (or, more specifically, the existence of gravitational waves) and the principles of quantum theory,[12] and, therefore, the need for a quantum theory of gravity. Thus, he writes that

> [A]s a result of the internal-atomic movement of electrons, atoms must radiate not only electromagnetic but also gravitational energy, if only in minuscule amounts. Since this cannot be the case in nature, then it appears that the quantum theory must modify not only Maxwellian electrodynamics but also the new theory of gravitation (Einstein 1916, 696).[13]

In this case Einstein is clearly troubled by the potential clash between the theoretically predicted gravitational radiation combined with the empirically observable stability of atoms: *any* moving mass (even the electrons in atoms) will radiate gravitational energy. In other words, something like Planck's law of radiation would have to be found for gravitation in order to account for the stability. He repeated this claim again in 1918, stating that "an improved version of quantum theory would lead to changes in the gravitational theory" (Einstein 1918, 167).[14]

[11]A little later it would also come to be understood that there is a "formal interaction" between general relativity and quantum objects stemming from the peculiar nature of fermions: including objects with half-integer spins imposes a variety of constraints on the spacetime structure, and therefore on the gravitational field (resulting in a slightly modified theory of gravitation). This was a rather slow lesson.

[12]As Kragh has pointed out, the version of quantum theory that Einstein would have been thinking about at this early phase of general relativity's development was the Bohr-Sommerfeld theory—see Kragh (2000, 965). Einstein would have been particularly impressed with the way the Sommerfeld theory integrated (special) relativity and quantum theory. Helmut Rechenberg claims (though doesn't provide a source) that Sommerfeld published his results after Einstein informed him that the general relativity would not modify the results in any appreciable way (Rechenberg 1995, 160).

[13]"Gleichwohl müssten die Atome zufolge der inneratomischen Elektron-enbewegung night nur electromagnetische, sondern auch Gravitations-energie ausstrahlen, wenn auch in winzigem Betrage. Da dies in Wahrheit in der Natur nicht zutreffen dürfte, so scheint es, dass die Quantentheorie nicht nur die Maxwellsche Elektrodynamik, sondern auch die neue Gravitationstheorie wird modifizieren müssen."

[14]By 1919 he was already going down the path of unitary field theories that would mark much of his later work: "there are reasons for thinking that the elementary formations which go to make up the atom are held together by gravitational forces" (Einstein 1919, 191). As Stachel notes (1998, 526), this marks a reversal in the priority given to the two theories, general relativity and quantum theory. Whereas prior to 1919 he believed that the latter might lead to modifications in the former; here general relativity (coupled with the electromagnetic field) is now being used to *explain* the quantum structure of matter. We can surmise that it was as a result of the work by others on general relativity and its unification with electromagnetism. Max Born writes that Einstein, up until 1920, was still very concerned with the relation between quantum and relativity. Einstein wrote him: "I

This looks like a potential *empirical* motivation for pursuing quantum gravity. However, as Gorelik correctly points out, whilst atomic radiation (computed along the lines of Maxwell's theory) leads to the collapse of the atom in (order of) 10^{-10} s (a fact inconsistent with observations), atomic *gravitational* radiation, computed using Einstein's formula, has a collapse time of the order of 10^{37} s. Therefore, there would in fact be no empirical inconsistency as a result of gravitational radiation and we should not be puzzled by the stability of atoms in this case.

Gorelik (1992, 365) argues that an "analogy with electrodynamics" lay behind this comment of Einstein's. This analogy was a persistent feature of early research on quantum gravity—see below. One must also bear in mind that the issue of absorption and emission of radiation must have occupied a central place in his thinking at the time of writing, for his paper on the emission and absorption of radiation in quantum theory appeared very shortly afterwards—replete with the statement that "it seems no longer doubtful that the basic idea of quantum theory must be maintained". What is remarkable, given what we know of the certainty he professed about general relativity, is that he openly considered the possibility that the quantum theory would demand some kind of "modification" (what we would now refer to as a quantum correction) of general relativity![15]

However, similar claims were made intermittently over the next decade or so, though nothing amounting to a serious attempt to construct a full-blown quantum theory of gravity was undertaken. These claims were primarily from German (or German speaking) physicists. For example, as early as 1919, Arthur von Haas writes (on the basis of "unification" ideals) that:

> Arguably, one of the most important future tasks of the axiomaization of physics is the implementation of quantum theory in the system of the general theory of relativity. (Hass 1919, 749)[16]

---

always brood in my free time about the quantum problem from the standpoint of relativity. I do not think the theory will have to discard the continuum. But I was unsuccessful, so far, to give tangible shape to my favourite idea, to understand the quantum theory with the help of differential equations by using conditions of over-determination ..." (Born 1955, 257: from their private correspondence). By 1926 he was "toiling at deriving the equations of motion of material particles regarded as singularities from the differential equations of general relativity" (ibid., p. 258).

[15]This openness of Einstein to the possibility of a quantum theoretical modification of general relativity would not last for long, of course, and was already beginning to sour at this stage. His taste for quantum theory soon soured to the extent that towards the end of his life he was searching for ways to reproduce quantum mechanical phenomena using a purely classical field theory. Suraj Gupta (who developed a special-relativistic theory of quantum gravity in the 1950s) has a different (inverted) interpretation of Einstein's underlying reasons for distrusting quantum mechanics: "Because his theory is different from other field theories, he tried to construct unified field theories and because he could not see how his theory in the curved space could possibly be quantized, he criticized quantum mechanics" (Gupta 1962, 253).

[16]"Eine der wichtigsten Zukunftsaufgaben, die in dieser hinsicht der physikalischen Axiomatik gestellt ist, ist wohl die Einfügung der Quantentheorie in das System der allgemeinen Relativitätstheorie."

Though he doesn't explicitly name the individual constants associated to the ingredient theories (viz. c, $\hbar$, G—see the next section), it is reasonable to surmise that this is what Haas had in mind in the following passage:

> The main task of the axiomatization of physics will be the problem concerning the integration of the universal constants of physics. Also the solution of this question may be expected to reveal deeper knowledge of the relations, only intimated by Hilbert, holding between gravity and electricity, and of a further integration of these relations with the quantum hypothesis. (ibid., p. 750)[17]

This interpretation is strengthened by the fact that Haas went on to consider the various possible combinations of other constants in other contexts, investigating the way they demarcate domains (Hass 1938).

Quantum theory was invoked several times (in discussions of general relativity, and unified field theories) to mark some kind of boundary of the *applicability* of a theory.[18] Einstein himself expressed just this view, in a lecture entitled "Ether and the Theory of Relativity" at the University of Leyden, May 5th 1920. This address is interesting for many reasons, historical and philosophical. For our purposes it is interesting because Einstein once again speculates on the possible restrictions that the quantum theory might place on general relativity:

> Further, in contemplating the immediate future of theoretical physics we ought not unconditionally to reject the possibility that the facts comprised in the quantum theory may set bounds to the field theory beyond which it cannot pass.

Indeed, we can find several examples of Einstein expressing this kind of sentiment. Inasmuch as his comments (here and in his 1916 paper) have been investigated by historians, they has tended to be in the context of the study of gravitational waves. It is true that gravitational waves are naturally involved here, but since Einstein is considering the possibility that the radiation of such waves is quantized, we ultimately have what can also be seen as heralding the beginning of quantum gravity.[19]

Perhaps the most famous interplay between gravity and quantum prior to 1930 was Bohr's usage of general relativity to argue against Einstein's "photon in a box" critique of his interpretation of quantum mechanics, at the 1927 Solvay Congress. As Oskar Klein explains:

---

[17]"Aufgabe der physikalischen Axiomatik sein wird; es ist das Problem des Zusammenhanges zwischen den universellen Konstanten der Physik. Auch die Lösung dieser Frage darf vielleicht erhofft werden von einer tieferen Erkenntnis der von *Hilbert* erst angedeuteten Beziehungen zwischen Gravitation und Elektrizitat und von einer Verknüpfung dieser Beziehungen mit der Quanten-hypothese."

[18]For example, Goldstein and Ritter note how Weyl adopts this position in his *Raum, Zeit, Materie* (Goldstein and Ritter 2003, 104).

[19]The beginnings of quantum gravity are usually traced back to a 1930 paper of Léon Rosenfeld's; however, there was, aside from Einstein's remarks, quite a lot of activity dealing with the general problem of quantum gravity, i.e. concerning the joint treatment of quantum and gravity. Though Rosenfeld's paper was, so far as I know, the first paper to apply the then newly developed methods of quantum *field* theory to the problem, thus treating the gravitational field like the successfully quantized electromagnetic field.

> We know from BOHR's account how ingeniously EINSTEIN defended his standpoint—the essential incompleteness of the quantal description of nature—and how BOHR refuted every one of his arguments with more than ingenuity. What impressed us younger people most was, I think, the "Einstein box," where BOHR successfully turned general relativity theory against EINSTEIN. ... And still EINSTEIN, who accepted all defeats with the utmost fairness but without changing his basic view, may have felt that on the side of the quantum physicists the importance of the general relativity claim in the search for the laws of the microworld was usually underestimated. (Klein 1955, 117).

Einstein used quantum theory and special relativity to try to circumvent the Heisenberg relations. Bohr used a combination of quantum theory and general relativity in order to eliminate the inconsistency that Einstein derived. As Christian Møller recalls:

> Well I remember of course the excitement when Bohr was able to beat Einstein with his own weapon. That was at a Solvay meeting; Einstein invented a way of showing that quantum mechanics was not consistent. He proposed to determine the energy of the photon which had come out of the box by weighing the box before and afterwards. Then Bohr could show that if one takes Einstein's formula for the rate of a clock in a gravitational field then it comes exactly to making the thing consistent again. And Gamow even made a model of this box with a spring and clock and shutter, which opened at a certain time and closed again at a certain time. Møller [http://www.aip.org/history/ohilist/4782.html]

We can see from this brief look at the early days of the quantum-gravity interface that there was a real desire to join the two theories together and "complete the revolution".[20] Moreover, there was a general belief that constructing such a theory would be "business as usual". That is, it was generally assumed that there would be no *special* difficulty in quantizing the gravitational field. The earliest attempts to bring these theoretical frameworks together involved the same methods as had and would be used for the other fundamental interactions.[21]

11.3 Planck Scale Pragmatism

The issue of *defining* quantum gravity is itself fraught with some historical difficulties. The notion has changed as other areas of physics (and mathematics and cosmology) have advanced. Ashtekar and Geroch, in their review of quantum gravity, characterize quantum gravity as "some physical theory which encompasses the principles of both quantum mechanics and general relativity" (Ashtekar and Geroch, 1974, 1213). This leaves a fair amount of elbowroom for the form such a theory might take.

---

[20] Though this barely skims the surface of a deep vein of early work on quantum gravity. For a more detailed, thorough study of the very earliest research on quantum gravity, see Rickles (forthcoming-a).

[21] As Abhay Ashtekar puts it, the methodology was "to do unto gravity as one would do unto any other physical field" (Ashtekar 2005, 2). As is becoming clear after decades of intense effort, gravity is not like any other force, at least not in terms of its formal representation, nor, many believe, in terms of how it is (or ought to be) conceptualized.

We can, however, say with certainty at what scales quantum gravitational effects would be expected to manifest themselves. This follows from the fact that there is a unique way to mix the fundamental constants that characterise the "ingredient" theories so as to generate units of (*L*)ength, (*M*)ass, and (*T*)ime. From general relativity we have the gravitational (or Newton) constant G_N (equal to $6.67 \times 10^{-11} m^3/\text{kg s}^2$), characterising the scale at which generally relativistic effects matter, and from quantum field theory we have *c* (the velocity of light *in vacuo*) and , Planck's constant of quantum action. These combine to give us:

$$L_P = \sqrt{\frac{\hbar G_N}{c}^3} = 1.616 \times 10^{-35} m \tag{1}$$

$$T_P = \sqrt{\frac{\hbar G}{c}^5} = 5.59 \times 10^{-44} \text{sec} \tag{2}$$

$$M_P = \sqrt{\frac{\hbar c}{G_N}} = 2.177 \times 10^{-5} g \tag{3}$$

At these scales, all three physical theories are expected to play a role, and (if we accept that general relativity is a theory of spacetime geometry) it is this scale that we expect quantum geometry to dominate. Curiously, these units were discovered by Planck almost three decades before quantum field theory was discovered, and almost two decades before general relativity was completed (and six years before special relativity): (Planck 1899). Planck was interested in producing *universal* descriptions of the world, that could even be understood by extraterrestrial civilisations! For this reason he pursued a set of natural scales that would make no reference to such local circumstances as the size of the Earth or aspects of its orbit and rotation.[22]

A kind of (quite understandable) pragmatism guided the early neglect of quantum gravity research. The scales at which phenomena would be apparent were known then to be well out of reach of direct tests.[23] Though Dirac believed quite firmly

[22]See Gorelik (1992) for more on the curious discovery of these units and their subsequent propagation into early quantum gravity research. Note that by the mid-1950s the notion of the Planck length was understood by those working on the so-called canonical approach as a measure of the fluctuations of spatial geometry. For those working along spacetime covariant approaches, the Planck length marked a natural boundary to the wavelengths of quantum fields. See §4 for more on these two approaches.

[23]The characteristic "Planck length" is computed by dimensional analysis by combining the constants that would control the theory of quantum gravity into a unique length. As shown above, this is $l_p = \sqrt{\hbar G/c^3} = 1.6 \times 10^{-33} cm$: a minuscule value, making gravity (effectively) a "collective phenomenon" requiring lots of interacting masses. That quantum gravitational effects will not be measurable on individual elementary particles is, therefore, quite clear: indeed, the Planck energy is $\sqrt{\hbar c^5/G} = 10^{22} MeV$! Bryce DeWitt devised rigorous arguments to show this to be the case: the gravitational field itself does not make sense at such scales. He showed that the static field from such a particle (with a mass of the order 10^{-20} in dimensionless units) would not exceed the quantum fluctuations. The static field dominates for systems with masses greater than 3.07×10^{-6}. The

that general relativity and quantum theory would have something to say to each other (and indeed did important work on the subject), he nonetheless accepted the pragmatic argument:

> Since the time when Einstein's general theory of relativity first appeared, various more general spaces have been proposed. Each of these would necessitate some modifications in the scheme of equations of atomic physics. The effects of these modifications on the laws of atomic physics would be much too small to be of any practical interest, and would therefore be, at most, of mathematical interest. (Dirac 1935, 657)

However, Oskar Klein (describing his own approach as a contribution to "an intimate alliance of the two fundamental viewpoints of present physics, that of complementarity and that of relativity" (Klein 1955, 117)) describes and rejects this pragmatist attitude:

> Now, it is very usual to regard the point of view of general relativity as insignificant in quantum theory because the direct effects of gravitation in ordinary atomic phenomena are very small.This, however, may easily be the same kind of fallacy, which it would have been to regard the electron spin as unimportant for the formulation of the laws of chemical binding, because the direct interaction between spin magnetic moments is, in general, negligible compared with chemical binding energies.
>
> [W]e shall tentatively take the point of view that general relativity is fundamental for the formulation of the laws of quantum field theory and that the demand of an adequate formulation of other invariance claims, e.g. that of gauge invariance, should be regarded as an indication of the need for a natural generalization of the relativity postulate. [ibid, p. 98]

As I suggested above, a second factor behind the neglect was that the early views on quantum gravity were tightly bound to the quantization of the electromagnetic field in quantum electrodynamics. It was thought that there would be no *special* puzzles caused by quantizing the gravitational field, since surely one classical field is much like any other. For example, in their famous paper marking the birth of QED Heisenberg and Pauli wrote:

> We might also mention, that quantization of the gravitational field, which also appears to be necessary for physical reasons, may be carried out by means of an analogous formalism to that applied here without new difficulties. (Heisenberg and Pauli 1929, 3)[24]

These days we have a few more physical, quasi-empirical reasons to think that a quantum theory of gravity is necessary. General relativity is now (thanks to the singularity theorems) firmly believed to generically predict spacetime singularities.

---

gravitational field is from this viewpoint an "emergent" "statistical phenomenon of bulk matter" (DeWitt 1962, 372). An earlier version of this viewpoint was suggested by van Dantzig (1938). The idea that gravity is emergent, has gained in popularity recently: see Novello et al. (2002) for a review.

[24]"Erwähnt sei noch, daβ auch eine Quantelung des Gravitationsfeldes, die aus physikalischen Gründen notwendig zu sein scheint, mittels eines zu dem hier verwendeten vollig analogen Formalismus ohne neue Sehwierigkeiten durchführbar sein dürfte." Note that Heisenberg and Pauli explicitly mention the remark of Einstein's from 1916, along with Klein's 1927 paper on five-dimensional quantum theory, in a footnote attached to this passage.

It is thought that our own universe may have emerged from such a singularity (="the big bang", and may wind up in another (= "the big crunch"). It is also thought that they may exist within our Universe inside black holes. General relativity does not apply to singular situations, so a theory of quantum gravity is expected to tell us what happens here. Such reasoning was not open to the earliest researchers on quantum gravity since inasmuch as they were understood at all, singularities were thought to be fictional. Nor did the big bang model (and the notion of a big crunch) exist during the initial phases of quantum gravity research—even when it was conjectured, it was not taken up easily).

Another piece of information that suggests the need for a quantum theory of gravity came from the consideration of quantum field theory on (fixed) black hole spacetimes. Hawking discovered that in such a semi-classical theory (QFT coupled to a *classical* gravitational field), black holes emit radiation and can evaporate (= "The Hawking Effect"). However, the semi-classical theory is not sufficient to analyse all aspects of the process, since the "end point" falls outside. There are several possibilities for the final stage: a (most likely Planck-scale) remnant, unitary evolution (not to be had in the purely semiclassical theory), or total evaporation (and, therefore, information loss).[25]

We might also mention the predicted value of the cosmological constant (the energy of empty space) made by quantum field theory, on the basis of the zero-point modes.[26] The observed value for the energy density comes out very close to zero: $\rho \simeq 10^{-30} gcm^{-3}$. This is a very long way from quantum field theory's prediction. One way to bring this value down is by imposing a cutoff at the Planck length, ignoring those modes that have wavelengths smaller than this, or by turning on the interactions between the vibrational modes.

These other reasons would take some time, and required advances in cosmology and astrophysics, amongst other things. The nature of the problem of quantum gravity adapted itself to these new conditions, by setting itself new puzzles (such as the conditions surrounding the big bang and within the interior of a black hole) and by utilizing any new data as targets that a respectable approach must hit. The construction of renormalized quantum field theory, and the renormalization group, would also stimulate and radically modify new work on quantum gravity. Certainly, by the 1950s, it was no longer believed that the quantization of the gravitational would be a matter of course. We give a brief review of some of the strategies adopting in

---

[25]The fact that black holes are thought to radiate implies that they possess an entropy too. Bekenstein computed this as $k_B \frac{A}{AL_P^2}$ (i.e. the entropy goes up a quarter as fast as the black hole's horizon, or surface area, goes up). This result offers a number that the latest approaches to quantum gravity are expected to be able to derive—many of them are indeed able to do so.

[26]The energy spectrum of an harmonic oscillator, namely $E_N = (N+1/2)\omega$, has a non-zero ground state in quantum mechanics. This is the zero-point energy, standardly explained by reference to the uncertainty principle (i.e. there's no way to freeze a particle). In the context of (free) quantum field theory the field is understood to be an infinite family of such harmonic oscillators, and as a result the energy density of the quantum vacuum is going to be infinite on account of the nonzero contribution from each vibrational mode of the fields being considered.

response to the problem of quantum gravity in the days following the development of quantum field theoretical techniques (and the additional problems that quantum field theory brought with it).

11.4 The Slow and Difficult Birth of Quantum Gravity

> Although the quantum description of the gravitational field has many points of similarity to conventional quantum field theory, it nevertheless seems incapable—or capable only with difficulty—of incorporating certain conventionally accepted notions. [Bryce DeWitt [1], p. 330]

Bryce DeWitt was one of the first people to write a doctoral thesis on quantum gravity, which he did under the (somewhat minimal) supervision of Julian Schwinger at Harvard University in 1950, a time when quantum gravity research was still very unfashionable. He wrote the words above in 1962. Though they are expressed with an air of obviousness, they encode within them more than four decades of struggle to try and treat the gravitational field and quantum theory within in the same framework.

Between 1950 and 1962 gravitation research underwent a significant transformation, as a result of several (often interdependent) factors. DeWitt himself dropped gravitation research after his doctoral work, and worked instead on detonation hydrodynamics (specifically computer simulations). This work would in fact turn out to be highly applicable in general relativity in the subfield of numerical relativity. DeWitt became a pioneer in numerical relativity, and discovering ways of programming aspects of relativity. A technique he developed at Livermore—multidimensional (two and three dimensions in DeWitt's example) Lagrangian hydrodynamics (DeWitt 1953)—was later used in the gravitational 2-body problem and black hole simulations.[27] One might think of this as a flow of ideas from his war work into gravitational research. However, the flow worked both ways: DeWitt was able to develop his ideas on higher dimensional Lagrangian hydrodynamics on account of his general relativistic background with general coordinate invariance and Jacobians. However, there is no doubt at all that his experience with computing influenced his thinking enormously.

DeWitt was involved in a a chain of interesting events relating to the history of quantum gravity. He was a physicist emerging right at the end of the second world war. Moreover, he did his postgraduate work at Harvard, where there was a close connection between the students' work and military applications—many did their "work experience" on military projects. Those who had done work for the military developed a certain "number crunching" mindset. There is little doubt that this *modus operandi* filtered in to the work that resulted after the war. DeWitt's work especially was highly computational.

---

[27] See DeWitt (1982) for DeWitt's own reminiscences about his time at Livermore working on this approach, and its later relevance—see also Smarr 1984, 13.

However, there were other key circumstances that contributed to the fortunes of quantum gravity research (bringing DeWitt back into the fold in a central way), involving (amongst other things) DeWitt's marriage to a French mathematical physicist (Cécile Morette, a great organizer of conferences and people, as well as a great mathematical physicist herself), various interventions in relation to funding opportunities by John Wheeler, an off-hand submission to an essay competition, and industrial, military, and government support. One can see in this story the attempts of various parties to produce a convergence of interests. This wasn't always possible. Scientists desire freedom to pursue whatever research project they desire with sufficient resources to pursue their goals, and the industrial and government sources have a more diverse set of goals and interests, include potential technological applications (leading to financial gains and power gains), prestige, or perhaps understanding of some aspect of the world.

In the case where the scientists have as their goal the "navigation among the potentialities proffered by nature" John Stachel has described this process of convergence as one of "negotiation" (Stachel 1994, 143). Any account of scientific discoveries must take account of this milieu, though Stachel is quick to point out that this does *not* imply the neglect of nature. Though there is a certain amount of elbow room in scientific discoveries, and so the evolution of scientific research and the nature of the theories that result, all of this this must be in accord with the "the potentialities proffered by nature".[28] This is not strong social constructivism, then. The contingency is very heavily constrained by nature.

Stachel goes on to give an alternative possible scientific history, in which a different theory of gravitation was "discovered" that was perfectly in accord with nature's potentialities (since it matches Einstein's version on all relevant observables). He borrows the example from Feynman who asked: "Suppose Einstein never existed, and the theory [of GR] was not available" (cited in Stachel (1994, 146)).[29] Could one replicate "the physics" of Einstein's gravitational theory using what other (non-geometrical) tools were available, namely special relativistic quantum particle theory? The answer is Yes,[30] as several people had already suspected before Feynman posed his question. One uses the fact that the gravitational interaction

[28]Indeed, he gives as a very apt example, the U.S. Air Force's support of "anti-gravity" projects: no amount of support of any calibre could generate such a phenomenon. In this sense, the goals of the U.S. Air Force were not in accord with the potentialities of nature: no amount of coaxing was able to bring it about. I might add that there was even a convergence between government and industry (in the form of Roger Babson, a wealthy businessman who was searching for a gravity shield). For more on these and related aspects of the history of general relativity, see: Goldberg (1992), Kaiser (1998), Kennefick (2007), Kaiser (2000), Rickles (forthcoming-c).

[29]This question was asked in the context of a pivotal conference in the history of gravitational research, including quantum gravity research: *Conference on the Role of Gravitation in Physics*. See DeWitt and Rickles (2011) for the report of this conference, and a description and assessment of the conference.

[30]Although the matter is not as straightforward as Stachel (and Feynman) suggest. For details on the subtleties involved, see Wald (1986). Note also that Stachel suggests the analysis takes place in the context of quantum field theory; in fact the analysis involves the particle picture only.

has observed qualitative properties that can be encoded into field quanta (named gravitons) with specific properties:

- Obeys inverse-square law—and so is *long range*
- Is always attractive
- Macroscopicly observable
- Couples to all massive objects with equal strength independently of their constitution
- Causes a red shift
- Bends light around the Sun
- Causes a correction (relative to Newton's theory) in the perihelion of Mercury

One then assigns properties to the exchange particle in a somewhat bespoke fashion. We can see immediately that the particle must be massless in order to satisfy the long-range requirement (and also to get the right value for the bending of starlight around the Sun). The fact that gravitational effects can be seen at macroscopic scales means that the particle must have integer spin. A more complex argument is required for the attractiveness properties. We will skip this here (but see Weinberg (1964) for the full argument), and simply note that a spin-2 particle is demanded in order to have *universal* attraction that couples in the right way to matter. The particle must also self-interact by virtue of universality—it is this that causes the nasty divergences in the quantum theory at high frequencies since it leads to graviton-graviton coupling.

Hence, had certain contingent factors been otherwise, we might have had a very different theory of gravitation.[31] Though it is clear that there isn't an unlimited supply of empirically adequate alternatives, and they are often very hard to construct. Inasmuch as one approach could be rationally (or logically, in Duff's terms) justified, so could the other. The selection of Einstein's geometrical approach is based on reasons outside of the standard cluster of empirical factors. In this case, we have the pragmatism mentioned earlier, coupled with the mere temporal precedence of the geometrical approach.

The two approaches are in fact jointly pursued to this day. Steven Weinberg's textbook on gravitation and cosmology uses the Lorentz invariant particle physics approach. The approach matches his training as a particle physicist. The division into two communities (the geometric relativists, and the analytical particle physicists) is a genuine phenomena that has deeper ramifications. There goes, side by side with the division, attitudes with respect to what are deemed relevant, important, and interesting questions. Feynman in particular was of the opinion that the

[31] Michael Duff makes a very similar point very clearly in his discussion of the approach to quantum gravity that follows this "alternative path" (covariant quantization): "the historical development of a physical theory and its logical development do not always proceed side by side, and logically, the particle physicist has no strong a priori reason for treating gravity as a special case" (Duff 1975, 79).

particle physicist's approach to general relativity involved a healthy rejection of philosophical issues:

> The questions about making a "quantum theory of geometry" or other conceptual questions are all evaded by considering the gravitational field as just a spin-2 field nonlinearly coupled to matter and itself ... and attempt to quantize this theory by following the prescriptions of quantum field theory, as one expects to do with any other field. (Feynman 1972, 377)

The general relativists by contrast show a deep engagement with conceptual issues, having to do with the nature and existence of space, time, change, and so on.

In fact, many early approaches, including Birkhoff's flat-space approach, were rejected because they did not meet the requirements set by these latter tests. But, still it is very possible that had quantum field theory been to hand earlier, and had quantum gravity been seen as more of a pressing problem (e.g. if the pragmatic argument was absent, or there was a greater desire for unification for the sake of unification), the flat space, special relativistic approach to gravity might have superseded Einstein's on account of its greater amenability in terms of quantizability, and its formal coherence with the rest of physics.

We backtrack in the rest of this section, to investigate the earliest work on quantum gravity. The aim here is to highlight the motivations behind construction, selection, and rejection. The key tools are the use of simplification techniques and analogies (with successful, superficially related theories).[32]

11.4.1 Flat Space Approaches

Simplification often characterises the earliest work in some field. One might find toy models, for example. Or, in cases where one has a non-linear theory, the use of linearisation techniques. General relativity is a non-linear theory: gravity couples to energy-momentum, and the gravitational field has energy-momentum, therefore gravity gravitates. This is part and parcel of the equivalence principle. The non-linearities lead to many (but by no means all) of the complications that are faced in quantum gravity. Rosenfeld, in his 1930 work, attempted to quantize the linear theory. This, as most acknowledge, is a preliminary exercise. One would attempt to account for the nonlinearities by adding quantum corrections.

George Temple (1936) introduced the perturbative method into GR, whereby the metric tensor is expanded in powers of the gravitational coupling constant (Temple 1936). The linear expansion is, as mentioned, much easier to quantize: waves of a particular frequency ν are simply quantized according to Einstein's relation $E = \hbar\nu$

[32]These two often come together as a package. For example, one of the simplification techniques (discussed below) is to linearize the theory, so that the quanta of the theory do not interact and self-interact. This is the case in quantum electrodynamics, the only successfully quantized theory in earlier times. Hence, the simplifying move and the analogy move produce an equivalent result.

(where the energy packets $\hbar\nu$ are the gravitons[33])—the higher order terms are the problematic ones, since they determine graviton interactions (including self-interactions). Solutions of the (unquantized) linearized field equations correspond to weak gravitational radiation in empty space. The quantized radiation would correspond to a small number of gravitons propagating in empty space.

In 1939 Pauli and Fierz, in a general study of the quantization of fields (Fierz and Pauli 1939), also employed the linear approximation of general relativity, and only considered this linear field in interaction with the electromagnetic field.[34] This approach was important for future developments in quantum gravity research, however, it suffered from an inability to recover the perihelion in the classical limit (when coupled to matter).[35] Given the desire to have a theory of gravity that was in step with the other forces, the approach was, nonetheless, developed further.

Suraj Gupta was the first to explicitly split the metric tensor apart, into a flat Minkowskian part and the residue. The residue was conceived as a gravitational field potential, and would represent the gravitational interaction. Hence, the theory amounts to a specially relativistic quantum field theory of gravitation. Gupta tackled the problem in two stages: first he considered the linear theory (as Pauli and Fierz had done). This has the problem that there are negative energy states, with no physical counterpart. He then, in a second paper, considers the gravitational field interacting with the full energy-momentum tensor.

Belinfante and Swihart developed a version of this approach (for which they claim priority over Gupta: (Belinfante and Swihart 1954, 2). They initially attempted a quantization of Birkhoff's theory of gravity but were unable to find a Lagrangian that gave Birkhoff's equations of motion—a fact they interpreted as the inability of the theory to satisfy the reciprocity (i.e. action-reaction) principle linking gravity and matter.

The earliest approaches to quantum gravity in this sense (i.e. in the sense of quantization of the gravitational field) were, quite naturally, pursued by those who had

[33]According to Stachel, this particle was coined "graviton" by Blokhintsev and Gal'perin (1934). There is another usage of the term in 1935 by Sir Shah Sulaiman (a mathematician and high-court judge) who put forward a competing theory to Einstein's which postulates the existence of gravitons on which the pull of gravity depends (*Science News Letter*, November 16, 1935, p. 309).

[34]In this paper, we also find for the first time the idea that gravity corresponds to a massless, spin-2 field, so that the particle carrying the force would be massless and spin-2 (note that the presence of spin-2 particles implies that a theory containing them would, ceteris paribus, be generally covariant). Thus, they write: "for vanishing rest-mass, our equations for the case of spin 2 go over into those of the relativity theory of weak gravitational fields (i.e. $g_{\mu\nu} = \delta_{\mu\nu} + \gamma_{\mu\nu}$, neglecting terms of order higher than the first in $\gamma_{\mu\nu}$); the "gauge-transformations" are identical with the changes induced in $\gamma_{\mu\nu}$ by infinitesimal co-ordinate transformations" (Fierz and Pauli 1939, 214)—see also Fierz (1939); Fierz and Pauli (1939) (especially 6 from the latter).

[35]Birkhoff later developed a theory of gravitation based on flat spacetime (Birkhoff 1943), and this was quantized by his student Moshinsky (1950). However, it suffered from the same empirical problems that Pauli and Fierz's theory faced. Note that the covariant approach is not the only approach to involve flat space. The ADM (Arnowitt, Deser, Misner) approach (a canonical approach: see below) also involves flat space quantization (the flatness is in this case asymptotic).

skills in quantum field theory. However, even as early as 1938 Jacques Solomon argued that the standard field quantizations methods would fail for *strong* gravitational fields, for then the approach ceases to give a good approximation to Einstein gravity (Solomon 1938, 484).[36] Alternative approaches were suggested as the extent of the problems facing quantum field theory became ever more apparent. In particular, it was argued that quantum field theory might be better adjusted to fit general relativity, rather than the other way around:

> Obscurity is about to come to an end. Quantum field theory has now reached a stage where the emergence of new points of view alone will lead to real progress. A considerable step forward was made immediately after the second world war when Schwinger in this country and Tomanaga in Japan, and with them several other investigators, introduced consistently (special) relativistic procedures into the quantum theory of the electromagnetic field. Further progress on a fundamental level will very likely be brought about by the introduction of general-relativistic approaches into quantum field theory. This opinion is based on the fact that general relativity gives us a deeper understanding of the nature of fields and their relationships to particles than has been achieved anywhere else in theoretical physics. This understanding will be preserved by any theory that will maintain the principle of equivalence (similarity of gravitation with inertial effects, such as centrifugal "forces"), even though it may deviate in its specific details from the general theory of relativity as it was originally conceived by Einstein. (Bergmann 1953, 112)

The linearization approach was recognized to be a provisional step up the ladder. Even at this early stage of development (and though there were alternative approaches), we can see rational moves in operation guiding the evaluation of the linearization approaches.

11.4.2 The Electrodynamical Analogy

Formal analogies between general relativity and Maxwell's theory of electromagnetism misled quantum gravity researchers for many years.[37] As Bryce DeWitt nicely put it, at the time of the first studies of quantum gravity "[q]uantum field theory had ... scarcely been born, and its umbilical cord to electrodynamics had not yet been cut" (DeWitt 1970, 182).[38] Moreover, renormalized QED was not yet constructed, and so the divergences could not even be properly conceptualized, yet

[36]As Stachel notes (1998, 561), Solomon, along with Matvei Bronstein (another early maverick in quantum gravity), were both casualities of the war, of Hitler and Stalin respectively.

[37]However, in many cases the analogies were necessary to get a foot hold on general relativity. For example, in the context of gravitational radiation the notion of electromagnetic radiation offered many essential clues. Felix Pirani is unequivocal about the utility of this radiation analogy (partial though it is): "*Some* analogy has to be sought, because the concept of radiation is until now largely familiar through electromagnetic theory, and one cannot define gravitational radiation sensibly without some appeal to electromagnetic theory for guidance" (Pirani 1962, 91).

[38]However, the formulation of Maxwell's theory, championed by Mandelstam, using path-dependent variables (holonomies) was very productive, leading the way to loop gravity—see Mandelstam (1962, 353) and also Gambini and Pullin (1996).

alone resolved. Note, the analogy to electrodynamics even pervaded the early naming of the theory of quantum gravity, which was labeled "quantum gravidynamics" (a name long since discarded).

Among the very earliest studies was a paper by Leon Rosenfeld in which he undertook a (tree-level, lowest perturbative order[39]) computation of the gravitational self-energy of a photon. It was known that the electron's interaction with its own field (the electron's electromagnetic "self-energy") suffered from divergences (attributed to its non-vanishing mass), and, as was to be expected, Rosenfeld's calculation revealed (quadratic) divergences for the gravitational case too.[40] This pointed to a generic problem facing any field theory.[41]

The electrodynamic analogy already began to break down at the classical level, as a result of the investigation into gravitational radiation and the possibility of its measurement. The measurement of electromagnetic radiation involves third derivatives of position—one measures a fluctuating "jerky" force (accelerations alone would give a *steady* force). Forces are represented by gradients of potentials, in this case the force is given by the first spatial derivative of the potential, so that the radiation must be given by the derivative of this, namely the second spatial derivative of the potential telling us how the force is changing. In general relativity, on the other hand, the situation is more complicated. Radiation would be given by the third derivative of the potential—the second derivative would describe a static gravitational field (i.e. curvature).

A second difference concerns the difference in the respective charges of electromagnetism and gravitation. In the case of the electromagnetic field, there are both

[39]In field theory one seeks to calculate the *amplitude* for occurrence of processes. The perturbative approach, where one expands some quantity in powers of the coupling constant of the theory, offers the standard methodology. Feynman developed a fairly mechanical diagrammatic notation for doing these computations. There are two types of diagram: those with (one or more) closed loops and those without closed loops. The latter are known as tree diagrams, and they are the simplest to evaluate since the (4-momenta of the) external lines determine the (4-momenta of the) internal lines, with no need to perform integration over the internal momentum variables. By contrast, diagrams with closed loops have internal lines that are not determined by the external ones. For each loop there is a four-fold integration to be performed (each involving integration over the independent momentum variables).

[40]Though Rosenfeld was one of the earliest quantizers of gravity, by 1966 he was less convinced that there was a problem of quantum gravity (Rosenfeld 1966). Or at least, in his mind the problem had been radically misconceived as a mathematical one (involving the necessity of unification on the basis of formal inconsistencies) instead of an empirical one. For Rosenfeld that absence of empirical clues meant that one could only ever probe quantum gravity from the "epistemological side", which implied that the considerations could not establish the conformity of any such investigations to the world of phenomena: "no logical compulsion exists for quantizing the gravitational field" (p. 606). Peter Bergmann called the problem one of "esthetic unease" (Bergmann 1992, 364).

[41]DeWitt later performed a gauge-invariant, Lorentz covariant version of Rosenfeld's computation (using the tools of renormalization that had only recently been showcased at the first Shelter Island conference). DeWitt showed that contra Rosenfeld, the analysis revealed the necessity of charge renormalization, rather than a non-vanishing mass).

positive and negative charges, so one can get neutralising effects. In the case of gravity there are only "positive" charges (or at least just one kind of charge, all with the same sign).

A third difference concerns the non-linearity of gravitational interactions. Any quantization of the theory would result in quanta that interacted with each other, and *self*-interacted! This is quite unlike the situation in quantum electrodynamics, which turned out to be special in respect of its linearity. These blatant differences did not deter researchers, for the case of quantized Maxwell theory was one of the few available tools to guide theory construction, in spite of its imperfect fit.

The divergence problem (in standard non-gravitational quantum field theory) guided the development of the early work in very large part. Mandelstam was concerned with avoiding the use of an indefinite metric to construct a quantum field theory. Pauli and Källen had argued (in their study of the Lee model with cutoff (Kallen and Pauli 1955)) that there were a certain values $g > g_{critical}$ of the coupling constant at which the use of a Riemannian metric breaks down and the theory becomes non-renormalizable. Mandelstam (like many others) claimed that this was unphysical.[42]

However, in the 1960s, the electromagnetic analogy did bear some fruit in the detection of quantities of the "right sort" for classical and quantum gravity; namely, coordinate independent, gauge-invariant observables. If one can find these observables, then one could in principle turn them into operators and construct a Hamiltonian. In the electromagnetic case, one has the electric and magnetic fields at one's disposal. These are gauge invariant entities. However, if one then considers the behaviour of charges interacting with the fields, then operators associated with the particles are not gauge invariant: gauge transformations alter the phase of the particles. The solution is to consider path-dependent quantities known as holonomies (see Mandelstam et al. (1962, 347)):

$$\Phi(x, P) = \phi(x)e^{-ie\int_P^x dz_m A_m(z)} \tag{4}$$

Hence, rather than working with local particle operators, one works with these spread out quantities (and the electromagnetic field). This approach retains the covariance of the original theory. Mandelstam argues that, in this case at least, "there is a close analogy between the electromagnetic field and the gravitational field" (Mandelstam et al. 1962, 353). The development of Yang-Mills theory saw the development of a new analogy. Because of the closer similarities between Yang-Mills fields and the gravitational field (both are non-linear and have infinite dimensional gauge groups), Murray Gell-Man suggested to Feynman (in the late 1950s), who was becoming interested, that he attempt to quantize Yang-Mills theory first, as a preparatory exercise (Feynman 1972, 378). This led to groundbreaking work in the construction of quantum gauge field theories (of the kind that make up the current standard model of particle physics).

---

[42]The introduction of an indefinite metric to resolve divergence problems in quantum field theory was given by Dirac in 1942 (Dirac 1942).

11.4.3 Gravity as a Natural Cutoff

As explained above, much of the early work was characterised by a desire to construct a theory free of divergences: a finite theory. Or at least a theory with "controllable" infinities. There were several methods that were developed to achieve this, based around the introduction of fundamental length scales. The potential utility of introducing gravity into elementary particle physics, so as to eliminate divergencies, spurred on work on quantum gravity enormously. The most obvious strategy is to impose a cutoff. There were many suggestions that gravity might act in this way, as some kind of "regulator". The divergences in question were those of QED, and meson theories, which were still, pre-WWII, a somewhat mathematically murky territory. The problem concerned the transitions between quantum states, during which time (a very short time, determined by the uncertainty relations) energy conservation is violated. The great hope for introducing gravity into elementary particle physics was that it would terminate the wavelengths before they get to the problematic high-energy (ultraviolet) wavelengths.

Pauli makes several comments to this effect, including the following remarks in a letter addressed to Abrikosov, Khalatnikov, and Pomeranchuk:

> I was very interested in *Landau's* remarks on the possibility of a connection of the cut-off moment of quantum electrodynamics with *gravitational* interaction (his article "on quantum theory of fields" in the Bohr-festival volume). It appeals to me, that the situation regarding divergencies would be fundamentally changed, as soon as the light-cone itself is not any longer a *c*-number equation. Then every given direction is space-time would have some "probability to be on the light-cone", which would be different from zero for a small but finite domain of directions. I doubt, however, that the *conventional* quantization of the $g_{\mu\nu}$-field is consistent under this circumstances. (Zürich, 15 August 1955; in von Meyenn (2001, 329))

He followed up on this same line in his comments after a talk by Oskar Klein:

> It is possible that this new situation so different from quantized theories invariant with respect to the LORENTZ group only, may help to over- come the divergence difficulties which are so intimately connected with a c-number for the light-cone in the latter theories. (Pauli in Klein (1956, 6928))

However, as with the linearization approaches, and those based on the electromagnetic analogy, the divergences in the gravitational case were more complicated than had been used to, for precisely the reasons Pauli alludes to:

> [I]t must constantly be borne in mind that the bad divergences of quantum gravidynamics are of an essentially different kind from those of other field theories. They are direct consequences of the fact that the light cone itself gets shifted by the non-linearities of the theory. But the light-cone shift is precisely what gives the theory its unique interest, and a special effort should be made to separate the divergences which it generates from other divergences. (DeWitt 1962, 374).

In his PhD thesis Bryce DeWitt, under the supervision of Julian Schwinger, sought to revisit Rosenfeld's work on the computation of gravitational self-energies. DeWitt would also revisit the idea of Landau that gravity might act as a natural regulator (DeWitt 1964). Though Landau didn't explicitly mention the Planck scale (he placed the location of the cutoff much higher), Pauli appeared to think that Landau had quantum gravitational effects in mind. It is clear that if there is a fundamental length, below which quantum field theoretic processes cannot operate, then one has what Landau sought. DeWitt was able to confirm that (at lowest order of perturbation) when gravity is included, the self-energies of charged particles (and the gravitons themselves) remain finite (though often very large).

More indirect, however, was Peter Bergmann's method of utilising the fact that the gravitational field equations determined particle trajectories free of any notions of divergences. He believed this would follow from the analysis of Einstein, Hoffmann and Infeld, according to which the assumption of geodesy for a free particle's motion was redundant, since it already could be seen to follow (by a method of successive approximation) from the field equations alone.

I might also note here that ultimately string theory emerged from the divergences problems facing quantum field theories of fields other than the electromagnetic field (particularly the strong interaction). In particular, since the perturbative approach breaks down when the coupling constant determining interactions strengths is high (as in strong interaction physics), alternative approaches were sought in the late 1950s and throughout the 1960s. One of the more popular of these approaches combined Heisenberg's S-matrix theory with dispersion theory. The S-matrix is a tool to encode all possible collision processes. Heisenberg suggested that one take this to embody what was relevant about the physics of collision processes. In particular, all that was observable were the inputs and outputs of collision processes, observed when the particles are far enough apart in spacetime to be non-interacting, or free. This back box approach to physics was very much inspired by the Copenhagen philosophy. The dispersion relation approach to physics tried to construct physical theories on the basis of a few central physical axioms, such as unitarity (conservation of probabilities), Lorentz invariance, and causality (effects can't precede causes). These two approaches were combined, by Geoff Chew amongst others, so that the focus was on the analytic properties of the S-matrix. One model for the S-matrix, incorporating some other principles thought to be involved in strong interaction physics, was the Veneziano model. This used the Euler beta function to encode the various desirable properties of the S-matrix. The model was found to be generated by a dynamical theory of strings. (See Cushing (1990) for a detailed historico-philosophical account of the early development of string theory.)

Developing the cutoff idea, and the idea that there might be a minimal (fundamental) length, leads one quite naturally into the idea that space and time might not be continuous, but better modelled instead by a discrete lattice or similar structure. This was suggested by several people. In a paper from 1930 Ambarzumian and Iwanenko (1930) argued for the introduction of a spatial lattice structure for physical space as a way of eliminating the infinite divergences from the self-energy of the electron. The basic idea was that the existence of a minimal length would imply a maximal frequency (567).[43] Alfred Schild (1948) investigated the properties of such

[43] I might add that this rich background of work on the mixture of geometry and quantum theory provides a nice background out of which Matvei Bronstein's work emerged—see Gorelik (1992).

a discrete lattice in order to see if it would break essential symmetries. In particular, he was responding to the objection that discrete theories would violate Lorentz invariance.[44] He wasn't able to devise a model to preserved all such symmetries, but enough to provide a plausible candidate for a background for a physical theory. Here again we find constraints operating on the various approaches; in this case the Lorentz symmetry of the classical theory.

Another discrete approach, of David van Dantzig (1938, 1955), was motivated by a combination of general covariance (as expressed in Einstein's "point-coincidence" argument) and the definition of observability in such a theory. He argued that in a generally covariant theory the observable things will be coincidences: events. van Dantzig argues that in order to not introduce unmeasurable structure into the interpretation or formulation of one's theory, one should dispense with the existence of a four-dimensional continuum, in favour of a discrete manifold of events.[45] Peter Bergmann describes the approach as one of "constructing 'spaces' that have certain topological properties similar to those of point spaces in the large but do not possess 'points' as elementary constituents" (Bergmann, following a talk of Wigner's talk: (1955, 226). The general approach lives on in several of the current approaches, including causal set theory and dynamical triangulations.

11.4.4 General Relativisation of the Dirac Equation

As with the electrodynamical analogy that led to a flow of ideas from quantum field theory to general relativity; so (at around the same time, though somewhat earlier) there was a "geometrical analogy", responsible for a flow from general relativity to quantum theory. Following Einstein's early remarks about the need for some kind of relationship between gravitation and the quantum theory, much of the work in the field of quantum gravity (up until the 1930s) concentrated on bringing the quantum theory in a form conducive to integration with (classical) general relativity.

George Temple argued for a modification of the Dirac equation (of the electron) on pain of "abandoning the theory of relativity" (Temple 1928, 352). Temple's approach was to construct a system of wave equations which possessed "all the advantages as Dirac's equations and which shall be tensorial in form in accordance with the general theory of relativity" (ibid.). In a slightly different way, both Fock and Weyl also attempted to merge the Dirac equation with the geometry of general relativity. Their strategy (discovered independently) was to modify the structure of the manifold so as to allow for spin—by adding (local) spinor structures. Fock desired (and thought he'd achieved) a "geometrization of Dirac's theory of the electron and its subsumption within general relativity" (Fock 1929, 275). Together with

[44]Rafael Sorkin would later defend the causal set approach from the same charge.

[45]I should point out that van Dantzig steers clear of positivism. He notes that "it is not sufficient to take only observed events; we have to add to these also possibly observable, hence fictitious events" (Comments after a talk of Wigner's: (1955, 224). I take it by "fictitious" he means counterfactual: i.e. they *could* be observed.

Iwanenko, they labeled their theory "quantum linear geometry". The basic idea was to modify the geometry in such a way as to include the properties of the Dirac matrices. They suggest introducing a linear differential form $ds = \Sigma\gamma_\nu dx_\nu$ that when squared would deliver the standard Riemann interval ds^2.

Fock believed that this approach could lead to solutions of the most pressing problems in quantum field theory at that time (negative-energy solutions and the ubiquitous divergences). Though he devised a near identical theory to Fock, Weyl would later distance himself from the geometrization programme. See Scholz (2005) for a nice account of this episode.[46]

The mathematician Dirk Struik was invited to MIT by Norbert Wiener. Struik had experience in the mathematics of general relativity, and had assisted in the development of parts of the differential geometric and group theoretic aspects. Wiener had met Struik on a visit to Göttingen—*cf.* Rowe (1989, 23). Together they worked on a unified theory of general relativity, electromagnetism, and quantum theory. The methodology was as above: to subsume quantum theory (in this case, specifically Schrödinger's wave mechanics) within general relativity.

Wigner, perhaps more than anyone else (save Weyl), recognized the importance of symmetry in quantum theory. Rather than focusing specifically on making the Dirac equation generally relativistic, Wigner adopted the method of showing how quantum mechanics itself was not really in any conflict with what he saw as the two basic principles of general relativity; namely that "coordinates have no independent meaning" and "that only coincidences in space-time can be observed directly and only these should be the subject of physical theory" (Wigner 1955, 219). Wigner appears to have something like Kretschmann's "point-coincidence" objection to the principle of general covariance in mind when he writes that "[t]his observation is so stringent that, properly considered, every physical theory conforms with it ... this is true also of the present day quantum mechanics" (ibid.).

His approach was to formulate a version of quantum mechanics without coordinates, using just field components. Fred Hoyle was in agreement with Wigner, calling the use of coordinates "a psychological survival from the Newtonian era" (Wigner 1955, 224):

> Now that we realize that coordinates are no more than parameters that must be eliminated in determining relations between observables, it becomes natural to ask whether we are using the most advantageous parameters, or even whether any such parameters are necessary. (ibid.)

This introduces another distinction between the approaches: there are those that seek to quantize the gravitational field, and those that seek to general relativize quantum theory. Again, these two broad categories can be found in the current crop of approaches to the problem of quantum gravity.

---

[46]The general relativization of other physical systems continued for some time. John Wheeler ran a seminar on Dirac's Equation in general relativity as part of his course on advanced quantum mechanics (in 1955)—this was transcribed by Charles Misner.

11.4.5 Canonical Versus Covariant

Two other distinct lines were clear very early on (post-1930): that involving quantizing the full metric using canonical quantization methods, and that of covariant quantization of a perturbation on a flat spacetime. The former approach involves the Hamiltonian formulation of general relativity in which the canonical "configuration" variable is the spatial metric (on a spatial hypersurface) and the equations of motion determine how it evolves with respect to an arbitrary slicing of spacetime into space and time. In modern terms, the latter (covariant quantization) approach amounts to the derivation of the Feynman rules for the propagators and the vertexes describing the quanta of the theory. In other words, the approach proceeds by constructing Feynman diagrams. These to hand, one can then ascertain whether the theory is renormalizable at various loops. The classical (or "tree level" or "zero loop") case involves the computation of the scattering of classical gravitons via the vertices and the propagators. The quantum case builds in closed loops to the tree diagrams. The Feynman diagrams describe the possibilities of breaking apart, forming loops, and joining together. Carrying out this procedure led Feynman to the discovery of "ghosts", a compensatory (unphysical) field needed purely to render quantum Yang-Mills field theories consistent.[47]

There are in fact several ways to achieve a covariant approach. Bryce DeWitt used Rudolph Peierls (coordinate independent) version of the Poisson bracket to define the commutator in terms of Green's functions. DeWitt was able to quantize the gravitational field without restricting himself to flat spacetime. The method, known as the "background field method", invoked physical degrees of freedom (in this case, a stiff elastic medium, like a physical ether, and a field of clocks) to localize points, and there allow for the localization of physical quantities. The gravitational field is taken to interact with this background field.[48] Computationally, the approach marked a great advance. However, the background is unphysical in most realistic cases—but this highlights DeWitt's (much like Feynman, and other physicists with war experience) focus on getting the job done, and computing numbers, over conceptual issues.

The canonical (Hamiltonian) approach was pursued in slightly different ways by several schools. Bergmann was the first to apply canonical quantization methods to non-linear covariant field theories (Bergmann and Brunings 1949). The Hamiltonian

[47]In 1974 't Hooft and Veltman were able to show that Einstein's theory of gravity (without matter and in 4D) was finite at one loop; however, adding matter to the theory destroyed this ('t Hooft and Veltman 1974). In 1985 Goroff and Sagnotti later did the two loop computation and found that the theory was non-renormalizable (Goroff and Sagnotti 1985).

[48]It is possible that this approach was developed by DeWitt using material he'd had to master for his work in higher-dimensional Lagrangian hydrodynamics, in which one has to consider a mesh that forms a dynamical background for the materials one is studying.

formulation of classical general relativity was perfected by Dirac, in 1958 (Dirac 1958).[49] The basic idea is expressed by DeWitt as follows:

> A canonical theory looks at spacetime as a sequence of 3-dimensional slices, each characterized by its intrinsic 3-geometry. The slicing is, of course, not unique. However, if two 3-geometries are chosen, one may try to solve the "Sandwich Problem": find those spacetimes (4-geometries) which can have these 3-geometries as slices. (DeWitt 1970, 186)

The four dimensional spacetime (diffeomorphism) symmetry of general relativity is clearly broken (at least superficially) in this approach. The symmetry is canonically rendered using four constraint functions on the chosen spatial manifold, a scalar field (known as the scalar constraint) and a vector field (known as the diffeomorphism constraint). These have the effect, respectively, of pushing data on the slice onto another nearby (infinitesimally close) slice and shifting data tangentially to the slice. In a canonical approach (to field theories) one writes theories in terms of fields and their momenta. Spacetime covariant tensors are split apart into spatial (tangential) and temporal (normal) components. This naturally obscures general covariance, but the theory is generally covariant despite surface appearances. The general covariance of the Einstein equations, reflecting the spacetime diffeomorphism invariance of the theory, is encoded in constraints.[50] Taken together, when satisfied, these constraints are taken to reflect spacetime diffeomorphism invariance; together they tell us that the geometry of spacetime is not affected by the action of the diffeomorphisms they generate. This job is done by two, of course, since the diffeomorphism constraint deals with aspects of the spatial geometry and the Hamiltonian constraint deals with aspects of time. Imposing both delivers the desired full spacetime diffeomorphism invariance.

One then quantizes the theory, including the quantization of the constraints, so that quantum states are annihilated by the full Hamiltonian constraint (a combination of the scalar and diffeomorphism constraints): $\hat{H}\Psi[g] = 0$. This equation (known as the Wheeler-DeWitt equation) contains all of the dynamics in quantum geometrodynamics. The quantum states (the wave-functionals Ψ) depend only on the 3-metric, not on time—hence, the solutions represent stationary wavefunctions. In fact, the diffeomorphism constraint implies that the quantum states are only dependent on the 3-geometry rather than the metric—that is to say, the states

[49]Donald Salisbury has investigated the early history of the canonical quantization formalism, with its introduction of "constraints" (on which see below): (Salisbury 2007, 2009). He traces crucial details back to Rosenfeld. However, from the post-Rosenfeld development he omits from his story crucial work by Paul Weiss (a student of Max Born and Paul Dirac): (1938a, 1938b). The pre-Rosenfeld story involves Emmy Noether's work on the identities generated by the general covariance of general relativity.

[50]Hamiltonian constraints, in general relativity, constitute an infinite set of relations holding between the canonical variables of the theory (the spatial metric and its conjugate: $(g_{\mu\nu}, p^{\mu\nu})$). Any choice of physical variables must satisfy these constraints on some initial hypersurface. The constraints are taken to generate (infinitesimal) coordinate transformations of the initial hypersurface. Since the theory is independent of coordinate transformations (on pain of underdetermination), these are taken to constitute gauge transformations. Dirac developed a general framework for such constrained systems, classically and quantum mechanically.

are invariant under diffeomorphisms of Σ thanks to the diffeomorphism constraint. However, the geometrodynamical (so expressed) approach ran out of steam due to technical difficulties. New variables based on a canonical transformation of the phase space of general relativity led to a more tractable formulation, but not until the mid-1980s.

***

The historical review presented above is, of course, incomplete, and rather rough. However, it contains elements that ought to be of interest to philosophers, historians, and sociologists. QG research shows quite clearly many cases in which none of the standard predictivist models of scientific evaluation are operating. Novel predictions are not an issue; certainly not in the earliest research. The pragmatic argument was trundled out time and again to show that novel testable predictions were out of the question, but research continued in spite of this consensus.[51] Yet the fact that "casualties" have occurred shows that evaluation *is* nonetheless in operation. It also reveals, even in the earliest work, a great diversity in competing approaches, many of which are empirically indistinguishable (i.e. the canonical versus covariant approaches) yet conceptually quite distinct and with their own distinct set of internal problems. As we saw with Stachel's example of the geometric vs spin-2 field representation of gravity, this can lead to genuine underdetermination: the geometric understanding of general relativity that prevails might easily have been replaced by a very different highly non-geometric one. The reasons are not experimental or observational: it is more a matter of timing—*cf.* Cushing (1994).

From the earliest phase of quantum gravity research to the present day, their is a vast array of approaches, most of them highly distinct. They are constrained by much the same factors: the facts we know about the world already, basic physical principles, and mathematical consistency. However, analogy and simplification also play a crucial role. The methods of simplification, and the analogies chosen are determined in large part (and quite naturally so) by the traditions of the practitioners: by the tools they have to hand and by the range of theories that they are acquainted with. This has elements of both Andy Pickering's and Peter Galison's approaches to the development of theory. I'll draw some of these points of contact out in the final two sections.

11.5 The Right Job for the Tools?

Very early on in the history of quantum gravity there emerged distinct paths that were trodden as a result of the differing backgrounds of the actors. With different sets of tools come different points of interest and different research questions. For

[51] There are, on the other hand, approaches which are inconsistent with old evidence. For these, we can at least adopt some empirical criterion: accommodationism—though as a *criterion* it is still applicable only to some episodes; other cases will violate it if an approach has other things going for it.

example, in the canonical (Hamiltonian) quantization approach one is concerned with such global features as the wave-function of the universe, and the domain space of such a wave-function. The construction and understanding of the configuration space is a largely classical problem. By contrast, the covariant approach is more "local", focusing on graviton scattering and other typically quantum field theoretic quantities (*cf.* DeWitt 1967, 1239). These are entirely different ways of approaching a subject, with distinct commitments, but without empirical distinction.

David Kaiser notes how, in his lectures at Caltech, Feynman introduced general relativity for "physics students ... who know about quantum theory and mesons and the fundamental particles, which were unknown in Einstein's day" (Feynman's *Lectures on Gravitation*; cited in Kaiser (2005, 329). We have seen that the particle physicist's approach is conceptually very different from Einstein's own geometrical approach. Feynman, though not the first to write general relativity in this way, sought to tailor the presentation of the theory to the needs of his students.[52] The approach was designed to mesh with the modern particle physicist's "multi-field" mindset. With the profusion of new particle types thrown up by the latest generation of particle accelerators, students were used to thinking about large numbers of fields (associated with the particles). Feynman's approach was to treat "the phenomena of gravitation" as the addition of "another field to the pot" (ibid.); as nothing special, much as many of the very earliest researchers had done. Those trained in the "geometrical way" naturally baulk at this particle physics approach. The divisions between the two ways of approaching gravity are felt strongly by the two sides—one can find the genuine animosity involved in the division in today's debate over background independent versus background dependent approaches (see Smolin (2006) for a discussion of this debate).[53]

[52]Recall, as I said earlier, that there is not unlimited freedom in how one can do this: the principles going into the theory, together with the observational data it would need to account for, constrain the form of the theory very tightly indeed.

[53]David Kaiser (2007) has argued in a similar way that particle cosmology (roughly: general relativity combined with elementary particle physics) emerged in the 1980s for external reasons. In particular the cold war bubble burst resulting in a cessation of particle physics funding opportunities. The upshot of this was that there were a bunch of physicists who needed to get their funding from somewhere. The funding, of course, is (inter alia) to enable to construction of equipment that can test their theories and generate new phenomena to stimulate new theories. This predicament forces the ploughing of new research avenues. In this case the hybridization of cosmology and particle physics. The joining of these subjects is not entirely accidental: cosmology involves (in certain areas) extreme high-energy phenomena, of just the kind that could function as a laboratory for particle physicists! Moreover, the two fields had some history together: the union is not entirely novel. However, it was certainly professionalized in the period Kaiser studies. Background conditions (external factors) can, then, quite clearly be seen to impact institutions which in turn impact scientific thought. I mention this case study since it impacts directly on the development of quantum gravity. Cosmological scenarios could be exploited to generate phenomenology for quantum gravity theorists. Not only this, the interaction between particle physics and cosmology resulted in the training of a generation of new scientists with strong skills in both quantum field theory and general relativity, ideal for quantum gravity research.

There are several interesting institutional changes that enter in to this argument, that trigger several phases of research inquiry. The earliest workers on quantum gravity had strong backgrounds in both general relativity and quantum theory, and had made strides in both, and often had written textbooks in both fields. But as quantum theory became more complex, and engaged with experiments more, it occupied more and more of physicists' attentions. It had a larger share of mathematical and conceptual problems. General relativity was, from very early on, seen to be essentially finished and very hard to apply in realistic situations. Hence, the physicists' interests and toolkits shifted more and more into quantum theory, and this shift can be seen in the evolution of approaches to the quantum gravity problem. It wasn't really until John Wheeler and Peter Bergmann began working on quantum gravity in the 1950s, and gathered high quality students around themselves that the geometrical approaches began to make their mark once again. But by this stage there were very strong disciplinary splits, marked out by the existence of distinctive "schools".

Of course, as I said above, though these institutional factors contribute a great deal to the nature of the research that is done in quantum gravity, and can go a long way in explaining why certain physicists hold some particular approach, there is not much latitude in the possibilities that theories can take. The available tools are one constraint among many others. I finish by saying something about the notion of constraints, and the role they play in establishing scientific beliefs, and accounting for changes in science. I am particularly concerned with finding some kind of constraint that can play a role functionally similar to that played by experiments in guiding theoretical developments.

11.6 Convergence and Constraints

Ian Hacking notes that the convergence on some feature or result (of a theory) can prove very convincing (in evaluative terms) in cases where the convergence comes about through quite different instruments and experiments, using quite distinct physical principles:

> we are convinced because instruments using entirely different physical principles lead us to observe pretty much the same structures in the same specimen. (Hacking 1983, 209)

If we keep finding some similar behaviour in a wide variety of conditions, then we are prone to believe that the behaviour is a universal feature, not some artefact of a model or experiment. Certain results crop up in multiple formalisms and in the context of quite distinct investigations in quantum gravity research. For example, the divergences, or in later work, the existence of quantum geometry. This is especially important in the absence of experiments. Indeed, in the absence of experiments it is perfectly natural to expect *theoretical* considerations to play a more central role.

One can usefully view this increase in the importance ascribed to theoretical considerations by looking at it through the lens of Peter Galison's notion of "constraints" (Galison 1995). Constraints are very much the life-blood of science. They serve to minimize the latitude one has in theory construction. The satisfaction of constraints can in itself act as an evaluative measure. In the absence of experiments and observation, new kinds of constraints must come to the fore, to guide theorizing.[54] The black hole entropy value I mentioned earlier functions, in some sense, as a constraint almost (but not quite) like an experimental constraint: it provides a number that the approaches must be able to derive. The newer approaches are tested against this constraint, and when an approach is able to satisfy it, it is seen to have passed a "test". It thus provides new material for constructing theories, or working out the possibilities of old theories. Renormalizability too acted as a crucial constraint in the post-WWII years (*cf.* Galison 1988). Weinberg describes the importance of this constraint:

> [I]t seemed to me to be a wonderful thing that very few quantum field theories are renormalizable. Limitations of this sort are, after all, what we most *want*; not mathematical methods which can make sense out of an infinite variety of physically irrelevant theories, but methods which carry constraints, because these constraints may point the way towards the one true theory. In particular, I was very impressed by the fact that [QED] could in a sense be *derived* from symmetry principles and the constraint of renormalizability; the only Lorentz invariant and gauge invariant renormalizable Lagrangian for photons and electrons is precisely the original Dirac Lagrangian. (Weinberg 1980, 1213; cited in Galison 1995, 22)

The original perturbative approach to quantum gravity was rejected because it conflicted with this constraint. By contrast, string theory was given credence because it offered the prospect of a finite theory. However, it was then found (in its early years) to violate the constraint that there be no quantum anomalies in the theory (i.e. symmetries that are in the classical theory but broken at the quantum level). The subsequent satisfaction of this constraint (by choosing suitable gauge groups) provided almost as significant a degree of motivation for renewed interest in the theory as a successful physical experiment. There is little doubt as to the power of constraints of this sort in the minds of physicists.

However, as with experiments, we shouldn't place too much weight on them: they are rarely decisive. As I mentioned in the string theory case; these constraints (if followed too rigidly) can lead one to drop a theory prematurely, only to be found at a later date to satisfy it. I mentioned earlier in fn.9 that the lessons of the Bohr-Rosenfeld analysis of measurability of the electromagnetic field were taken to transfer over to the gravitational case (yet another example of the analogical reasoning so prevalent in quantum gravity research).[55] The idea was that the gravitational field would *necessarily* have to be quantized if it were coupled to another quantized field, or to quantized matter. This belief spurred on physicists in the early days. However, in the 1957 Chapel Hill conference (*On the Role of Gravitation*

---

[54]The classic tests of GR functioned as constraints on the early quantum gravity approaches in a more or less standard way (i.e. empirical adequacy). What is more interesting is where we have approaches that match up with respect to all pre-existing data, but that don't make any novel testable predictions beyond this old data. In this case, any decisions must be based on theoretical considerations.

[55]Note that such analogies can themselves be interpreted as constraints, for one is essentially making a claim that two systems are sufficiently similar so that the (well-known) constraints that apply to one will most likely apply to the other.

in Physics), Rosenfeld argues that the analysis he performed with Bohr does not translate into the gravitational case. The crucial disanalogy is that one cannot (even theoretically) find a measuring instrument that would not generate perturbations in the measurement result: this is due to the equivalence principle. In the electromagnetic case the fact that there are both positive and negative charges allows one to control the perturbations. The electromagnetic field can be shielded.

This supposed necessity (suggested by the thought experiment) previously functioned as a constraint on quantum gravitational theorizing. However, John Wheeler was willing to suggest (following Rosenfeld's remarks) that perhaps the measurement problem for quantum gravity could be ignored for the present and than one place more emphasis on "the organic unity of nature" as a key constraint (Wheeler 1957, 83). Hence, though the development of theory demands constraints to guide it at any one time; the constraints it makes use of don't have to be constant. Quantum gravity provides a very useful episode in which non-experimental constraints can be seen to evolve as the wider (theoretical and experimental) context evolves around it (and under quantum gravity research's own internal dynamics).

For example, in the very earliest approaches, there was no quantum field theory available, so what would become an important constraint (renormalizability) was absent. Once QED was constructed, however, one could get a handle on computable aspects of quantum gravity, and compare them to what had been done in QED. Rosenfeld's computation of the self-energy of the graviton was such an example. The development of renormalisability led to an easily applicable criterion to decide whether a theory was worth pursuing. Interestingly, the constraint of renormalizability played a lesser role once the tools of renormalization *group* theory had been assimilated.

General invariance became a constraint itself. Other important constraints include unitarity (probability conservation), Lorentz invariance, and causality. These combined in an interesting way (with known resonance data from particle collision experiments) to lead to string theory in the late 1960s. Taken together, these principles can home in on a very small number of possible candidate theories. Indeed, for a long time it was believed that they could work in tandem to produce a *unique* theory, though this view is less popular today. Whether the constraints can force uniqueness or not, it is true that they reduce the freedom one has in theory construction, and this is crucial, for without them one would have infinite freedom! A new field known as quantum gravity phenomenology, currently in the early stages, is developing in order to provide additional data to further constraint the possible theories of quantum gravity.

The notion of constraints seems to offer some promise in exposing the innards of the black box I began with, in a way that might be conducive to philosophers. The constraints are at work on both constructive and evaluative levels. The interest to sociologists enters through the fact that different communities are determined by their different trainings (with different toolkits), and this difference spills over into a difference over what constraints ought to be respected (renormalizability versus general invariance, for example). Only a close investigation of the historical details can reveal which constraints guided some particular theory choice. The constraints will often be sociological as well as mathematical and empirical.

11.7 Conclusion

Quantum gravity research constitutes an ideal and novel historical episode that should appeal to historians, philosophers, and sociologists of science alike. The absence of possible experiments and experimental anomalies that usually drive the development of the field expose an entirely different set of inner workings than we are used to seeing in science. One can see how a range of virtues (such as unification, beauty, and so on) beyond "the usual suspects" can guide both the construction and justification of theories. One also sees the strong role played by analogies, which continued to be pursued despite the knowledge that the analogy was far from perfect. Methodologically, what the development of quantum gravity reveals is that what is deemed appropriate will depend upon what constraints are available at the time, and this is prone to changes of a great variety of sorts. I have argued that the framework of *constraints* provides a useful tool with which to prise open the black box that contains the development of quantum gravity. However, quantum gravity itself provides a tool with which to see the operation and evolution of theoretical constraints that are often overpowered by experimental constraints.

Acknowledgements My thanks to audiences at the 2nd &HPS conference at the University of Notre Dame in 2009, and the History of Quantum Theory conference at the University of Sydney in 2008. My thanks also to the editors of this volume both for the invitation to contribute, and for their comments. I gratefully acknowledge the Australian Research Council for funding an Australian Research Fellowship, under which this work was carried out.

References

Ambarzumian, V., and D. Iwanenko. 1930. "Zur Frage nach Vermeidung der unendlichen Selbstrückwirkung des Elektrons". *Zeitschrift für Physik A: Hadrons and Nuclei* 64(7–8): 563–67.

Ashtekar, A. 2005. "The Winding Road to Quantum Gravity". *Current Science* 89(12): 2064–74.

Belinfante, F.J., and J.C. Swihart. 1954. A Theory of Gravitation and its Quantization. Research Report, Purdue University.

Bergmann, P.G. 1953. "Review of The Theory of Relativity by C. Møller". *The Scientific Monthly* 76(2): 112–4.

Bergmann, P.G. 1992. "Quantization of the Gravitational Field, 1930–1988". In *Studies in the History of General Relativity*, edited by J. Eisenstaedt and A.J. Kox, 364–6). Boston, MA: Birkhaüser.

Bergmann, P.G., and J.H.M. Brunings. 1949. "Non-Linear Field Theories II. Canonical Equations and Quantization". *Reviews of Modern Physics* 21: 480–7.

Birkhoff, G. 1943. "Matter, Electricity and Gravitation in Flat Space-Time". *Proc Natl Acad Sci USA* 29(8): 231–9.

Blokhintsev, D.I., and F.M. Gal'perin. 1934. "Gipoteza Neitrino i Zakon Sokhraneniya Energii". *Pod Znamenem Marxisma* 6: 147–157.

Born, M. 1955. "Physics and Relativity". In *Fünfzig Jahre Relativitätstheorie, Bern, July 11–16, 1955*, edited by A. Mercier and M. Kervaire, 244-60. *Helvetica Physica Acta, Suppl. 4.* Basel: Birkhaüser Verlag.

Burian, R.M. 2002. "Comments on the Precarious Relationship Between History and Philosophy of Science". *Perspectives on Science* 10(4): 398–407.

Curiel, E. 2001. "Against the Excesses of Quantum Gravity: A Plea for Modesty". *Philosophy of Science* 68(S1): S424–S41.

Cushing, J.T. 1990. *Theory Construction and Selection in Modern Physics: The S Matrix.* Cambridge: Cambridge University Press.

Cushing, J.T. 1994. *Quantum Mechanics: Historical Contingency and the Copenhagen Hegemony.* Chicago, IL: The University of Chicago Press.

DeWitt, B.S. 1953. A Numerical Method for Two-Dimensional Lagrangian Hydrodynamics. Livermore Report: UCRL-4250.

DeWitt. B. 1962. "The Quantization of Geometry". In *Gravitation: An Introduction to Current Research,* edited by L. Witten, 266–328. New York: Wiley.

DeWitt, B.S. 1967. "Quantum Theory of Gravity. III. Applications of the Covariant Theory". *Physical Review* 162(5): 1239–56.

DeWitt, B.S. 1970. "Quantum Theories of Gravity". *General Relativity and Gravitation* 1(2): 181–9.

DeWitt, B.S. 1982. "The Early Days of Lagrangian Hydrodynamics at Lawrence Livermore Laboratory". In *Numerical Astrophysics*, edited by J.M. Centrella, J.M. LeBlanc, and R.L. Bowers, 474–81. Boston: Jones and Bartlett Publishers, Inc, 1985.

DeWitt, B.S. 1964. "Gravity: A Universal Regulator?". *Physical Review Letters* 113(3): 114–8.

DeWitt, C., and D.P. Rickles. 2011. The Role of Gravitation in Physics: Report from the 1957 Chapel Hill Conference (Co-edited with Cecile DeWitt). Max Planck Research Library for the History and Development of Knowledge, Volume 5. http://www.edition-open-access.de/sources/5/index.html

Dirac, P.A.M. 1935. "The Electron Wave Equation in de Sitter Space". *Annals of Mathematics* 36(3): 657–69.

Dirac, P.A.M. 1942. "Bakerian Lecture. The Physical Interpretation of Quantum Mechanics". *Proceedings of the Royal Society of London* A 180(1): 1–40.

Dirac, P.A.M. 1958. "The Theory of Gravitation in Hamiltonian Form". *Proceedings of the Royal Society of London* A 246: 333–43.

Dirac. P.A.M. 1983. "The Origin of Quantum Field Theory". In *The Birth of Particle Physics,* edited by L. Brown and L. Hoddeson, 39–55. Cambridge: Cambridge University Press.

Duff, M.J. 1975. "Covariant Quantization". In *Quantum Gravity: An Oxford Symposium,* edited by C.J. Isham, R. Penrose, and D.W. Sciama, 78–135. Oxford: Clarendon Press.

Einstein, A. 1916. "Naherungsweise Integration der Feldgleichungen der Gravitation". *Sitzungsberichte der Koniglich PreufiischenAkademie der Wis-senschaften Berlin* XXXII: 688–96.

Einstein, A. 1918. "Über Gravitationswellen". *Sitzungsberichte der Koniglich Preufiischen Akademie der Wissenschaften Berlin:* 154–67.

Einstein, A. 1919. "Spielen Gravitationsfelder im Aufbau der materiellen Elementarteilchen eine wesentliche Rolle?" *Sitzungsberichte der Koniglich Preufiischen Akademie der Wissenschaften Berlin* 1: 349–56. Translated as: "Do Gravitational Fields Play an Essential Part in the Structure of the Elementary Particles of Matter?" In *The Principle of Relativity,* edited by A. Einstein, A. Lorentz, H. Weyl, and H. Minkowski, 189–98. Dover Publications, 1932.

Einstein, A. 1940. "Considerations Concerning the Fundaments of Theoretical Physics". *Science* 91(2369): 487–92.

Feynman, R.P. 1972. "Quantizing the Gravitational and Yang-Mills Fields". In *Magic Without Magic,* edited by J. Klauder, 377–408. New York: W. H. Freeman.

Fierz, M., and W. Pauli. 1939. "On Relativistic Wave Equations for Particles of Arbitrary Spin in an Electromagnetic Field". *Proceedings of the Royal Society of London. Series A, Mathematical and Physical Sciences* 173(953): 211–32.

Fock, V. 1929. "Geometrisierung der Diraschen Theorie des Electrons". *Zeitschrift für Physik* 52: 869–77

Fock, V., and D. Iwanenko. 1929. "Quantum Geometry". *Nature* 123: 838.

Galison, P. 1988. Multiple Constraints, Simultaneous Solutions. *PSA: Proceedings of the Biennial Meeting of the Philosophy of Science Association. Vol. 1988, Volume Two: Symposia and Invited Papers:* 157–63.

Galison, P. 1995. "Context and Constraints". In *Scientific Practice,* edited by J.Z. Buchwald, 13–41. Chicago: University of Chicago Press.

Goldberg, J.M. 1992. "US Air Force Support of General Relativity: 1956–1972". In *Studies in the History of General Relativity,* edited by J. Eisenstaedt and A.J. Kox, 89–102. Boston, MA: Birkhaüser.

Goldstein, C., and J. Ritter. 2003. "The Varieties of Unity: Sounding Unified Theories 1920–1930". In *Revisiting the Foundations of Relativistic Physics*, edited by A. Ashtekar et al., 93–149. Dordrecht: Kluwer.

Gorelik, G. 1992. "The First Steps of Quantum Gravity and the Planck Values". In *Studies in the History of General Relativity,* edited by J. Eisensaedt and A.J. Kox, 364–79. Boston, MA: Birkhaüser.

Goroff, M.H., and A. Sagnotti. 1985. "Quantum Gravity at Two Loops". *Physics Letters B* 160(1–3): 81–86.

Gupta, S. 1962. "Quantum Theory of Gravitation". In *Recent Developments in General Relativity,* 251–258. New York, NY: Pergamon Press.

Hacking, I. 1983. *Representing and Intervening: Introductory Topics in the Philosophy of Natural Science*. Cambridge: Cambridge University Press.

Hass, A. von. 1919. "Die Axiomatik der Modernen Physik". *Naturwissenschaften* 7(41): 744–50.

Hass, A. von. 1938. "The Dimensionless Constants of Physics". *Proceedings of the National Academy of Sciences of the United States of America* 24(7): 274–6.

Heisenberg, W., and W. Pauli. 1929. "Zur Quantenelecktrodynamik der Wellenfelder". *Zeitschrift für Physik* 56(1–2): 1–61.

Hoddeson, L. 2002. "Toward a History-Based Model for Scientific Invention: Problem-Solving Practices in the Invention of the Transistor and the Development of the Theory of Superconductivity". *Mind & Society* 3: 67–79.

Humboldt, W. von. 1822. "On the Historian's Task". *History and Theory* 6(1) 1967: 57–71.

Kaiser, D. 1998. A ψ is just a ψ? Pedagogy, Practice, and the Reconstitution of General Relativity, 1942–1975". *Studies in History and Philosophy of Modern Physics* 29: 321–38.

Kaiser, D. 2000. *Making Theory: Producing Physics and Physicists in Postwar America.* Unpublished PhD Thesis, Harvard University.

Kaiser, D., ed. 2005. *Pedagogy and the Practice of Science*. Cambridge, MA: MIT Press.

Kaiser, D. 2006. "Whose Mass Is It Anyway? Particle Cosmology and the Objects of Theory". *Social Studies of Science* 36: 533–64.

Kaiser, D. 2007. "When Fields Collide". *Scientific American* 296 (June): 62–9.

Kennefick, D. 2007. *Traveling at the Speed of Thought: Einstein and the Quest for Gravitational Waves*. Princeton, NJ: Princeton University Press.

Klein, O. 1955. "Quantum Theory and Relativity". In *Niels Bohr and the Development of Physics*, edited by W. Pauli, L. Rosenfeld, and V. Weisskopf, 96–117. London: Pergamon Press.

Kragh, H. 2000. "Relativity and Quantum Theory from Sommerfeld to Dirac". *Annalen der Physik (Leipzig)* 9(11–12): 961–74.

Mandelstam, S., D. Bohm, C. Moller, W. Kundt, and A. Lichnerowicz. 1962. "Quantization of the Gravitational Field [and Discussion]". *Proceedings of the Royal Society of London, Series A* 270(1342): 346–53.

Moshinsky, M. 1950. "On the Interactions of Birkhoff's Gravitational Field with the Electromagnetic and Pair Fields". *Physical Review* 80(4): 514–9.

Novello, M., M. Visser, and G. Volovik, eds. 2002. *Artificial Black Holes*. Singapore: World Scientific.

Kallen, G., and W. Pauli. 1955. "On the Mathematical Structure of T. D. Lee's Model of a Renormalizable Field Theory". *Kgl. Danske Vidensk. Selsk. Mat.-Fys. Medd.* 30: 3–23.

Pearce Williams, L. 1975. "Should Philosophers be Allowed to Write History?". *British Journal for the Philosophy of Science* 26: 241–53.

Pinch, T.J. 1992. "Opening Black Boxes: Science, Technology and Society". *Social Studies of Science* 22(3): 487–510.

Pirani, F. 1962. "Survey of Gravitational Radiation Theory". In *Recent Developments in General Relativity*, 89–105. New York: Pergamon Press.

Planck, M. 1899. "Über Irreversible Strahlungsvorgänge". *Sitzungsberichte der Preussische Akademie der Wissenschaften Berlin* 5: 440–80.

Rechenberg, H. 1995. "Quanta and Quantum Mechanics". In *Twentieth Century Physics, Volume I,* edited by L.M. Brown, A. Pais, and B. Pippard, 143–248. Bristol: Institute of Physics Publishing.

Rickles, D.P. (forthcoming-a). *The Development of Quantum Gravity, Volume 1: 1916–1956.* Oxford: Oxford University Press.

Rickles, D.P. (forthcoming-c) The Institute of Field Physics, Inc.

Rickles, D.P. (forthcoming-b). *A Biography of String Theory: From Dual Models to M Theory.* Springer.

Rosenfeld, L. 1966. "Quantum Theory and Gravitation". In *Selected papers of Leon Rosenfeld,* Reprinted in R.S. Cohen and J. Stachel (Eds.), 599–608. Dordrecht: Reidel, 1979.

Rovelli, C. 2000. "The Century of the Incomplete Revolution: Searching for General Relativistic Quantum Field Theory". *Journal of Mathematical Physics* 41(6): 3776–801.

Rowe, D.E. 1989. "Interview with Dirk Jan Struik". *The Mathematical Intelligencer* 11(1): 14–26.

Salisbury, D.C. 2007. Rosenfeld, Bergmann, Dirac and the Invention of Constrained Hamiltonian Dynamics. http://arxiv.org/pdf/physics/0701299.

Salisbury, D.C. 2009. "Leon Rosenfeld and the Challenge of the Vanishing Momentum in Quantum Electrodynamics". *Studies In History and Philosophy of Science Part B: Studies In History and Philosophy of Modern Physics* 40(4): 363–73.

Schild, A. 1948. "Discrete Space-Time and Integral Lorentz Transformations". *Physical Review* 73(4): 414–5.

Scholz, E. 2005. "Local spinor structures in V. Fock's and H. Weyl's work on the Dirac equation (1929)". In *Geometrie au Vingtieme Siecle, 1930–2000*, edited by D. Flament, 284–301. Paris: Hermann.

Smarr, L. 1984. "The Contribution of Bryce DeWitt to Classical General Relativity". In *Ahead of His Time: Bryce S. DeWitt. Essays on the Quantum Theory of Gravity in Honor of his 60th Birthday*, edited by S. Christensen, 1–20. Bristol: Adam Hilger.

Smolin, L. 2006. "The Case for Background Independence". In *The Structural Foundations of Quantum Gravity,* edited by D. Rickles, S. French, and J. Saatsi, 196–239. Oxford: Oxford University Press.

Solomon, J. 1938. "Gravitation et Quanta". *Journal de Physique et de Radium* 9: 479–85.

Stachel, J. 1994. "Scientific Discoveries as Historical Artifacts". In *Trends in the Historiography of Science*, edited by K. Govroglu, J. Christianidis, and E. Nicolaidis, 139–48. Dordrecht: Kluwer Academic Publishers.

Stachel, J. 1998. "The Early History of Quantum Gravity (1916–1940)". In *Black Holes, Gravitational Radiation and the Universe,* edited by B.R. Iver and B. Bhawal, 525–34. Dordrecht: Kluwer Academic Publishers.

Temple, G. 1928. "The Tensorial Form of Dirac's Wave Equations". *Proceedings of the Royal Society of London. Series A* 122(789): 352–7.

Temple, G. 1936. "Gauss's Theorem in General Relativity". *Proceedings of the Royal Society of London. Series A* 154(789): 354–63.

't Hooft, G., and M. Veltman. 1974. "One Loop Divergencies in the Theory of Gravitation". *Ann. Inst. Henri Poincare* 20(1): 69–94.

Wald, R.M. 1986. "Spin-Two Fields and General Covariance". *Physical Review D* 33(12): 3613–25.

Weinberg, S. 1964. "Photons and Gravitons in S-Matrix Theory: Derivation of Charge Conservation and Equality of Gravitational and Inertial Mass". *Physical Review* 135(4B): B1049–B56.

Weinberg, S. 1980. "Conceptual Foundations of the Unified Theory of Weak and Electromagnetic Interactions". *Science* 210: 1212–8.

Weiss, P. 1938a. "On the Hamilton-Jacobi Theory and Quantization of a Dynamical Continuum". *Proceedings of the Royal Society of London* A169(936): 102–19.

Weiss, P. 1938b. "On the Hamilton-Jacobi Theory and Quantization of Generalized Electrodynamics". *Proceedings ofthe Royal Society ofLondon* A169(936): 119–33.

Wigner, E. 1955. "Relativistic Invariance of Quantum Mechanical Equation". In *Funfzig Jahre Relativitatstheorie, Bern, July 11–16, 1955,* edited by A. Mercier and M. Kervaire, 210–26. *Helvetica Physica Acta, Suppl. 4*. Basel: Birkhaüser Verlag.

Wheeler, J. 1957. "Comments after Rosenfeld's Remarks". In *Conference on the Role of Gravitation in Physics*, edited by C. DeWitt, WADC Technical Report 57–216. Wright-Patterson Air Force Base, Ohio.

van Dantzig, D. 1938. "Some Possibilities for the Future Development of the Notions of Space and Time". *Erkenntnis* 7: 142–6.

van Dantzig, D. 1955. "On the Relation Between Geometry and Physics and the Concept of Space-Time". In *Fünfzig Jahre Relativitätstheorie, Bern, July 11–16, 1955*, edited by A. Mercier and M. Kervaire, 48–53. *Helvetica Physica Acta, Suppl. 4*. Basel: Birkhaüser Verlag.

von Meyenn, K. 2001. *Wissenschaftlicher Briefwechsel mit Bohr, Einstein, Heisenberg u.a. Band IV, Teil III: 1955–1956 / Scientific Correspondence with Bohr, Einstein, Heisenberg, a.o. Volume IV, Part III: 1955–1956*. Springer.

Chapter 12
History and Philosophy of Science at Work: Making Regenerative Medicine Research Better

Jane Maienschein

History and Philosophy have found various ways to be friends over the millennia, and so have History of Science and Philosophy of Science. History gives particulars that ground interpretations in stories and make them real. Philosophy gives analysis and connections to general themes that carry us beyond the particulars. Together, they offer perspective that can be more valuable and richer than either alone. Thus, it is hard to disagree with the claim that history and philosophy of science have much to offer each other.

Of course, there is also a tradition of philosophy and history of science pointing in different directions, drawing on different methodologies, and taking different measures of successful investigation. Historians tell stories, recount particulars of the individual episode before them, and often resist any efforts at generalization. What matters is people and places. Furthermore, in recent decades history of science as a field has been dominated increasingly by cultural history, where it is the context and culture that matter more than the science itself. Mere "internalists" who concentrate on the logic and methodology of science have been reviled. Meanwhile, philosophers have sought just the generalizations that some historians have rejected. They have examined just the internal logic and reasoning that some historians have eschewed in favor of contextualization.

Notwithstanding the tension that has appeared in diverse ways, some historians and philosophers have remained friends and have worked hard to overcome tensions and to draw on different methodologies and different values to achieve deeper and richer understanding of the nature and context of scientific practices. Despite disciplinary differences, historians and philosophers meet together in their annual professional meetings. An energetic group has organized a series of workshops on &HPS to integrate history and philosophy of science and promote the synergies. Collaboration and communication can work, and I offer a case study in favor of that claim.

Here I take a particular example and offer it as evidence in favor of the stronger claim that, in drawing on both history of science and philosophy of science together,

J. Maienschein (✉)
Arizona State University, Tempe, AZ, USA
e-mail: maienschein@asu.edu

S. Mauskopf, T. Schmaltz (eds.), *Integrating History and Philosophy of Science*, Boston Studies in the Philosophy of Science 263, DOI 10.1007/978-94-007-1745-9_12,

it is possible to make science better. Making this case requires understanding what it means to assert that the science is better as well as how we can know. I do not pretend to have an argument completely worked out for this claim. Yet I propose that the particular example of regenerative medical research allows a strong demonstration that at the very least goes far toward making such an argument.

The discussion starts with a description of regenerative medical research today, and its context of claims that this is an exceptionally productive research field ripe for translation from bench to bedside and that it ought to be pursued energetically and with significant public investment. I discuss what is meant by regenerative research, by NIH's translational imperative, and by the political and ethical as well as scientific contexts in which this research exists.

The second section asks about the philosophical analysis of underlying assumptions of regenerative research, including metaphysical assumptions about what is being regenerated and epistemological assumptions about how we know what works. This section brings us discussion of form and function, preformation and epigenesis, determinism and adaptation, and wholes and parts. This philosophical analysis uncovers general themes and shows how they are playing out in this particular case. Philosophical inquiry brings analysis and reflection to the often and perhaps necessarily over-enthusiastic presentation of the research. As a result, we see that researchers are making several assumptions that are limiting the scope of their research and have pointed the research in limited directions while missing others that might be at least as productive.

Third comes an exploration of the lines of research that have led to regenerative research today. This historical tracing of particulars shows a number of research questions and approaches that were set aside, not understood, or otherwise ignored or lost to current researchers. There is value in recovering them. We can learn from examining paths not taken or ways of working long cast off. And we can gain a much wider and richer picture of the research today by placing it in historical context.

Finally comes the section addressing the "so what" questions about any project: so what if scientists today ignore history and philosophy? So what if they could learn more through this study; do they really need to? Well, no, it is certainly not necessary that each scientist study history and/or philosophy. But I argue that history and philosophy can nonetheless make science better as a whole. The task here is to show what such a claim might mean and in what sense it might be true.

12.1 Regenerative Research: The Science in Context

"Regenerative medicine" covers a diversity of research approaches, but the term has been shaped by the National Institutes of Health and leading institutes in the U.S. and elsewhere. The Whitehead Institute in Cambridge, Massachusetts, gives a widely-shared definition: "Regenerative medicine: Seeks to understand how and why stem cells, whether derived from human embryos or adult tissues, are able to develop into specialized tissues, and seeks to harness this potential for tissue-replacement therapies that will restore lost function in damaged organs" (see Whitehead 2008). Irving Weissman's Institute for Stem Cell Biology and

Regenerative Medicine has a similar focus, so that his team feels that "The Stanford Stem Cell Biology and Regenerative Medicine Institute is at the forefront of a groundbreaking approach to biomedical research and patient care. This approach aims to harness the power of stem cells—master cells from which all specialized cells and tissues in our bodies are derived—to target and remedy the root causes of today's most devastating diseases" (Stanford 2008). In fact, the entire California Institute for Regenerative Medicine makes clear that their mandate is to promote and develop stem cell research (CIRM 2008) .

All these institutes, and nearly all of the many, many others with those words in their names, emphasize the goal of drawing on stem cell technologies to restore function for clinical purposes. A few also mention regeneration of structures, in order to recover lost function, but the focus remains on function and is nearly always stated in terms of clinical application, even when the research being carried out includes basic developmental biology and related fields.

Amidst the vast number of research publications and polemics available, two summary sources are particularly helpful in providing insight into how researchers see the field developing. The first is the NIH report "Regenerative Medicine 2006." This 65-page compilation of six chapters summarizes work on embryonic stem cells, bone marrow stem cells, nervous system repair with stem cells, genetically modified stem cell experimental therapies, and intellectual property issues surrounding stem cell research. A 2007 addition entitled "Mending a Broken Heart" addresses stem cells in cardiac repair. Each chapter offers an overview of the dominant research areas and the implications for regenerative research agenda, which very much reflects NIH priorities at the time (NIH 2006).

Notice that the report is entitled "Regenerative Medicine" and that every chapter concerns stem cell research. It follows the 2001 report on "Stem Cells and the Future of Regenerative Medicine" (NIH 2001) and reflects NIH packaging of stem cell research in terms of regeneration and the other way around. We see a similar pattern in California's Proposition 71 and its implementation. Regenerative medicine can surely be more than stem cell research, but they have become linked and nearly synonymous for some purposes and contexts. It is worth understanding why. Similarly, stem cell research is about far more than just regenerating function but has gotten packaged as applied regenerative research for clinical purposes. Again, it is worth understanding why. And it is not enough to say knowingly, "Ah, it's political." We need a better sense of the research and the political climate including a look at NIH and research funding, at public expectations of publicly-funded science and the scientific community's expectations for their research, and of the stem cell research and regenerative applications actually being carried out. Let us take each in turn.

12.2 NIH Mandate

The National Institute (at first just singular) of Health began in 1930 with the Ransdell Act. A mix of advocates argued that the U.S. Congress needed to take responsibility and fund research leading to health improvements. They began targeting specific diseases, which led to multiple institutes, each for a favorite disease that has gained sufficient advocacy (see Starr 1982).

The NIH mission has remained focused on health. Since 2002 it has also emphasized translational research, dedicated to "translating" scientific research carried out at the laboratory benchside into clinical applications at the bedside. In response to political pressure to make the applications more quickly and more visibly, the new NIH Director Elias Zerhouni developed a "roadmap" to facilitate translation (NIH 2008; see Maienschein et al. 2008). The agency also began funding Clinical and Translational Science Centers (CTSCs) and a number of targeted major projects.

Yet the precise expectations and interpretations remain unclear. As Declan Butler suggests: "Ask ten people what translational research means and you're likely to get ten different answers." Butler points to the new rhetoric as beginning with the first appearance of the term "translational research" in PubMed in 1993, following research on the BRCA1 and other cancer genes (Butler 2008, 841).

The push for translation arose from opportunity as the genome project produced new knowledge apparently available for application and public demand for health results. Zerhouni felt that "There was a widening gap between basic and clinical research" that needed to be addressed. An unsigned editorial in *Nature* agrees that this is still true, acknowledging that "Some researchers complain that an emphasis on translation swings the pendulum too far towards applied science at the expense of basic research, but this concern has little foundation. In fact, what is worrying is the extent to which biomedicine in the past few decades has swung so far toward pure science" (Editor 2008).

Stem cell research, with its public promises of significant clinical applicability, has become a poster child for translational research. The slogan "regenerative medicine" works well for public interests, NIH translational needs, and a growing research community's interests. This is particularly ironic, since the specific research area that many consider most promising uses human embryonic stem cells and is currently limited by President George W. Bush's Executive Order on August 9, 2001 restricting use of federal funds for just that research (White House 2001). Regenerative medicine, translational imperative, and stem cell research are all tied together by the accidents of history and politics (see Maienschein et al. 2008).

California gives us a site where the intersection has played out most forcefully and to greatest immediate effect. Proposition 71, passed by the state's voters in November 2004 and signed into law by a supportive Republican Governor Arnold Schwarzenegger, led to the establishment of the California Institute for Regenerative Medicine (CIRM). The proposition emphasized stem cell research in particular. The campaign was well documented, especially by the *Washington Post, New York Times, Science* and *Nature*, and brought a parade of Nobel Prize winners and Hollywood celebrities to endorse the call for publicly-funded stem cell research. The packaging in newspaper and television ads emphasized the treatments that would result and directly linked stem cell research to predicted clinical results. Bills in the U.S. Congress were first oriented toward opposing such research and especially cloning, but shifted to supporting stem cell research by 2005. The Stem

Cell Research Enhancement Acts explicitly to allow such research did pass in 2006 and 2007, but President Bush vetoed—the first veto of his presidency.

The stage was set for a program of regenerative medical research in the twenty-first century, with broad support from the public and the scientific community. Such research is being done energetically in California and some other states, by privately funded research institutes, and especially in other countries that have not had the same ethical and political debates as the U.S. Many researchers report that a rising percentage of publications of stem cell research are coming from countries other than the U.S., including some that have invested heavily in this area (Owens-Smith and McCormick 2006).

12.3 Expectations of Science

It is worth a brief reflection on the social contract concerning science and the implications for regenerative medicine. There is little point in rehearsing the well-worn paths of bioethics and policy debates, but there are other relevant factors shaping the scientific context and therefore the science. One of these concerns what the public expects of science. Clearly when the NIH was established, Congress and the public supporters expected research and results that would improve health and cure disease. The National Science Foundation was established in 1950, with the mission "to ensure that the United States maintains leadership in scientific discovery and the development of new technologies." As Vannevar Bush proposed, NSF would pursue new knowledge that would lead to useful results someday in some way, while the NIH was expected to focus on solving disease problems (Bush 1945).

The current push for translation and particularly for regenerative medicine puts a greater emphasis on outcomes—often particular defined outcomes and preferably achieved quickly. The public voting for the California initiative wanted not some vague promises or future applications, but to cure Parkinson's disease and diabetes, among others. They wanted researchers to engineer stem cells to regenerate particular lost functions. A clearly desirable goal, this strong direction for research may nonetheless not suit the way scientific research institutions work best and it may distort the types of research done. While this distortion may be a good thing in some ways, it clearly shifts the emphasis and ways of working.

In the short run, the focus on results of particular kinds benefits established researchers already working on stem cell science. Yet many of these researchers are excellent developmental biologists with strong track records who were working on other problems. It remains to be seen whether, how, and to what extent the current demand for results of particular kinds impacts the research enterprise beyond adding lots of funding to new directions. It also remains to be seen how the public reacts when it becomes clear that researchers are not able to deliver on all the promises made during political campaigns. The scientific community is surely already making wonderful discoveries, but just as surely many will be surprising and not the outcomes that had been predicted. We may not figure out ways to get stem cells to produce dopamine neurons in the brain to repair Parkinson's losses, for example, but we may be able to engineer cells to prevent or control the disease in other ways before it does the damage. Researchers—and the entire research network—are likely to deliver on the loosely understood social contract that public investment will produce some clinical results. Those just may not be the particular results that the public campaign emphasized and that the public supporters imagined.

12.4 Stem Cell Research and Regenerative Applications

So where are we with regenerative medicine? The NIH website does a good job of updating research statutes and pointing to work from within and outside the NIH. The individual institutes do an excellent job of presenting their research programs and the results that have been published. Not surprisingly, they do less presentation of research in progress or research results that have likely significant proprietary value that they (and their funders) want to protect.

The 2006 NIH report provides a useful survey, and it is worth noting the precise message delivered and the particular language chosen. Like most discussions of regenerative medicine, it starts with Prometheus. Chained to a rock, this mythical Greek spends every day with an eagle eating his liver. Yet every night the liver regenerates, "enabling him to survive." Furthermore, as the introduction to the NIH report puts it, "The scientific researchers and medical doctors of today hope to make the legendary concept of regeneration into reality by developing therapies to restore lost, damaged, or aging cells and tissues in the human body" (NIH 2006, i). This interpretation of the myth is instructive. Usually, the tale is presented in terms of the punishment of Prometheus for his having given fire to humans. As a result, he must endure having his liver eaten again and again, suffering for his act. Yet here, we get an uplifting tale: look, livers can regenerate, and today's researchers can help make this happen. Isn't this great! The interpretation tells us much about the optimistic expectations for regenerative medicine even when, as the report makes clear, very few clinical applications have yet been established.

The first full chapter of the report on regenerative medicine is entitled "Embryonic Stem Cells" and spells out what they are, what they can do, how they are most effectively cultured (growing best in media that have proven problematic for human clinical use), why pluripotency is so desirable but potentially problematic, and possibilities for genetic manipulation. This sets the stage for more clinically applied work.

Chapter 2 discusses hematopoietic stem cells from bone marrow. The essay points to the post WWII attempts to restore blood supplies to patients with leukemia and other diseases resulting from irradiation. Hematopoietic stem cells are the only stem cells known to provide consistent stem cell therapy, having been used and their efficacy proven since 1959 as clinical transplantation. This is a case of taking cells that are already somewhat differentiated and known to give rise reliably to blood cells and transplanting them to a patient. Translation starts with transplantation in this case, and the history of transplantation research from developmental biology interests with the clinical applications in ways to which we will return. This is also a case of transplanting "adult" stem cells, meaning cells taken from older-than-embryonic stages. This chapter, by Jos Domen, Amy Wagers, and Irving L. Weissman, concludes with a realistic assessment of the challenges as well as hopes that "After more than 50 years of research and clinical use, hematopoietic stem cells have become the best-studied stem cells and, more importantly, hematopoietic stem cells have seen widespread clinical use. Yet the study of HSCs remains active and continues to advance very rapidly. Fueled by new basic research and clinical discoveries, HSCs hold promise for such purposes as treating autoimmunity, generating tolerance for solid organ transplants, and cancer therapy. However, many challenges remain" (Domen et al. 2006, 28). The 180 papers cited give ample indication of activity in this research field.

Chapter 3, by David Panchision, looks at research on nervous system repair. Quite a number of diseases and conditions would benefit from regeneration of neural function. Until the 1990s, it was generally assumed that nerve cells stop developing in adults, so that the best hope was in limiting damage or retaining existing neural networks. Then it became clear that at least some neurons differentiate in adults, perhaps from residual stem cells. "These findings are exciting because they suggest that the brain may contain a built-in mechanism to repair itself. Unfortunately, these new neurons are only generated in a few sites in the brain and turn into only a few specialized types of nerve cells. Although there are many different neuronal cell types in the brain, we are now optimistic that these new neurons can 'plug in' correctly to assist brain function." These findings give increased hopes for getting cells to do what is needed for degenerative diseases. "For this reason, a huge effort is underway to develop new treatments, including growth factors that help the remaining dopamine neurons survive and transplantation procedures to replace those that have died." But we should not do just any research at any cost. Rather, "it is the current task of scientists to bring these methods from the laboratory bench to the clinic in a scientifically sound and ethically acceptable fashion" (Panchision 2006, 35, 37, 42).

Chapter 4 looks at gene therapy as related to stem cell research, especially using stem cells as a vehicle for such genetic manipulation. Chapter 5 addresses Intellectual Property issues, especially given the international nature of the research and the complexities of funding.

Chapter 6 was added in 2007 and looks at cardiac repair. Though diseases gain great attention with poster cases like Parkinson's Michael J. Fox or spinal cord injury's Christopher Reeve, or with poignant stories such as the degeneration of Alzheimer's or the failed insulin function of juvenile diabetes, in fact heart disease is the most common in the U.S. and many other countries. Cardiac disease is the number one cause of death in the U.S. and apparently has been in every year starting in 1900 with the exception of 1918, where influenza surpassed it.

Researchers are exploring diverse ways to repair heart muscle cells, and so far heart transplantations have been the most successful of regenerative approaches—in the few cases where hearts are available for transplant and not rejected. Some trials with transplanted cells, including myocardial progenitor stem cells, seemed to lead to differentiation into heart cells of several types. Yet it now seems more likely that "transplanted stem cells release growth factors and other molecules that promote blood vessel formation (angiogenesis) or stimulate ‘resident’ cardiac stem cells to repair damage. Additional mechanisms for stem-cell mediated heart repair, including strengthening of the post-infarct scar and the fusion of donor cells with host cardiomyocytes, have also been proposed” (Goldthwaite 2006, 58).

Whatever the mechanism, a major remaining challenge is getting the cells delivered to the functional site, and another is timing the cell delivery effectively. For cardiac repair, experimentation on multipotent or progenitor adult stem cells may hold at least as much promise as with pluripotent embryonic stem cells. Indeed, using such adult cells may avoid some of the risks of teratoma formation and other problems of undifferentiated pluripotent cells that have too much potency. So while such research holds great promise, “the use of these cells in this setting is currently in its infancy—much remains to be learned about the mechanisms by which stem cells repair and regenerate myocardium, the optimal cell types and modes of their delivery, and the safety issues that will accompany their use. As the results of large-scale clinical trials become available, researchers will begin to identify ways to standardize and optimize the use of these cells, thereby providing clinicians with powerful tools to mend a broken heart” (63).

What we learn from the report as a whole is that all of stem cell research is just beginning. While we have a half century of experience with hematopoietic stem cell transplants, decades of study of mouse stem cells, and several decades of experience with a select handful of organ transplants, we have actually made tremendous progress in understanding more and more details of developmental biology but not much progress with clinical applications—not yet. And what we have learned has often challenged or contradicted previous assumptions, as we will consider in the next sections.

Another more recent set of publications appeared May 15, 2008 in *Nature*. Intended for an audience of researchers, these reports are more technical and detailed, but they also show that progress is occurring quickly on many fronts. One area of considerable promise concerns induced pluripotent stem (iPS) cell lines. Here we see changing assumptions. Early stem cell research focused on being able to take embryonic stem cells from blastomeres, because these were the ones that exhibited pluripotency, and culture them in such ways that the particular culture medium determined what kinds of cells they become. But considerable study of developmental processes and how growth factors shape differentiation has begun to show how already differentiated cells can be re-differentiated and even act (at least as far as researchers can tell) very much (though not precisely) like pluripotent embryonic stem cells. This is tremendously exciting research because it may reduce the need for embryonic cells and it shows a great deal about the complex of factors that shape development (Zon 2008, 311).

What had been called “cell fate” and “determination” has now been joined by ideas of “differentiation,” “de-differentiation,” “redifferentiation,” and “reprogramming.” Development is once again an exciting dynamic process, as it was around the early twentieth century, rather than a matter of playing out inherited deterministic preformationist programs. This trend is good for biology.

Language matters in all this flux of discussion—and especially as researchers and different areas at the bench, the bedside, and in the public try to communicate effectively and reliably. It is hard, for example, to get away from the idea of cells as being programmed. Senior editor Natalie DeWitt, in introducing the section on regenerative medicine, starts her column by invoking regeneration: "Although some of our cells have the innate ability to replenish themselves—and, by doing so, to repair ageing and injured tissues and organs—most of the body's cells form the specialized cell type they are destined for and then go into lock down." She then moves to "the field of programming" which she locates as beginning with John Gurdon's work on frog cloning by nuclear transplantation from early somatic cells. Then a paragraph on Prometheus and creation as scientists discover "how to create new sources of such cells in a Petri dish." The last paragraph is virtually a statement of the NIH translational mission: "The articles in this Insight explore the promises and challenges of the next era of regenerative medicine—and how to use the information gained from the study of model organisms and cell culture to eventually heal" (DeWitt 2008).

In the fifty rich pages that follow, we learn about such research as cell therapies, molecular pathways, variability even within cell lines, specific genetic factors and knock out-knock in technologies, and what is meant by self-renewal of embryonic cells. Some of the work thought to be relevant is about regeneration, other research is about generation gone wrong with production of teratomas and immune system reactions/rejections. We learn about successes with adult cells that are not de-differentiated or re-differentiated by rather caused to differentiate in ways other than expected. To move to clinical successes, however, we will need to establish definitively both that the cells targeted are actually causing the effects claimed and that they do so in stable and predictable ways.

We see a diversity of approaches that involve basic developmental biology carried out in the lab. Researchers have to get stem cells, isolate and culture them, then make them do what is wanted, sometimes with genetic modifications. Transcription and growth factors are critical to facilitate differentiation of the "right" sort. Researchers internationally are busily studying all aspects of these processes, in humans, mice, and other organisms. And the NIH provides a valuable summary of current research at http://stemcells.nih.gov/research/current.asp, while other countries and institutes provide their own summaries.

The science in these papers is tremendously exciting, both for the promises of possible clinical applications and for its direct emphasis on what we learn about development and the basic research before clinical applications are in sight. Assumptions are shifting, and researchers are acknowledging that differentiation is much more complex and fascinating than the impression given in the public debates about California's research initiative or proposed Congressional legislation.

12.5 Philosophical Analysis

Given the burgeoning body of research and ambitions for medical applications, what can we learn by bringing the tools of philosophical analysis to bear? For our purposes here, I will set aside the vast bioethics discussion of stem cell research and its social and policy contexts. With some notable exceptions, this discussion has been rather myopically focused on a few standard issues and has started with basic assumptions that are highly contested. Unfortunately, only a few of those eager to enter debates about stem cell research have made serious attempts to understand deeply the science involved. Instead, they have latched on to well-worn issues of personhood, identity and autonomy, and worries about whether people should be allowed to donate or even sell eggs or embryos for research or whether that constitutes exploitation. Fortunately, a few scholars have taken up more challenging and new issues, such as the moral and scientific status of chimeras or the implications of induced Pluripotent Stem Cell research (Robert and Baylis 2003; Robert 2004a; Robert 2006).

Let us begin here by recognizing that stem cell research already exists—on embryonic stem cells from human blastocysts and a range of stem cells from other sources. Let us acknowledge that the work is both possible and that some researchers are already experimenting with such things as chimeras made up of cells from more than one individual and even more than one species. Whatever the moral status, the science of regenerative medicine is underway in the U.S. and perhaps especially elsewhere. We are already repairing and replacing cells and tissues in a diversity of ways. Let us then turn to analyzing the work and let others debate the ethics and policy and make the laws.

Philosophical analysis is useful because it allows us to ask some of the questions that researchers are not asking. In some cases, it just has not occurred to them to ask and in other cases the underlying assumptions are so strong that the answers seem clear. Let us focus on four sets of questions to get at different parts of stem cell research. In each case, I will examine the driving questions and assumptions, and will also challenge existing views where appropriate. Key areas of interest focus around:

1. Metaphysical issues. What are stem cells, and how do they work? Is there, for example, such a thing as "stemness" and if so what is it and what does it do? What do toti-, pluri-, multi-, and uni-potency and progenitor status mean? This leads to other questions, such as whether if stem cells are the sources of new cells in the body, then does manipulating stem cells change the autonomy or identity of that body?
2. What does "regenerative" really mean? Does something become regenerative because it actually regenerates—and in what sense? Regeneration of the same part, of the same function, or of some replacement function that "works" even if in a different way? Does regenerative medicine involve repair of structure or function—or replacement with some others? And how—through genetic engineering, injection, transplantation? By causing new differentiation of something previously undifferentiated, or de-differentiated and re-differentiated? Is programming involved—necessarily?
3. Epistemologically, what counts as an explanation of the regenerative phenomena? Does regenerative success result from the presence of particular genes, transcription factors, function, assumptions about stemness, or what? How do we demonstrate/confirm such claims?

4. And how can we develop new knowledge when we cannot directly observe the regenerative processes? What role do assumptions about model systems play, or about the behavior of cells as they are necessarily transplanted from in vitro to in vivo settings?

In the end, we see that the research community is making a number of assumptions that may be wrong. It matters because putting wrong cells in the wrong places or in such a way that they start to do wrong things could be doing degenerative rather than regenerative medicine. Obviously that would be problematic. In more detail, then:

1. What are stem cells and how do they work? Immediately after James Thomson and John Gearhart brought human stem cells to public attention in 1998, numerous versions of definitions arose. Stem cells are those that both remain undifferentiated and also retain their capacity for self-renewal. Stem cells are never totipotent, meaning that they never have the capacity to become an entire organism. They can be—as embryonic stem cells taken from blastocysts distinctly are—pluripotent, meaning that they have the capacity to become any kind of cell. At least that's the assumption.

In fact, there is not any way to prove absolutely that any given cell or even group of cells has this capacity to become any from among all the possible kinds of cells. Here is one assumption already worth uncovering and examining. Perhaps there are factors making some stem cells from the same cell line pluripotent and others only multipotent, that is capable of becoming any of several but not any from among all the types of cells. Perhaps all the cells in the same cell line cultures in the same medium and derived from the same blastocyst are not, in fact, the same. This would be very valuable to know, in which case more detailed studies of the nature, causes, and effects of the diversity would be potentially useful. In fact, there is growing evidence of such differences.

Multipotent or unipotent stem cells are self defined, and progenitor stem cells seem equally so, in that they are apparently destined to become a particular kind of cell for which they are the progenitors. But what makes a cell determined enough to count as a progenitor but not differentiated enough to count as a whatever-type-it-is-cell already? Considerable work is being done on developmental regulators and factors allowing self-renewal or guiding differentiation (for example, see Zon 2008, 308).

Early assumptions still very much adopted a traditional view of development that the arrow of differentiation goes in only one direction, and that genetic control (or programming) with some input from environmental signals shapes the nature and tempo of the differentiation process. Recent accumulating evidence challenges that assumption and suggests that "reprogramming" is not just a rarity brought about by such interventions as cloning technology. Such reprogramming, de-differentiation or already differentiated cells, and re-differentiation based on new conditions seem to happen much more commonly than thought until quite recently (Robert et al. 2006).

Philosophers would find it easy to say something like "well, we could have told you scientists to question your deterministic assumptions." Hindsight is easy. What do we have to add now? First, what do we gain—and lose—by holding so tightly to the programming metaphor? And also what is gained—and lost—by imagining a developmental arrow in only one direction, or by thinking in terms of single cells and their particular environments rather than more complex systems?

Perhaps normal development involves something that acts functionally like programming, with information sources captured somewhere in the DNA and/or the material biochemical structure of the embryo and subsequent developmental stages. Perhaps when cells are taken out of context, whether when pluripotent embryonic stem cells are removed from a blastocyst or adult stem cells from bone marrow, perhaps their entire functioning is reset. Perhaps they are no longer "programmed" at all (if they ever really were). Perhaps we should expect rather than be surprised if cells that apparently are the same behave differently, because just perhaps they are not really programmed in any very deterministic way. Perhaps they are responding much more to environmental cues or to signaling among cells and the surrounding medium than we thought possible. Furthermore, perhaps the arrows of differentiation flow both ways—or many ways, with more and less differentiation at different points in the cycles of cell division (as seems likely) or different densities of cells—but in deterministic ways. Perhaps individual cells have "minds" of their own and can "choose" different behavior and developmental pathways based on random choice, availability of necessary growth factors, or relationships with neighbors.

It could be quite useful to take up theoretical developmental biology that draws on new metaphors, explores new ways of thinking about the "social" interactions among cells, and looks beyond genetic transcription factors to include other environmental conditions. For example, Jason Scott Robert has articulated a vision of creativity in development that is directly on point here (Robert 2004b). He theorizes that from a single-cell stage, organisms adopt, construct, process, and regulate developmental resources of various sorts dispersed throughout the organism and its environment. Accordingly, development is a semi-autonomous, creative, self-constitutive process engaged in by the developing organism. Perhaps past evolutionary adaptations to changing conditions are relevant. Scott Gilbert's call for a robust ecological evolutionary developmental biology (or eco-evo-devo) sounds compelling here. Such dynamics systems approach might prove informative for explaining why hematopoietic transplants work something and not others, or why some cells become self-replicating and cancerous in some contexts and not others (Gilbert 2001).

One question remains that philosophers like to worry about and biologists usually do not, but that might matter here. What if a patient develops diabetes and we replace the function that produces insulin; then he develops leukemia and we replace the bone marrow and hence introduce new hematopoietic stem cells; then he develops cardiac disease and we replace multiple kinds of heart muscle cells or even the heart itself. Then brain cells with dopamine-producing neurons to control Parkinson's, and so on. As with the philosophers' concern about Aristotle's ship in which each part is replaced with new parts, we ask about identity: is it the same person after all those changes? What if the stem cell parts come from other individuals and carry different genetic materials, and hence make the resulting individual a genetic chimera: does that matter? Is this the same person? What if we genetically engineered stem cells: would that make a difference?

Is there a point at which regenerative medicine goes so far that it is more generative of something new and different than regenerative of something that already exists and just needed repair? If there is such a point, where is it—and how do we know? And does it matter? Such questions hold philosophic interest, certainly, but they also raise the serious possibility that perhaps not all medical intervention is desirable for practical as well as moral reasons. If we change enough, with or without genetic engineering, do we make the new whole unable to function "properly"?—or in a way that we consider a successful medical result? This is not just an ideal abstract and not just an ethical question.

2. Related to the last concern is the broader question what we mean by "regenerative." The term suggests regeneration, which suggests re-generating of something. That is one interpretation, and some of the stem cell research is oriented toward the goal of regenerating lost function (getting cells in the right place to do the right thing) and in some cases lost function and structure. Regenerative medicine includes more than this, though, as with hematopoietic stem cells that may produce new blood cells but in different combinations than the original and may get more or different capacities than they had before. Or even further there are examples of repair of damaged cells with something different—skin that is scarred but covers the wound, for example. Or engineered prostheses to replace function in different ways than the original, such as a wheelchair or pacemaker in the heart or shunts to reduce pressure in the brain. Transplantation, starting with heart, skin, and bone marrow transplants, have had the longest history of success but also limitations that have proven instructive. It remains to be seen how other forms of replacement, repair, or regeneration will work.

3. Epistemologically, what counts as an explanation of regenerative phenomena, and how can we demonstrate this? This is a very difficult question that has few definitive answers as yet, though it is becoming clear that a mix of genes, transcription factors, growth factors, environmental stimuli, and interactions with other cells are all relevant. All contribute to causing differentiation to go as it does. And it requires a complex interactive systems theory to explain how and why the various causal factors interact (Robert et al. 2006).

One challenge is testing a theory. Even where it is clear that particular factors like presence of a transcription factor are associated with an effect, and even where the effect did not occur before and does now, this does not give a very robust causal explanation. Yes, that factor may have been a necessary but probably not a sufficient condition. Causation is often difficult to demonstrate, of course, but this is a case where each cell line is different and perhaps even individual cells within the same lines are different because of interactions with the other cells. To some extent, cultures of exactly defined conditions yield similar-seeming cells. But when they are transplanted into different environments, it is very difficult to establish which factors made a difference.

4. Part of the problem is inability to observe directly the regenerative process. We can see cells in a dish, but only indirectly the results when they are transplanted to a new site to produce regenerative results. And in the new setting, cardiac cells may seem to produce beats like myocardium is supposed to—but how can we tell? How can we tell epistemologically? And how can we even do the research in humans that would require transplanting cells into patients when we have no way of determining the safety or efficacy of such a transplantation before we try it? We can test in animal models, as usual in medicine, but one of the things we know clearly is that stem cells develop differently in different environments (which is, after all, why we value them—we believe that we can cause them to differentiate in defined desired ways by culturing them in particular ways). Therefore, we should expect that these cells in animal models will precisely NOT behave as they would in humans. The animals therefore do NOT serve as very useful models for purposes of stem cell research (Robert 2004a, b; see Robert et al. 2006).

We are left with ethical concerns about the extent to which we are comfortable creating chimeras of human and animal cells, but equally challenging are the epistemological barriers. Researchers can begin to address them, but only by carrying out a great deal of fundamental work in developmental, cell, and molecular biology. Much of this work holds little immediate promise of translation. And much will require exploring creative new theoretical approaches to complex systems and getting well beyond metaphors and deeply-engrained assumptions about programming and uni-directional differentiation.

12.6 History of Regeneration Research: Tangled Threads

Regenerative medicine has grown out of many different lines of research, addressing different questions with different methods and approaches and with different goals in mind. Alejandro Sanchez-Alvarado provides a nice introductory lecture from the perspective of a current leading research on regeneration (Sanchez-Alvarado 2008). I likewise find a rich set of traditions informing research today. It is worth looking at some of those (see Maienschein 2009, for more detailed discussion).

Arguably, the first true stem cell research in the modern sense was Ross Granville Harrison's experiment on nerve fiber development (Harrison 1907). Harrison's particular interest was in nerve development. He wanted to show how nerve fibers grow and hypothesized that they reach out by protoplasmic outgrowth. He denied the alternative popular view that organisms develop their nervous systems because of preformed bridges. For Harrison, arguing and pointing to more and more beautiful silver nitrate solutions like those of Camillo Golgi or Santiago Ramon y Cajal would not settle the question.

Harrison sought to do that definitive settling of the question with a crucial experiment. He cut neuroblast cells (neural stem cells) out of the frog and placed them into an artificial culture medium, in this case a hanging drop of frog lymph. Out grew a beautiful nerve fiber, just as in the case of normally developing frog nerve fibers. The experiment had answered the question about development, had produced the first successful tissue culture ever, and had (in retrospect) demonstrated the capacity of neural stem cells to differentiate in an artificial medium. The work began while Harrison was at the Johns Hopkins University and continued after he moved to Yale in 1907. There, under the influence of bacteriologists, he refined the methods and achieved even greater success with his cultures (Maienschein 1991).

Harrison moved on to other questions about embryology and left it to others such as Alexis Carrel at the Rockefeller Institute to carry tissue culture to new applications and to clinical translations (Landecker 2007; Stapleton 2004). Meanwhile, other researchers looked at the biological phenomena of regeneration. Thomas Hunt Morgan was one of those, whose 1901 book *Regeneration* provided a summary of centuries of work to that point. Morgan's 316 pages described the experimental and theoretical studies of regeneration, beginning with a retrospective review of earlier studies on a diversity of organisms such as hydra, worms, frogs, and planarians. The book grew out of a series of lectures presented at Columbia University, where Morgan was a faculty member. And they drew on his earliest reflections presented in a lecture at the Marine Biological Laboratory in Woods Hole (Sunderland 2007; Morgan 1901).

Morgan had come to see regeneration as a way to understand normal development but also reflecting special circumstances and in some cases special capacities. Regeneration was, in many ways, a problem of growth and replacement of material form that took place with the guidance of "tensions" within the organism itself.

Lewis Wolpert analyzes Morgan's "ambivalence" reflected in his tensions hypothesis and notes that at times Morgan's interpretation seems very like the gradient theory of Charles Manning Child, which Wolpert himself favors. At times, Morgan seemed to embrace gradients and rely on them to explain polarity and other phenomena. Yet at other times, Morgan reverted to alternative accounts. As Wolpert notes, Child's own views are very difficult to understand at times, so it is perhaps not surprising that Morgan and his contemporaries largely ignored Child and his extensive studies of planarians and gradients (Wolpert 1991; Mitman and Fausto-Sterling 1992). At that time, neither gradients nor tensions provided much of a tractable research program for getting at how either gradients or tensions actually work or what causes them. Saying that gradients cause regeneration but that we have no idea how gradients are established does not get us very far, as Morgan explained.

Another important line of research that led to what was considered an embryological "gold rush" came in the 1920s and concerned "induction" and the role of what Hans Spemann identified as the "organizer." Could it be that whatever chemical forces and factors cause induction are causing differentiation and therefore generation? This was an obvious question, but the researchers most focused on studying induction saw normal development and differentiation as the real prize. Regeneration was a curious phenomenon, yes, but it was not clear how planarians or earthworms or hydra generating missing parts of themselves would reveal much about normal development. Re-generation might well be different processes or depend on different causes than induction and generation in normal development.

One line of research pursued especially by Harrison, Spemann, and their many students involved transplanting all sorts of frog parts: limb buds and eye vesicles, and also bits of "organizer" material from the dorsal lip of the blastopore. Or they tried other non-frog and non-organic materials to see which ones induced differentiation and how. Still others transplanted nuclei into egg cells, leading to Robert Briggs' and Thomas King's cloning of frog's eggs in 1951 and John Gurdon's later demonstration that even later developmental stages could be cloned. Transplantation and cloning raised serious questions about the capacities and limits of cells to become dedifferentiated, redifferentiated, and differently differentiated, for example. Underlying assumptions about what was possible and not possible kept researchers from pushing further in the direction of cloning later stage or somatic cells, and it is intriguing to reflect on why those assumptions were made (see Maienschein 2003).

After spending decades studying genetics in fruitflies and winning a Nobel Prize, Morgan returned to regeneration as a foundational problem in embryology. It became clear that he had never really given up his fascination with regeneration. In *Embryology and Genetics* in 1934, he noted there are really two different forms of regeneration. "In one the new structure develops by a remodeling of the old materials; in the other the new structures are formed out of new materials that are derived from the old part" (Morgan 1934, 164). Morgan was well aware that "location" may be everything, but that merely invoking it explains nothing. Perhaps the new cells contact with other cells and there is some physical or chemical influence, or perhaps the old cells act as "determiners" or "organizers" for the new cells.

Others have taken a more theoretical rather than experimental approach. Paul Weiss, for example, asserted in his 1939 *Principles of Development* that "Regeneration is the repair by growth and differentiation of damage suffered by an organism" and that the processes "are fundamentally of the same nature and follow the same principles as the ontogenetic processes" (Weiss 1939, 458). Some cases involve structural organ or cell repair, other cases bring physiological repair of function. As Donna Haraway has persuasively shown, Weiss turned to the concept of "fields" as a physical way of producing pattern (Haraway 1976, chapter 5). Like Morgan, Weiss concluded that "Here the basic problem of development—how parts which have not been so before become different from one another—rises again in its full important, and if it were for no other reasons, an *epigenetic* view of development would have to be postulated on the strength of regeneration phenomena alone" (Weiss 1939, 478).

Others began to seek chemical explanations for differentiation and development. Early in the last century, Jacques Loeb invoked what he called the "mysterious Fernwirkung." By mid century Joseph Needham summarized a number of such approaches in his 1942 edition of his *Biochemistry and Morphogenesis.* After reminding readers of the variety of studies on regenerating body parts and tissues, he places the discussion in the context of problems of determination, differentiation, cell competence, the power of "organ districts" (areas giving rise to particular organs), renewed cell pluripotency, and questions about abnormal cancers and normal regenerations. He concluded that "Regeneration is a repetition of ontogenesis in so far as the organ districts involved are the same, but the processes of necessity are somewhat different. There is probably a more restricted set of competences in the reacting material, but within the limits of the organ district in question the material is certainly undetermined" (Needham 1942, 447).

About the same time, Jean Brachet took up the question of biochemical effects in his *Chemical Embryology.* Brachet emphasized the formation of blastema, or undifferentiated cells around the edges of the wound. These cells seemed to make possible the development of new cells and differentiation of the right sort to cause regeneration of the original form rather than just new formation of something quite different. Brachet acknowledged that "Whether these cells arise from a migration of adjacent cells or whether they come from a dedifferentiation of more complex cells still is a controversial question" (Brachet 1950, 429).

In 1956, C. H. Waddington took the discussion further. In chapter 14 of his *Principles of Embryology,* he concluded that "The evidence suggests that to some extent at least the formation of a regeneration blastema involves a true dedifferentiation" (Waddington 1956, 306). Yet in other types of regeneration there might be special cells that had never differentiated in the first place. After discussing the role of fields, Waddington attempted to capture the reactions mathematically and asked his readers to focus again, as Child had, on metabolic activity.

Charles Bodemer's 1968 textbook *Modern Embryology* captures the considerable further thinking up to that point. It was clear that regeneration occurs widely, that reconstitution occurs in some cases, that rather little was known about what causes differentiation, redifferentiation, or dedifferentiation, but that these processes were fundamental to understanding developmental biology.

This abbreviated and quite selective review shows that different lines of research considered different aspects of regeneration within the study of development. The path to stem cells and regenerative medicine was neither direct nor clear. Yet there are other stories to tell. For example, WW II stimulated interest in medical reconstructions of lost and injured functions that led to research on regeneration of function through replacement. Sometimes prostheses can do the job, such as with limbs or vision enhancements, but they clearly are not regenerative in any very robust sense. Other medical cases in the 1950s led to discovery of the regenerative capacities of hematopoietic stem cells in bone marrow. These special cells seemed more unique than typical of anything else, and their capacities and limitations surely led many researchers to concentrate exclusively on them and ignore other kinds of cells that did not have the same abilities. It was easy to assume that such stem cells were very rare and special.

Yet another extremely important line of research led to identification, isolation, culturing, and establishing of stem cell lines. This happened in mice, growing out of work at the Jackson Laboratory with the 129 strain of mice that generated teratomas. The regular appearance of these tumors raised the question why, as well as how that mouse strain could help reveal processes of development. Martin Evans and Gail Martin took up the study and by 1981 cultured cell lines from the inner cell mass of mouse blastocysts, thereby generating the first embryonic stem cell lines (Evans and Kaufman 1981; Martin 1981).

This and subsequent successes with mouse cell lines in turn raised questions about whether culturing human embryonic stem cells might also be possible. James Thomson at the University of Wisconsin-Madison finally succeeded in developing non-human primate stem cell lines and then also human stem cell lines cultured from embryonic stem cells on layers of mouse feeder cells, all in work published in 1998 (for publications and context, see http://ink.primate.wisc.edu/~thomson/publications.html). At the same time, John D. Gearhart at Johns Hopkins used the same basic approach to culture cells from human fetal tissue. Their contributions, as has been well documented, served as a starting point for the push to find ways to culture stem cells for clinical purposes and can be seen as the public starting point for the contemporary Regenerative Medicine movement.

William Haseltine is given credit for the term, including with the short-lived name of a new journal *e-biomed. The Journal of Regenerative Medicine* and in an opening speech at the first regenerative medicine conference in December 2000. The term gained traction, and in 2001, an article in *Nature Biotechnology* noted that the new "'Regenerative medicine' encompasses the broad range of disciplines—and companies—working toward the common goal of the repair or replacement of cells, tissues, and organs." Furthermore, "Regenerative Medicine promises a more permanent solution than current pharmaceutical 'fixes', and with the launch of a few products in this class, it is moving from the realms of science fiction to the surgery" (Petit-Zeman 2001). A search of titles in PubMed shows an increasing number of articles using the term, with a sharp escalation upward in recent years.

12.7 Conclusions: How History and Philosophy of Science Help Make Science Better

We still have a lot more questions than answers about the deep processes of differentiation, dedifferentiation, and regeneration, for example, or of the relations of genetics and development. Yet we know a lot more about the biological processes than we did a century ago. And what we know is very exciting in its prospects for regenerative medicine, though the clinical results will likely be quite different than the public, or even more researchers, now imagine. We are also beginning to recognize the deeper, richer, and older lines of research that have gone into the promising regenerative medicine programs today. We can agree with the eminent cell biologist Edmund Beecher Wilson, who sounded an optimistic note about cell biology in 1896: "We cannot foretell its future triumphs, nor can we repress the hopes that step by step the way may be opened to an understanding of inheritance and development" (Wilson 1896, 330).

What the history and philosophy of science can each do, in large part and in their different but synergistic ways, is to remind researchers of the fact that Wilson pointed to, namely, that of any particular line of research "we cannot foretell its future triumphs." One of the least surprising aspects of good science is that it is often surprising. The breakthroughs often come in different ways and different places then we expected. Historical particulars can show examples of paths not taken, mistaken assumptions that were misleading, failing to ask the important questions that would have stimulated discovery. Philosophical analyses help probe sacred assumptions, articulate questions, and suggest connections unexplored. Taken together, history and philosophy of science help by adding perspective and insight, stimulating the researcher to challenge assumptions, and to seek different models or methods for exploring the questions at hand.

In the case of regenerative medicine, it has been useful for public purposes and accepted by researchers that translation, stem cells, and regeneration hold tremendous promise for valuable applications. It has been too easy to fall into simplistic pictures of how development works—taking stem cells into culture and transplanting them into the "right" culture medium can seem like a straightforward process. The arrows can seem to point directly from dishes of cells to repaired or replaced structure and function.

We see, in fact, that the research and engineering challenges of regenerative medicine are more complex. Accepting more complex understanding of the science and medical applications will help make the research better. Even if it makes the political sales job harder in the immediate future, honesty should pay off with the public trust in the long run. And history and philosophy of science are useful in getting across the message that success will depend on networks, complex systems of developing organisms and of developing scientific and medical research networks.

References

Brachet, Jean. 1950. *Chemical Embryology*. New York, NY: Interscience Publishers. (Original French edition 1944).

Bush, Vannevar. 1945. The Endless Frontier: A Report to the President. Washington, D.C.: United States Government Printing Office.

Butler, Declan. 2008. "Crossing the Valley of Death". *Nature* 453: 840–42.

CIRM. 2008. http://www.cirm.ca.gov/ (accessed July 10, 2008).

DeWitt, Natalie. 2008. "Regenerative Medicine". *Nature* 453: 301.

Domen, Jos, Amy Wagers, and Irving L. Weissman. 2006. Bone Marrow (Hematopoietic) Stem Cells. In NIH, pp. 18–34.

Editor. 2008. "To Thwart Disease, Apply Now". *Nature* 453: 823.

Evans, Martin J., and Matthew H. Kaufman. 1981. "Establishment in culture of pluripotential cells from mouse embryos". *Nature* 292: 154–56.

Gilbert, Scott. 2001. "Ecological Developmental Biology: Developmental Biology Meets the Real World". *Developmental Biology* 233: 1–12.

Goldthwaite, Charles A. 2006. Mending a Broken Heart. Stem Cells and Cardiac Repair. In NIH 2006 (2007 update), 57–65.

Haraway, Donna. 1976. *Crystals, Fabrics, and Fields. Metaphors of Organicism in Twentieth-Century Developmental Biology*. New Haven, CT: Yale University Press.

Harrison, Ross Granville. 1907. Observations on the Living Developing Nerve Fiber. *Anatomical Record* 1: 116–18.

Landecker, Hannah. 2007. *Culturing Life. How Cells Became Technologies*. Cambridge: Harvard University Press.

Maienschein, Jane. 1991. *Transforming Traditions in American Biology, 1880–1915*. Baltimore, MD: Johns Hopkins University Press.

Maienschein, Jane. 2003. *Whose View of Life? Embryos, Cloning, and Stem Cells*. Cambridge: Harvard University Press.

Maienschein, Jane, et al. 2008. "The Ethos and Ethics of Translational Research". *The American Journal of Bioethics* 8: 43–51.

Maienschein, Jane. 2009. Regenerative Medicine in Historical Context. *Medicine Studies: An International Journal for the History, Philosophy and Ethics of Medicine & Allied Sciences* 1: 33–40.

Martin, Gail. 1981. "Isolation of a pluripotent cell line from early mouse embryos cultured in medium conditioned by teratocarcinoma stem cells". *Proceedings of the National Academy of Sciences* 78: 7634–8.

Mitman, Gregg and Anne Fausto Sterling. 1992. Whatever Happened to Planaria? C.M. Child and the Physiology of Inheritance. In *The Right Tool for the Right Job: At Work in Twentieth-Century Life Sciences*, edited by A.E. Clarke and J.H. Fujimura. Princeton, NJ: Princeton University Press.

Morgan, Thomas Hunt. 1901. *Regeneration*. New York, NY: Macmillan.

Morgan, Thomas Hunt. 1934. *Embryology and Genetics*. New York, NY: Columbia University Press.

Needham, Joseph. 1942. *Biochemistry and Morphogenesis*. Cambridge: Cambridge University Press.

NIH. 2001. *Stem Cells: Scientific Progress and Future Research Directions*. Department of Health and Human Services. June 2001. http://stemcells.nih.gov/info/scireport/2001report

NIH. 2006. *Regenerative Medicine*. Department of Health and Human Services. http://stemcells.nih.gov/info/scireport/2006report.htm (with 2007 update added).

NIH. 2008. http://nihroadmap.nih.gov/overview.asp (accessed July 21, 2008).

Owens-Smith, Jason and Jennifer McCormick. 2006. "An International Gap in Human ES Cell Research". *Nature Biotechnology* 24: 391–92.

Panchision, David. 2006. "Repairing the Nervous System with Stem Cells". In NIH 2006, 35–43.

Petit-Zeman, Sophie. 2001. "Regenerative Medicine". *Nature Biotechnology* 19: 201–06.

Robert, Jason Scott. 2004a. "Model Systems in Stem Cell Biology". *BioEssays* 26: 1005–12.

Robert, Jason Scott. 2004b. *Embryology, Epigenesis, and Evolution: Taking Development Seriously*. New York, NY: Cambridge University Press.

Robert, Jason Scott. 2006. "The Science and Ethics of Making Part-Human Animals in Stem Cell Biology". *FASEB Journal* 20: 838–45.

Robert, Jason Scott and Françoise Baylis. 2003. "Crossing Species Boundaries" (Target Article with Commentaries and Response.). *The American Journal of Bioethics* 3: 1–13.

Robert, Jason Scott, Jane Maienschein, and Manfred Laubichler. 2006. "Systems Bioethics and Stem Cell Biology". *Journal of Bioethical Inquiry* 3: 19–31.

Sanchez-Alvarado, Alejandro. 2008. Lectures on "The Problem of Regeneration. Part I. A Brief (Natural) History of Regeneration," http://www.ascb.org/iBioSeminars/sanchez/sanchez1.cfm (accessed July 21, 2008).

Stanford. 2008 http://stemcell.stanford.edu (accessed July 10, 2008).

Stapleton, Darwin, ed. 2004. *Creating a Tradition of Biomedical Research: Contributions to the History of the Rockefeller University*. New York, NY: Rockefeller University Press.

Starr, Paul. 1982. *Social Transformation of American Medicine*. New York, NY: Basic Books.

Sunderland, Mary. 2007. Unpublished. Regenerating and Cloning. Paper presented at the workshop on the History of Cloning, Max Planck Institute for the History of Science, Berlin.

Waddington, Conrad Hal 1956. *Principles of Development*. New York, NY: Macmillan.

Weiss, Paul. 1939. *Principles of Development*. New York, NY: Henry Holt and Company.

Whitehead Institute. 2008. www.whitehead.mit.edu/news/ontopic/stem_cells/stemcells_glossary.html (accessed July 10, 2008).

White House. 2001 http://www.whitehouse.gov/news/releases/2001/08/20010809-2.html (accessed July 21, 2008).

Wilson, Edmund Beecher. 1896. *The Cell in Development and Heredity*. New York, NY: Macmillan.

Wolpert, Lewis. 1991. "Morgan's Ambivalence: A History of Gradients and Regeneration". In *A History of Regeneration Research, Milestones in the Evolution of a Science,* edited by Charles E. Dinsmore, 1–217. Cambridge: Cambridge University Press.

Zon, Leonard I. 2008. "Intrinsic and Extrinsic Control of Haematopoietic Stem-Cell Self-Renewal". *Nature* 453: 306–13.

Chapter 13
Social Epistemology of Stem Cell Research: Philosophy and Experiment

Melinda Bonnie Fagan

13.1 Introduction

When it comes to social aspects of our knowledge-generating practices, history and philosophy of science seem starkly opposed.[1] I argue that this opposition stems from an assumption of normative/descriptive dualism. This dualism polarizes the study of scientific inquiry into two mutually exclusive, yet co-dependent, projects: description of our actual scientific practices and their results, or abstract examination of epistemic ideals detached from our practices. If we must choose between describing the historical unfolding of our scientific practices, or elaborating abstract epistemic ideals, an integrated history and philosophy of social epistemology of scientific inquiry is precluded. I show that this dualism can be overcome, by explicating a conception of the epistemic ideal of scientific objectivity from the social aspects of our scientific practices. This ideal of objectivity is both normative and engaged with the historical unfolding of experimental inquiry. It is thus a first step toward an integrated social epistemology of scientific inquiry, to be elaborated by further historical and philosophical study.

This essay is organized as follows. I begin by setting out the problem for social epistemology of scientific inquiry (Section 13.1) and then introduce a framework for its resolution (Section 13.2). This thin framework, based on robust consensus in philosophy of social action, is then fleshed out with a study of the recent history of blood stem cell research. I first provide some background for this case study (Section 13.3), then describe a key episode in the search for the hematopoietic stem cell (HSC, Section 13.4). Results of this socio-historical study (Section 13.5) combine with normative requirements of the social action framework to yield an integrated ideal of scientific objectivity (Section 13.6).

[1]For a range of perspectives, see (Hollis and Lukes 1982; Labinger and Collins 2001; Zammito 2004).

M.B. Fagan (✉)
Rice University, Houston, TX, USA
e-mail: mbf2@rice.edu

S. Mauskopf, T. Schmaltz (eds.), *Integrating History and Philosophy of Science*,
Boston Studies in the Philosophy of Science 263, DOI 10.1007/978-94-007-1745-9_13,

13.2 Normative/Descriptive Dualism and Social Epistemology of Scientific Inquiry

I assume the following: (1) the distinction between knowledge and opinion is a *sine qua non* of epistemology; (2) any adequate epistemology of scientific inquiry must explicate the distinction between scientific knowledge and opinion in a way that relates to our scientific practices; and (3) any adequate *social* epistemology of scientific inquiry must do so in a way that engages significant social aspects of those practices.[2] These are minimal adequacy conditions, evidently. But, given normative/descriptive dualism, (3) cannot be met.[3] Consider a thoroughly descriptive account of scientific inquiry: epistemic standards distinguishing knowledge from opinion are of a piece with our scientific practices. There is no absolute, stable or invariant standard outside the historical contexts in which these practices emerge and (for a time) persist.[4]

Knowledge is distinguished from opinion in virtue of satisfying epistemic standards resulting from complex and highly contingent social negotiations. As social structures, values and interests change over time, epistemic standards for scientific knowledge change in correlated ways. Although they may be described, these changes cannot be epistemically evaluated. Any standard for such evaluation would transcend the variable socio-historical contexts in which our scientific practices occur. But, on this view, there are none. In contrast, normative epistemology (in the Anglophone analytic tradition) requires such standards: epistemic ideals that prescribe our practices independently of the historical course of scientific inquiry.[5] Normative/descriptive dualism effectively segregates history and philosophy of science, with respect to the social aspects of scientific inquiry. On the descriptive approach, social epistemology of scientific inquiry is a form of historical investigation. On the normative approach, it is continuous with epistemology in the analytic tradition. Normative/descriptive dualism thus polarizes social epistemology of scientific inquiry, along the familiar fault-lines that divide social epistemology and science studies.

---

[2]If an account fails to distinguish knowledge from opinion then it is outside the scope of epistemology, falling instead into another domain (e.g., social science, psychology, philosophy of mind). If an account fails to engage our scientific practices, then it does not concern scientific inquiry as we practice it, though it may address other epistemological issues (e.g., analysis of knowledge, dynamics of epistemic authority, characteristics of ideal or finished science). If an account fails to engage social aspects of our practices, then it is outside the scope of social studies of science.

[3]Neither can (2), though I shall not argue the point here, as the controversy over social aspects of scientific knowledge is more severe and entrenched.

[4]See, e.g. Barnes and Bloor (1982), Fleck (1979[1935]), Fuller (1988), Rouse (1996), Kusch (2002). Of course, there may be invariant epistemic standards common to all scientific contexts (though empirical evidence for this is not encouraging). But invariance merely partitions epistemic standards into "same" and "different" relative to contexts being compared. This distinction does not provide a basis for epistemic evaluation across socio-historical contexts. Generality as such does not clarify the distinction between scientific knowledge and opinion.

[5]See, e.g. Goldman (1999), Kitcher (2001).

Yet an adequate social epistemology of scientific inquiry needs both. On a thoroughly descriptive approach, the distinction between scientific knowledge and what is accepted as such within a particular socio-historical context collapses (cf. Barnes and Bloor 1982, 27). Scientific knowledge is thus identified with beliefs accepted as knowledge at a particular time/interval by the scientific community (or an authoritative portion thereof)—that is, opinion accepted as authoritative. The distinction between knowledge and opinion is just the difference between authoritative and non-authoritative opinion. Such a distinction can be drawn for any cognitive enterprise: law, religion, politics, philosophy, art, etc. So there is nothing, on a descriptive view, to distinguish scientific inquiry from other social practices that involve belief and opinion—that is, most human activities. The descriptive approach therefore fails to explicate the distinction between knowledge and opinion in a way that engages our scientific practices. It engages, instead, a much broader domain of human social action, within which our scientific practices recede into the broader social fabric. Thoroughgoing descriptivism effaces the epistemic significance of scientific inquiry.[6] This approach, on its own, is therefore inadequate for social epistemology of scientific inquiry (though it might be defended on its own terms as epistemology of human social endeavor).

The normative approach is vitiated in a complementary way. Though epistemic ideals such as rationality and objectivity have long been considered characteristic of scientific inquiry, their relationship to our actual practices is difficult to specify. Social constructivist critiques of normative epistemology highlight this difficulty. How can epistemic standards distinct from our practices nonetheless exert "prescriptive grip" on them? Is it not less mysterious to dismiss them as pretty fictions to dazzle the uninitiated and inspire novices? This would make normative epistemology "no more than an empty play on words or an epistemology of the imagination" (Fleck 1979[1935], 21). In response, normative epistemologists offer case studies of epistemic ideals in scientific practice: intuitively clear cases of epistemic success and failure in historical or contemporary scientific inquiry are shown to conform to some idealized epistemic standard (e.g., Kitcher 1993; Friedman 1999). But such case studies are not a satisfactory rebuttal. Scientific inquiry is not "pre-packaged" into cases for philosophical consumption. "*Exogenous*" epistemic ideals engage our actual practices only in conjunction with further assumptions specifying which aspects of scientific episodes are epistemically relevant. So case studies in normative epistemology show (at best) that an epistemic ideal applies to our scientific practices relative to a partition of those practices into epistemically relevant and irrelevant aspects. But this is question begging. Normative social epistemologists have themselves persuasively argued against the individualistic assumption that social aspects of scientific inquiry are epistemically irrelevant (e.g., Longino 1990; Kitcher 1993; Kornblith 1994; Goldman 1999; Solomon 2001). Parallel arguments can be made for

[6]This is not a merely theoretical worry. In recent sociology of science, attention has in fact shifted to science as policy, as regulation, as a strand of political economy, or as the epistemic face of the modern state (e.g., Knorr Cetina 1999; Drori et al. 2003; Jasanoff 2005; Frickel and Moore 2006). The work of scientists themselves has largely vanished from sociological discussion.

the social interactions and negotiations that establish epistemic standards in actual practice.

The descriptive approach presents an incisive challenge for normative epistemology of science—not by proving that epistemic standards must be socially constructed (as noted above, this is not even an adequate alternative), but by highlighting the arbitrary and *ad hoc* nature of the epistemic relevance relations that underwrite the application of ideal epistemic standards to our scientific practices. Many different specifications of epistemic relevance are possible, and absent a principled rationale for selecting among them the associated epistemic ideals lack normative force for our practices.[7] Application to our scientific practices thus relativizes epistemic ideals to one partition among many. A "relevance partitioning principle" would of course underwrite application of ideal epistemic standards to our practices. But it does not seem that any such is available. Any attempt to identify such a principle will face the same difficulties as epistemic ideals themselves. *A priori* arguments in support of one partition over another cannot settle the matter, since what is at issue is the application of epistemic ideals to our practices, not the cogency of those ideals as such. On the other hand, appeal to contemporary acceptance of certain aspects of certain episodes as exemplary concedes to descriptivism. If epistemic ideals apply to our scientific practices only in virtue of our acceptance of certain exemplars, then the distinction between epistemic standards and local criteria of acceptance collapses in practice. So, despite the availability of philosophical case studies, "exogenous" epistemic ideals fail to prescribe our scientific practices, being relative to an arbitrary distinction, or collapsing into descriptivism.

Normative/descriptive dualism thus poses a dilemma for epistemology of scientific inquiry: either epistemic standards distinguishing scientific knowledge from opinion are dependent on social structures, values or interests, or they are not. If they are, then epistemology collapses into description of socio-historical facts, and scientific inquiry recedes into the complex and dynamic pattern of the broader social and cultural fabric. If epistemic standards are independent of social structures, values or interests, then epistemology proposes abstract ideals with no clear prescriptive relation to our actual practices of scientific inquiry. An adequate social epistemology of scientific inquiry requires an epistemic standard that can bridge the gap between abstract ideals and our pervasively social scientific practices. But it is difficult to see how such a standard can be articulated. Neither historical nor philosophical case studies (as discussed above) can be of help here. In fact, the dilemma for social epistemology of scientific inquiry is posed by these two kinds of case studies. The descriptive approach is underwritten by studies of many different disciplines and historical contexts, which robustly indicate that our scientific practices are suffused with social interactions and sociological influences.[8] Selection of topics for inquiry,

---

[7]Recognition of this arbitrariness is one plausible motivation for endorsements of pluralism in social epistemology and philosophy of science (e.g., Goldman 2002; Kitcher 2004; Kellert et al. 2006).

[8]For example: Collins (1975), Knorr-Cetina (1981), Latour and Woolgar (1979), Shapin and Schaffer (1985), Pickering (1995), Collins (1998), Knorr Cetina (1999). See Shapin (1982),

development and implementation of methods and evidential standards, and acceptance of scientific claims are all social practices, involving multiple individuals and (unavoidably) their various interests. Further historical case studies demonstrating this often-reproduced empirical result are otiose, and (as noted above) unmotivated. On the other hand, philosophical case studies showing that "exogenous" epistemic ideals apply to our scientific practices are vitiated by question-begging assumptions as to which aspects of our practice are epistemically relevant. Further illustrations that selected aspects of a selected scientific episode conform to an independently-obtaining epistemic ideal are of no help. To move beyond both sorts of case study, a new framework for examining scientific inquiry is needed.

13.3 Social Action Framework

Philosophy of social action provides a framework that is compatible with both normative and descriptive approaches to scientific inquiry, and so accommodates their diverse case studies to one another. Its core is a thin but robust consensus concerning practical reasoning and action, deeply entrenched in everyday and technical explanations of human activity, and explicit in recent philosophical accounts of social action. This consensus view is simply that human action can be understood and explained in terms of "fit" between goals and means (instrumental rationality weakly construed).[9] This widespread commitment to means-end reasoning in explanation and understanding of human action entails two constraints on agents' goals and means so understood. A goal must, at minimum, engage an agent's motivation such that her intentional action may ensue, the latter being subject to assessment in terms of instrumental rationality. To take an action to be a means to a goal is to include it in a plan for achieving that goal. This is not to stipulate that intentions or actions must be instrumentally rational, only that they fall within the scope of instrumental rationality, to be understood and explained in terms of "fit" between goal and means. This is an exceedingly weak constraint, but it does rule out two cases: (1) one's having a goal which cannot be achieved no matter what one does;

Golinski (1998) for detailed surveys of the relevant literature; (Knorr Cetina 1999, 263 note 1; Bloor 2004, 919–20), for citation lists.

[9]Appeal to inquirers' goals and means is explicit in Latour's actor-network theory (1987) and Pickering's mangle of practice (1995). In Shapin's interest model (1975), the tie between broad socio-cultural interests and inquirers' goals and means remains partly implicit, yet underwrites the explanatory force of these accounts. Several influential sociological accounts discuss the goals and means of individual inquirers in light of "the end of science" conceived as the *telos* of a social structure (Merton 1973) or the expression of a mood characteristic of a "thought-style" (Fleck 1979[1935]). Others (Collins 1975; Knorr Cetina 1981) focus on the social organizations and epistemic practices that structure scientists' means-end reasoning; the latter provides the starting point for such laboratory studies, and is presupposed by them. Means-end reasoning underpins the philosophy of political naturalists like Rouse (1996) and Fuller (1988), as well as their accounts of scientific inquiry. The same goes for naturalistic epistemologists, such as Hull (1988), Goldman (1999), Kitcher (2001). The goals and means of scientific communities also figure in Solomon's (2001) and Longino's (2002) social accounts of scientific rationality and knowledge (respectively).

and (2) one's taking as means actions that could not be included in a coherent plan specifying how one's goal might be achieved. These constraints impose two necessary (though insufficient) conditions for instrumentally rational action: achievability (for goals) and coherence (for means).

Recent philosophical accounts of social action extend these minimal preconditions for instrumental rationality to activities involving multiple interacting participants (Gilbert 1989; Searle 1990; Bratman 1999; Kutz 2000; Miller 2001; Tuomela 2005). Though there are deep differences among them, all these accounts endorse the idea of a shared goal achieved by multiple participants acting according to their parts.[10] For practical reasoning in social action, what is at issue is not what an individual can do, but what multiple individuals can accomplish together: a shared goal. In social as well as individual action, an agent is committed to a plan that includes her intended action as a part. What distinguishes participant means from means taken in individual action is that the plan necessarily involves others' actions as well. This entails an additional requirement of coherence: that one's means be coordinated with those of other participants. Social action is understood and explained in terms of the connection between shared goals that participants hope to accomplish together, and the coordinated means by which they try to do so. This entails two requirements: that the shared goal of a given social action be achievable, and the means taken to it coordinated among participants. These minimal preconditions for instrumental rationality therefore provide a minimal consensus framework, compatible with both normative and descriptive perspectives on scientific inquiry.

Of course, the framework of goal-oriented social action subject to constraints of instrumental rationality, as such, does not illuminate the social epistemology of scientific inquiry. Fleshing out this minimal framework requires empirical study of scientific episodes. I focus next on one such episode: the effort to isolate and characterize blood stem cells.[11]

13.4 Preliminaries

The search for blood stem cells (HSC[12]) emerged from the confluence of cell biology, genetics and radiation research in the mid-twentieth century. In the early 1960s, it coalesced around a new experimental approach: the spleen colony assay. A key result in 1988 led to a developmental turn for the field, which has had important

[10]For example, if my climbing partner and I share the goal of climbing Half Dome, then we are each committed to trying to get to the top as a duo. Accordingly, we plan and execute our climb by coordinating actions, e.g., taking turns to lead and belay. Each of us participates in social action aimed at the shared goal of reaching the top together. In contrast, everyone who plans to climb Half Dome has the *same* goal, in the sense that all plan to reach the same place. But would-be climbers do not all *share* this goal. We are not all trying to reach the top together. My partner and I aim to reach the top together, but whether or not any of the others also do so is not our concern. If their goals figure at all in our plans, it is only as a background condition, like inanimate objects or weather.

[11]Fagan (2007) includes a preliminary version of this study, as well as the social action framework.

[12]The acronym is for "hematopoietic": literally, "blood-making".

ramifications for stem cell and cancer biology. Historical study of this episode, focused on shared goals and coordinated means of participants, fleshes out the minimal social action framework for a representative experimental success. Importantly, this is not a case study of the sort critiqued in the previous section. It is not a philosophical case study illustrating the application of an exogenous epistemic ideal for scientific knowledge; no such ideal has been proposed. Nor is it an historical case study illustrating that epistemic standards change in response to changing social structures, interests or values. Clearly the evidential standards we use do vary and change; no further case studies are needed to establish this point. But I do not identify these standards with that which distinguishes scientific knowledge from opinion. So my narrative description of this episode does not reinforce the normative/descriptive divide and thereby polarizing social epistemology of science. One might insist that it is a case study of scientific inquiry in the social action framework, but the latter is sufficiently thin to make that a rather trivial exercise. It is the empirical study within this minimal framework that yields a substantive result.

The HSC episode was selected for three reasons. First, it is well suited to examining the relation of epistemic standards for scientific knowledge and social aspects of scientific practice within a framework of shared goals and coordinated means. The search for HSC exhibits a complex social structure, is explicitly goal-oriented, and is recognized by practicing scientists as including several important successes. In all three respects it is typical of contemporary experimental biomedicine.[13] So the search for HSC is a representative episode in which the features of interest for social epistemology of scientific inquiry in a social action framework are evident, but not peculiarly exaggerated. My second reason for selecting the HSC episode is methodological. The account that follows is based on the published record, interviews with participating researchers, and my own experience as a graduate student in the Weissman lab (1994–1997). The last familiarized me with this episode of immunology research and provided access to many of the social interactions involved. However, I was not directly involved in the search for HSC and did not participate in the episode described here. Personal experience allowed me to understand published sources and to obtain interviews with participants more efficiently than would have been possible otherwise.

Third, the HSC episode is significant for understanding the history of immunology and stem cell research. HSC occupy a distinctive role in the immune system and in our understanding of it. Though diverse cells are involved in immune function, all develop from a common precursor type, localized (in adults) to bone marrow. Most blood and immune cells live only a few days or weeks, and do not divide with sufficient rapidity to replenish themselves. But one or a few HSC can completely reconstitute an immune system that functions over the long-term, continuously dividing into progeny that differentiate into all the cells of the immune system.[14]

[13]For example, this episode is one of nineteen singled out by the editors of *Immunological Reviews* as "turning points in modern immunology" (Koretzky and Monroe 2002).

[14]This is why bone marrow transplants are clinically effective: to treat leukemias (for example), the entire immune system is ablated with radiation or chemotherapy, and then completely reconstituted by a bone marrow transplant containing a few HSC.

As the beginning of the developmental history of the immune system, HSC provide an inclusive starting point for explaining and understanding its diverse mechanisms and our experimental manipulations thereof.[15] Examining the epistemic history of HSC research provides an illuminating view of immunology more generally. Furthermore, HSC are the best understood and most facilely manipulated of all stem cell types, providing a standard for characterizing other stem cells (embryonic, neurogenic, tumorigenic) in clinical and laboratory settings. So tracing the development of evidential standards for isolating HSC and other stem cell types sheds light on the epistemology of stem cell research more generally. The search for HSC is thus a representative episode of recent experimental biomedicine, of interest to both historians and philosophers of science, in which social aspects of scientific inquiry play an important role, and which my own academic history has put me in a position to study.

A bit more should be said about interview methodology. The aim of interviews was to identify and characterize social interactions recognized by participants as crucial in the search for HSC, and to reveal participants' attitudes toward these interactions (see below for details). Specifically, I sought to understand how interviewees conceived of their research activities in relation to those of other scientific inquirers, within and among laboratories and research communities, and the impact of these interactions (if any) on achievement of research goals. Eleven interviewees were selected to provide a range of perspectives on the search for HSC. All but two are or were at one time members of the Weissman lab. Their periods of involvement with the search for HSC range from two to three years to more than three decades, and from the late 1960s to the present day. Their roles were diverse: graduate student, laboratory manager, medical student, post-doctoral fellow, principal investigator, and technician. Their subsequent career trajectories also vary widely, and include academic research, clinical research, and industry. The description emerging from these multiple interviews is therefore robust to these diverse participants' perspectives.

Qualitative research interviewing (Merton et al. 1956; Briggs 1986; Seidman 1998; Zuckerman 1977) was used to allow participants' attitudes to emerge rather than imposing interviewer's assumptions via leading questions. Interview guides and biographies were prepared in advance for each subject. Interviews focused on the search for HSC, and tended to proceed chronologically; otherwise discussion was unstructured, and ranged in duration from 75 minutes to two hours. These discussions were recorded on tape during visits to subjects' laboratories, and supplemented by one or more of the following: a tour of laboratory facilities, further informal discussions with lab personnel, attendance of lab meeting. Laboratory visits contextualized the taped interviews in two ways. First, they provided information about interviewees' current setting and style of working, and framed their attitudes

[15] Embryonic stem cells play an analogous role in understanding and explaining organismal development. This is not to say that HSC are foundational for immunology in the sense of providing first principles for theories (modern immunology arguably has no such principles), nor that HSC encapsulate the whole of the subject in a kind of "meta-preformation".

toward past interactions in terms of contemporary roles and projects. Second, these engagements with interviewees' current working environments provided an opportunity to discuss the relation between the search for HSC and their current projects, eliciting interviewees' attitudes toward scientific success over time. Along with the published record, participant interviews and visits yielded a fine-grained account of the social structure of the search for HSC.

13.5 Search for HSC

Though the existence of HSC was inferred in Owen's 1945 study of blood group genetics in bovine twins, the search for these elusive cells began over a decade later.[16] A key interim discovery occurred in 1951, when radiation biologists observed that lethally-irradiated mice could be "rescued" by bone marrow transplantation. High levels of radiation destroy the immune system, which is ordinarily fatal. But mice given lethal doses survived if later injected with bone marrow cells from a donor of the same inbred strain. Transplanting bone marrow cells effectively transplanted a functioning immune system. Spurred by the obvious clinical applications, biomedical researchers began to systematically investigate "radiation rescue" in mice. The search for HSC grew out of this research program, focused around a new experimental system: the spleen colony assay. The assay was invented by medical biophysicists at the Ontario Cancer Institute, who noticed that, after about two weeks, "rescued" mice developed nodules on their spleens.[17] Each nodule was found to be a colony or clone descended from a single donor bone marrow cell, containing all the blood cell types known at the time. Cell preparations from these nodules could rescue lethally irradiated mice and produce splenic colonies in their turn. HSC were defined in terms of the capacities of colony-forming cells: (1) radiation rescue by immune reconstitution; (2) multipotency (differentiation into multiple blood cell types); and (3) self-renewal (maintaining these capacities beyond the life span of a single blood cell). So defined, HSC could be detected only in retrospect, after these capacities had been realized. A quantitative version of the spleen colony assay measured the "colony-forming units" of a given "rescue" preparation—by this means, HSC were shown to be rare cells, comprising >0.1% of bone marrow in adult mice.[18] At this point the experimental goal became clear: prospectively enrich bone marrow cell preparations for HSC, using the spleen colony assay to measure enrichment.

This goal was shared by a diffuse network of hematologists, medically-trained experts on blood cells. About a dozen hematological research groups took up the project in the mid-1960s (including the inventors of the spleen colony assay), modifying the core method by selecting donor cells from bone marrow by size, cell

[16]See Fagan (2007) for more detail on the origins of the search for HSC.

[17]Till and McCulloch (1961), Becker et al. (1963), Siminovitch et al. (1963).

[18]By present estimates, well-supported by experimental and clinical data, the frequency is considerably lower, approximately 0.0005 in whole bone marrow.

cycle state, density, and/or surface phenotype.[19] Competition to isolate HSC was tempered by regular meetings, at which representatives of the main hematological groups compared the results of their different variations on the spleen colony assay and attempted to assemble them into a consensus account of the HSC phenotype.[20] Through such interactions, the different lab groups formed a more inclusive group, with the shared goal of prospectively isolating HSC. The hematologists' search for HSC proceeded by division of labor: each group used a somewhat different method, the (quantitative) spleen colony assay provided a common standard for evaluating these variations, and results were pooled and compared at regular meetings to arrive at consensus. Experimental protocols lengthened, as methods of enrichment were concatenated. By the mid-1980s, the community had made considerable progress: up to 200-fold enrichment of HSC from mouse bone marrow (Visser et al. 1984). This was the state of play when the search was abruptly transformed.

The transformative event was a widely-publicized announcement by the Weissman lab at Stanford University in 1988 that the search for HSC in mice was over—murine HSC had been purified and characterized (Spangrude et al. 1988).[21] This result (two decades in the making) emerged from a distinct line of research that differed in three important ways from the hematological search for HSC. First, the shared core of the Weissman group's search was not an experimental assay system, but the shared goal of characterizing mechanisms of immune cell development in terms of surface phenotype, movement and function of single cells. Second, being uncommitted to any particular experimental method, the Weissman group made opportunistic use of whatever new techniques were available. Importantly, they enjoyed early access to fluorescence-activated cell sorting technology (developed at Stanford in the Herzenberg laboratory) as paying members of a "shared FACS users group."[22] Third, instead of dispersed division of labor, the lab served as a center for continuous and cumulative collaboration in the search for HSC. Research interests within the lab were diverse, and its shifting membership was free to pursue whatever project they chose. The result was a loose and shifting assemblage of lines of inquiry, many of which concerned blood cell development.

---

[19] See reviews in Watt et al. (1987), Spangrude (1989), Visser and van Bekkum (1990).

[20] For example, annual meetings of the Midwest and Southern "Blood Clubs" (mid-1980s); annual symposia on Molecular Biology of Hematopoiesis (1985–1989). The main groups seeking HSC were in Toronto, Manchester, Melbourne, and more diffusely distributed in the Netherlands and the Eastern US (primarily NYC).

[21] See Fagan (2007) for further details on the Weissman group's search and the 1988 result.

[22] FACS is a method for rapidly sorting cell populations, one cell at a time, according to level of surface expression of particular molecules, which are detected by specific binding of antibodies conjugated with fluorescent tags. The aim is to separate functionally distinct but morphologically similar cells without killing them, so purified cell populations can be used in further experiments. The first FACS apparatus was developed at Stanford University in collaboration with Becton Dickinson (Bonner et al. 1972; Herzenberg et al. 1976; Keating and Cambrosio 2003; Herzenberg and Herzenberg 2004).

In the early 1980s, three such projects were coordinated into a focused search for HSC. In this search, three kinds of interaction were crucial: multiple collaborations within the Weissman lab, participation in Stanford's shared FACS users' group, and a collaboration with a West German laboratory. The upshot was a cell labeling and tracking experiment that characterized and isolated a cell population 2,000-fold enriched for HSC function. This was the result announced in *Science* (and the popular press) in 1988. Though the paper had only three authors, the result depended on decades of sustained collaboration involving dozens of researchers, in and outside the Weissman lab. Initially controversial, the Weissman group's result is now recognized as a significant contribution to modern immunology; a representative biomedical success.

Interestingly, their recognized accomplishment was *not* isolation of HSC, nor even greater enrichment of HSC from mouse bone marrow. Methodological diversity among the hematologists yielded divergent assessments of how the Weissman group's result compared to earlier work, preventing clear consensus among those most concerned with this issue.[23] Arguably, the Weissman group did not improve on extant HSC purification protocols. Divergent assessments of the issue persist today. Moreover, within a year all interested parties agreed that the Weissman group's cell population was heterogeneous, and various groups (including Weissman's) began working to characterize more finely-grained cell populations.[24] So the success of 1988 did not consist in the isolation of a pure blood stem cell, and at least arguably, not even of HSC enrichment relative to other available methods.

What did the 1988 success consist of? It had two components: articulation of a new model[25] of blood cell development coordinating HSC capacities with cell phenotype; and, just as important, a new direction and impetus for the search for HSC as a project of cellular immunology. Both resulted from distinctive features of the Weissman group's search for HSC. As noted above, the hematological HSC

[23]Though all agreed that HSC are pluripotent, self-renewing and responsible for radiation rescue, different groups took different aspects of HSC as primary. Responses to the 1988 result thus discriminated between experimental methods and emphases. Many hematologists conceived of HSC primarily in terms of spleen or in vitro colony formation, while the Weissman group defined HSC in terms of the correlation between in vitro colony formation and in vivo immune reconstitution. The dispute has not been resolved; within the Weissman lab, the 10-fold greater enrichment was and is recognized as success.

[24]Lemischka et al. (1986), Visser, in (Radetsky 1995, 91; Spangrude 1989 (interview of 12/4/2006); Müller-Sieburg (interview of 4/6/2007)).

[25]"Model" is the term used by HSC researchers, and, increasingly, by philosophers of science as a generalization of "theory", which admits non-linguistic representations as well as more traditional theories amenable to axiomatic presentation (e.g., Giere 1988; Longino 2002). Models in this sense are representations of subjects of inquiry, in which mathematical laws or idealized causal or formal relations are satisfied. Theories may be thought of as "families" of models and associated similarity claims. Models in science represent parts of the world under investigation in particular respects and degrees, which vary depending on available techniques and the purposes for which those models are constructed. Techniques and purposes, in turn, vary widely across disciplines, fields, and socio-historical contexts. Improvements to a model strengthen or extend the similarity claims associated with it, according to the standards of the relevant research community.

research community proceeded by aggregating the results of different groups working to isolate it. Their methods were diverse, not coordinated by a single model of cell development or physiology. In contrast, the Weissman group coordinated the defining capacities of HSC with cell surface characteristics at the single-cell level, in a way that readily extended to humans and to other developmental stages. The 1988 result amounted to an improved model of the development and function of the immune system, and (via developmental analogy) gestured toward future clinical applications. This fit with the biomedical goal of immumology: knowledge of the immune system and treatment of infectious disease, autoimmune disorders, and cancers (Paul 1983; Kuby 1994; Paul 2003). The tie to cellular developmental immunology gave new direction and impetus to the field of HSC research. The widely-publicized *Science* paper thus created a new interface between previously distinct lines of inquiry. Controversy then ensued over their different methods and standards for isolating HSC. Yet the occasion of this controversy became the most enduring aspect of the Weissman group's success. The 1988 model itself was superseded within a year, but the Weissman group's distinctive method (coordinating surface phenotype with developmental potential and immune function at the single-cell level) was emulated and modified by many other groups, and the search for HSC drew on rapid advances in cellular immunology throughout the 1990s. The two aspects of success were interdependent: announcement of the coordinated model initiated the interface of experimental hematology and developmental immunology, and the standards according to which that model counted as improved emerged from that interface, endorsed by the more inclusive HSC community.

Other successes followed: further refinement of bone marrow cells by self-renewal capacity (Morrison and Weissman 1994), reconciling apparently incompatible models of blood cell development (Kawamoto et al. 1997; Kondo et al. 1997), and ramifications of HSC research for stem cell and cancer biology, realized in new interfaces between these fields (Dontu et al. 2003). Continued pursuit of HSC led in turn to further new interfaces with neurobiology, developmental biology, evolutionary biology, and cancer research. Concomitantly, models of blood cell development became increasingly robust and detailed. In these various ramifications of the search for HSC, two aspects of success are robustly recognized by participating researchers: improved models of cell development, and coordination of groups or individuals with different goals. This pattern recurs at different levels of social organization: within a single lab, among different lab groups, and across fields and disciplines.[26]

13.6 Scientific Success

The two aspects of the 1988 success are both *coordinations*: of HSC function and cell phenotype in an improved model of blood cell development, and, at the social level, of the search for HSC with other successful lines of inquiry. As noted above, this pattern of recognized success recurs throughout the episode, at different levels

[26]Full results in (Fagan 2008, under review).

of social organization. This recurring pattern is also seen in other episodes of experimental inquiry, including those described in influential science studies texts: Boyle's experiments with the air-pump, isolation and characterization of thyrotropin releasing factor (hormone) at the Salk Institute, gravitational wave research in Italy and the US, Pasteur's anthrax vaccine, the Wassermann test for syphilis.[27] So diverse socio-historical case studies indicate that this pattern of recognized scientific success (improved models, and new interfaces) is robust across various disciplines and historical contexts.

This is, of course, an empirical result, not a normative thesis about scientific success. Yet this pattern of success displays some kinship with normative epistemic ideals. Models are recognized as improved when they increase the scope, consistency, precision or accuracy of scientific accounts of the empirical world. Coordination of diverse lines of inquiry via new interfaces similarly recalls epistemic virtues long associated with scientific knowledge: consistency, coherence, unification.[28] But to identify epistemic ideals with the two aspects of recognized scientific success would be to conflate the normative and descriptive aspects of scientific inquiry. Models are improved relative to epistemic standards of groups, which vary widely over time and across fields, and in response to social interactions and structures. Formation of new interfaces is a highly contingent matter, influenced by available technology, world events, and mere coincidences as well as the (often unexpected) results of lines of inquiry themselves. So the way in which epistemic ideals of consistency, coherence and unification are cashed out in any particular new interface depends on many contingencies, as well as the standards for evaluation of models accepted by the groups in question.

The social action framework captures the normative/descriptive relation in terms of (minimal) instrumental rationality: the two aspects of success are participant means to the shared epistemic goal of scientific inquiry, scientific knowledge. That is, recognized successes (improved models and new interfaces) are recognized as contributions to scientific knowledge: provisional, partial, and unavoidably enmeshed in human social interactions, yet also directed toward a further end. In this way, the social action framework accommodates normative and descriptive

[27]Shapin and Shaffer (1985, 3–7, 30–31), Shapin (1996, 96), Latour and Woolgar (1979, 106), Collins (1998, 299), Latour (1983, 260–64), Fleck (1979[1935], 14–19).

[28]New interfaces between distinct lines of inquiry can arise in three ways: a single line of inquiry divides into two (or more) distinct branches; two distinct lines of inquiry merge to become one; and two distinct lines of inquiry remain distinct, but alter their relation to one another. All three are a means to (though not a guarantee of) greater consistency, coherence or unification of scientific knowledge. Division of a single line of inquiry into two disambiguates the goals and means at work within a line of inquiry, reconciling inconsistencies between apparently incompatible models and thereby organizing inquiry more efficiently. Merging of two distinct lines of inquiry to form a new, more inclusive group is, roughly speaking, the converse of division of labor. The subject matter of distinct lines of inquiry is seen to connect, such that models previously thought unrelated are seen as relevant to one another. Conflict and controversy ensue, precisely because of the new connection; coordination is achieved (at least in some cases) via these apparently antagonistic social interactions. Lines of inquiry are then seen as more coherently organized, contributing via different roles to a larger project with a more inclusive shared goal.

approaches to social epistemology of scientific inquiry without identifying scientific knowledge with the outcome of successful scientific episodes, nor dismissing as irrelevant the socially-enmeshed epistemic standards used in actual scientific episodes. The final task is to specify this end or shared goal, using the results of empirical study of our scientific practices. One such result (generalizing from the HSC episode) is that the two aspects of success are coordinated in our scientific practice. Models count as improved according to the epistemic standards of some particular group pursuing a line of inquiry. These standards are unavoidably enmeshed in that group's socio-cultural context. Models meeting the standards of different lines of inquiry are brought into critical contact by formation of new interfaces between research groups. New epistemic goals and standards for their achievement are then negotiated. Models that satisfy standards of improvement in one context are thus given critical scrutiny from a new (though not wholly unrelated) perspective. And so on, as inquiry continues. The two aspects of success are thus coordinated means for ongoing successful inquiry—iterations of model-construction and interface-formation.

13.7 Scientific Objectivity

Having characterized these means by socio-historical study of scientific practice, one may ask: what must the shared goal of our scientific practices be like, given the means taken to it? Here the minimal requirements for social action come into play. Recall that these are preconditions for conceiving scientific inquiry in terms of instrumental rationality.[29] If scientific practices can be understood and evaluated in terms of instrumental rationality, then they have a shared goal that is achievable by coordinated participant means: construction of improved models and coordination of lines of inquiry via new interfaces. Historians, sociologists and philosophers of science, as well as scientists themselves, do try to understand scientific inquiry in this way (among others). So these requirements, though minimal, have quite broad prescriptive force. Applied to the descriptive account of scientific success in the social action framework, these minimal constraints on social action explicate the distinction between scientific knowledge and opinion.

Scientific knowledge, the shared goal of our scientific practices, must be achievable by the coordinated means taken to it. There is of course no guarantee of success, nor can a comprehensive plan be detailed. This uncertainty is in part due to the fact that formation of new interfaces between distinct lines of inquiry is highly contingent, as is the negotiation of new epistemic standards brought into critical contact thereby. A consequence of this socially-enmeshed contingency is that the pattern of formation of new interfaces cannot be specified in advance. If formation of new interfaces is unpredictable in advance, then the epistemic standards resulting from

[29]This is not to say that scientific inquiry is instrumentally rational; only that it may be understood and evaluated in these terms (i.e., the fit between goals and means). This account is neutral regarding further requirements for instrumental rationality.

such new interfaces are likewise unpredictable. So the standards to which a successful model will be held accountable cannot be specified in advance. Scientific knowledge (conceived as the aim of inquiry) must be such as to possibly result from the coordinated means taken to it. Thus, minimal constraints on instrumentally rational social action require that the shared epistemic goal of scientific inquiry be achievable by the interplay of construction of improved models and formation of new interfaces, where the epistemic standards successful models are to satisfy cannot be specified in advance. More simply: scientific knowledge must be such as to possibly satisfy epistemic standards not specifiable in advance. This is, admittedly, a very thin characterization of scientific knowledge, but it suffices to rule out "knowledge by agreement" as the goal of scientific inquiry.

This is an important, albeit negative, result: knowledge that is so only in virtue of the epistemic standards of specifiable groups in particular socio-historical contexts, is not an achievable shared goal of our scientific practices. So the distinction of knowledge from opinion is not drawn by epistemic standards accepted within particular socio-historical contexts. One may grant that our scientific practices are pervasively social, and indeed that social interactions are necessary for epistemic success in all but the most fragmentary and circumscribed episodes of scientific inquiry, without identifying scientific knowledge with authoritative opinion in particular contexts. Put more positively, this result specifies the ideal of scientific objectivity in relation to our scientific practices. Our scientific practices, conceived as social action satisfying prerequisites for instrumental rationality, aim at knowledge that is so in virtue of satisfying epistemic standards that are not limited to any specifiable group. Such knowledge is "objective" in a sense long associated with the epistemic distinctiveness of scientific inquiry, but hotly contested in recent studies of science (e.g., Longino 1990; Daston and Galison 1992; Boghossian 2006). Objective knowledge, in this sense, is knowledge independent of the opinions of any single individual or group of individuals. In the terminology used here, that anyone (or any specifiable group) accepts a model as scientific knowledge does not make it so. The relevant epistemic standard does not depend on features specific or idiosyncratic to particular groups of inquirers (and, *a fortiori*, individual inquirers).

This result is not to be confused with an analysis of the concept of objectivity. It is, rather, an explication of an epistemic ideal implicit in our scientific practices, brought out by framing descriptive socio-historical narratives of scientific inquiry in terms of social action theory. This conception of scientific objectivity is normative, in two senses. First, it is required for understanding scientific inquiry in terms of means-end reasoning. Second, it specifies an epistemic ideal that allows for principled epistemic evaluation of our scientific practices. To be sure, this thin conception of scientific objectivity does not allow for all the epistemological critique of science one might want. It is, rather, a starting point from which more substantive epistemic ideals could be elaborated.[30]

---

[30]For example, this minimal account could ground or warrant Longino's social epistemic norms for reliable empirical knowledge (1990, 2002).

13.8 Summary and Conclusion

The account proposed here bridges the gap between social aspects of scientific practice, on the one hand, and epistemic ideals of scientific knowledge, on the other, moving beyond the historical and philosophical case studies that frame the dilemma for social epistemology of scientific inquiry. It moves beyond historical case studies of the social aspects of scientific inquiry by embedding them in a social action framework that entails minimal normative requirements. Descriptive socio-historical accounts in turn yield a robust two-part account of scientific success that approximates traditional epistemic ideals of scientific knowledge. This two-part account of scientific success unifies diverse socio-historical case studies of scientific inquiry and so characterizes participant means to the shared epistemic goal of these practices. The minimal constraints for social action then specify the epistemic ideal of scientific objectivity. So my account also goes beyond philosophical case studies illustrating the application of an "exogenous" epistemic ideal. The problem of principled application to our scientific practices (Section 13.1) does not arise. Instead, the distinction between scientific knowledge and opinion is explicated by engaging with the social aspects of our scientific practices from the outset. The resulting epistemic ideal of scientific objectivity is genuinely normative and provides a starting point for elaborating further social epistemic norms for scientific inquiry. So it may play a grounding and framing role for philosophical case studies of scientific inquiry as well. All this is a matter for further study. This essay has shown only that the dilemma for social epistemology of scientific inquiry posed by normative/descriptive dualism can be overcome, integrating history and philosophy of science. It is through understanding our scientific practices as social action that we gain purchase on epistemic ideals of *our* scientific inquiry, rather than idealized abstractions.[31]

[31] I thank the editors for the opportunity to contribute to this collection. An earlier version of this paper was presented at &HPS1 (University of Pittsburgh, 10/12/08); I thank the participants, especially my co-presenters Hasok Chang and Theodore Arabatzis, for valuable questions and comments. This project has also benefited greatly from discussions with Colin Allen, Jordi Cat, Elihu Gerson, Tom Gieryn, Sander Gliboff, James Griesemer, Elisabeth Lloyd, Jutta Schickore, and Fred Schmitt, and from comment from audiences the University of California at Davis, the University of California at Santa Cruz, the University of Western Ontario, and the 2007 meeting of the International Society for the History, Philosophy, and Social Studies of Biology (ISHPSSB, University of Exeter, 7/29/07). The empirical portion of the study was made possible by the generous participation of Laurie Ailles, Arlene Bitmansour, Samuel Cheshier, Robert Coffman, Tony DeTomaso, George Gutman, Leonore and Leonard Herzenberg, Libuse Jerabek, Motonari Kondo, Sean Morrison, Jerry Spangrude, Christa Müller-Sieburg and Irving Weissman. Financial support was provided by a Doctoral Dissertation Improvement Grant from the National Science Foundation (SES-0620993) and a Dissertation Year Fellowship from the College of Arts and Sciences at Indiana University.

References

Barnes, Barry and David Bloor. 1982. "Relativism, Rationalism and the Sociology of Knowledge". In *Rationality and Relativism*, edited by M. Hollis and S. Lukes, 21–47. Oxford: Blackwell.

Becker, A.J., E.A. McCulloch, and J.E. Till. 1963. "Cytological Demonstration of the Clonal Nature of Spleen Colonies Derived from Transplanted Mouse Bone Marrow Cells". *Nature* 197: 452–54.

Bloor, David. 2004. "Sociology of scientific knowledge". In *Handbook of Epistemology*, edited by Illka Niiniluoto, Matti Sintonen, and Jan Wolenski, 919–62. Dordrecht: Kluwer.

Boghossian, Peter. 2006. *Fear of Knowledge: Against Relativism and Constructivism*. Oxford: Oxford University Press.

Bonner, W.A., H.R. Hulett, R.G. Sweet, and L.A. Herzenberg. 1972. "Fluorescence Activated Cell Sorting". *Review of Scientific Instruments* 43: 404–09.

Bratman, Michael E. 1999. *Faces of Intention: Selected Essays on Intention and Agency*. (Cambridge Studies in Philosophy) Cambridge: Cambridge University Press.

Briggs, Charles L. 1986. *Learning How to Ask: A Sociolinguistic Appraisal of the Role of the Interview in Social Science Research*. Cambridge: Cambridge University Press.

Collins, Harry. 1975. "The Seven Sexes: A Study in the Sociology of a Phenomenon, or the Replication of Experiments in Physics". *Sociology* 9: 205–24.

Collins, Harry. 1998. "The Meaning of Data: Open and Closed Evidential Cultures in the Search for Gravitational Waves". *American Journal of Sociology* 104: 293–338.

Daston, Lorraine and Peter Galison. 1992. "The Image of Objectivity". *Representations* 40: 81–128.

Dontu, G, M. Al-Hajj, W.A. Abdallah, M.F. Clarke, and M.S. Wicha. 2003. "Stem cells in normal breast development and breast cancer". *Cell Proliferation* 36: 59–72, S1.

Drori, Gili S., John W. Meyer, Francisco O. Ramirez, and Evan Schofer. 2003. *Science in the Modern World Polity: Institutionalization and Globalization*. Stanford, CA: Stanford University Press.

Fagan, Melinda B. 2007. "The Search for the Hematopoietic Stem Cell: Social Interaction and Epistemic Success in Immunology". *Studies in History and Philosophy of Biological and Biomedical Sciences* 38: 217–37.

Fleck, Ludwik. 1979. *Genesis and Development of a Scientific Fact*, trans. F. Bradley and T. J. Trenn; edited by T.J. Trenn and R.K. Merton. Chicago, IL: University of Chicago Press (1st edition. published 1935, German).

Frickel, Scott and Moore, Kelly, eds. 2006. *The New Political Sociology of Science: Institutions, Networks, and Power*. Madison, WI: University of Wisconsin Press.

Friedman, Michael. 1999. *The Dynamics of Reason*. Stanford, CA: CSLI.

Fuller, Steve. 1988. *Social epistemology* (2nd edition, 2002). Bloomington, IN: Indiana University Press.

Giere, Ronald. 1988. *Explaining Science: A Cognitive Approach*. Chicago, IL: Chicago University Press.

Gilbert, Margaret. 1989. *On Social Facts*. Princeton, NJ: Princeton University Press.

Goldman, Alvin I. 1999. *Knowledge in a Social World*. Oxford: Oxford University Press.

Goldman, Alvin I. 2002. *Pathways to Knowledge: Private and Public*. Oxford: Oxford University Press.

Golinski, Jan. 1998. *Making Natural Knowledge: Constructivism and the History of Science*. Cambridge: Cambridge University Press.

Herzenberg, L.A. and L.A. Herzenberg. 2004. "Genetics, FACS, Immunology, and Redox". *Annual Review of Immunology* 22: 1–31.

Herzenberg, L.A., R.G. Sweet, and L.A. Herzenberg. 1976. "Fluorescence-Activiated Cell Sorting". *Scientific American* 224: 108–117

Hollis, M. and S. Lukes, eds. 1982. *Rationality and Relativism*. Oxford: Blackwell.

Hull, David L. 1988. *Science as a Process: an Evolutionary Account of the Social and Conceptual Development of Science*. Chicago, IL: University of Chicago Press.

Jasanoff, Sheila. 2005. *Designs on Nature: Science and Democracy in Europe and the United States*. Princeton, NJ: Princeton University Press.

Kawamoto, H., K. Ohmura, and Y. Katsura. 1997. "Direct Evidence for the Commitment of Hematopoietic Stem Cells to T, B, and Myeloid Lineages in Murine Fetal Liver". *International Immunology* 9: 1011–19.

Keating, P., and A. Cambrosio. 2003. *Biomedical Platforms: Realigning the Normal and the Pathological in Late-Twentieth-Century Medicine*. Cambridge: The MIT Press.

Kellert, Stephen H., Helen E. Longino, and C. Kenneth Waters. 2006. *Scientific Pluralism*. Minnesota Studies in Philosophy of Science, Volume XIX. Minneapolis, MN: University of Minnesota Press.

Kitcher, Philip. 1993. *The Advancement of Science*. New York, NY: Oxford University Press.

Kitcher, Philip. 2001. *Science, Truth and Democracy*. Oxford: Oxford University Press.

Kitcher, Philip. 2004. "The Ends of the Sciences". In *The Future for Philosophy*, edited by Brian Leiter. Oxford: Clarendon.

Knorr-Cetina, Karin. 1981. *The Manufacture of Knowledge: An Essay on the Constructivist and Contextual Nature of Science*. Oxford: Pergamon Press.

Knorr Cetina, Karin. 1999. *Epistemic Cultures: How the Sciences Make Knowledge*. Cambridge: Harvard University Press.

Kondo, M., I.L. Weissman, and K. Akashi. 1997. "Identification of Clonogenic Common Lymphoid Progenitors in Mouse Bone Marrow". *Cell* 91: 661–72.

Koretzky, G., and J. Monroe. 2002. "Introduction". *Immunological Reviews* 185: 5–6.

Kornblith, Hilary. 1994. "A Conservative Approach to Social Epistemology". In *Socializing Epistemology: The Social Dimensions of Knowledge*, edited by F.F. Schmitt, 93–110. Lanham, MD: Rowman & Littlefield.

Kuby, J. 1994. *Immunology*, 2nd edition. New York, NY: Freeman .

Kusch, Martin. 2002. *Knowledge by Agreement*. Oxford: Clarendon Press

Kutz, Christopher. 2000. "Acting Together". *Philosophy and Phenomenological Research* 61: 1–31.

Labinger, J.A., and H. Collins, eds. 2001. *The One Culture? A Conversation about Science*. Chicago, IL: University of Chicago Press.

Latour, B. 1983. "Give Me a Laboratory and I Will Raise the World". In *Science Observed*, edited by E. Mulkay, and K. Knorr Cetina, 141–70. London: Sage.

Latour, Bruno. 1987. *Science in Action*. Cambridge: Harvard University Press.

Latour, B., and S. Woolgar. 1979. *Laboratory Life: the Construction of Scientific Facts* (2nd edition, 1986). Princeton, NJ: Princeton University Press.

Lemischka, I.R., D.H. Raulet, and R.C. Mulligan. 1986. "Developmental Potential and Dynamic Behavior of Hematopoietic Stem-Cells". *Cell* 45: 917–27.

Longino, Helen. 1990. *Science as Social Knowledge: Values and Objectivity in Scientific Inquiry*. Princeton, NJ: Princeton University Press.

Longino, Helen. 2002. *The Fate of Knowledge*. Princeton, NJ: Princeton University Press.

Merton, Robert K., Marjorie Fisk, and Patricia L. Kendall. 1956. *The Focused Interview: A Manual of Problems and Procedures*. Glencoe, IL: Free Press.

Merton, Robert K. 1973. *The Sociology of Science*. Chicago, IL: University of Chicago Press.

Miller, Seumas. 2001. *Social Action: a Teleological Account*. Cambridge: Cambridge University Press.

Morrison, Sean J., and Irving L. Weissman. 1994. "The Long-Term Repopulating Subset of Hematopoietic Stem Cells Is Deterministic and Isolatable by Phenotype". *Immunity* 1: 661–73.

Owen, Ray. 1945. "Immunogenetic Consequences of Vascular Anastomoses Between Bovine Twins". *Science* 102: 400–01.

Paul, W.E. 1983. "Preface to Volume 1". *Annual Review of Immunology* 1: vii.

Paul, W.E., ed. 2003. *Fundamental Immunology*, 5th edition. Philadelphia, PA: Lippincott, Williams, and Wilkins.
Pickering, A. 1995. *The Mangle of Practice: Time, Agency and Science*. Chicago, IL: University of Chicago Press.
Radetsky, Peter. 1995. "The Mother of All Blood Cells". *Discover* 16: 86–93.
Rouse, Joseph. 1996. *Engaging Science: How to Understand its Practices Philosophically*. Ithaca, NY: Cornell University Press.
Searle, John. 1990. "Collective Actions and Intentions". In *Intentions in Communication*, edited by P. Cohen, J. Morgan, and M. Pollack, 401–15. Cambridge: MIT Press.
Seidman, Irving. 1998. *Interviewing as Qualitative Research*, 2nd edition. New York, NY: Teacher's College Press .
Shapin, Steven. 1982. "History of Science and Its Sociological Reconstructions". *History of Science* 20: 157–211.
Shapin, Steven. 1996. *The Scientific Revolution*. Chicago, IL: University of Chicago Press.
Shapin, Steven, and Simon Schaffer. 1985. *Leviathan and the Air-pump*. Princeton, NJ: Princeton University Press.
Siminovitch, L., E.A. McCulloch, and J.E. Till. 1963. "The Distribution of Colony-Forming Cells Among Spleen Colonies". *Journal of Cellular and Comparative Physiology* 62: 327–36.
Solomon, Miriam. 2001. *Social Empiricism*. Cambridge: MIT Press.
Spangrude, G.J. 1989. "Enrichment of Murine Hematopoietic Stem-Cells: Diverging Roads". *Immunology Today* 10: 344–50.
Spangrude, G.J., S. Heimfeld, and I.L. Weissman. 1988. "Purification and Characterization of Mouse Hematopoietic Stem Cells". *Science* 241: 58–62.
Till, J.E., and E.A. McCulloch. 1961. "A Direct Measurement of the Radiation Sensitivity of Normal Mouse Bone Marrow Cells". *Radiation Research* 14: 213–22.
Tuomela, Raimo. 2005. "We-Intentions Revisited". *Philosophical Studies* 125: 327–69.
Visser, J.W.M., J.G.J. Bauman, A.H. Mulder, J.F. Eliason, and A.M. de Leeuw. 1984. "Isolation of Murine Pluripotent Hemopoietic Stem Cells". *Journal of Experimental Medicine* 59: 1576–90.
Visser, J.W.M., and D.W. van Bekkum. 1990. "Purification of Pluripotent Hematopoietic Stem Cells – Past and Present. *Experimental Hematology* 18: 248–56.
Watt, S., D.J. Gilmore, J.M. Davis, M.R. Clark, and H. Waldmann. 1987. "Cell-Surface Markers on Haemopoietic Precursors: Reagents for the Isolation and Analysis of Progenitor Cell Subpopulations". *Molecular and Cellular Probes* 1: 297–326.
Zammito, John H. 2004. *A Nice Derangement of Epistemes*. Chicago, IL: University of Chicago Press.
Zuckerman, Harriet. 1977. *Scientific Elites: Nobel Laureates in the United States*. New York, NY: Free Press.

Index

S. Mauskopf, T. Schmaltz (eds.), *Integrating History and Philosophy of Science*, Boston Studies in the Philosophy of Science 263, DOI 10.1007/978-94-007-1745-9,

Q

R

MIX
Papier aus verantwortungsvollen Quellen
Paper from responsible sources
FSC® C105338

If you have any concerns about our products,
you can contact us on
ProductSafety@springernature.com

In case Publisher is established outside the EU,
the EU authorized representative is:
Springer Nature Customer Service Center GmbH
Europaplatz 3, 69115 Heidelberg, Germany

Printed by Libri Plureos GmbH
in Hamburg, Germany

Distinguished Figures in Mechanism and Machine Science

HISTORY OF MECHANISM AND MACHINE SCIENCE

Volume 7

Series Editor

MARCO CECCARELLI

Aims and Scope of the Series

This book series aims to establish a well defined forum for Monographs and Proceedings on the History of Mechanism and Machine Science (MMS). The series publishes works that give an overview of the historical developments, from the earliest times up to and including the recent past, of MMS in all its technical aspects.

This technical approach is an essential characteristic of the series. By discussing technical details and formulations and even reformulating those in terms of modern formalisms the possibility is created not only to track the historical technical developments but also to use past experiences in technical teaching and research today. In order to do so, the emphasis must be on technical aspects rather than a purely historical focus, although the latter has its place too.

Furthermore, the series will consider the republication of out-of-print older works with English translation and comments.

The book series is intended to collect technical views on historical developments of the broad field of MMS in a unique frame that can be seen in its totality as an Encyclopaedia of the History of MMS but with the additional purpose of archiving and teaching the History of MMS. Therefore the book series is intended not only for researchers of the History of Engineering but also for professionals and students who are interested in obtaining a clear perspective of the past for their future technical works. The books will be written in general by engineers but not only for engineers.

Prospective authors and editors can contact the series editor, Professor M. Ceccarelli, about future publications within the series at:

LARM: Laboratory of Robotics and Mechatronics
DiMSAT – University of Cassino
Via Di Biasio 43, 03043 Cassino (Fr)
Italy
E-mail: ceccarelli@unicas.it

For other titles published in this series, go to
www.springer.com/series/7481

Marco Ceccarelli
Editor

Distinguished Figures in Mechanism and Machine Science

Their Contributions and Legacies, Part 2

Springer

Editor
Marco Ceccarelli
DiMSAT
University of Cassino
Italy
ceccarelli@unicas.it

ISBN 978-94-007-3110-3   e-ISBN 978-90-481-2346-9
DOI 10.1007/978-90-481-2346-9
Springer Dordrecht Heidelberg London New York

Softcover reprint of the hardcover 1st edition 2009

Printed on acid-free paper

Springer is part of Springer Science+Business Media (www.springer.com)

Series Preface

This is the second volume of a series of edited books whose aim is to collect contributed papers within a framework that can serve as a collection of persons in MMS (Mechanism and Machine Science). This is a continuation of the first volume that was published in 2008, again combining very ancient and very recent scholars in order to give not only an encyclopaedic character to this project but also to emphasize the significance of MMS over time.

This project has the characteristic that the papers illustrate, by recognizing persons and their scientific work, mainly technical developments in the historical evolution of the fields that today are grouped in MMS. Thus, emphasis is also given to biographical notes describing efforts and experiences of people who have contributed to the technical achievements whose technical survey is the core of each contributed paper. This second volume of the project has been possible thanks to the invited authors who have enthusiastically shared in this initiative and who have spent time and effort in preparing the papers.

The stand-alone papers cover the wide field of the History of Mechanical Engineering with specific focus on MMS. I believe that readers will take advantage of the papers in this book and future ones by supplying further satisfaction and motivation for her or his work (historical or not).

I am grateful to the authors of the articles for their valuable contributions and for preparing their manuscripts on time. Also acknowledged is the professional assistance by the staff of Springer Science + Business Media and especially by Miss Anneke Pot and Miss Nathalie Jacobs, who have enthusiastically supported this project with their help and advice in the preparation of the first and second book.

I am grateful to my wife Brunella, my daughters Elisa and Sofia, and my young son Raffaele. Without their patience and understanding it would not have been possible for me to work on this book and the dictionary project.

Cassino, October 2009 Marco Ceccarelli (Editor)

Contents

AL-JAZARI (1136–1206)

Lotfi Romdhane and Saïd Zeghloul

Abstract Ismail Al-Jazari, Arab inventor who is remembered for his design of water-raising machines and many unusual clocks and automata. His book entitled: "Book of Knowledge of Ingenious Mechanical Devices" presents a whole range of devices and machines, with a multiplicity of purposes.

From all the presented machines, two are most remarkable: his famous elephant clock, which was by far the most sophisticated clock at that time and his invention of a water pump using a crank-slider-like system, which was the first known machine to use a crank.

Biographical Notes

Al-Jazari is by far the most famous mechanical engineer of the pre-medieval era. His name is

"الشيخ رئيس العمل بديع الزمان أبو العز إبن إسماعيل إبن الرزاز الجزري",

"Al-Shaykh Rais Al-Amal Badii Al-Zaman Abu Al-Izz Ibn Ismail Ibn Al-Razzaz Al-Jazari".

This long name is actually a combination of his title and his real name. The first three titles "Al-Shaykh Rais Al-Amal" indicate that he was a notorious chief engineer, "Badii Al-Zaman" the best of all times, "Abu El-Izz" the well respected, "Ibn Ismail" sun of Ismail (his father), "Ibn Al-Razzaz" his grandfather, and "Al-Jazari"

L. Romdhane (✉)
Laboratoire de Génie Mécanique, Ecole Nationale d'Ingénieurs de Sousse,
Technopole Sousse 4054, Tunisia
e-mail: lotfi.romdhane@gmail.com

S. Zeghloul
Université de Poitiers, Laboratoire de Mécanique des Solides,
CNRS UMR 6610, Boulevard Marie et Pierre Curie, Téléport 2,
BP 30179, 86962, Futuroscope Chasseneuil Cedex, France
e-mail: said.zeghloul@lms.univ-poitiers.fr

M. Ceccarelli (ed.), *Distinguished Figures in Mechanism and Machine Science: Their Contributions and Legacies, Part 2*, History of Mechanism and Machine Science 7,
DOI 10.1007/978-90-481-2346-9_1, © Springer Science+Business Media B.V. 2010

Fig. 1 Al-Jazari

indicates that his family came from Al-Jazira, which is the region in northern Syria and Iraq between the two rivers Tigre and Euphrates (Fig. 1).

He lived most of his life away from his homeland in Diyar Bikr in Upper Mesopotamia, which is in present-day southern Turkey. He was in the service of three consecutive Artuqid rulers: Nur al-Din Muhammad ibn Arslan (1174–1185), Qutb al-Din Sukman ibn Muhammad (1185–1200) and Nasir al-Din Mahmud ibn Muhammad (1200–1222). The only known information about his life is that given in his book. He completed writing his book in early 1206 and it took him 7 years to complete it. He recorded that when he started writing his book he had more than 25 years of experience in building machines. Hence, historians concluded that he started his active life in around 1175. Several aspects of his life made him different: the first one is that in the contrary of most scientists (علماء oulama in Arabic) at that era of Islamic civilization, he was interested only in inventing and building machines. Indeed, most of the scientists of that time were pluri-disciplinary, in the sense understood nowadays. Some of the most famous scientists, like Al-Farabi, Ibnou-Sina (Avicenna), Al-Khawarizmi, to name a few, were excelling in several disciplines going from poetry to theology to philosophy to science to medicine. Al-Jazari, however, was a true mechanical engineer and he excelled in his job as a chief engineer in the court of the Artuqid rulers.

His Work

Al-Jazari finished his single contribution in 1206. His book is entitled: ال جامع ب ين ال ع لم وال عمل ال ناف ع ف ي ص ناعة ال ح يل i.e., "Al Jamaa byna al ilm wal amal ennafii fi sinaat al hiyal", which was translated in several ways (Fig. 2).

Fig. 2 First page of Al-Jazari's book (The Topkapi Palace Museum, Istanbul, N° A 3472)

The first attempt to translate this book was by Wiedmann and Hawser, who translated parts of it into German in the first quarter of the twentieth century (Hill 1998). The most known translation of this book, however, is by Hill and is entitled "The Book of Knowledge of Ingenious Mechanical Devices" (Hill 1974). Others preferred translating it as "A Compendium on the Theory and Practice of the Mechanical Arts" (Al-Hassan 1976). Some sources refer to the same book simply as "Automata" (Chapius and Droz 1958). All these attempts tried to capture the meaning of this long and very eloquent title. If we analyze the key words of the title we can find the word "ilm علم ", which means science or knowledge in its broadest sense. "amal ennafii العمل النافع" can be translated as the useful work or power, but from the meaning of the whole title, it is probably intended for practice or the practical aspect of design or simply engineering. Al-Jazari tried by allying science and practice to design machines called here "hiyal الحيل", which is the plural of «hyla حيلة » meaning an artifact or a tricky way to get something done.

Al-Jazari's book presents a whole range of devices and machines, with a multiplicity of purposes. What is interesting to note is the considerable degree of engineering skill required to manufacture these devices, and the use of delicate mechanisms and sensitive control systems. In this respect, Al-Jazari can be called a true design engineer more than a scientist. Dimargonas, in his recent book on mechanical design (Dimargonas 2001) said: "…thus, Al-Jazari's work is really a study in systematic machine design!"

Al-Jazari's book is divided in six types (أنواع Anwaa, which is the plural of Nawâ نوع type in Arabic) totaling 50 different devices (أشكال Ashkal, plural for Shakl شكل: : geometric form in Arabic):

- Ten different Clocks
- Ten designs of Automata vessels dispensing wine and water for drinking sessions
- Ten designs of Water dispensers for ritual ablution (oudhouالوضوء in Arabic) and bloodletting devices
- Ten fountains and musical automata
- Five different designs of water raising machines
- Five miscellaneous machines like instruments for measuring spheres and locks

From all the presented machines, two are most remarkable: his famous elephant clock, which wasby far the most sophisticated clock at that time, and his inventions of a water pump using a crank-slider-like system. Even though, only a few of his inventions can be classified as utilitarian machines, Al-Jazari still departed from his predecessors when he presented these inventions. Indeed, despite him living in the court of his ruler, he was also listening to the needs of the people. The fifth chapter of his book presents five different, useful inventions mainly to raise water for irrigation and domestic purposes. This attitude was unusual at that time since most of the scientists (*Oulama* علماء) were at the service of the caliphs and their concern is mostly motivated by responding to the need of these rulers. Moreover, in writing his book, he tried to use simple Arabic and to clarify as much as he can the way his machines work and he even gave precise information on how to build the different parts of these machines. At that time, this way of writing was unusual, since most of the scientists (*Oulama* علماء) were more interested in impressing their peers by writing their books using a sophisticated language hardly understandable by common people, hence limited to the elite.

The rest of his book, but an important part of it, presents another type of engineering that is different from the utilitarian technology described in the fifth chapter. This type of technology is usually called "fine technology", since its distinguishing features derived from the use of delicate mechanisms and controls. Some of these devices had obvious practical uses: water clocks were used in astronomical observations and were also erected in public places. Others gave amusement and aesthetic pleasure to the members of royal circles, which led him to invent the first programmable humanoid robot in 1206. Al-Jazari's robot was a boat with four automatic musicians that floated on a lake to entertain guests at royal drinking parties (Margaret 2006; Franchi and Güzeldere 2005). Some of these devices functioned as curiosities.

Review of his Main Work on Mechanism Design

To show the ingenuity of the proposed designs, we will be mainly interested in water-raising machines, some of which are original and can still be seen in several contemporary machines. By far the most important contribution of Al-Jazari to mechanism design is the crank-slider system and his use in a machine. This first description of a crank-slider mechanism came 300 years before Western Engineers, e.g., Francesco di Giorgio Martini and Leonardo Da Vinci, who used the crank-slider-type mechanism in the fifteenth century (Al-Hassan 1976). Ramelli also used the crank-connecting rod in a pump in his book of 1588 (Teach and Ferguson 1994).

We also present some of his inventions in designing clocks. To show the ingenuity of Al-Jazari who, according to several historians (Ören 2001), laid down the foundation of automatic mechanical control as perceived today.

It is also interesting to note that the degree of precision in the description of his devices allowed people with no background in mechanics, to understand how these machines should work.

Water Raising Machines

Throughout history, the supply of water for domestic and irrigation purposes has always been a vital consideration. Earlier designs can be found since the antiquity. The water-raising technology was of great benefit to the people of that time and it also laid the foundation for the machinery of irrigation and the machine design, in general.

Despite being at the service of the Artuqid rulers, Al-Jazari was sensitive to the needs of the common people and he proposed in his book five different designs of water raising machines. These machines are intended to help people in everyday life and to alleviate some of their burden. Two different machines, which are mostly an improvement of the most ancient water raising machine, called the "Shaduf", are first presented. This type of machine lies on a lever system with a counterweight at one end to help fill a bucket, attached at the other end, and lifting it. This system is still in use nowadays in some Middle Eastern countries.

Al-Jazari suggested the use of a hollow tube as a lever instead of the beam and a scoop at the end instead of a bucket, so when it is raised the water can flow inside it to the irrigation system. This system made raising water easier and faster.

Al-Jazari's third machine was an improvement of the "Saqiya". The classical design of the "*Saqiya*" is an animal-powered mechanism of two perpendicular wooden gears. An animal is harnessed to the vertical axis of the horizontal wheel. The second gear is driving a vertical wheel with clay pots attached to its circumference (pots-wheel). As the vertical pots-wheel rotates, the pots are dipped into the water one by one. As each pot reaches the top, its water pours into a wooden sluice.

The first idea of Al-Jazari is to take advantage of the power of the water, in the form of a flow, and to use it to drive the system. He presents his invention (Fig. 3) where he does not eliminate completely animals from the driving system but rather he designs a complementary system to help the animal. Hence, depending on the flow of the water, the system can be fully autonomous, but if the flow is not enough (during the dry season, for example) the animal can still drive the system. This way of thinking drove most of the Al-Jazari's inventions.

The second idea of Al-Jazari is to use a pots-chain instead of a wheel with pots as it was known at that time (Fig. 3). The chain-like system solution allowed the use of the "Saqiya" in places like wells, where the classical design could not be used due to the limited space. Moreover, the proposed design can be adapted easily for different heights by simply changing the length of the chain.

Therefore, Al-Jazari improved the well-known design of "Saqiya" in two ways: the first one was by adding a water wheel to help driving the system in an autonomous way, and the second one was by replacing the wheel with pots, in the classical design, by a chain-like system containing the pots.

The way this machine was sketched is also very interesting. Al-Jazari tried to make his drawing as clear as possible by trying to show the 3D aspect of the

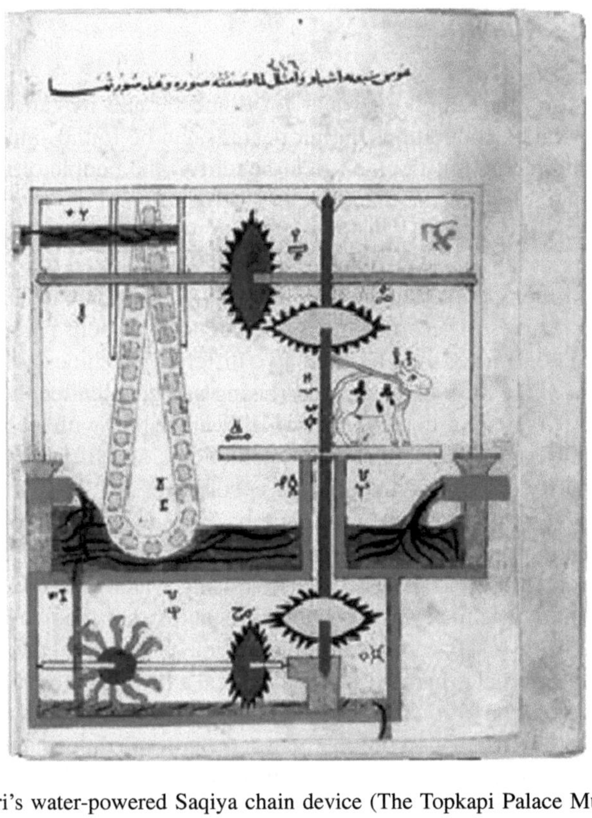

Fig. 3 Al-Jazari's water-powered Saqiya chain device (The Topkapi Palace Museum, Istanbul, N° A 3472)

machine. Indeed, one can notice that the gears, which are circular, are drawn as some sort of ellipses. Depending on the direction of the great axis of the ellipse, one can know the plane containing the gear. For example in Fig. 3 the gears with the vertical axis are drawn with "ellipses" with horizontal great axes and similarly for the horizontal axes gears. Moreover, the chain-like system of pots is represented skewed to show that it is contained in the vertical plane perpendicular to the one of the drawing, but still shows enough details. This representation is, indeed, one first attempt to use projection to represent 3D systems. Al-Jazari also used also letters and some special symbols to help him in the description of the system in very much the same way as in patents today.

The fourth machine used a flume beam (a hollow beam or a tube) and was animal powered. The tube oscillates back and forth thanks to a couple of gears driven by an animal and a crank-slider system. The scoop at the end of the tube is filled with water when it is at its lowest position and when it is raised, the water can flow inside the tube to the irrigation system. The idea of the tube here is very similar to the one used in the first two machines (Shadhuf). This shows how Al-Jazari is using modular designs and then improving on them each time, which attests to his practical sense of optimizing the design process. What is original, however, in this invention is his idea of converting a continuously rotating motion to an oscillating one using an inversion of the crank-slider mechanism. Several historians (Hill 1998; Bertrand 1969) considered this design as the first one to use a crank-slider mechanism in a machine. Indeed, even though the adopted solution is probably not the best, according to our contemporary knowledge, this design opened a whole era for more complex machines going up to the steam engine and the internal combustion engines. The idea of having a pin-in-a-slot type joint will be used again in his fifth invention (Fig. 4).

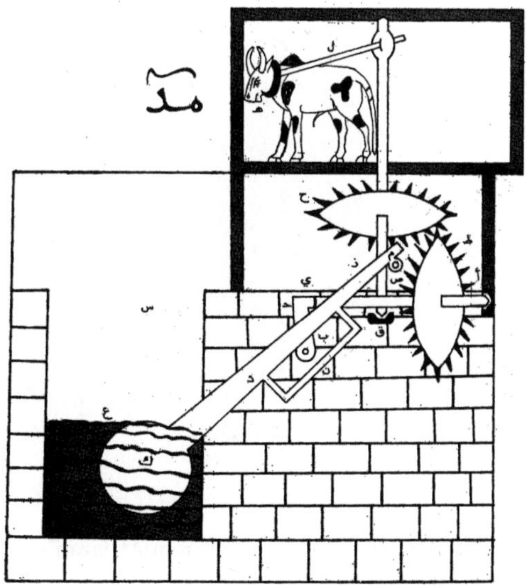

Fig. 4 Fourth water raising machine by Al-Jazari from Al-Jazari's manuscript

The fifth machine, presented by Al-Jazari, took this idea of converting different types of motion further by proposing what we nowadays call a piston pump. The translation motion of the piston is obtained by transforming a continuously rotating motion through a tricky system of gears and links.

This machine, a twin-cylinder, water-driven pump, represents the first attempt to use suction and discharge to lift water to a certain height. Up until that time (twelfth century), all the water-raising machines relied on filling some type of container and lifting it to empty it at the desired height. The machine proposed by Al-Jazari went away from this solution to try to force water through a pipe, which allowed him to lift water to heights never reached before with a relatively small-size machine. This invention contains the most features of relevance for the development of mechanical design, and is very similar to contemporary pumps (Fig. 5).

Three main novelties can be noted in this design. The first one, and probably the most important, is the use of eccentricity to convert the continuously rotating

Fig. 5 The reciprocating pump from Al-Jazari's manuscript (Topkapi Palace Museum Library, Ahmet III 3472)

motion to an oscillating motion. Moreover, through a pin-in-slot type joint, Al-Jazari succeeded in getting the necessary reciprocating motion of the pistons.

The second novel aspect in this machine is the use of valves as a means to allow the suction and the discharge process in the reciprocating pump. Several contemporary machines are still using this design in controlling the suction and the discharge phases (steam engine, internal combustion engine, piston pumps…). All the historians agree that this invention was not known before by any of his predecessors.

The last aspect is the use of a large water wheel to make the system completely autonomous. Some historians mentioned that Al-Jazari also used animals to drive this machine, mainly during the drought seasons where the level of the water is low.

Clocks

Al-Jazari also mentioned many amusing, yet practical, devices, including several unusual clocks which were much more elaborate than those developed before. They were sometimes driven by weights. As the weight fell, it encountered resistance from a float riding in a water tank. Others involved a bucket that tipped over when filled with water, hitting a ratchet that moved the clock ahead (Fig. 6).

a

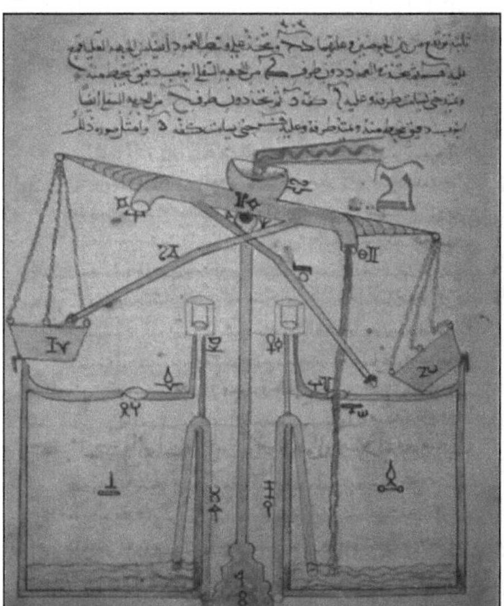

b

Translation: (a vertical pole) located between the two basins is used to sustain the horizontal tube in the middle. On top of the tube at the middle a funnel and the two ends of the tube are opened. At the bottom side of the tube, a small tube emerges from it and extends until it reaches the bucket. Also it gets out of the tube another small tube and it extends until it reaches the bucket. And I represent all in the figure.

Fig. 6 (**a**) Water clock by Al-Jazari (Süleymaniye library, Istanbul, Turkey), (**b**) translation of the text

Water clocks

This water clock uses two tipping buckets that are filled alternatively, making the horizontal tube tip from one side to the other. The flow of water is therefore switching from one tank to the other after a fixed period of time required to fill one bucket. A similar design was proposed by Al-Jazari as a musical instrument. The flow of water in the tank chases the air inside it and forces it to go through a whistle (Nadarajan 2007; Hill 1974).

The interesting idea in this design is the use of the small tubes to fill the two buckets. The oscillation of the tube triggers a system (not represented) to indicate the time at constant intervals. This clock can work as long as the water is flowing.

Candle Clocks

Al-Jazari also describes candle clocks which all worked in a similar manner. Each design specified a large candle with a uniform cross-section and known weight. The candle was installed inside a metal sheath, to which a cap was fitted. The cap had a hole in the centre, around which was an indentation on the upper side. The candle, whose rate of burning was known, had its wick going through the hole. The bottom of the candle rested in a shallow dish that had a ring on its side and was connected through pulleys to a counterweight. As the candle burned away, the weight pushed it upward at a constant speed. The automata are operated from the dish at the bottom of the candle by letting a small sphere drop in a container, which is suspended to the arm of the robot. Under the weight of the ball, the arm of the robot lowers, indicating the hour. The container descends until it hits the floor, which makes it to tip and to let the ball out (Fig. 7).

A second design is presented in Fig. 8, where the scribe sitting on the left is rotating as the candle goes up. A set of pulleys and a cable with a counterweight allows the transmission of the motion between the candle and the scribe. The tip of his pen indicates the hours and the minutes, which are engraved on a circular scale. The set of spheres on the right is made such that one sphere drops from the falcon's beak at the top of each hour.

Weight Clocks

The most famous of Al-Jazari's clocks is his three-meter-high elephant clock. Using weights dropped from its top at regular intervals, an intricate mechanism is triggered.

The clock features a perforated submersible bowl, a device that is at the heart of the clock mechanism. The small orifice in the submersible bowl is carefully

a

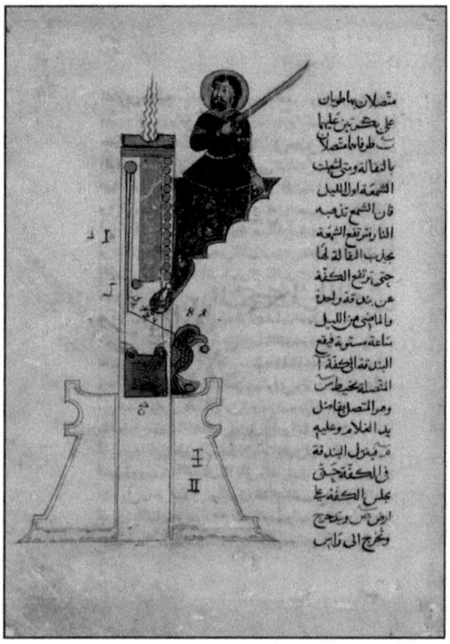

b

Translation:
The candle is attached by two strings through two pulleys to a weight. As the candle burns away it gets lighter and it lifts off under the effect of the weight, which lets one small ball to drop in a container suspended by a string to the arm of the robot. Under the weight of the ball, the arm of the robot lowers, indicating the hour. The container tips when it gets in contact with the floor and the ball rolls out of the mouth of the bird.

Fig. 7 (**a**) Candle clock of Al-Jazari (Smithsonian Institute), (**b**) translation of the text

Fig. 8 A candle clock from Al-Jazari's manuscript

calibrated to produce the correct rates of flow. This rate of flow determines the time at which the clock strikes at hourly intervals. As the bowl fills with water, it sinks and a set of pulleys and a cable with a counterweight allows the transmission of the motion to the scribe sitting on the back of the elephant. The tip of his pen indicates the hours and the minutes, which are engraved on a circular scale. When the bowl reaches the bottom, it releases a ball from the castle, which falls onto a channel mechanism that rotates a dial to reveal the time. The ball then travels to the falcon's beak and is then dropped into the mouth of a snake, which then swings due to the weight of the ball and drops the ball into a vase. The motion of the snake pulls on the bowl and empties it. The ball then disappears into the elephant's body and it activates the Mahout sitting on the head of the elephant to strike the cymbals. The snake swings back to his original position due a counterweight. At this time, the bowl is again floating and it starts to sink and the whole process repeats itself as long there are balls on top of the elephant (Fig. 9).

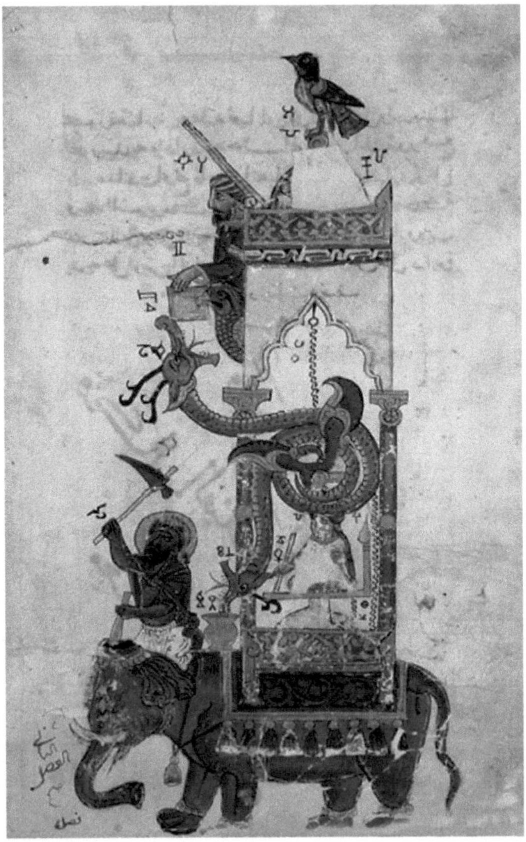

Fig. 9 Elephant clock (NY Metropolitan Museum)

Miscellaneous

Other chapters of Al-Jazari's work describe fountains and musical automata which are of interest mainly because the flow of water in them alternated from one large tank to another at hourly or half-hourly intervals. Several ingenious devices for hydraulic switching were used to achieve this operation (Rosheim 1994). These revolutionary machines owed him the title of the father of robotics (Chapius and Droz 1958; Nocks 2007).

One of Al-Jazari's machines was a boat with four automatic musicians that floated on a lake to entertain guests at royal drinking parties. It had two drummers, a harpist and a flautist (Fig. 10). The heart of the mechanism is a rotating cylindrical beam with pegs (cams) protruding from it. These just bump into little levers that operate the percussion. The point of the model is to demonstrate that the drummer can be made to play different rhythms and different drum patterns if the pegs are moved around (Fig. 11) (Ören 2001). Al-Jazari is also an inspiration for a musical robot live-coding system bearing his name, and his work is acknowledged in several contemporary books on robotics and mechanical design (Ceccarelli 2004; Dimargonas 2001).

Al-Jazari also noted a number of practical joke devices in his manuscript. Some were trick drinking vessels that appeared to contain water but could not be emptied. Others looked empty but produced water when tipped over.

In producing these not-so-useful inventions, Al-Jazari was typical of his age.

Fig. 10 Musical Automata from Al-Jazari's manuscript

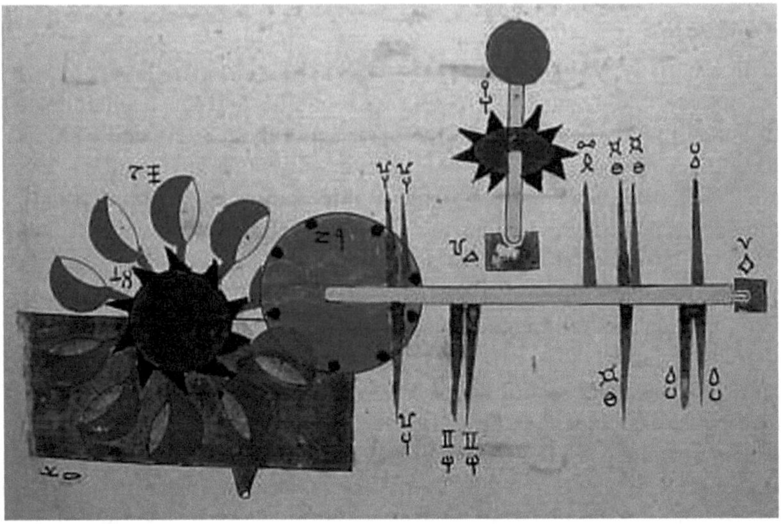

Fig. 11 Cam mechanism from Al-Jazari's manuscript

On the Circulation of his Work

Many copies of this book are conserved in a number of museums worldwide such as Top Kapi in Istanbul, and the Museum of Fine Arts in Boston, the museum of the Louvre in France and Oxford's Bodleian Library (Hill 1998).

British engineer Donald Hill, who has a special interest in Arab technology, writes: "It is impossible to over-emphasize the importance of Al-Jazari's work in the history of engineering; it provides a wealth of instructions for design, manufacture and assembly of machines" (Hill 1974).

About al-Jazari's book, Sarton says that "this treatise is the most elaborate of its kind and may be considered the climax of this line of Moslem achievement" (Sarton 1975).

Al-Jazari's book was translated into English by Hill (1974). It is a complete translation of his work, using mainly MS Graves 27 at the Bodleian Library, Oxford, UK (Hill 1998).

It is worth mentioning that the written Arabic heritage in mechanical technology began with the Banu Musa book, written in the nineth century. Banu Musa described a hundred ingenious devices. They have made contributions to technological refinements and new applications. However, Saliba (2007) compared the work of Al-Jazari and the work of Banu Musa and he concluded: ".... when their works (Banu Musa) are compared to the work of Al-Jazari, we note a remarkable maturity in the latter's work that is nowhere to be found in the works of Banu Musa. With Al-Jazari, we begin to notice discussions regarding the real function of mechanical devices, and real appreciation of their significance as tools that did not only fulfill daily functions for the society but that they were also tools that could demonstrate the way in which the natural physical principles worked" (Saliba 2007).

Many of the ideas employed in the construction of ingenious devices were useful in the later development of mechanical technology and they had a tremendous influence on the history of the mechanical design of machines. His utilitarian inventions were also used in the entire Arab world at that time and spread all the way to Spain. It is believed that the most likely location for the transfer of information between the East and the West was Spain where Muslims and Christians coexisted for eight centuries (Cigola and Gallozzi 2000; Cigola 2006). On the other hand, with the Crusades, the West met the East and the transfer of scientific and technical knowledge from the Eastern world was accentuated.

We can also note that Al-Jazari introduced several individual components like conical valves and locks. Conical valves appeared later in Leonard da Vinci's book (1495) and as a float-controlled regulator for steam boilers in England in 1784 (Mayer 1970).

Al-Jazari developed techniques to build his machines such as casting in closed mold boxes, which reappeared in Europe in 1500 (Smith 1968; Baer 1984).

In his book, Rosheim (2006) mentioned the book of Al-Jazari as one of the sources that might have influenced Leonardo da Vinci in building some of his clocks. This information was also reported in (Needham 1965) in his study on the influence of China and the Medieval Arabic civilization on the West.

Taqi Eddin, in the sixteenth century, also proposed a six-cylinder pump most likely inspired from the pump, which was invented by Al-Jazari 300 years before (Al-Hassan 1976).

Al-Jazari's third water-raising machine allowed lifting water higher by simply increasing the length of the chain and without drastic changes in the whole design. However, with classical designs of the "*Saqiya*", increasing the height requires changing the whole wheel for one having greater diameter, which is a much more difficult task.

One such machine was located on the River Yazid in Damascus (thirteenth century) (Fig. 12) and is thought to have supplied the water needs of a nearby hospital.

Several prototypes of Al-Jazari's machines have been built recently. During the Islam Festival in England in 1976, three of Al-Jazari's machines were shown. The first one was a monumental water clock, which is now exhibited in Asten in the Netherlands. Other prototypes can be seen in Frankfurt, Germany and in Aleppo, Syria. A full-scale prototype can also be seen in a mall in Dubai (Fig. 13) and a very elaborate simulation of the same system can be found on the web (Fig. 14).

In addition to the elaborate explanations and illustrations, Al Jaziri's book has an artistic and a historical value. All illustrations in the book are in color and they have great artistic merit. Historians believe that great artists lived at the court of the Artuqids who helped Al-Jazari in illustrating his book. These illustrations were also used by historians to study the clothing styles of men and women in Diyar Bikr in the thirteenth century (Akurgal and Hilber 1980; Raby 1985; Turner 1996).

Fig. 12 Vestige from a Saqiya in Damascus, Syria http://www.flickr.com/photos/damascene/80436428/

Fig. 13 Al-Jazari's clock at Dubai's Ibn Battuta Mall

Fig. 14 Simulation of the elephant clock (http://communication.ucsd.edu/aaytes/transliteracies/cezeri2.htm)

Modern Interpretation of Main Contributions

His inventions of water-raising machines have been in use since the thirteenth century. Indeed, the idea of having a chain of pots, in building the Saqiya, is still used nowadays in some Middle Eastern countries. Figure 15 shows a recent sketch of a Saqiya and the built system in the south of Egypt. We can notice that the use of two cogwheels with different diameters, as shown in Fig. 15, was not proposed by Al-Jazari.

In all his machines, Al-Jazari uses only cogwheels with identical diameters; hence we can conclude that gears are used in order to transmit movement between perpendicular axes rather than as a speed multiplier.

The fourth raising machine proposed by Al-Jazari (Fig. 4) is the first one made of only rigid bodies. The kinematic diagram of this machine is shown in Fig. 16. The dimensions are chosen as close as possible to the ones proposed in the manuscript. In Fig. 16a, we can see the lowest position of the tube where the bucket is being filled with water, and in Fig. 16b we have the second extreme position of the tube being emptied in the irrigation system. Figure 17 shows a planar kinematic

Fig. 15 (**a**) A recent sketch of a Saqiya as proposed by Al-Jazari, built in the south of Egypt, (**b**) the built system (http://www.saqiya-luxor.com/)

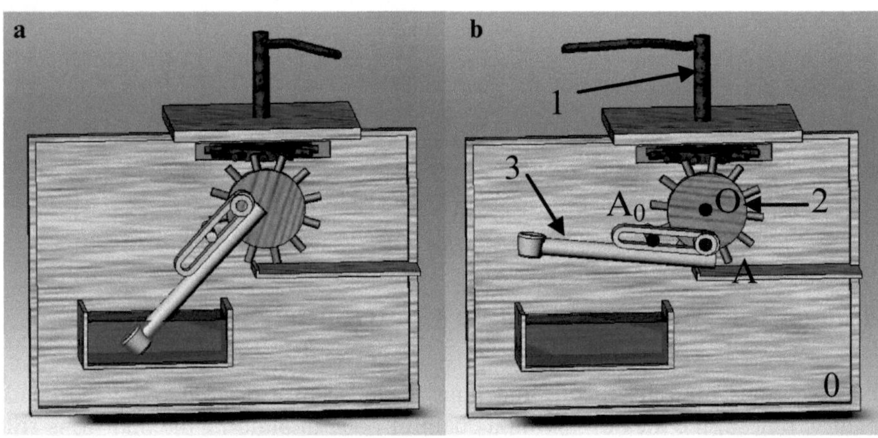

Fig. 16 a 3D CAD model of Al-Jazari's fourt water raising machine. (**a**) Lower position, (**b**) upper position

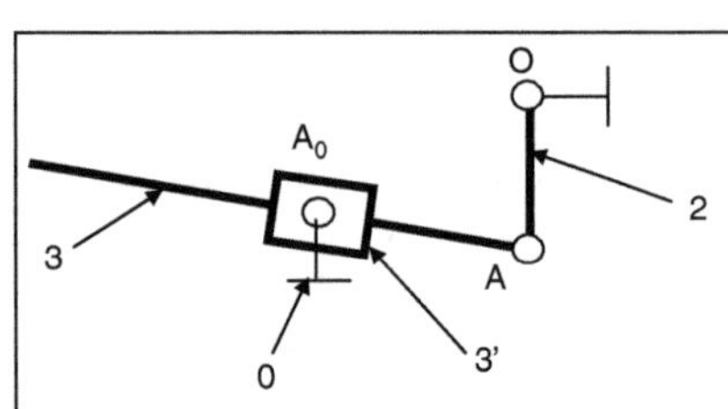

Fig. 17 A planar kinematic diagram of Al-Jazari's water-raising machine

diagram of the proposed machine. The pin-in-a-slot joint was replaced by two lower kinematic pairs. Therefore, the proposed mechanism is an inversion of the well-known crank-slider mechanism. Indeed, most historians agree that this is the first use of a crank-slider system in a machine and is probably the first use of the pin-in-a-slot joint also (Al-Hassan 1976; Teach and Ferguson 1994; Hill 1998; Bertrand 1969).

The drawback of this machine is its limited height to which it can lift water. This limitation inspired Al-Jazari to move away from the classical solution of raising the water and to invent the pumping solution. His fifth machine (Fig. 5) is a twin cylinder pump where the crank and the pin-in-a-slot joint are again used to build this system. Figure 18 shows a 3D CAD model of the system. The planar mechanism made of bodies 0-2-3-4 is the one transforming the continuous rotation into a reciprocating motion of the pistons (4). Figure 19a shows a kinematic diagram of the mechanism when the joint in B is modeled by a hinge joint.

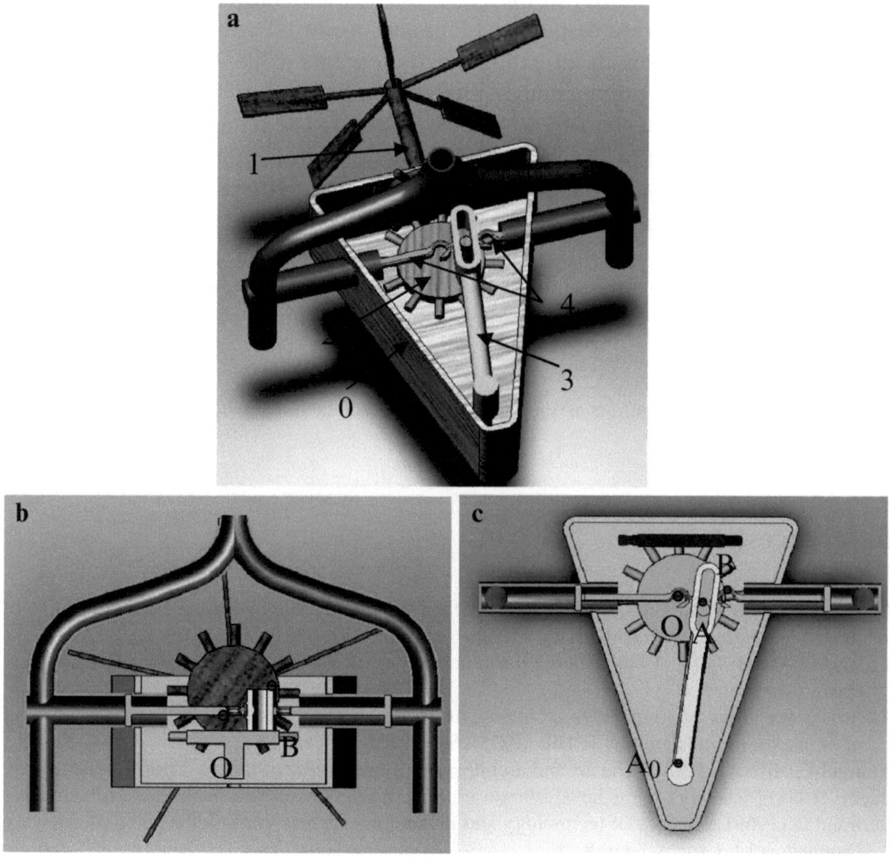

Fig. 18 (**a**) 3D CAD model of Al-Jazari's pump, (**b**) a vertical cut, (**c**) an horizontal cut

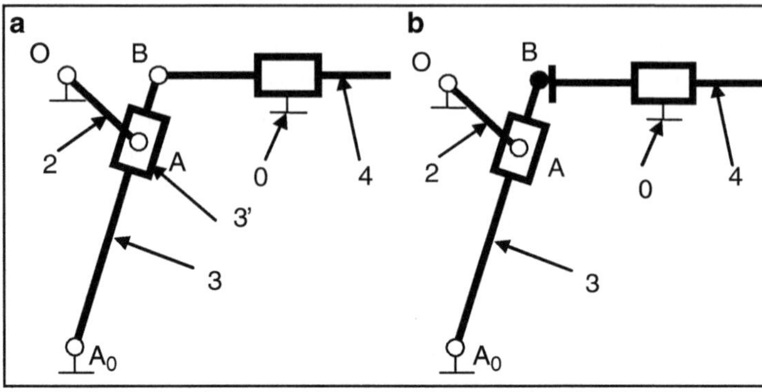

Fig. 19 A planar kinematic diagram of Al-Jazari's pump. (**a**) A hinged solution, (**b**) the adopted solution

One can see that the mobility of the system is 0 and no motion is allowed in this case. Al-Jazari was aware of this problem and proposed a unilateral joint made of a revolute joint but with an exaggerated clearance (Fig. 19). Therefore, this can be modeled by a point contact joint as shown in Fig. 19b. This practical solution allowed the system to work despite the existence of a backlash when inversing the direction of the motion of the pistons. Since the input speed is not important, the behavior of the system was deemed satisfactory by Al-Jazari. Nowadays, this mechanism has a connecting rod to allow the correct functioning.

This type of mechanisms is not far from a design point of view, from mechanisms taught at engineering schools nowadays, and its similarity with the quick-return mechanism is striking.

Acknowledgement The authors wish to thank Samir Lahouar for his help in drawing some of the presented models.

References

Al-986 AY (1976) Taqi Eddin and Arabic Mechanical Engineering. Institute for the History of Arabic Science, Aleppo University, Aleppo

Al-Hassan AY, Hill DR (1986) Islamic technology – an illustrated guide. Cambridge University Press, Cambridge

Al-Jazari, *The Book of Knowledge of Ingenious Mechanical Devices: Kitáb fí ma'rifat al-hiyal al-handasiyya*, translated by P. Hill (1973), Springer Science.

Akurgal E, Hilber L (1980) The art and architecture of turkey. Rizzoli, New York

Baer E (1984) Metalwork in medieval islamic art. State University of New York Press, Albany, NY

Bertrand G (1969) A history of technology and invention, progress through the ages, ed. Maurice Daumas, vol I. Crown, New York

Bunch BH, Hellemans A (1993) The timetables of technology: a chronology of the most important people. Simon & Schuster, New York

Ceccarelli M (2004) Fundamentals of mechanics of robotic manipulation (intelligent systems, control and automation: science and engineering). Springer, New York
Chapius A, Droz E (1958) Automota: a historical and technological study. Neuchatel: Editions du Griffon, France
Cigola M (2006) Greek and Muslim automata: influence of Greek knowledge in medieval Islamic science. In: Proceedings of the 2nd International Conference on Ancient Greek Technology, Athens, 2006
Cigola M, Gallozzi A (2000) Drawings of automata in middle age. In: Ceccarelli M (ed) Proceedings International Symposium on History of Machines and Mechanisms, Cassino, Italy, 11–13 May 2000
Dimargonas AD (2001) Machine design: a CAD approach. Wiley-IEEE, New York, p 14
Franchi S, Güzeldere G (2005) Mechanical bodies, computational minds. MIT Press, Boston, MA
Hill DR (1998) Studies in medieval Islamic technology [Texte imprimé]: from Philo to Al-Jazari – from Alexandria to Diyar Bakr. David A. King, Aldershot (GB)
Losano Mario G (2003) Automi d'Oriente. "Ingegnosi meccanismi" arabi del XIII secolo. Medusa Edizioni, Italy
Margaret A (2006) Mind as machine: a history of cognitive science. Oxford University Press, Oxford
Mayer O (1970) The origins of feedback control. MIT Press, Cambridge
Nadarajan G (2007) Islamic automation: a reading of Al-Jazari's the book of knowledge of ingenious mechanical devices (1206). MediaArtHistories/MIT Press, Cambridge, MA), pp 163–178
Needham J (1965) Science and civilization in china, Vol IV-2. Cambridge University Press, London, pp 435–481
Nocks L (2007) The Robot: the life story of a technology. Greenwood, West Port, CT
Ören (2001) Advances in computer and information sciences: from Abacus to Holonic agents. Special Issue on Artificial Intelligence of Elektrik (Turkish Journal of Electrical Engineering and Computer Sciences – Published by TUBITAK – Turkish Science and Technical Council), 9:1, 63–70, T.I. 2001
Raby AJ (ed) (1985) The art of Syria and the Jazira. Oxford University Press, Oxford
Rosheim ME (1994) Robot evolution: the development of anthrobotics. Wiley-IEEE, New York
Rosheim ME (2006) Leonardo's Lost Robots. Springer-Verlag, Berlin, p 117
Saliba G (2007) Islamic science and the making of the European. MIT Press, Boston, MA
Sarton G (1975) Introduction to the history of science, vol II. Krieger, New York
Seo MH, HaHn Y, Lee J, Nam M (2000) Clock-work mechanisms of the automatic striking cepsydra Jagyungnu. In: Ceccarelli M (ed) Proceedings International Symposium on History of Machines and Mechanisms, Cassino, Italy, 11–13 May 2000
Smith CS (1968) The early history of casting moulds and the science of solidification. In: Mullins WW, Shaw MC (eds) Metal transformations. Gordon & Breach, New York
Teach GM, Ferguson ES (1994) The various and ingenious machines of Agostino Ramelli. Dover, New York
Turner HR (1997) Science in medieval Islam: an illustrated introduction. University of Texas Press, Austin
Turner J (1996) The Grove Dictionary of art. Oxford University Press, New York, p 322
Ungerer A (1930) Horloges monumentales. Edition des archives alsaciennes d'histoire de l'art, Strasbourg
http://communication.ucsd.edu/aaytes/transliteracies/cezeri2.htm
http://en.wikipedia.org/wiki/Al-Jazari
http://www.history-science-technology.com/Notes/Notes%203.htm
http://news.independent.co.uk/world/science_technology/article350594.ece
http://www.muslimheritage.com/day_life/default.cfm?ArticleID=188&Oldpage=1
http://www.muslimheritage.com/topics/default.cfm?ArticleID=466
http://www.pawfal.org/index.php?page=AlJazari
http://www.shef.ac.uk/marcoms/eview/articles58/robot.html
http://www.knowledgehunter.info/wiki/Robot

Ivan Ivanovich Artobolevski (1905–1977)

Olga V. Egorova and Nikolai V. Umnov

Abstract Ivan Ivanovich Artobolevski devoted his life to the science of machines and to the creation of a theoretical base for the construction of modern and future machines which could improve man's work and release him from tiresome, monotonous and unhealthy operations. He received enormous respect and authority in his Motherland and great recognition abroad as a talented scientist, organizer and also as a statesman.

Biographical Notes

Ivan Ivanovich Artobolevski (Fig. 1) was born in Moscow on October 9, 1905 to the family of a faith teacher of Moscow Imperial Commercial School.

His father, Ivan Alexievich Artobolevski, was elected MA professor of theology at Moscow Agricultural Institute (Timiryazev Agricultural Academy nowadays). Artobolevski's house was always full of scientists and philosophers, and among them the elder friend of his father, the famous historian in Russia – Vasily O. Kluchevsky who, in those days, worked as a professor at Moscow State University named after Michael V. Lomonosov (MGU). The atmosphere of creative work and the relation with well-known natural scientists such as D.N. Prianishnikov, S.A. Zernov and N.A.Kulagin et al. couldn't but influence on the interest of Ivan Ivanovich towards the acquisition of knowledge, the learning of natural science and technology. Artobolevski passed his first examination in 1915 when he went on to

O.V. Egorova (✉)
Department of Theory of Machines and Mechanisms, MGOU: Moscow State Open University, 129805 P.Korchagin Street, 22, Moscow, Russia
e-mail: cuba2006@inbox.ru

N.V. Umnov
IMASH: Institute for Engineering Science named after A.A.Blagonravov, P.O. Box 101990 Maliy Haritonevski per., 4, Moscow, Russia
e-mail: n_umnov@mail.ru

M. Ceccarelli (ed.), *Distinguished Figures in Mechanism and Machine Science: Their Contributions and Legacies, Part 2*, History of Mechanism and Machine Science 7, DOI 10.1007/978-90-481-2346-9_2, © Springer Science+Business Media B.V. 2010

Fig. 1 Ivan Ivanovich Artobolevski (1905–1977)

the Third Moscow Classical Gymnasium. Later, in 1921, under the influence of the outstanding teacher and professor A.F.Fortunov, who inculcated in his pupil love for mathematics, he went on to the faculty of mechanical engineering at the Timiriazev Agricultural Academy and successfully graduated from it in 1924.

It is interesting that his governess taught the little Ivan French and he advanced so much in his studies that, being a young boy, was able to read French books in their original form. This knowledge was useful to him in the future when he became a well-known scientist.

After the October revolution, in 1918 by reason of the separation of the church from the state and the extinction of the chair of theology, Artobolevski's father was dismissed and, in spite of being an excellent teacher and scientist, he was forced to accept the condition of a parson in a parochial church.

Naturally the family was in poverty and, besides, they had to quit their service apartment. The future scientist had to begin his labour activity combining studies with work as a librarian, a laboratory assistant, a preparer in the chair and a land surveyor's assistant.

The first and the main of I. Artobolevski's teachers were V.P. Goriachkin and N.I. Mertsalov. The founder of the new, for that time, science – "The agricultural mechanics", academician Goriachkin exerted great influence on further scientific activities of Ivan Ivanovich who worked under his leadership as a mechanics laboratory assistant, and obtained a lot of knowledge and experience in the field of the dynamics of machines and experimental investigations. Thanks to his chef, Professor Mertsalov, Artobolevski became interested in the theory of mechanisms, and studied projective geometry and spherical trigonometry.

According to his teachers' advice he attended a complete course at the physicomathematic faculty of Moscow State University.

It's not casual that after having obtained a ponderable stock of fundamental knowledge, even being very young he solved, a number of complicated problems in the field of the theory of flat and space mechanisms of agricultural machines.

Ivan Artobolevski was an excellent teacher. His lectures were always full of the latest information about the achievements in the international science and technology. He skilfully organized the educational process and promoted the introduction of modern methods of teaching. In 1940, he published a textbook for students, postgraduates and teachers of the mechanic- mathematical faculty of MGU, which became the basis of the university course "Theory of Mechanisms and Machines". Artobolevski constantly improved the courses he taught, and his textbooks gained international recognition and were repeatedly republished abroad.

Since he was young he had loved science, and in spite of his enormous lack of time, he always found time for literature, theatre, music and art. As a time cultured and educated person, he was keen on architecture, he collected books on art and painting. He preferred Russian artists and painters.

His acquaintance with one of the greatest intelligencia, A.V. Lunacharski, had influence on the development of young Ivan's artistic taste, because he visited his house and often met many outstanding personalities of literature, music and art there. One of his favourite writers was and continued to be up to his last days, the Russian writer Fedor M.Dostoevski. The scientist's wife, Olga Nikolaevna Artobolevski, being a pianist and singer (Fig. 2) promoted his approach to art personalities.

Fig. 2 Ivan I.Artobolevski and his wife Olga Nikolaevna

The most prominent Soviet and foreign scientists, musicians, artists could be seen with their hospitable family. They were friends with many opera singers and artists of drama theatres, painters, writers and sportsmen. They arranged home concerts with the participation of famous artists in their country house as well as in their apartment in Moscow. The great Russian tenor Ivan S. Kozlovski, being their friend, sang his arias and romances at these parties. It is not causal that the list of Artobolevski's friends mentioned in his autobiography, amazes by the abundance of known and unknown names.

All the colleagues and students noticed his remarkable self-control. Artobolevski never lowered himself to blasphemy, even in the circle of people sharing the same ideas, although sometimes the lack of a basis of the attacks of his detractors who insisted on trying to prove that a man with such origins (from the family of a theologist) he couldn't be a good mathematician, gave reason for this. What is this? Is it a character, an education and good breeding, or a strong part of Christian moral inculcated in his childhood? I think, all together.

One more fact in his biography arouses admiration. He voluntarily joined the army in 1941. Since the beginning of the Second World War all the collaborators of the Institute for Engineering Science (IMASH) had been evacuated to deepest Kazan (Russia), but the correspondent-member of the Academy of Sciences of the USSR Artobolevski, went to the war as a volunteer. Fortunately, he had been there only three weeks because, on the orders of the higher leadership of the country, all scientists, including him, were taken from the front. But the deed itself shows his immense conscience and patriotism, his true love to his Motherland.

In his mortal life, Artobolevski was a very interesting person, he liked classical music, art, read a lot, was fond of sports, played tennis, cultivated flowers and loved animals. He lived an interesting but sometimes uneasy life, especially in his youth because of his social origin. But he continued working and believed that the awful injustices towards him would stop one day. And so it happened.

His multiphase scientific activity was highly appreciated by the Academy of Sciences of the USSR. In 1936 he had a scientific degree of Doctor in engineering science conferred on him; in 1939 he was elected a correspondent-member of the Academy and in 1946, at the age of 40, an academician on mechanics. Due to his great services in the development of national science and social activity, Artobolevski was awarded with the highest title of the Soviet Union – the Hero of Socialist Labour and the highest award – the Lenin Order (during his life he had been awarded this six times!). He achieved a lot and all his achievements received the highest assessment in his Motherland.

The academician also gained enormous recognition abroad. In 1967, the Institute for Mechanical Engineering of Great Britain awarded him James Watt Gold Medal (Fig. 3) – the highest award in the world given to engineers. Earlier this award was conferred on the greatest mechanics and engineers of the world – G.Ford, A.Stodol, and G.Taylor.

Artobolevski dedicated a lot of his efforts and energy to the integrity of the specialists of different countries, who worked in the field of the theory of machines, and to the International Federation on Theory of Machines and Mechanisms

Fig. 3 Ivan I.Artobolevski in London being awarded with the James Watt Gold Medal

(IFToMM), as he understood perfectly well the importance of international contacts for Soviet and world science.

This Federation was founded in 1969 and Ivan I.Artobolevski was unanimously elected its first President. He held the post for two periods and then became its past-president up to his final days. This is the greatest international recognition of a man who had dedicated all his life to chosen science.

As his President, he directed the work of the sessions of general assemblies, the meetings of their execution councils, participated in the work of organizing committees of the international magazine IFToMM, drawing more and more Soviet and foreign specialists into the work of the Federation. It is here where he used the French language which he spoke to perfection.

Artobolevski also carried out much work for the International Organization of Scientific Workers (IOSW) and he was elected its vice-president in 1965. This organization was founded in 1945 by the great French man, Frederick Jolio Curie, who could direct the efforts of scientists towards the struggle for the peace and happiness of all people, against nuclear war and the use of other means of mass destruction. In 1959 Ivan Artobolevski was awarded the Jolio Curie Silver Anniversary, medal by the International Peace Council.

Aacademies of science of many countries elected Artobolevski as a honorary academician and honorary doctor. He was awarded many local and international honours.

Artobolevski's life is a vivid example of the untiring scientific, social and state activity of a scientist who did a lot to promote scientific, technological and engineering programs and the flourishing of mankind. And he achieved everything through his work and talent.

A Brief History of IFToMM

IFToMM was founded as the International Federation on Theory of Machines and Mechanisms in Zakopane, Poland, on September 29, 1969, during the Second World Congress on the Theory of Machines and Mechanisms (TMM) (Fig. 4).

The main promoters of the IFToMM World Federation were Academician Ivan I.Artobolevski (USSR) and Prof. Erskine F.R.Crossley (USA), whose principal aim was to overcome the obstacles of the time of the Cold War by developing international collaboration on TMM science for the benefit of the world.

IFToMM started as a family of TMM scientists among whom we may identify the IFToMM founding fathers who signed or contributed to the foundation act (Fig. 5), with the initial 13 Member Organizations (countries), in the persons of: Academician Ivan I. Artobolevski (USSR), Prof. Erskine F.R. Crossley (USA), Prof. Michael S. Konstantinov (Bulgaria), Dr. Werner Thomas (GFR), Prof. B.M. Belgaumkar (India), Prof. Kenneth H. Hunt (Australia), Prof. J. Oderfeld (Poland), Prof. Jack Phillips (Australia), Prof. George Rusanov (Bulgaria), Prof. Wolfgang Rössner (GDR), Prof. Zènò Terplàn (Hungary), Prof. Jammi S. Rao (India), Prof. Giovanni Bianchi (Italy), Prof. Adam Morecki (Poland), Nicolae I. Manolescu (Rumania), Leonard Maunder (UK), Douglas Muster (USA), Ilic Branisky (Yugoslavia).

The foundation of IFToMM was the result of the intense activity in stimulating and promoting international collaboration, more that never in the past, as started in the late 1950s, as is documented by several letters that are stored in the IFToMM Archive.

Fig. 4 1969 IFToMM Congress (IFToMM Archive)

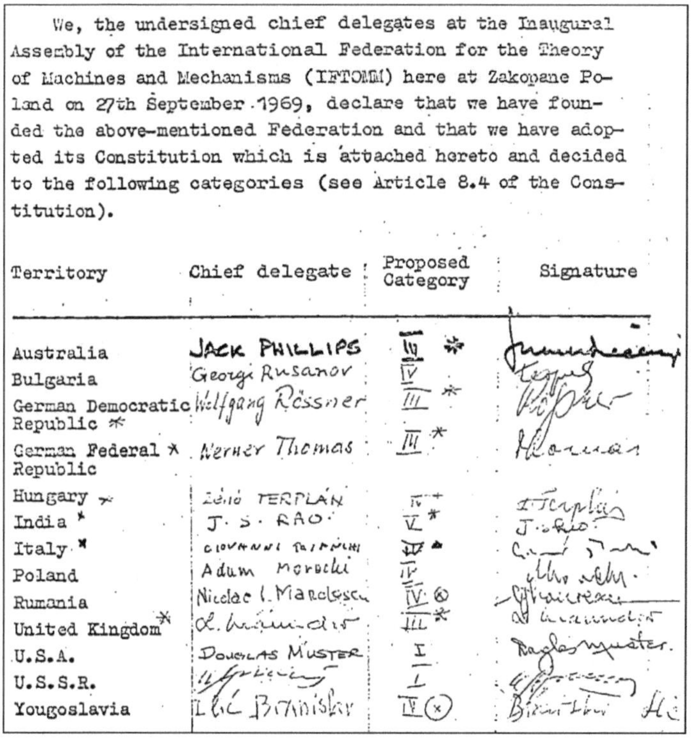

We, the undersigned chief delegates at the Inaugural Assembly of the International Federation for the Theory of Machines and Mechanisms (IFTOMM) here at Zakopane Poland on 27th September 1969, declare that we have founded the above-mentioned Federation and that we have adopted its Constitution which is attached hereto and decided to the following categories (see Article 8.4 of the Constitution).

Territory	Chief delegate	Proposed Category	Signature
Australia	JACK PHILLIPS	III *	[illegible]
Bulgaria	Georgi Rusanov	IV	[illegible]
German Democratic Republic *	Wolfgang Rössner	III *	[illegible]
German Federal * Republic	Werner Thomas	III *	[illegible]
Hungary *	Zénó TERPLÁN	IV *	[illegible]
India *	J. S. RAO.	V *	[illegible]
Italy *	GIOVANNI [illegible]	IV *	[illegible]
Poland	Adam Morecki	IV	[illegible]
Rumania	Nicolae I. Manolescu	IV ⊗	[illegible]
United Kingdom *	[illegible]	III *	[illegible]
U.S.A.	DOUGLAS MUSTER	I	[illegible]
U.S.S.R.	[illegible]	I	[illegible]
Yougoslavia	Ilić Branislav	IV (×)	[illegible]

Fig. 5 IFToMM Foundation Act, 1969 (IFToMM Archive)

The first World Congress was held in 1965 in Varna, Bulgaria, during which the foundation of IFToMM was planned and it was later agreed to during the Second World Congress on TMM in Zakopane, Poland. The Congress series was immediately recognized as the IFToMM World Congress and in 2007 there were celebrations of the 12th anniversary with the participation of the 48 Member Organizations.

The IFToMM community has continuously grown over time and the TMM has evolved to large engineering science, including new disciplines. This has brought, in 2000, to the amendment of the name of the IFToMM Federation to the IFToMM International Federation for the Promotion of Mechanism and Machine Science and a change in the name of TMM to MMS (Mechanism and Machine Science), in order to emphasize the modern mission of the IFToMM community.

IFToMM activity has grown in many aspects and nowadays there are 13 Technical Committees, five Permanent Committees, five Journals and one affiliated journal "Mechanism and Machine Theory".

The IFToMM community evolved from a purely family character of a few beginners and founders to a worldwide scientific community through the following generations:

- 1950s–1979 First generation: founding fathers and their friendly colleagues up to the fourth IFToMM World Congress in Newcastle upon Tyne in 1975 with Prof. Maunder as Congress Chair
- 1980–1995 Second Generation: pupils and people educated by the founding fathers and their friendly colleagues; up to the ninth World Congress in Milan in 1995 with Prof. Rovetta as Congress Chair
- 1996–today Third Generation: educated people within the framework of the IFToMM and within IFToMM activity with 48 national organizations as IFToMM members

Knowing the History of IFToMM[1] can also promotea greater awareness of community identity and significance. More information on the IFToMM and its activities can be found on the website: http://www.iftomm.org.

List of Main Works of Ivan Artobolevski

Bibliography includes more than 1,000 published works. The main are:

- "On the Structure of Space Mechanisms" (1935)
- "Theory of Space Mechanisms" (1937)
- "The Experiment of Structural Analysis of Mechanisms" (1939)
- "Fundamentals of Unique Classification of Mechanisms" (1939)
- "Synthesis of Mechanisms" (1944)
- "Acoustic Dynamics of Machines" (1969)
- "Machines of 20th century" (1974)
- "The Mechanisms in Modern Engineering Design" (1970–1975)
- Main Textbooks for institutions of higher education: "Theory of Mechanisms and Machines" (1940), "Theory of Mechanisms" (1940, 1945, 1952, 1953, 1965, 1967, 1975, 1988 editions)

A Review of Main Works by Ivan Artobolevski

The circle of Artobolevski's interests was very wide. He worked in the field of the structure and kinematics of plane and space mechanisms, was engaged to the analysis of force, the theory of balancing of machines, the investigation of the movement of mechanisms, the calculation of mass of fly-wheels. He also wrote works on an approximate and exact synthesis of mechanisms and, together with his colleagues, he elaborated on the basis for automatic machines.

[1] Details on the History of IFToMM can be found in the first chapter of the Proceedings of the first International Symposium on the History of Machines and Mechanisms HMM2000 (published by Kluwer) in which all the past IFToMM Presidents have outlined their historical perspective of IFToMM in contributed papers with references.

Fig. 6 Title pages of the works by Ivan Artobolevski

Several 1,000 printed works belong to his pen (Fig. 6), among which, the most important are "Theory of Space Mechanisms" (1937), "Structure, Kinematics and Kinetostatics of Multilinking Plane Mechanisms" (1939), "Theory of Mechanisms and Machines" (1940), "Synthesis of Mechanisms" (1944), a reference book in four volumes "Mechanisms" (1947–1951), "Theory of Mechanisms for Reproduction of Plane Curves" (1959), "Synthesis of Plane Mechanisms" (1959), "Theory of Mechanisms and Machines" (1975) and an edition in five volumes "The Mechanisms in Modern Engineering Design" (1970–1975) which is considered as the encyclopaedia of all ancient and modern mechanisms and includes the description, systematization, classification and the main properties of about 5,000 mechanisms of different types and purposes.

Structure and Classification of Mechanisms

In 1939, Artobolevski's investigations in the field of structure and classification of mechanisms finished with the publication of the monograph "Fundamentals of Unique Classification of Mechanisms". This work gave wide possibilities for the further development of the theory of machines and mechanisms because it made the construction of the methods of the kinematic and kinetostatic analysis of mechanisms considerably easier. Thus, the problem concerning the investigation of the mechanism of any complexity comes down to the analysis of separate groups from which the mechanism is composed. In the presence of a strict scientific classification, the projection of the mechanism becomes easier.

Besides the structural classification of mechanisms, Ivan Artobolevski suggested a functional classification in which the mechanisms are divided into uniform groups which depend on the functions performed by them in machines and devices (for example, the mechanism of clamps, feed, regulators, brakes, etc.) and also the structure-constructive classification of the mechanisms on types of links and kinematic pairs (of lever, tooth, friction, shaft and others). Both classifications are stated in a multivolume reference book which represents a systematic collection of descriptions of mechanisms applied in different branches of machine and device engineering. The reference book had several editions and was widely used by designers while projecting machine and devices.

Kinematics and Kinetostatics of Plane Mechanisms

In 1939 Artobolevski finished his investigations in the field of kinematic, and kinetostatics with the publication of the monograph "Structure, Kinematics and Kinetostatics of Multilinking Plane Mechanisms". Later on, he carried out additional investigations on kinematics and managed to set up some new loci peculiar to plane mechanisms that have a great significance. Two of these new loci, relating to the definition of velocities and accelerations of plane mechanisms, were named by him as analogues to the circles of Bress and Jagir. In the article "Geometrical Methods of Solution of Some Problems on Theory of Mechanisms" (1947), he showed that all the problems bound up with the definition of acceleration could be solved with the help of the construction of the plans of velocities only.

Space Mechanisms

The results of Artobolevski's investigations on the theory of space mechanisms are summarized in the monograph "Theory of Space Mechanisms" (1937). In this monograph, Ivanovich examined the problems of the kinematics of a solid body during the different tasks of its motion, gave the axoid theory and, finally, exposed the methods of kinematic analysis of space mechanisms according to his own classification. This work is built up on the principles of geometrical school with wide utilization of the method of right-angled projections.

Dynamics of Mechanisms

One of the first works of Artobolevski on dynamics is dedicated to the solution of the problem of the motion of mechanisms under the influence of set forces when these forces are the functions of the arrangement of the links. For the first time this problem was solved by Ivanovich with the help of the method proposed by him which led him to determine the velocities and accelerations of the links of a mechanism which is under the influence of set forces. The problem on the motion of mechanisms is closely related to the problem of the calculation of fly masses for machine units which are under the influence of constantly changing forces. He managed to show that the approximate method of calculation of fly masses proposed by Radinger has some essential deficiencies and it gave serious deviations in the dimensions of the calculated fly masses. Using Lagrange's second-order equation, Artobolevski proposed a geometrical method of determination of derivatives from masses reduced to the functions of a turn angle. Further investigations on questions about the motion of a machine unit under the influence of set forces, are related to the creation of experimental methods of determining the mechanical parameters of a machine. At the Institute for Engineering Science of the AS of USSR (now RAS), Artobolevski created the Laboratory of Machine Dynamics where he worked out not only the recommended methods of experimental investigation of machines, but also many original means of registration and calculation of mechanical parameters of machines. Artobolevski also paid great attention to another highly important problem of the dynamics of machines, the balancing of the mechanism on the base. He solved the problem of balancing of mechanisms with complex kinematic diagrams. While solving it, he proposed some original methods of balancing four-bar chain mechanisms with the help of counterbalances and also of complex mechanisms with the help of additional toothed gearing which equilibrate the forces of inertia of different orders.

During the last years of Artobolevski's direct participation, one more trend appeared in the modern machine dynamics. The methods of acoustic dynamics of machines allowed him to solve problems on elimination of harmful high-frequency vibrations and lowering the level of noise. Nowadays, the acoustic dynamics of machines has developed into a huge scientific program which has not only scientific-engineering significance but also social significance as well. The level of vibrations and the noise in machines characterizes not only the dynamic tension of the machines but also the work conditions in the everyday life. Artobolevski was far-sighted here giving enormous support to some scientists who had been working on this problem.

Synthesis of Mechanisms

In the field of projection of plane mechanism, Artobolevski suggested geometric methods of synthesis with the help of centroid and interrounded curves for the most general cases of the task of plane motion. The result of these investigations is published in the monograph "Synthesis of Plane Mechanisms" (1939) and in the book "Synthesis of Mechanisms" (1944). In the monograph "Theory of Mechanisms for the Reproducing of Algebraic Curves" (1959), some methods were showed that made possible the synthesis of mechanisms for drawing and rounding not only conic sections but also other spreading curves. This work is the first big generalization in the world literature in the so-called theory of exact synthesis of mechanisms.

All the achievements of the synthesis of mechanisms were summed up in the book "Synthesis of Plane Mechanisms" (1959) which he wrote together with some of his colleagues. This book widely covers the different methods of synthesis of mechanisms that it can be considered, by right, as a work of an encyclopaedic nature.

Theory of Automatic Machines

As far back as in the thirties, Artobolevski, together with a group of scientists, began working on a summary monography dedicated to the general theory of automatic machines. The first two volumes were published under the title "Methods of Analysis of Automatic Machines" (1945, 1949). The methods of structural, kinematic and kinetostatic analysis of automatic machines and the method of their power valuation are stated in the book. The elaborated methods of projection of complex systems of machines of automatic operation which include technological machines and apparatus transporting and informative machines, engines and also auto-operators and industrial robots, obtained a scientific base.

Manipulators, Robots and Walking Machines

Artobolevski was one of the first in the Soviet Union to attract the attention of specialists to the problems of robots, manipulators and walking machines. In his works with his students, not only the basis of the theory of mechanic systems of manipulators were laid but also possibilities were shown for their further development. Artobolevski made much of the development of this problem in the USSR, formulating the most important scientific problems in that field. He, together with his pupils, worked out the methods of kinematic and dynamic analysis of executive organs (grips) of manipulators and robots, which possess high mobility and suggested criteria for integral valuation of their geometry and kinematics. Problems were discussed on the development of investigations directed towards the creation of robots with elements of artificial intellect.

History of Science and Technology

Artobolevski paid much attention to the questions of the history of science and technology. Fighting for the historical justice in world science, he defended, with arguments, the priority of Russian scientists. His works are interesting and instructive, dedicated to the analysis of the scientific legacy of Mikhail V. Lomonosov, Ivan P. Kulibin, P.L. Chebyshev, N.E. Zhukovski, L.V. Assur, S.I. Vavilov and also of his teachers I.P. Goryachkin and N.I. Mertzalov. At the same time, he always underlined the valuable contribution made by Russian scientists Chebyshev, Vishnegradski, Petrov and others to the formation of a scientific school on machines and its influence on the development of a world scientific school on the theory of machines and mechanisms.

He also carried out a very detailed investigation dedicated to Leonardo da Vinci's legacy. He showed how closely the works on physics and mechanics of this genius were related to the theory of different mechanisms and machines, devices and their use in practice. Leonardo da Vinci, in his works, appears before us as a theoretic scientist, an outstanding engineer, constructor and innovator, the one who worked out theories of transmission and projected hydro-technical constructions, built excavating machines, weaving-looms and different mechanisms for military affairs, invented devices of different types to polish the lens of optic devices and many other things. It is not casual that in 1968, Artobolevski was elected a honourable member of the International Academy of History of Sciences (Paris).

During the last years, together with Alexei N.Bogoljubov, he wrote a capital work "Leonid Vladimirovich Assur" in which it was proved that some questions on the structure of mechanisms were worked out by Assur more wider and deeper in comparison to the works of Burmester and Grubler.

A great number of Artobolevski's scientific books and articles are translated into many languages and are published in dozens of countries. This testifies to perpetuate his memory not only in Russia but far across its boundaries as well.

On the Circulation of Works

Among the personalities of science and engineering who greatly contributed to the scientific and engineering progress in the world, academician Ivan Ivanovich Artobolevski played a very important role. He founded the Soviet (Russian) school on the theory of machines and mechanisms, made great generalizations in the theory of machines and left a rich legacy: monographies, reference books, textbooks, scientific and publicistic articles in magazines and newspapers.

Artobolevski came to the conclusion which he later elaborated on, that a harmonious system for the classification of mechanisms existed, the beginnings of which he discovered in the works of Assur. This classification permitted him to create a strict scientific base for the theory of machines and mechanisms and it made considerably easier the elaboration of the general methods of calculation of mechanisms. This theory also became the base of the fundamental course "Theory of Machines and Mechanisms" (1940) distinguished by the strictness and harmony of the exposition of material. Afterwards, this work, with some changes and supplements, was republished repeatedly in the Soviet Union and abroad and was a great success. Artobolevski was constantly engaged to the development of the science "Theory of Machines and Mechanisms" (TMM) and wrote not only a new chapter of science important for some branches of industry but also became the main ideologist of new tendencies.

Knowing perfectly well the international literature on the TMM, he kept up a wide correspondence with the greatest scientists of different countries and even knew some of them personally, having met with them abroad.

In his letters to his colleagues, he exchanged some new scientific ideas on the science of machines and mechanisms. Such communication turned out to be very fruitful and on this basis scientific seminars on TMM appeared in different cities of the Soviet Union, such as Leningrad (now Saint Petersburg), Kiev, Odessa, Tbilisi, Kaunas, Alma-Ata and also in some fraternal countries such as Bulgaria, Poland and Czechoslovakia.

Being a scientist and statesman, Ivan Artobolevski had to travel much and he gave lectures for the common people. He always took an active part in the organization by holding seminars, discussions, conferences, symposiums and congresses in the field of TMM, being at the head of scientific councils and a member of the editing and publishing councils.

The scientific activity of Artobolevski is indissolubly connected with teaching. Yet being a student of Timiryazev Academy, he began teaching at the chair of agricultural machines. In 1929 he was elected professor and obtained the Chair of Technological Mechanics at the Moscow Chemical and Technological Institute named after Dmitry Mendeleev. Later he gave lectures at the Moscow Institute for Chemical and Mechanical Engineering, Air-Military Academy named after N.E.Zukovski, Moscow State University, and became at the chair of the "Theory of Machines and Mechanisms" Department at the Moscow Aviation Institute (Fig. 7).

At different periods of time, he gave courses on TMM, the balance of air and sea engines, the theory of agricultural machines, the theory of Space mechanisms, the

Fig. 7 A group of professors of "Theory of machines and mechanisms" department, Moscow Aviation Institute, 1948

basis of the theory of automatic machines, the theory and calculation of mill machines and the theory of revolving vibrations of the shafts of engines.

Artobolevski was an excellent lecturer, he gave lectures with inspiration and it seemed that he enjoyed them. He always had a great audience and his lectures were attended not only by students but also by teachers.

He helped young scientist in their investigation, and he did that with pleasure and consideration. About 100 scientific workers defended their candidate thesis under his leadership and 30 pupils became Doctors of Science and professors.

Modern Interpretation of Main Contributions to Mechanism Design by Ivan Artobolevski

Ivan I.Artobolevski wrote a great four-volume work "Mechanisms" which he had been working on for 10 years. The first edition, which was published in 1947–1952, contained a description of about 4,000 mechanisms applied by modern science, and was then the most complete directory. Engineers, thus received a theoretical basis for designing machines.

In 1808, the "Course of Construction of Machines" ('Essai sur la composition des machines') by Lanz and Betancourt was published – one of the first history textbooks of engineering and machines. It has a table where the authors included all known mechanisms at the end of the eighteenth – the beginnings of nineteenth. For one and a half centuries the quantity of mechanisms has increased almost 30 times! At the end of the 1960s, Artobolevski started to work on a new edition of the directory. Its first two volumes named "Mechanisms in Modern Engineering Design" (Fig. 6), were published in 1970 and 1971, contained 2,228 lever mechanisms, and in the second volume there appeared a description of 123 mechanisms made by Artobolevski himself: for example, a link-lever ellipsograph, the mechanism for plotting and rounding of ellipses and a hyperboloid, a link-lever hyperbolograph. The experience creating of these mechanisms allowed Artobolevski to later develop on the general theory of their synthesis. Such mechanisms have found application at processing surfaces of the second level, including surfaces of cylinders having in sections, curves of the second order. In 1937, the Commission for Engineering Science in the Academy of Sciences of the USSR was created, transformed then into the USSR Academy of Science Institute for Engineering Science (now RAS Institute for Engineering Science, named after A.A.Blagonravov), where Artobolevski created a number of laboratories and supervised the Department of "Theory of Machines and Mechanisms", and later the Department of "Mechanics of Machines". Here, in that Institute, Ivan Ivanovich concentrated the basic scientific activity during his lifetime.

As time passed, a number of problems appeared which earlier didn't ever enter the theory of mechanisms and machines: vibrating, vibrating-percussive systems, the theory of machines with variable structure and parameters, account of elasticity of parts and backlashes in kinematic pairs, the theory of systems of automatic machines, etc. These problems are actual today. They have not only scientific and technical, but also social values connected with studying complex system "man–machine–environment". Not casually, Artobolevski always spoke about two aspects of vibration: on the one hand it is necessary to escape from fluctuations and to achieve their elimination, and on the other hand, fluctuations in machines can be useful and could be used for conducting various technological processes. Under his initiative and under his management in RAS Institute for Engineering Science, named after A.A.Blagonravov a conference was held on vibrating machines with the purpose of attracting the attention of scientists and engineers to this problem. Since the 1950s of the twentieth century and until today a lot of new offers have appeared on use of a vibrating principle by means of which such technological operations as condensation, separation and crushing of materials, drilling of rocks and removal of pressure casting are carried out, sorting and transportation of fine details and many other things. Artobolevski was never a scientist of one idea. He took great interest in everything new and constantly put new tasks before his pupils and colleagues and actively developed new sections of the Theory of Machines, such as acoustic dynamics of machines; the theory of walking machines; the theory of automatic systems; and also synthesis multiple parameter, multicriterial systems in general. It is an important merit that Artobolevski was the first to restore the seditious word "Robot" after a very long-term prohibition of cybernetics in the Soviet Union. During his direct participation in the Institute of Surgery named after A.V.Vishnevski, the laboratory of medical cybernetics was created, successfully solving problems of the diagnosis of various diseases. Together with A.E.Kobrinski, he gave a lecture "Some problems of construction of systems known as robots" at a conference on the basic problems of theory of machines and mechanisms.

Under Artobolevski's initiative with the 1967 All-Union, symposiums concerning the theory and principles of mechanisms of robots and manipulators on which possible directions of their further development were discussed, were also started on a regular basis. He was the first to show the huge value of these machines in the future not only from the scientific and technical point of view, but also from the social and humanistic point of view. Artobolevski spoke of robotics as being one of the advanced directions of progress. "It is very important, that these systems", – he wrote, "allow the release of human beings from the work in zones of raised radioactivity, harmful and dangerous to health, from dust content, gassed conditions, zones of heat and pressure". His words were prophetical. It is expected, that the robots of the subsequent generations possessing elements of an artificial intellect, can replace man in extreme situations: in space, under water at greater depths, under the ground, in the struggle against terrorism, having left for man only the functions of supervision, management and introduction a complex system. Under Artobolevski's initiative, a Scientific Council on the theory and principles of mechanisms of robots and manipulators was also created, and he was its chairman from 1973 to 1977. He attached huge significance to the coordination of scientific activity of scientists, engineers, teachers, to collective discussion of achievements of a science and its prospects since 1937 and up to the end of his life he had been a permanent head of the seminar on theory of machines and mechanisms, and the editor-in-chief of works of this seminar (160 collections have been published).

Mechanics of machines was the first applied science, as the result of the industrial revolution of the eighteenth century. The two following centuries brought with them a stream of information and duplication of sciences, and a number of sciences about machines in particular, and at the present time has reached a three-value figure. But, until mankind uses machines, the theory of machines and mechanisms (TMM) in the sum of human knowledge will play one of the most important roles. In one of the publicistic clauses, addressed to the youth, Artobolevski wrote: "Most important in the education of scientific youth is the development of independence and creative ideas. To learn and learn creativity – it is possible. Most important for a creative person is his skill to observe a natural phenomenon, to know the mechanism of processes, and to transfer this knowledge to the world of techniques".[2]

During his last years, Artobolevski wrote one of his greatest works: "Successes of the Soviet school on theory of machines and mechanisms", (Fig. 8), in which he

К 60-летию ВЕЛИКОЙ ОКТЯБРЬСКОЙ СОЦИАЛИСТИЧЕСКОЙ РЕВОЛЮЦИИ

академик

И. И. АРТОБОЛЕВСКИЙ

УСПЕХИ СОВЕТСКОЙ ШКОЛЫ ТЕОРИИ МАШИН И МЕХАНИЗМОВ

АЛМА-АТА

ММ 77

ВСЕСОЮЗНЫЙ СЪЕЗД

ВСЕСОЮЗНЫЙ СЪЕЗД
по теории машин и механизмов

Москва — 1977

Fig. 8 Title page of the last (life time) publication of Ivan Artobolevski: "Successes of the Soviet school on theory of machines and mechanisms"

[2]Artobolevski I.I. "Be not afraid to dare", "The Komsomolskaya Pravda" (the newspaper), 11 May 1974.

summed up the activity of all Soviet schools on the theory of machines actually created by him. It was his last publication. His scientific ideas and the published works help acceleration of scientific and technical progress. His life deserves to be studied in detail and to become an example for new generations of modern engineers and scientists.

Acknowledgements The authors wish to thank all the employees of IMASH, the Library of IMASH, the Publishing House "Nauka" and Elena Atrobolevski in particular. It is also important to mark the role of the Academician Konstantin Vasilievich Frolov[3] and his efforts in holding the 2005 meeting dedicated to the 100th anniversary of the birth of Academician Ivan Ivanovich Artobolevski.

References

Academician Ivan Ivanovich Artobolevski, under reduction of Dubrovski V.A., Collected Papers, Nauka, Moscow, 2007 (Papers of the meeting of the Scientific Council of RAS Institute for Engineering Science named after A.A. Blagonravov and of the Incorporated Scientific Council of the Russian Academy of Science on a complex program "Machine industry", devoted to the 100-anniversary from birthday of the academician I.I.Artobolevski, Moscow, 12–13 October 2005)

Artobolevski II (1956) About machines of vibrating action. Publishing house of AS of the USSR, Moscow, 47 p (Together with A.P.Bessonov and A.V. Shljahtin)

Artobolevski II (1945) Course on theory of mechanisms and machines. OGIZ, San Petersburg, Moscow, 450 p

Artobolevski II (1977) Dynamics of machine units on limiting modes of movement. Nauka, Moscow, 326 p (Together with V.S. Loshinin)

Artobolevski II (1961) Great Russian scientist .V.Lomonosov. In: M.V.Lomonosov – To the 250th anniversary of his birthday. Moscow, pp 1–21

Artobolevski II (1979) Introduction to acoustic dynamics of machines. Nauka, Moscow (Together with J.I. Bobrovitski and M.D. Genkin)

Artobolevski II (1952) Leonardo da Vinci as the designer Info of S of the USSR, . No. 6, pp 891–902

Artobolevski II (1971) Leonid Vladimirovich Assur (1878–1920). Nauka, Moscow, 264 p (Together with A.N. Bogoljubov)

Artobolevski II (1975) Mechanisms in modern techniques. Cum and frictional mechanisms. Mechanisms with flexible parts. Nauka, Moscow, . 4, 448 p

Artobolevski II (1976) Mechanisms in modern techniques. Hydraulic, pneumatic and electric mechanisms. Nauka, Moscow, . 5, 848 p

Artobolevski II (1970) Mechanisms in modern techniques. Lever mechanisms. Nauka, Moscow, .I, 608 p

Artobolevski II (1971) Mechanisms in modern techniques. Lever mechanisms. Nauka, Moscow, T.2, 1007 p

Artobolevski II (1973) Mechanisms in modern techniques: gear mechanisms, vol. 3. Nauka, Moscow, 576 p

Artobolevski II (1951) Mechanisms, vol. 4. Publishing house of AS of the USSR, San Petersburg, Moscow, I–IV, 1176 p

[3] Unfortunately, the academician Konstantin V. Frolov passed away on 18 November 2007.

Artobolevski II (1949) Mechanisms, vol. 3. Publishing house of AS of the USSR, San Petersburg, Moscow, 1044 p
Artobolevski II. Mechanisms, vol. 2. Publishing house of AS of the USSR, San Petersburg, Moscow, 900 p
Artobolevski II (1947) Mechanisms, vol. 1. Publishing house of AS of the USSR, San Petersburg, Moscow, 887 p
Artobolevski II (1945) Methods of the analysis of automatic machines: structure and analysis, part 1. Publishing house of AS of the USSR, San Petersburg, Moscow, 120 pages (Together with S.I. Artobolevski, G.A. Shaumjan and V.A. Judin)
Artobolevski II (1949) Methods of the analysis of automatic machines: kinematic and kinetostatic analysis, part 2. Publishing house of S of the USSR, Moscow, 251 pages (Together with S.I. Artobolevski, G.A. Shaumjan and V.A. Judin)
Artobolevski II (1954) Modern condition of the theory of machines and its immediate tasks. Publishing house of S of the USSR, Moscow, 40 p
Artobolevski II (1957) Modern scientific and technical revolution & distribution of political and scientific knowledge. Znanie, Moscow, 24 p
Artobolevski II (1949) Nikolay Ivanovicha Mertsalov. Life and activity, 1866–1948 // Seminar on , 1949, om 7. Ed. 26. pp 5–17
Artobolevski II (1973) Problem book on theory of mechanisms and machines. Nauka, Moscow, 256 p (Together with B.V. Edelshtein)
Artobolevski II (1947) Problem book on theory of mechanisms and machines. Gostechizdat, San Petersburg, Moscow, 196 p (Together with V.A. Zinovjev and B.V. Edelshtein)
Artobolevski II (1956) Problems of engineering science in development of engineering and technological processes in connection with automation. Publishing house of AS of the USSR, Moscow, 33 p (Together with A.A. Blagonravov, V.I. Dikushin and V.S. Kulebakin)
Artobolevski II (1969) Some problems of creation of walking machines // Info of S of the USSR, 2, pp 44–51 (Together with N.V. Umnov)
Artobolevski II (1967) Soviet school of the theory of machines for 50 years // Mashinovedenie, 6, pp 5–11
Artobolevski II (1939) Structure, kinematics and kinetostatics of iterative flat mechanisms. ONTI, San Petersburg, Moscow, 232 p
Artobolevski II (1944) Synthesis of mechanisms. Gostechizdat, San Petersburg, Moscow, 387 pages (Together with Z.S. Bloh and V.V. Dobrovolski)
Artobolevski II (1959) Synthesis of plane mechanisms. Fizmatguz, Moscow, 1084 pages (Together with N.I. Levitski and S.A. Cherkudinov)
Artobolevski II (1939) Synthesis of plane mechanisms. Publishing house of AS of the USSR, San Petersburg, Moscow, 42 p
Artobolevski II (1951) The outstanding Soviet scientist and public figure – Sergey Ivanovich Vavilov. Pravda, Moscow, 48 p
Artobolevski II (1946) The Russian inventor and designer Kulibin. Molodyaya Guardiya, Moscow, 24 p
Artobolevski II (1975) Theory of mechanisms and machines, 3rd edn. Nauka, Moscow, 640 p
Artobolevski II (1940) Theory of mechanisms and machines. Gostechizdat, San Petersburg, Moscow, 762 p
Artobolevski II (1938) Theory of mechanisms and machines: dynamics of mechanisms, part 2, 2nd edn. MICHM, Moscow, 366 p
Artobolevski II (1938) Theory of mechanisms and machines: kinematics of mechanisms, part 1, 1st edn. MICHM, Moscow, 213 p
Artobolevski II (1959) Theory of mechanisms for reproduction of plane curves. Publishing house of AS of the USSR, Moscow, 255 p
Artobolevski II (1965) Theory of mechanisms. Nauka, Moscow, 776 p
Artobolevski II (1974) Scienziati e tecnologi wntemporanei. Arnoldo Mondadori Ed., Milano, p 52
Artobolevski II (1964) Who's who in Soviet science and technology. Compiled by I. Telberg, 2nd edn., revised and enlarged by A. Dmitriev and V.G. Telberg. Published by Telberg book Corp., New York, p 12

Artobolevski II (1973) Who's who in the world, 2nd edn., 1974–1975. Chicago, Marquis who's who, p 45
Artoholevski II (1962) Dictionary encyclopedic sht, vol. 1. Ed. politica, Bucuresti, pp 207–208
Bessonov AP (1979) Ivan Ivanovich Artobolevski. Znanie, Moscow
Bogoljubov AN (1975) Soviet school of machine mechanics. Nauka, Moscow
Ceccarelli M (2004) Classifications of mechanisms over time. Proceedings of International Symposium on History of Machines and Mechanisms, HMM2004. Kluwer, Dordrecht, pp 285–302
History of Mechanics in Russia, Monograph, under reduction of Bogoljubov A.N. and etc (1987). Naukova Dumka, Kiev, pp 319–321
Kuzmenko RI, Finashina GN (1975) Ivan Ivanovich Artobolevski, notes to bibliography of USSR scientists, 2nd edn. Nauka, Moscow
Warm welcome for scientists, Vechernyaya chronicle, 8/IX, No 30583, p 9, photo
Watt medal, Chart, mech. Engineering (1967) vol. 14, no. 4, pp 183–184
W lle H (1969) Bilduugavorlanf fiir die Auigaben von Morgen.B Neues Oeutschland, 20/VI, S. 1–2

Ludwig Burmester (1840–1927)

Teun Koetsier

Abstract Ludwig Burmester was a nineteenth century German mathematician. His *Lehrbuch der Kinematik, Erster Band, Die ebene Bewegung* (Textbook of Kinematics, First Volume, Planar Motion), of 1888, contains the first far-reaching attempt at a synthesis of theoretical kinematics and kinematics of mechanisms. His most influential technical contribution is the Burmester theory which deals with four or five discrete positions of a moving plane. The goal of the Burmester theory is the synthesis of mechanisms. In the present paper, Burmester's work is discussed against the background of the development of the theory of machines and mechanisms in the nineteenth century.

Biographical Notes

Ludwig Ernst Hans Burmester, Fig. 1, was born on May 5, 1840, as son of a gardener in the village of Othmarschen near Hamburg in Germany. He died on April 20, 1927, as a respected professor emeritus of Descriptive Geometry and Kinematics at the University of Munich and a member of the Bavarian Academy of Sciences. Between these dates lies a rich and varied professional life dominated by Geometry and Kinematics.

At the age of 14, Burmester became an apprentice in the workshop of a Hamburg precision mechanic. He was allowed to go to the Polytechnical Preparatory School in Hamburg run by Otto Jensen. Jensen was an excellent man who was not only a great teacher of mathematics but good at convincing rich citizens of the city of Hamburg to support talented children like Ludwig Burmester. Because Burmester thought that there was a good future in telegraph machines, he left Hamburg and went to Berlin where Siemens & Halske were building such machines. Yet, in the end his desire to study prevailed.

T. Koetsier (✉)
Department of mathematics, Faculty of Science, Vrije Universiteit, De Boelelaan,
P.O. Box 1081NL-1081HV, Amsterdam, The Netherlands
e-mail: t.koetsier@few.vu.nl

M. Ceccarelli (ed.), *Distinguished Figures in Mechanism and Machine Science: Their Contributions and Legacies, Part 2*, History of Mechanism and Machine Science 7,
DOI 10.1007/978-90-481-2346-9_3, © Springer Science+Business Media B.V. 2010

Fig. 1 Ludwig Burmester (1840–1927) when he was a professor in Dresden (Courtesy of Dr. Klaus Mauersberger, T. H. Dresden)

In 1862 he enrolled at the Polytechnical School in Dresden in the department for future teachers of mathematics, science and technology. He attended the classes of Oskar Schlömilch on several areas of mathematics. It is remarkable that he seems to have acquired a considerable knowledge of geometry that later dominated his activities, all by himself. In 1864, Burmester received his diploma in Dresden with honours.

In the next years, Burmester studied in Göttingen where he obtained his doctor's degree in 1865.

In 1868, Burmester married Gabriele Schallowetz (August 4, 1848–November 15, 1919). Out of the marriage a daughter, who died very early and three sons were born. After 1865, Burmester worked for 4 years as a secondary school teacher in Lodz in Russian Poland.

In 1870, Burmester had to leave Lodz after the Russians had closed down the secondary school he was teaching in. In 1872, Burmester's luck changed, he became professor of descriptive and synthetic geometry in Dresden, where he had become Privatdozent after his Habilitation in 1871. In 1875, he started teaching projective geometry and in 1879 he started to teach kinematics. Much of his work in this period is related to kinematics. One of his earliest publications after his appointment in Dresden in 1873 combines a growing interest in kinematics with geometrical questions concerning projection and shadows. On 1 October 1887, Burmester was appointed to the chair of descriptive geometry and kinematics in Munich. For the events leading to his appointment, see Hashagen (2003, pp. 500–506). In Munich, where he lived in the Kaulbachstraβe 83, Burmester was highly respected. In 1905, he became a member of the Bavarian Academy of Sciences. In the last years of his life he developed a great interest in the mechanical engineering side of cinematography. He lectured on this subject and he studied different methods for guiding film in a projector.

Burmester died in 1927 of a heart attack. He was buried next to his wife in the cemetery of Kreuth near Lake Tegernsee. The grave is still there. Two necrologies appeared, written by Finsterwalder (1927) and Müller (1930). They show that one of the giants of nineteenth century German kinematics had died.

Burmester's Work

Burmester's Göttingen thesis is titled *Ueber die Elemente einer Theorie der Isophoten* (About the elements of a theory of isophotes). It deals with the representation of lighted surfaces in three-dimensional space. Isophotes on a surface are lines of constant light intensity: they are the sets of points for which the cosine of the angle between the light direction and the surface normal has a constant value. It is remarkable that Burmester's treatment of the subject is entirely analytical. The contrast with Burmester's later works, in which his method is more and more synthetic, is considerable.

In 1871, his first book appeared in Leipzig: the *Theorie und Darstellung gesetzmäßig gestalteter Flächen* (Theory and representation of well-defined surfaces). It was reprinted in 1875. The book is about the subject of his dissertation, the theory of isophotes.

Interesting is Burmester's distinction between lines of "true light intensity", the isophotes, and lines of "observed light intensity", which he called isophengs. They are the sets of points for which the product of, on the one hand, the cosine of the angle between the light direction and the surface normal, and, on the other hand, the cosine of the angle between the direction of the eye and the surface normal, has a constant value. Figure 2 is an example taken from (Burmester 1871).

Burmester was interested in all aspects of the theory of perspective as well. In 1883, in Leipzig he published his book *Grundzüge der Reliefperspective* (The Main Features of Relief Perspective), dealing with the principles of relief perspective. In ordinary perspective, a half space is mapped on one plane of projection; in relief perspective, a half space is mapped on the space between two parallel planes. Burmester built several models in order to demonstrate the effect of his theory (See Fig. 3).

In 1884, Burmester published *Grundlehren der Theaterperspective* (Foundations of Theater Perspective). It is said that he was involved in the design of the decor for some of Wagner's opera's.[1]

It is remarkable that both the subject of isophotes and the subject of relief perspective have recently attracted interest in the area of computer aided geometric design (Lordick, 2005).

---

[1]Cf. http://www.mathematik.tu-muenchen.de/~kaplan/fakul/node16.html. I have not been able to check this. Wagner died in 1883, before Burmester came to München. Yet the Germans took theatrical design very seriously and it is quite possible that during Wagner's last years or after his death, Burmester was involved in the scenography in Bayreuth.

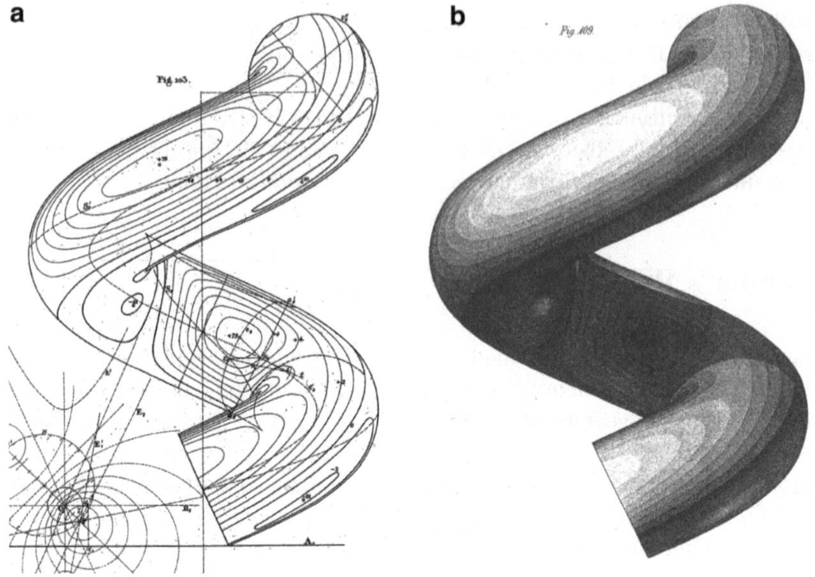

Fig. 2 Figures from Burmester's book on isophotes: (a) lines of equal light intensity; (b) a shadow representation

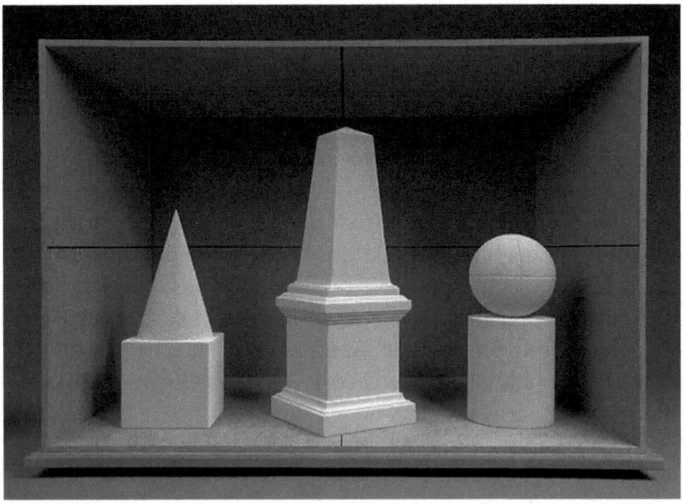

Fig. 3 A reconstruction by Dr. Daniel Lordick (TU Dresden) of one of Burmester's relief perspective models (Lordick 2005) (Courtesy of Dr. Lordick)

The turn to kinematics was crucial in Burmester's life. Before the 1960s of the nineteenth century, kinematics of mechanisms was hardly a coherent discipline and results from theoretical kinematics were not applied systematically to the kinematics of mechanisms. In the second half of the nineteenth century, things changed.

Kinematics enjoyed great popularity among both mechanical engineers and mathematicians. The work of the German engineer Franz Reuleaux (1829–1905) was in particular very influential. Reuleaux's *Theoretische Kinematik* (a theoretical treatise on kinematics of mechanisms, in spite of its title. See below), published in book form in 1875, paved the way for further mathematization of kinematics of mechanisms. On the other hand mathematicians also played an essential role. Ludwig Burmester was a prominent example.

Burmester's most important contribution to kinematics is undoubtedly his *Lehrbuch der Kinematik, Erster Band, Die ebene Bewegung* (Textbook of Kinematics, First Volume, Planar Motion). It was originally published in three installments; the first two installments appeared in 1886 in one volume of 560 pages and the last installment appeared at the end of 1887. In 1888 the three installments appeared in one volume. In 941 pages accompanied by an atlas with 863 figures presented Burmester gave a survey of everything he knew about planar kinematics, which included many of his own results (Fig. 4).

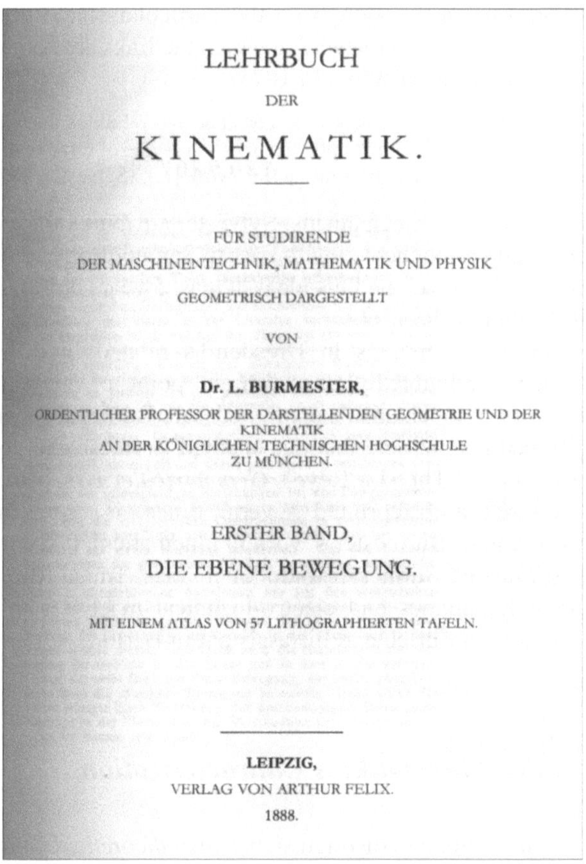
LEHRBUCH

DER

KINEMATIK.

FÜR STUDIRENDE

DER MASCHINENTECHNIK, MATHEMATIK UND PHYSIK

GEOMETRISCH DARGESTELLT

VON

Dr. L. BURMESTER,

ORDENTLICHER PROFESSOR DER DARSTELLENDEN GEOMETRIE UND DER KINEMATIK

AN DER KÖNIGLICHEN TECHNISCHEN HOCHSCHULE

ZU MÜNCHEN.

ERSTER BAND,

DIE EBENE BEWEGUNG.

MIT EINEM ATLAS VON 57 LITHOGRAPHIERTEN TAFELN.

LEIPZIG,

VERLAG VON ARTHUR FELIX.

1888.

Fig. 4 Title page of Burmester's *Lehrbuch der Kinematik* (1888)

In 1890, he and the other mathematicians from München were involved in the foundation of the *Deutsche Mathematiker-Vereinigung* (Association of German Mathematicians).

Review of Main Work Related to Mechanism Design

The Background

With the completely new classification of mechanisms by means of kinematical chains in the 1860s, Reuleaux had introduced a very abstract point of view in kinematics of mechanisms. For a mathematician, this boiled down to viewing a planar mechanism as a collection of coinciding Euclidean planes all moving (with one degree of freedom) with respect to each other. This was a major and not trivial step forward which helped to turn kinematics of mechanisms into a more coherent discipline. It involves two related elements: considering the frame of a mechanism as a link and, moreover, abstraction from the particular shape of the links in a mechanism and concentration on the way in which the links are connected. Without mentioning Reuleaux, Sylvester wrote in 1875:

> The true view of the theory of linkages is to consider every link as carrying with it an indefinitely extended plane and to look upon the question as one of relative motion [...] Fix any of these planes and the linkage becomes a link-work [...]. (Sylvester, 1875)

In the investigation of bar-mechanisms, Sylvester attributed the extension of the consideration to the planes connected with bars to Samuel Roberts. Yet the priority belongs to Reuleaux who developed his ideas at least 5 years before Roberts turned to kinematics at the end of the 1860s.

Burmester's turn to kinematics in Dresden is related in different ways to Reuleaux' influence. Undoubtedly, the abstract point of view introduced by Reuleaux in the theory of mechanisms combined with Burmester's great interest in the recently developed geometrical theories enabled Burmester to do his kinematical work. As for geometry, Theodor Reye's *Geometrie der Lage* (Projective geometry) of 1866, revised and reprinted several times, is representative of the geometrical background of Burmester's work in kinematics). It seems to have been Rittershaus who stimulated Burmester to do kinematics. Trajan Rittershaus (1843–1899), professor of kinematics and electro-technical machines in Dresden, had been Reuleaux' student in Zürich and his assistant in Berlin.

The Lehrbuch der Kinematik: A Characterization

Nowadays it is not uncommon to distinguish between *theoretical kinematics* dealing primarily with the general properties of motion and *kinematics of mechanisms* which concentrates on the kinematical properties of (classes of) specific mechanisms. As we will see below, Reuleaux used the term kinematics in a different way. We will use the modern terminology. Burmester's *Lehrbuch der Kinematik* contains the first far-reaching attempt at a synthesis of theoretical kinematics and kinematics of mechanisms. The planned second volume on spatial kinematics was never written. The content of the kinematical papers that Burmester wrote before the publication of his book, are all included in the book. In 1874 and 1875, he published on affine and equiform kinematics: the moving plane moves and is subjected to affine or equiform transformations as well, while the fixed plane remains Euclidean. The results return in the twelfth chapter of the *Lehrbuch der Kinematik*. In 1876 and 1877, he wrote two very original papers on what would later be called the *Burmester theory*. The Burmester theory returns in the ninth chapter of the book. In 1880 he published a systematic investigation of the instantaneous velocities of *n* rigid systems moving with respect to each other in the plane. Aronhold had discovered the three centers theorem when he prepared a course on theoretical kinematics in Berlin, which took place in 1866–67. Twenty years later, Kennedy would independently discover the same theorem. In Burmester's 1880, paper the full potential of the theorem in velocity analysis became clear. The results return in the seventh chapter of the book. It is remarkable that this paper exerted some influence in civil engineering. The results were used for the calculation of forces in a framework by means of the principle of virtual displacements (Kurrer 2002, pp. 521–525). After 1888, Burmester continued to publish on kinematics. In 1893, he thoroughly investigated a class of overconstrained mechanisms introduced by Kempe in 1878. He returned to affine kinematics, this was in 1902, a subject that he had studied in 1874 and 1875. In 1911 and in 1925, he published on acceleration in planar motion. The *Lehrbuch der Kinematik* consists of 12 chapters. It is a compilation of almost everything that was known on planar kinematics at that time. On the one hand, Burmester gave a survey of the results from theoretical planar kinematics. On the other hand, he systematically studied their application to practically all planar mechanisms that had been identified at the time. Examples of figures from the atlas are given in Figs. 5 and 6.

The Lehrbuch der Kinematik: Its Contents

Let us consider the structure of the *Lehrbuch der Kinematik* in more detail. The first chapter is devoted to the general properties of planar motion: velocity, the instantaneous center of rotation (the pole), the polhodes, the motion of three coinciding planes with respect to each other, construction of tangents and centers of curvature, with many examples of special motions.

Chapter two concerns the different cycloids and their properties.

Chapter three deals with cylindrical gears, cycloid gearing, involute gearing.

Chapter four deals with the gear in internal gear pumps (*Kapselraeder*). In this chapter, Burmester discusses all existing types of such pumps.

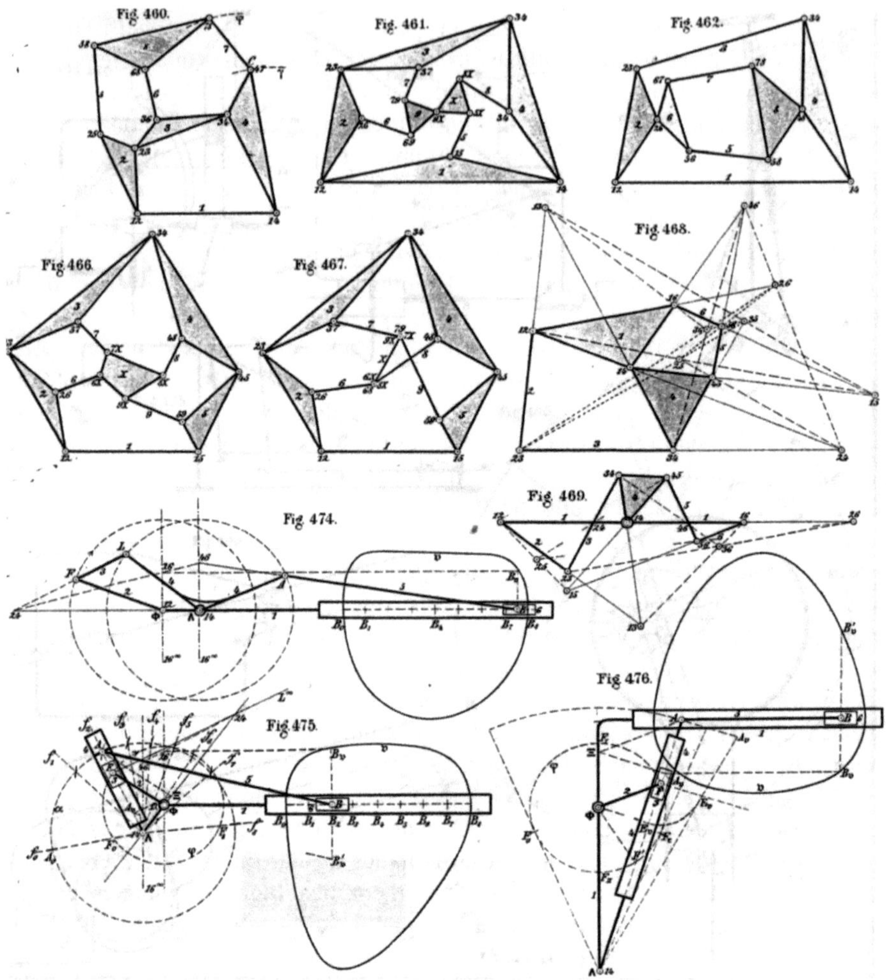

Fig. 5 Figures accompanying Burmester's seventh chapter of the *Lehrbuch der Kinematik* on compound planar mechanisms

He always moves from the very general to the specific. The theories given in the first chapter are applied in chapters two to four. In the fifth chapter, Burmester again starts from very general considerations, which are applied in chapters six and seven.

In the fifth chapter, he deals with the general theory of constraints (*Zwanglauf*): restrictions on the motion of a system. The final goal of this theory, which was defined by Reuleaux, is to determine the most favorable ways to realize a particular one-degree-of-freedom constrained motion. Reuleaux' notions of kinematic pair (consisting of two elements that constrain each other's motion) and kinematic chain

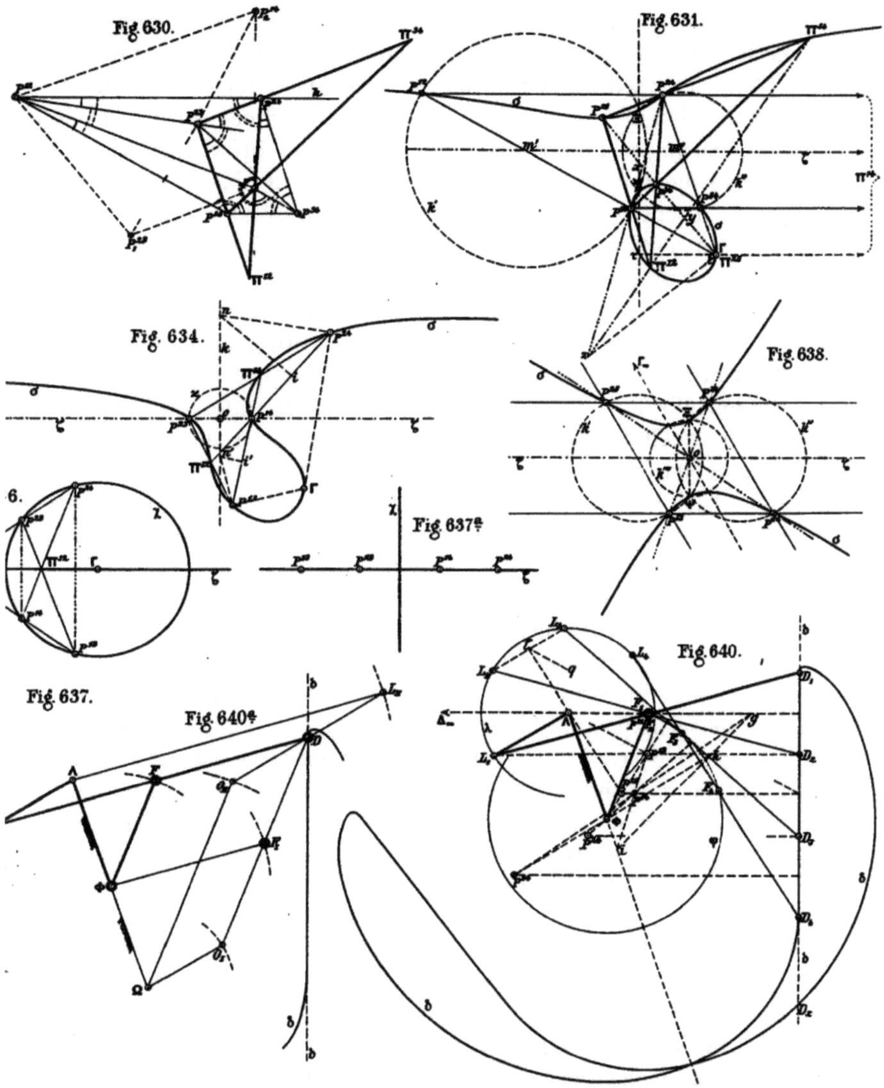

Fig. 6 Figures accompanying the treatment of the Burmester theory in the ninth chapter of the *Lehrbuch der Kinematik*. The curve σ in Fig. 634 is the centre point curve. In Fig. 638, σ consists of the line at infinity and a hyperbola

are the basic notions. There are, however, some terminological differences between Burmester and Reuleaux. Instead of Reuleaux' word 'kinematic chain', for example, Burmester uses the word 'mechanism', something Reuleaux would later interpret as an insult that revealed that Burmester had not understood the significance of Reuleaux' kinematics at all. For more details see the section on the reception of Burmester's work below.

The sixth chapter is very long, 122 pages. It deals with *simple planar mechanisms*. When the links (or elements including the one that is fixed, the frame) of a mechanism form a closed sequence and all elements execute a one-degree-of-freedom constrained motion with respect to each other, the mechanism is by definition a simple mechanism. First Burmester discusses the different planar four-bar linkages, Robert's theorem and the graphical velocity analysis of a planar four-bar linkage. Then he discusses special cases like the slider–crank mechanism and the Scotch-yoke mechanism with the graphical velocity analysis. He treats simple mechanisms with cams, with non-circular gear wheels and simple belt-driven mechanisms. There is also a section on locking and switching devices.

Burmester then proceeds in the seventh long chapter of 143 pages to compound planar mechanisms. In such mechanisms there will be at least one element that is connected through closure to more than two others. At the beginning of the chapter, Burmester discusses Grübler's results concerning constraints and results concerning the configuration of poles at a particular instant in the case of several coinciding moving planes. Then he turns to the discussion of all the compound mechanisms that he was familiar with. Wherever possible, Burmester applied the pertinent theoretical results to the many special mechanisms that he discusses. Watt's mechanism, Stephenson's mechanism and many others are treated. Figure 5 gives an idea of the variety of mechanisms that Burmester studied.

In the eighth chapter, Burmester deals with what he calls "guided mechanisms", like the pantograph. A one-degree-of-freedom kinematic chain is attached to the fixed plane by means of only one hinge. Then if we guide a point of one of the other links along a curve, we have a guided mechanism. In the same chapter, he discusses overconstrained mechanisms.

In the ninth chapter he deals with straight-line mechanisms. Also, here he develops a very general theory before it is applied. This time it is quite new: it is the theory nowadays called Burmester's Theory. Burmester's *Lehrbuch der Kinematik* is the culmination of a long development and is frequently based on papers that Burmester had written earlier. The treatment of the Burmester theory is based on the content of two papers from 1876 and 1877. Below, we will look at these papers in some detail.

In the twentieth chapter, of the *Lehrbuch der Kinematik*, he discusses slider controls for locomotives like Stephenson's control or Heusinger von Waldegg's control. In the eleventh chapter of 122 pages, Burmester gives an extensive theory of acceleration with applications to many simple and compound mechanisms. Also, here Burmester starts from the general theory and only then considers its applications. Finally in the twelfth chapter, Burmester deals with equiform and affine kinematics.

The Burmester Theory and the Burmester Points

The 1876 paper, which clearly represents Burmester's growing interest in applications of kinematics in mechanical engineering, consists of three parts. The paper is on the design of four-bar straight-line mechanisms taking into consideration the discrete positions of a moving plane. In 1876, Burmester first considered three discrete positions S_1, S_2 and S_3 of a moving plane and determined the points in the moving plane that are in those positions on a straight line. Right from the start, Burmester attacked the problem by means of projective geometry. By considering projective pencils of points and lines, he proved that the set of all points in S_2 that are in the three positions on a straight line, is a conic section. Because the circle points I and J are on it, it is a circle C_2 in S_2. This result was already known; it had been published without proof by Grouard in 1870. In his 1876 paper, Burmester went on to four positions, although he restricted himself to a proof of the theorem which says that, in general, precisely one point is in the four positions on a straight line.

In the 1877 paper, Burmester attacked the more general problem of the loci of points that are in a number of discrete positions on a circle. First he considered three positions and proved that the locus of the centers of the circles that are determined by triples of homologous points on three homologous lines is a conic section through the three poles. This theorem immediately yields: There exist either one or three circles that go through four homologous points on four homologous lines. The circles correspond to real points of the intersection of conic sections. By applying this theorem to the homologous lines of a pencil, the locus of centers of quadruples of four homologous points turns out to consist of the points of intersection of corresponding elements of two pencils of conic sections. Burmester analytically proved (the methods in the rest of the paper are synthetic) that this *Mittelpunktskurve* (centre point curve) is a circular curve of third degree. He also shows that it is a so-called focal curve, which is the locus of the foci of all conic sections that touch four given straight lines. Burmester then showed that the locus of points that are in four positions on a circle is also a focal curve, which he calls the *Angelpunktkurve* (pivotal point curve). In his book, he called this curve the "circle point curve". From the text of the 1877 paper, it is clear that Burmester's ideas were still being developed. For example, Burmester identifies the fixed plane with position 1, which means that he did not seem to realize that by considering the situation from the point of view of the moving plane, we are also dealing with four positions of the fixed plane with respect to the moving one and it is because of that the *Angelpunktkurve* must be of the same nature as the "centre point curve".

Finally, Burmester considered five positions and he found the points that are all five positions on a circle by intersecting two focal curves. Because from the nine points of intersection the circle points and three poles must be subtracted, he easily proved the existence of the four points that were later called the *Burmester points* by R. Müller.

One of the admirable characteristics of Burmester's work is the fact that he combined a great interest in theoretical results with an interest in applications. His 1877 paper on discrete position theory is no exception. The Burmester theory is immediately applied to Stephenson's link mechanism for controlling the steam valve of a locomotive.

Kempe's Focal Mechanism

In 1878, Alfred Bray Kempe published a paper which contains a remarkable over-constrained linkage (see Fig. 7). ABCD is a four-bar mechanism. If we choose arbitrary points S, R, V and U on the bars and connect them with bars of a fixed length to a joint P, we create, in general, a rigid structure. Kempe showed that P can be chosen in such a way that quadrilateral ASPV is similar to quadrilateral PRCU and quadrilateral BSPU is similar to quadrilateral PRDV. He, moreover, showed that for such points P, the mechanism has one degree of freedom. Obviously P moves in a circle with respect to AB. Kempe also showed that if the centers of this circle moves to infinity (on the extension of AB) and we remove the bars PR and PS (one has to remove them in fact, because their lengths become infinite), the mechanism becomes the straight-line linkage that Hart had found a year earlier.

It is easy to show that the similarity of the quadrilaterals implies that for the point P inside the quadrilateral we have:

$$\sphericalangle APB = \pi - \sphericalangle CPD.$$

In other words, P is a point from which opposite sides of the quadrilateral AB and CD are seen under angles that are each other's complement. Moreover, the converse holds as well: for all such points we can determine points S, U, R and V such that we get an over-constrained linkage consisting of a four-bar linkage divided into two pairs of similar quadrilaterals.

In 1893, Burmester wrote a long and very thorough paper about the mechanism. He showed that for a given quadrilateral, the locus of all points P that yield an over-constrained mechanism of this type is a circular curve of the third degree. Moreover, such points P are always the focus of a conic section that touches the sides of the quadrilateral. In other words, the third-degree curve is the set of all foci of the conic sections to which the four sides of the quadrilateral or their extension are tangent. Burmester introduced the term *focal mechanism* for this mechanism because, during the motion of the mechanism, the point P is always the focus of a

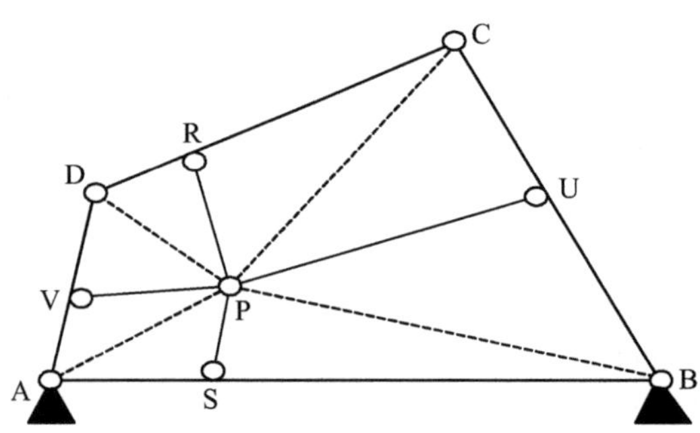

Fig. 7 Kempe's focal mechanism

conic section (changing during the motion) that touches the sides (or their extension) of the four-bar linkage.

Although beyond any doubt, Kempe discovered the mechanism, Wunderlich wrote a paper in 1968 in which he called it "Burmester's focal mechanism" without mentioning Kempe. Dijksman noticed Wunderlich's oversight and called the mechanism *Kempe's focal mechanism.* I prefer that name. The mechanism is also called the Kempe-Burmester focal mechanism (cf. Mallik et al. 1994, p. 121)

On the Reception of His Work

Background

Burmester's book is part of the development of scientific mechanical engineering in the nineteenth century. Before the nineteenth century machines had been studied traditionally in mechanics. When in the eighteenth century, rational mechanics was created, the investigation of machines had become an application of rational mechanics. Because of the difficulties involved, the results were limited. In the nineteenth century, it became clear that the geometrical aspect of machines and mechanisms could be investigated very successfully. Yet the investigation of machines remained applied mathematics or applied mechanics. There was no independent science of machines with a status comparable to the status of the established disciplines. Moreover, in the nineteenth century, the theory of machines was taught on the European continent at polytechnical schools, which in general, had a lower status than universities.

Scientific mechanical engineering was not born easily. For example, in 1877, J. Lüders wrote a booklet of 88 pages with the title *Wider Herrn Reuleaux* (Against Mr. Reuleaux) in which he described Reuleaux' *Theoretische Kinematik* as a book without any value, not written for the specialist but meant to impress the public at large. Others criticized Reuleaux as well because they failed to see the value of his work for mechanical engineering in practice.

In the second half of the nineteenth century, one can distinguish three different views of technology in Germany (Hensel et al. 1989, p. 167). The first view, represented by Franz Grashof (1826–1893), for a long time chairman of the Association of German Engineers (Verein Deutscher Ingenieure), defines technology in the tradition of French positivism: technology is applied natural science and applied mathematics. Burmester must have shared this view. In the second view, represented by Reuleaux, the machine is in the development of mankind, the essential element that determines man's relation with nature. Reuleaux can indeed be seen as one of the first philosophers of technology, who attempted to characterize the general development of mankind on the basis of the fundamental notion of machine. As a result of this position, he emphasized the need for an independent, unified theory of the machine. This theory would reserve a precise place for its application. The third view, represented by Alois Riedler (1850–1936), is opposed to the dominance of theory in the first two views. For Riedler, technology is more a socio-economic system of which theory is only one of the components. "Knowledge is a daughter of application, not the other way around", Riedler wrote (Hensel et al. 1989, p. 180). Words like 'reality', 'organisation' and 'labour' frequently occurred in his writings. He also opposed the specialization that resulted in curricula at technological universities that consisted of completely separate courses.

What Grashof, Burmester and also Reuleaux, however, had in common was the belief in the value of theoretical considerations and in deduction from the general to the particular. The difference was that Reuleaux defended a general theory of machines, independent of natural science and mathematics. Riedler was sceptical of all general theory. Another representative in Germany of the latter view was Th. Beck (1839–1917), factory owner and engineer from Darmstadt. Reuleaux had written (1875, p. 586): "What has 'advantage' to do with science?" Beck argued that the main task of mechanical engineering is to understand how to build a machine in such a way that it yields the greatest advantage (Hensel et al. 1989, p. 199).

When Reuleaux and Burmester wrote their books, the gap between the theory of machines and the practice of building, using and maintaining machines was still considerable. This created tensions and Reuleaux felt under attack from Riedler for not being practical enough, while, on the other hand, he felt that his work was not taken seriously enough by people like Grashof and Burmester. In this particular position, Reuleaux was unable to appreciate the value of Burmester's book.

Reuleaux' Severe Criticism of Burmester

In general, Burmester's book was well received. For example, in 1886, Schumann wrote a very positive review of the first two installments in the *Jahrbuch für die Fortschritte der Mathematik* and in 1887 he repeated his positive opinion with respect to the third installment. There was, however, one remarkable exception. In 1889 Reuleaux wrote:

> His [Burmester's – T. K.] book represents a phenomenon that probably has not yet occurred in our literature. As the main result of my investigation I must point out that *the book does not contain one new thought, not even a small one, in the area covered by the title of the book.* Yes, in order to honor the truth I must say that none of the laws of kinematics is treated completely correctly. Mr. Burmester turns out to be an amateur in kinematics (1894, p. LIII, italics are mine- T. K.).

This quotation is preceded by almost 20 (sic!) pages of severe criticism. Some of the criticism is undoubtedly justified. Wherever Burmester saw a possibility to apply his general theoretical considerations to a particular mechanism he did so, even if such an application seemed miles away from the practice of machine building. In this respect, Burmester's book was written clearly by a mathematician and not by an engineer. However, Reuleaux mainly criticized Burmester for using the word kinematics in an improper way. The above quotation in fact says: "the book does not contain any new contribution in the area of knowledge that I, Reuleaux, call kinematics; it may contain many new thoughts concerning other areas of knowledge".

Burmester defined kinematics in 1886 when he announced his book as follows:

> Kinematics, which encompasses the geometrical theory of motion and its application to machines, was born from the connection of geometry with the notion of motion. (Quoted by Reuleaux 1894, p. XXXII)

Reuleaux had defined kinematics as *Zwanglauflehre*, which can be translated as "the theory of constrained motion". Kinematics was, from Reuleaux' point of view, identical with the theory of mechanisms. Reauleux wrote that kinematics is

> The science that deals with the question how a machine should be composed such that the movements, i.e. changes of position, of the parts with respect to each other are completely determined (Reuleaux 1894, p. XVI)

The basic notions in kinematics are "kinematical pair", "kinematical chain" and "mechanism". In a kinematical pair, two elements are connected in such a way that the motion of two elements with respect to each other is constrained but keeps at least one degree of freedom. In a closed kinematical chain kinematical pairs are connected in such a way that one degree of freedom is left in the chain. By fixing one of the elements of a closed kinematical chain we obtain a mechanism. The mechanism becomes a machine when we use it to force the forces of nature do work by means of precisely defined movements.

Reuleaux distinguished between *theoretical and applied kinematics*. Applied kinematics dealt with the design of machinery in practice on the basis of theoretical kinematics.

It is clear that Reuleaux had defined kinematics as the core discipline in the new emerging science of machines that he envisaged. In Reuleaux' view of kinematics, the geometry of motion was most useful but it played only a supporting role in kinematics. Kinematics encompassed in this view much more than geometrical considerations.

Reuleaux' goal in life was the creation of a science of machines with a status comparable to the other sciences. That is why he gave his *Theoretische Kinematik* a deductive structure; he had tried to make the science of machines with kinematics at its heart as rigid as Euclid's *Elements*. In Reuleaux' perspective, Burmester did not simply use the word kinematics in a different way. No, Reuleaux saw Burmester's book as an attempt to reduce the science of machines to merely applied mathematics. In 1890 he wrote about Burmester and others:

> At the same time geometry imagined, that is its self-appointed imperialist protagonists imagined, to easily unharness the numerous recently derived kinematical theorems with the right of the conqueror and drive them like patient lambs into their sheepcote. (Reuleaux 1890, p. 248)

Reuleaux felt he had to defend the science of machines against enemies like Burmester. It seems that Reuleaux sent his criticism to the Rector and the Senate of the Technische Hochschule in Munich and later he even sent a copy of his 20-page criticism of Burmester's book to the Royal Bavarian Ministery of Education in Munich (Königliche Bayrische Kultusministerium) so that the ministry would see how incompetent a professor they had on their payroll. Burmester, well-positioned in Munich, Fig. 8, reacted to this with a public defense, which obviously did not satisfy Reuleaux. Actually, Burmester was not the real enemy. When he criticized Burmester, Reuleaux was professor in Berlin at the Technische Hochschule and he was at the top of his power. In 1890–1891, he was even rector of the Hochschule. In 1888, however, Alois Riedler had become professor at the Hochschule. Riedler had completely different views for mechanical engineering. He defended a curriculum with much less theory and more practical exercises. Riedler was a formidable opponent who clashed vehemently with Reuleaux. In the end, Riedler won. He succeeded in drastically reducing the number of classes devoted to mathematics and other theoretical subjects. In 1896 Reuleaux retired and soon kinematics was no longer an obligatory subject in Berlin.

H.–J. Braun characterized the situation in mechanical engineering in late nineteenth century Germany as a conflict about the correct method. That is what it was. Yet, one gets the impression that Reuleaux' self-image included infallibility which, in combination with his overambitious theoretical program and his willingness to fight his opponents any time anywhere at great lengths, turned the opposition between points of view into real battles. In particular, in Berlin the clash was very serious. Elsewhere things developed differently, as we will see below.

The Development in Dresden

It is interesting to compare the dramatic developments in Berlin with the situation in Dresden in the same period (Mauersberger 2001). Kinematics classes started in Dresden in 1870. They were given by Ernst Hartig (1836–1900) who was a specialist and pioneer with respect to measurement and test engineering. As we have seen, Burmester became professor for descriptive geometry in Dresden in 1872. In 1874, Reuleaux' pupil Trajan Rittershaus became professor of kinematics and electrotechnical machines. In the same period, Otto Mohr (1835–1918), specialist in graphical methods in statics and strength of materials, was professor of technical mechanics for the civil engineers. Moreover, in 1873, Gustav Zeuner (1828–1907) had become director of the Polytechicum in Dresden. Instead of fighting each other, these men stimulated each other in different ways. Although, in the course of time, the subject played a lesser role in the curriculum, for example, because dynamical aspects of machines became more important, in Dresden kinematics continued to be an important discipline. In 1894, Mohr had succeeded Zeuner, who had returned to teaching, as professor of technical mechanics. In 1900, the successor of Otto Mohr was Martin Grübler (1851–1935), famous for his contributions to the theory of constraints. A younger representative of the kinematical tradition in Dresden is Hermann Alt (1889–1945) who worked on the Burmester theory, as we will see below.

The Long-Term Reception of Burmester's Work

As for the long term reception of Burmester's work and in particular the Burmester theory, I will restrict myself. In the twentieth century, his work and the elaborations by others became an established part of modern kinematics. I will restrict myself to remarks about Italy, Germany and the United States.

Lorenzo Alievi (1856–1941) wrote his treatise *Cinematica della Biella Piana* (Kinematics of the plane coupler) in Rome in 1892. He published the book in Naples in 1895. It deals with the higher-order properties of the curvature of the four-bar coupler curve. Allievi was familiar with Burmester's *Lehrbuch der Kinematik* and with Schoenflies' *Geometrie der Bewegung* of 1886. Schoenflies had pointed out that Burmester's discrete position theory implied analogous, elegant results in instantaneous kinematics, for example, points of stationary curvature are on a curve of the third degree, the instantaneous analogue of Burmester's circle point curves. Moreover, the analytical methods of the differential calculus can be applied in instantaneous kinematics. Allievi drew the conclusion that the higher-order properties of the curvature of the four-bar coupler curve ought to be accessible to an analytical investigation. He started at the second-order level with the Euler-Savary equation and by differentiating, he reached higher-order results. This was a good idea. Half a century later, A. E. Richard de Jonge rightly praised Allievi's book (De Jonge 1943, p. 667).

In 1930, the German Rudolf Beyer published his *Technische Kinematik* (Technical Kinematics). Beyer starts his preface, next to a picture of Reuleaux, with the statement:

> In the spirit of Franz Reuleaux,
>
> with Burmester in mind,
>
> in the footsteps of Ferdinand Wittenbauer.

The first 147 pages are devoted to Reuleaux' theory of constraints in the spirit of the great man. The next 280 pages are written with Burmester in mind and the last 63 pages are devoted to dynamics in the footsteps of Wittenbauer. In the 1930s, Rudolph Beyer was the spokesman of the German kinematicians (Kerle 2007, p. 184) and the situation is perfectly clear. The quarrels of the second half of the nineteenth century were forgotten and Reuleaux and Burmester were both celebrated as great kinematicians. Reuleaux' lasting contribution turned out to have been his abstract approach of the machine: look at mechanisms and the problem of their classification in terms of kinematical pairs and kinematical chains. The bulk of the theory, however, at the time turned out to be, with Burmester in mind, the development of the general kinematical properties of motion in combination with their application to specific mechanisms.

The first half of the twentieth century is exactly the period in which kinematics was really applied in practice. In the 1990s, Riedler had been right: much of the theory that was taught at the Technische Hochschulen was not applied in practice. This had changed drastically in the 1930s. The nineteenth century graphical methods were really applied both in the analysis and the synthesis of mechanisms. Moreover, there were new developments. Worth mentioning is the fact that Herman Alt (1889–1954), professor in Dresden, further developed the Burmester theory (Mauersberger 2006).

In 1942 and 1943, A. E. Richard de Jonge wrote two papers in which he introduced the kinematics that had been developed in Europe to the American study of Mechanical Engineers in the United States. Writing about the US, De Jonge called "kinematic synthesis" the least known but most interesting branch of kinematics. The Burmester theory gets much attention in De Jonge's description. The author's intention was to wake up the Americans by pointing out that in Europe, in particular in Germany, but elsewhere as well, important work in kinematics had been done. In America hardly any published work existed. In the author's closure, he expressed his disappointment at the lack of interest for his papers.

However, the man who most successfully introduced modern kinematics in America was Ferdinand Freudenstein (1926–2006). When de Jonge published his papers on kinematics, he was adjunct professor at the Polytechnic Institute of Brooklyn. In 1954, he worked for the Reeves Instrument Corporation in New York. Freudenstein was enthusiastic about De Jonge's papers. He wrote:

> The greatest benefit of the use of the author's method is obtained when computing machines can be utilized. (Roth 2007, p. 242)

Through Freudenstein and his pupils, modern kinematics, including many of Burmester's ideas, conquered America. Roth (2007) contains Freudenstein's academic family tree at the time of his death. It consisted of 500 members. Worth mentioning in this context is a 1959 paper written by Freudenstein and Sandor in which he started an important series of papers on the vector-loop methods in synthesis and the modern development of the classical Burmester theory (Roth 2007, p. 175).

In 1954, De Jonge saw what was coming. In the course of the years the graphical methods were replaced by analytical methods and the solutions were calculated by means of computers.

Spherical and spatial analogues of the planar Burmester theory have been developed. In a good modern textbook like McCarthy's 2000 the theorem

Fig. 8 Ludwig Burmester when he was professor in Munich

that makes it possible to easily find the center points of RR-chains that correspond to four given positions of a body in the plane, is called *Burmester's theorem* (see p. 60). In 1967, Bernard Roth derived the spherical analogue of this theorem. McCarthy calls it the *Burmester-Roth theorem* (see p. 176). As for space, given five positions of a moving body, there are six fixed axes that can serve as central axes of CC-chains that can guide the body through given positions. McCarthy calls these lines the Burmester lines (p. 244).

The Burmester Curves

In German, *French curves* are called *Burmester curves*. A French curve is a template made out of wood or plastic composed of different curves. It is used in manual drafting to get smooth curves. It is not known exactly how the German Burmester sets that consist of three templates acquired this name. Yet it seems very probable that it was indeed Ludwig Burmester whose name was given to the templates because, in his *Grundzüge der Reliefperspective* (The Main Features of Relief Perspective) of 1883, Burmester remarked in a footnote:

> For exact mathematical drawings the suitably formed curved ruler is a useful aid like the circle and can be much recommended; because in a drawing made with ruler and compass the curves drawn by hand form an ugly contrast which greatly influences the beauty and the homogeneity of the drawing. […] Experience teaches that three suitably formed curved rulers are sufficient for drawing curves in almost all cases. (1883, p. 17)

Burmester made many drawings. The experience he mentioned must have been his own experience. The T.U. Dresden possesses a considerable collection of "Burmester curves" (See Fig. 9).

Fig. 9 Burmester curves (Courtesy of Dr. Klaus Mauersberger)

Fig. 10 A commercial set of Burmester templates (Courtesy of Faber-Castell)

It is possible that someone in Dresden or Munich took the set that Burmester used and started to produce it and sell it under the name *Burmester curves* (Fig. 10).

Acknowledgements I am very grateful to Frau Judith Winkler (Kreuth beim Tegernsee, the village where Burmester was buried) for her help and to Dr. Klaus Mauersberger (T.U. Dresden) for critically commenting on an earlier version of this paper.

References

Burmester's kinematical work

Burmester L (1873) Kinematisch-geometrische Constructionen der Parallelprojection der Schraubenflächen und insbesondere des Schattens derselben. Zeitschrift für Mathematik und Physik 18:185–203

Burmester L (1874a) Kinematisch-geometrische Untersuchungen der Bewegung affin-veränderlicher und collinear-veränderlicher ebener Systeme. Zeitschrift für Mathematik und Physik 19:465–495 (plus Table V)

Burmester L (1874b) Kinematisch-geometrische Untersuchungen der Bewegung ähnlich-veränderlicher ebener Systeme. Zeitschrift für Mathematik und Physik 19:151–170 (plus Table II)

Burmester L (1875) Kinematisch-geometrische Untersuchungen der Bewegung gesetzmässig-veränderlicher ebener Systeme. Zeitschrift für Mathematik und Physik 20:381–422 (plus Tables IV and V)

Burmester L (1876) Über die Geradführung durch das Kurbelgetriebe. Der Civilingenieur 22:597–606 (plus Table 30)

Burmester L (1877) Über die Geradführung durch das Kurbelgetriebe. Der Civilingenieur 23:227–250, 319–342 (plus Tables 10, 11, 14, 15, 16)

Burmester L (1878) Ueber den Beschleunigungszustand ähnlich-veränderlicher und starrer ebener Systeme. Der Civilingenieur 24:147–172

Burmetser L (1879) Ueber die Festlegung projectiv-veränderlicher ebener Systeme. Mathematische Annalen 14:472–497 (plus one Table)

Burmester L (1880a) Ueber das bifocal-veränderliche System. Mathematische Annalen 16:89–112 (plus Tables I and II)

Burmester L (1880b) Ueber die momentane Bewegung ebener kinematischer Ketten. Der Civilingenieur 26:248–285 (plus Tables 14–18)

Burmester L (1888a) Lehrbuch der Kinematik, Band 1, Die ebene Bewegung, Leipzig

Burmester L (1888b) Atlas zum Lehrbuch der Kinematik, Band 1, Die ebene Bewegung, Leipzig
Kinematische Flächenerzeugung vermittelst cylindrischer Rollung (1888). Zeitschrift für Mathematik und Physik 33:304–348
Berichtigung zu Buka's Bemerkungen (1888). Zeitschrift für Mathematik und Physik 33:190
Kinematische Untersuchungen der Mechanismen mit Bandbetrieb (1889). Der Civilingenieur 35:131–154
Geradführung und Proportionalität am Indicator (1890). Zeitschrift Deutscher Ingenieure 34:2 pages
Ueber die momentane Bewegung ebener Mechanismen (1890). Technische Blätter 22:1–18, 73–83
Burmester L (1893) Die Brennpunktmechanismen. Zeitschrift für Mathematische Physik 38:193–223
Burmester L (1902) Kinematisch-geometrische Theorie der Bewegung der affin-veränderlichen und starren räumlichen oder ebenen Systeme. Zeitschrift für Mathematik und Physik 47:128–156
Kinetographische Verwandtschaft ebener Systeme und räumlicher Systeme (1907). Münchener Berichte 37:17–32
Burmester L (1911) Konstruktion der Beschleunigungen bei zusammengesezten Mechanismen. Sitzungsberichte der Bayrischen Akademie der wissenschaften
Kinematische Aufklärung der Bewegung des Auges (1918). Münchener Berichte, pp 171–202
Burmester L (1925) Analysis der möglichen Beschleunigungszuständse eines komplan bewegten starren ebenen Systems. Zeitschrift für Angewandte Mathematik und Mechanik 5:510

Other works by Burmester

Burmester L (1871) Theorie und Darstellung gesetzmäßig gestalteter Flächen, Leipzig
Burmester L (1883) Grundzüge der Reliefperspective nebst Anwendung zur Herstellung reliefperspectivischer Modelle; als Erg. zum Perspectiv-Unterricht an Kunstakademien, Kunstgewerbeschulen u. techn. Lehranstalten, Leipzig
Burmester L (1884) Grundlehren der Theaterperspective, Allgemeine Bauzeitung, Heft 5–7

Other References

Allievi L (1895) Cinematica della biella piana. Regia Tipografia Francesco Giannini & Figli, Napoli
Bottema O, Roth B (1990) Theoretical kinematics. Dover, New York (first Edition in 1978)
Ceccarelli M (ed) (2007) Distinguished Figures in Machine Science, Springer
De Jonge AER (1942) What is wrong with kinematics and mechanisms? Mech Eng 64(April):273–278, and the author's closure, Trans ASME October:747–751
De Jonge AER (1943) A brief account of modern kinematics. Trans ASME August:663–683
Dijksman EA (1975) Kempe's (focal) linkage generalized, particularly in connection with Hart's second straight line mechanism. Mech Mach Theory 10:445–460
Dijksman EA (1976) Motion geometry of mechanisms. Cambridge University Press, New York
Finsterwalder S (1927) In memoriam Ludwig Burmester. Jahrbuch der Bayerischen Akademie der Wissenschaften, 29–35
Grouard A (1870) Extraits inédits des procès verbaux. L'institut : journal universel des sciences et des sociétés savantes en France et à l'étranger, Paris, pp 27–28
Hart H (1877) On some cases of parallel motion. Proc Lond Math Soc 8:286–289
Hashagen U (2003) Walther von Dyck (1856–1934). Mathematik, Technik und Wissenschafts organisation and der TH München. Franz Steiner Verlag, Stuttgart
Hensel S, Ihmig KN, Otte M (1989) Mathematik und Technik im 19. Jahrhundert in Deutschland. Soziale Auseinandersetzung und philosophische Problematik. Vandenhoeck& Ruprecht, Göttingen
Kempe AB (1878) On conjugate four-piece linkages. Proc Lond Math Soc 9:133–147
Kerle H. (2007) Kurt Hain (1908–1995), In (Ceccarelli, 2007), pp 183–215
Koetsier T (1989) The centenary of Ludwig Burmester's 'Lehrbuch der Kinematik'. Mech Mach Theory 24:37–38
Kurrer K-E (2002) Geschichte der Baustatik. Ernst & Sohn, a Wiley Company, Berlin, Germany

Lüders J (1877) Wider Herrn Reuleaux, Eine Würdigung der Wissenschaftlichen Arbeiten und Methoden des Herrn Professor Reuleaux. Verlag von E. Homann, Kiel

Lordick D (2005) Reliefperspektivische Modelle aus dem 3D-Drucker. IBDG (Informationsblätter der Geometrie), Heft 1/2005, Jahrgang 24, Innsbruck, pp 33–42

Mallik AK, Ghosh A, Dittrich G (1994) Kinematic analysis and synthesis of mechanisms. CRC Press, Boca Raton, FL

Mauersberger K (2001) Zur Geschichte der Getriebetechnik an der TH Dresden vor Lichtenheld. Wissenschaftliche Zeitschrift der TH Dresden 50:9–18

Mauersberger K (2006) Theory of mechanisms and machines at Dresden University of technology: A historical review. Dresden University of Technology, Dresden. http://technomag

McCarthy JM (2000) Geometric design of linkages. Springer, New York

Müller R (1930) In memoriam Ludwig Burmester. Jahresbericht der Deutschen Mathematiker Vereinigung 39(1):1–21. Abt. Heft1/4

Reuleaux F (1875) Theoretische Kinematik, Berlin (References are to the English Dover edition: Reuleaux 1963)

Reuleaux F (1890) Ueber das Verhältnis zwischen Geometrie, Mechanik und Kinematik. Zeitung des Vereins Deutscher Ingenieure 34:217–225, 243–248

Reuleaux F (1894) Der Konstrukteur, Ein Handbuch zum Gebrauch beim Maschinen-Entwerfen, Friedrich Vieweg und Sohn, Braunschweig, 4[e] umgearbeitete und erweiterte Auflage

Reuleaux F (1963) The kinematics of machinery, outlines of a theory of machines by Franz Reuleaux, with a new introduction by Eugene S. Ferguson. Dover, New York

Roth B (2007) In memoriam of Professor Freudenstein (1926–2006). Editorial in Trans ASME. J Mech Des 129:241–242

Reye KT (1886) Geometrie de Lage, Vorträge, Abtheilungen I und II, mit einer Aufgabensammlung, Hannover

Schoenflies A (1886) Geometrie der Bewegung, Leipzig

Sylvester JJ (1875) Without title. Nature 12:168

Wunderlich W (1986) On Burmester's focal mechanism and Hart's straight-line motion. J Mech 3:79–86

Henry George (Harry) Ferguson (1884–1960)

Gordon R. Pennock

Abstract This article reviews the life and significant contributions of Henry George (Harry) Ferguson, the pioneer of a system of farm mechanization which helped to revolutionize agriculture. Ferguson, see Fig. 1, was the inventor of the tractor that bears his name and which led to a partnership with Henry Ford and eventually to the Massey-Ferguson tractor. Ferguson was a philanthropist with a single-focus mission, the aim of which was agricultural production. During his lifetime it was this production which controlled the cost of living. As an employee of the Irish Department of Agriculture, during World War I, he supervised the operation and maintenance of the small number of tractors that were then in use. He came to the realization that efficient agriculture based on mechanization was the only solution to the problems of the world's food supply. He devoted his time and energy to the development of equipment which he believed would reduce the cost of food production and the cost of living. His passion to mechanize farming to the highest degree, occupied most of his working life. He was a man with a never-let-up personal drive and had a love for building and tinkering with machines. Ferguson received some 100 patents on improvements in carburetors for internal combustion engines, improvements relating to tractor plows, and the means for coupling agricultural implements to tractors. He was a brilliant engineer who brought about a major change in tractor design with his revolutionary linkage, the three point hitch, that allowed both tractor and implement to work as an integrated unit. Today, virtually all tractors are based in some manner on the unique ideas of Ferguson.

G.R. Pennock (✉)
School of Mechanical Engineering, Purdue University, West Lafayette, IN 47907-2088, USA
e-mail: pennock@ecn.purdue.edu

M. Ceccarelli (ed.), *Distinguished Figures in Mechanism and Machine Science: Their Contributions and Legacies, Part 2*, History of Mechanism and Machine Science 7,
DOI 10.1007/978-90-481-2346-9_4,

Fig. 1 Henry George (Harry) Ferguson (*c*.1930)

Biographical Notes

James Ferguson married Mary (Minnie) Bell on July 1, 1879, and over the next 24 years, 11 children were born into the family; namely, Joseph Bell Ferguson, James Patterson, Margaret Elizabeth, Henry George, Mary Agnes, Norman Priestly, William Edwin Persis, Sarah Ludley Agnew, John Victor Stanley, Edgar McKnight, and Hugh Fisher (www.proni.gov.uk). The children were born on their father's farm at Growell, in the parish of Dromore, 5 miles from Hillsborough, in County Down. Growell, a townland of scattered farmsteads and cottages, is some 15 miles south of the city of Belfast, the capitol of Northern Ireland. The Ferguson family were members of the Brethren, or Plymouth Brethren, and attended the Growell Gospel Hall. So Henry George, or Harry as he was always called, grew up under the teachings of the Holy Bible. However, he was to turn his back on this Christian influence and later in life he would acknowledge that he was, in fact, an agnostic.

It is interesting to note that Ferguson was born just a few months before Gottlieb Daimler received a patent for his motor-cycle and Karl Benz received a patent for his motor-car (Wymer 1961). By the time Harry was 1 year old, the Daimler motor-cycle and the Benz motor-car were the only petrol driven vehicles on the roads throughout Europe. Ferguson left school at the age of 15 and began working full-time with his father on the farm. His light build, however, proved most unsuitable for the heavy farm labor. In truth, he had no interest in the work. By the age of 17, he was preparing to immigrate to Canada. By most accounts he had the necessary paper work in his possession when the oldest brother Joe offered him the opportunity to serve an apprenticeship in his new business (Fraser 1972, 1973). Joe also had no interest in farm work and had started an apprenticeship, at the age of 15, as a maintenance mechanic with the firm of Combe Barbour, general engineers, making linen machines and other factory machines. Then in 1901, Joe joined with two other ex-apprentices from Combe Barbour, Stewart Hamilton and James Alexander McKee, to form the Stewart Partnership on the Shankill Road in North Belfast. This started as a small engineering business but soon concentrated on automobile and motorcycle repairs and, in fact, this was one of the first garages in the city of Belfast. A short time later, Joe opened his own garage on Little Donegall Street in Belfast and called his business, J.B. Ferguson and Company, Automobile Engineers. Later in life he was to own a car dealership in New York.

Belfast (from the Gallic, or Irish language – *Béal Feirste* – meaning sandy ford at river mouth) was originally a town in County Antrim (http://en.wikipedia.org./wiki/belfast). The County Borough of Belfast was created when it was granted city status in 1888 by Queen Victoria. The population within the city limits at that time, according to the 1881 census, was 208,122 (www.nirsa.gov.uk). The city is situated on the eastern coast and is flanked to the northwest by a series of hills, including the Cavehill which is thought to be the inspiration for Gulliver's Travels, a novel by the Irish writer Jonathan Swift. Belfast is a port city, located at the western end of Belfast Lough at the mouth of the River Lagan, making it an ideal location for the shipbuilding industry that once made it famous. When the Titanic was built there in 1912, Harland and Wolff was the largest shipyard in the world. This shipyard with its giant cranes, which could be seen from most parts of the city, is renowned for building many famous ships. The Olympic-class ocean liners which were a trio of the largest and most luxurious ocean liners of their time (namely the RMS Titanic, RMS Olympic and HMHS Britannic) were built there between 1908 and 1914. During and immediately after the Second World War, Belfast was an industrial city known around the world not only for the building of large ships, but for aircraft manufacture, the linen industry, and rope manufacturing. Not far from the shipyard is the aircraft factory, Short Brothers and Harland, which is renowned for the building of freight, and small passenger, aircraft. In the early 1960s, the factory would gain notoriety for its pioneering work with a vertical take-off and landing (VTOL) research aircraft known as the SC1. The Rolls-Royce Flying Bedstead proved the vertical lift capabilities of the RB108 engines, but only flew in England. Belfast is also the birthplace of many famous engineers and mathematicians, including, William Thomson, perhaps better known as Lord Kelvin (1824–1907), Osborne Reynolds (1842–1912), John Greenlees Semple (1904–1985), and John Stewart Bell (1928–1990).

So it was in this city, in the year 1901, that Harry Ferguson joined his brother as an apprentice mechanic in the car and cycle repair shop. They were very successful in both motor-cycle and car racing and established what was described by many as the best garage business of its kind within the city limits. Harry showed he had a natural gift for mechanics and engineering. In 1903, he built a motor-cycle to his own design and won a hill test to determine how its performance would compare with factory production motor-cycles. He also won several road races and trials. Then in 1907, he built and raced his own car. For most of his life, Harry promoted motor-cycle and car racing. His efforts led to the Stormont Road Races Act of 1932, which made possible the first Ulster Grand Prix on the roads of Northern Ireland. He also lobbied the Royal Automobile Club (RAC) to organize the famous Tourist Trophy (TT) motor cycle races (1928–1936). In later life, Harry would apply himself to the design of four-wheel drive cars. He designed the Ferguson Formula car which could transmit traction and braking torque equally to all four wheels. This resulted in the four-wheel drive, using controlled differentials to govern the speeds of the individual wheels. The differential had a limited range of motion only, the front wheels could travel up to 15% faster than the rear wheels (which is a condition required when turning a circle at maximum lock). The rear wheels could turn 5% faster than the front wheels, which was the margin to allow for deflation of a rear tyre. If there was a tendency for greater speed variations than these values, then the differential would lock and all four wheels would be driven at the same speed. The master differential allowed the use of the Maxaret anti-lock brake system that was developed by Dunlop, initially for aircraft wheels. Ferguson developed the first Formula 1 racing car with all wheel drive, the car was the Ferguson P99 (http://en.wikipedia.org./wiki/ferguson), the system included viscous coupling and the anti-lock braking system. The British race car driver Stirling Moss raced a car of this design in the 1960 and 1961 Formula 1 World Championships. Moss, one of the most successful Formula 1 racing car drivers until he was forced to retire in 1962, won the Gold Cup at Oulton Park with the P99. This, of course, was before four-wheel drive cars were excluded from the racing circuit. The P99 also won the British Hill Climb Championship on two occasions. The system, designed by Ferguson was later used in the Jensen FF car. This car had a four-wheel drive system with unequal front/rear torque splitting, a self-locking center differential, and the Maxaret anti-lock brake unit.

By the age of 25, Ferguson had designed and flown a monoplane (http://www.uftm.org.uk). Apparently his initial interest in aviation was galvanized by the excitement generated over Bleriot's crossing of the English Channel in July 1909. He attended the flying meetings at Rheims in August of that year and the Paris Air Exhibition in September. He started to build his first aeroplane immediately upon his return. Although Harry's main expertise was in the building and tuning of engines, he had studied all the aircraft on display in France in great detail, and this enabled him to mastermind the construction of the aircraft on the premises of J. B. Ferguson and Company Ltd., while his brother Joe was in America on motor-related business. Ferguson was the first Briton to build and successfully fly a powered aeroplane in Ireland, this was only 6 years after the pioneering flight of the Wright brothers. While early attempts to get airborne were unsuccessful, he did manage his first flight of this small monoplane on December 31, 1909, at Hillsborough. Ferguson faced the aeroplane into the wind, and with the power of an eight-cylinder engine, raised the aeroplane into the air to a height of about 12 ft. Apparently the aeroplane traveled a total distance of about 130 yards before Ferguson brought it back to earth for a safe landing. Ferguson also flew the monoplane at Magilligan Strand, County Londonderry. His flight at Dundrum Bay, just north-east of Newcastle, County Down, in August 1910, was officially recognized by the Aeroplane Club. He won a cash prize of £100 offered for a 2 miles flight, although Harry flew closer to 3 miles. A man who claimed to have helped Ferguson construct the monoplane was William John Sands (Flackes 1960) who lived on the Old Hillsborough Road, Lisburn, County Down. Due to his age at that time, however, it appears unlikely that Sands worked on the early version of the Ferguson aeroplane. Two friends, Joseph Martin and John L. Williams, however, were prominent throughout the design, building and flight testing of the aircraft. Joe lived in Dromore, County Down, and worked for J.B. Ferguson. He was an expert in building spoked wheels for motor cars. Sands was about 19 years old when he was employed by Ferguson, after serving an apprenticeship in the maintenance of machinery for the linen industry. He also operated a small car repair business on the Lisburn Road. Sands would end up working for Ferguson for more than 35 years, and their association would cover almost 50 years. They had a friendship which survived some very sharp disagreements. In fact, Sands would testify against Ferguson during the infamous lawsuit filed by Ferguson in January 1948 against the Ford Motor Company. Sands had come to believe that he was the true inventor of the Ferguson System (see the following section) and while Sands was an engineering genius in his own right, he was a designer and not an inventor. In the lawsuit against the Ford Motor Company, Ferguson was seeking damages in excess of $250 million on the grounds that Ford had conspired to destroy the Ferguson business and for alleged infringement of several Ferguson patents. The lawsuit was finally settled out of court some 4 years later, in April 1952, for less than $10 million in damages to Ferguson in payment of royalties on patented devices used by Ford after the break-up with Ferguson. These devices comprised, in large part, the Ferguson System.

Harry opened his own motor repair business in Belfast, in 1911, which he called the May Street Motor Company, Limited. Harry would eventually move the business to Lower Donegall Street East. Joe Martin went with Harry to May Street and John Williams also worked for Harry up to the time of his untimely death in 1940. Two other gentlemen who worked for Ferguson were Hugh Reid, who joined the company as a journeyman draughtsman in 1918, and Joseph Thompson, who started as an apprentice and, for many years, was manager of Harry Ferguson (Motors) Limited in Belfast. Subsequently, Thompson bought the business and it was renamed Thompson-Reid Limited. It is interesting to note that the internal combustion engine which was to play such an integral part in the automobile industry, in general, and tractors, in particular, was invented in 1885, just a few months after Harry was born. In 1913, Ferguson married a local girl called Maureen Watson in a registry office in Newry, County Down. The following year, as a loyalist, he was involved in gun running for the Ulster Volunteer Force. Then when the First World War broke out, the Irish Department of Agriculture offered him a wartime Governmental post. At this time, Harry was the agent for a 20 horse power American tractor, called the Overtime. The department asked Ferguson to supervise the operation and maintenance of the small number of tractors that were in use at that time to increase domestic food production. This could be described as the turning point in the life of Harry Ferguson. It was his introduction to the mechanization of agriculture, and the beginning of his focus on plow design. He came to the conclusion that efficient agriculture based on mechanization was the only solution to the problems of food supply around the world. He put all his energy into the development of equipment which he believed would bring down the cost of food production and the cost of living. William J. Sands was with Ferguson during the campaign to increase tractor usage and food production in Ireland during the First World War. He provided a clue to the origin of Ferguson's obsession with agricultural production in pointing out that Ferguson consulted with Dr. Wibberley of Queen's University, Belfast, who had written a book on farming along factory lines (Flackes 1960).

The Overtime tractor was a very heavy machine (it weighed over 2 tons) and dangerous to operate. Ferguson was of the opinion that only a tractor of his own design would do justice to his new plow. He came to the conclusion that the whole idea of making a tractor as one unit, and the implement as another, was entirely wrong. Ferguson designed and built a plow in 1917 which would increase efficiency in food production. His plow was to be coupled to the tractor by a three point linkage, that is, the three-point hitch (Rae 1980). It is interesting to note that Henry Ford began to work on developing a light, cheap and durable tractor as early as 1907, but did not arrive at a satisfactory design until about 1915. As fate would have it, the first Henry Ford tractor (named the Fordson) went into production in 1917. The plow that Ferguson developed, commonly referred to as the "Belfast plough", was first linked to the Model T Ford but was then designed specifically for the Fordson tractor. The man who translated Ferguson's plow ideas into metal was William John Sands. The new two-furrow implement was presented to the farming public at a demonstration in Coleraine, County Antrim. However, in 1918, the Ferguson plow was not at the level of perfection to which the designer aspired. He patented the duplex hitch in that year. However, patents were typically sacrificed because there was not enough money to keep them alive. Also, in that year of 1918, Harry and Maureen had a daughter Elizabeth (Betty). She was to be their only child. Betty would marry Anthony J. Sheldon, an employee in the Ferguson Company. They would have a daughter Sally (she would marry and become Mrs. Sally Fleming), and a son Jamie. The Ferguson Family Museum is currently operated by Jamie on the Isle of Wight (www.ferguson-museum.co.uk).

By 1920, the British farmer had still not decided that the tractor had come to stay. Ferguson and Sands traveled to the United States to visit Henry Ford and show him their, still from perfect, design. Ford offered Ferguson a position in his organization to develop the plow further but refused to provide him with manufacturing facilities. Preferring to remain independent, Ferguson refused the offer. The story of Ferguson's struggle throughout the 1920s and much of the 1930s is one of never giving up in face of repeated frustrations. Fortunately, he had the support of some fine men during this time, many of whom he had first met while taking technical courses at the Belfast Technological College. These men included Captain John L. Williams, Joseph Thompson, Ernest Hamilton Browne, Trevor Knox, Archibald Greer, John Chambers, and T. MacGregor Greer. In 1921, Ferguson took his plow to the Fordson tractor plant in Cork, Ireland, to carry out further tests with some of the Ford staff. The Fordson, as previously mentioned, was a light tractor with many of the same faults as the other light tractors of that period. However, little came from this interaction and Ferguson continued to focus on the development of his plow. He made an arrangement with the Shunk Company, Bucyrus, Ohio to manufacture the plow (Neufeld 1969). Then the business was shifted to Indiana and the Ferguson-Sherman Corporation of Indiana was formed in the town of Evansville, in the southern part of the state. This company was a sales organization which sold thousands of his hand-lift plows with automatic depth control. Some of the components for the Ferguson plow were manufactured at the near-by Vulcan Plow Company.

List of Main Works

During his life time, Ferguson would compile over 100 patents (http://www.google.com/patents; http://ep.espacenet.com) beginning in 1912, with improvements in and relating to carburetors for internal combustion engines, and continuing up through 1950. Most of the patents after 1915 would relate to improvements in the tractor, the plow and agricultural implements, for example, GB119883: Improvements relating to tractor ploughs; GB122703: Improvements relating to ploughs; CA200513: Plough; FR527705: Perfectionnements aux moyens d'accouplement des instruments aratoires; US1379399: Means for coupling agricultural machinery; CA207827: Plough; US1464130: Means for coupling agricultural implements; CA234363: Means of coupling agricultural implements to tractors; CA233226: Agricultural implement; US1501652: Agricultural implement; US1501651: Agricultural implement; CA236960: Agricultural implement; US1526972: Plow; and US1529425: Plow. However, the patents that would revolutionize farming were the patents for the Ferguson system, or the Ferguson hydraulic draft control system. The system was covered by numerous patents which dated from 1917 to 1939. Many of these patents owed a great deal to various members of the Ferguson team. However, it was Ferguson's vision which directed them towards specific solutions to the engineering problems.

Review of Main Works on Mechanism Design

The master patent, filed by Ferguson in 1925, was GB253566 and incorporated the principal of draft control whereby the depth of a ground-engaging implement is automatically controlled by reference to the effort, or draft, required to pull it. The movement of the linkage could be controlled by an electric, mechanical, or hydraulic device. This patent also covered draft control by means of the transmission torque of the tractor. Taken from this patent is Fig. 2 which is a schematic diagram of a disk harrow coupled to the hydraulic mechanism in conjunction with the draft control for angling the disks automatically.

The frame (46) is attached to the disk gangs by means of ball joints (47) and, at the forward end, is pivotally attached to the rocker (9). The rods (48), connected to the disk gangs and to the bell crank (49), serve to regulate the angle of the disk

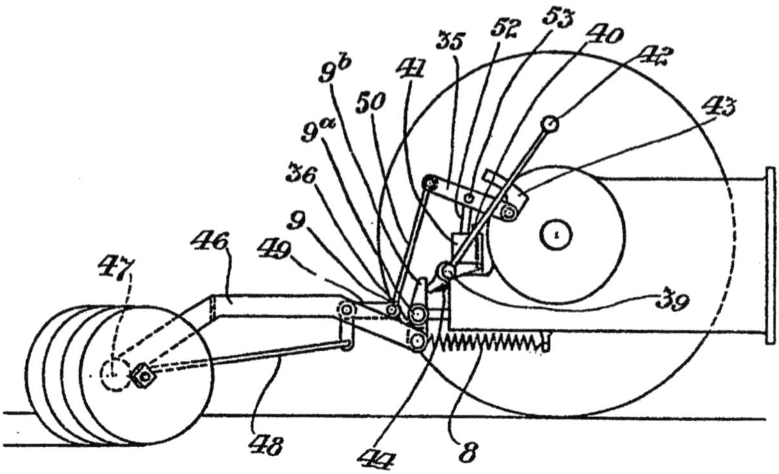

Fig. 2 A schematic diagram of a disk harrow with draft control

gangs. The bell crank is also coupled to the duplex links (35) by means of a rod (50). The duplex links, pivoted on the tractor, are actuated by the piston of the arm cylinder (41) through the piston rod (52) and the crossbar (53). The hand control lever (42) rotates the eccentric shaft (39) on which is mounted a bell crank, which is coupled to the valve (40). A quadrant (43) is provided to retain the hand control lever in any pre-determined position in order to adjust the working depth of the implement. The spring (8) is attached to the rocker (9) and the tractor frame and takes the pull of the implement. Any variation in the pull produces a movement of the rocker which is communicated to the bell crank (pivoted on the eccentric shaft (39) for hand control) which operates a valve (40) and in turn controls the pressure of the oil in the cylinder (41).

The Ferguson system was a system for the mechanization of agricultural operations with the aim and objective of enabling increased production of food and other farm commodities easily, economically and profitably, so that farm products could reach all the people at prices they could afford. The system was a combination of a mechanical linkage and a hydraulic mechanism for controlling the operation of farm implements. The original version of the Ferguson system is shown in Fig. 3 where A denotes the neutral position, B denotes the raised position, and C denotes the lowered position (Morling 1979). The essential features of the system are: (a) integration of the implements with the tractor; (b) hitch geometry to keep the front end of the tractor on the ground (i.e., weight transfer from the implement to the rear wheels of the tractor); (c) traction without excessive tractor weight (i.e., implement penetration without excessive implement weight); (d) continuous automatic control of the implement in the soil; (e) quick and simple attachment and detachment of the implements to and from the tractor; (f) protection of the implement from all normal hazards of operation (such as hitting rocks and other hidden obstructions); and (g) finger-tip and automatic hydraulic control and operation of the tractor and the

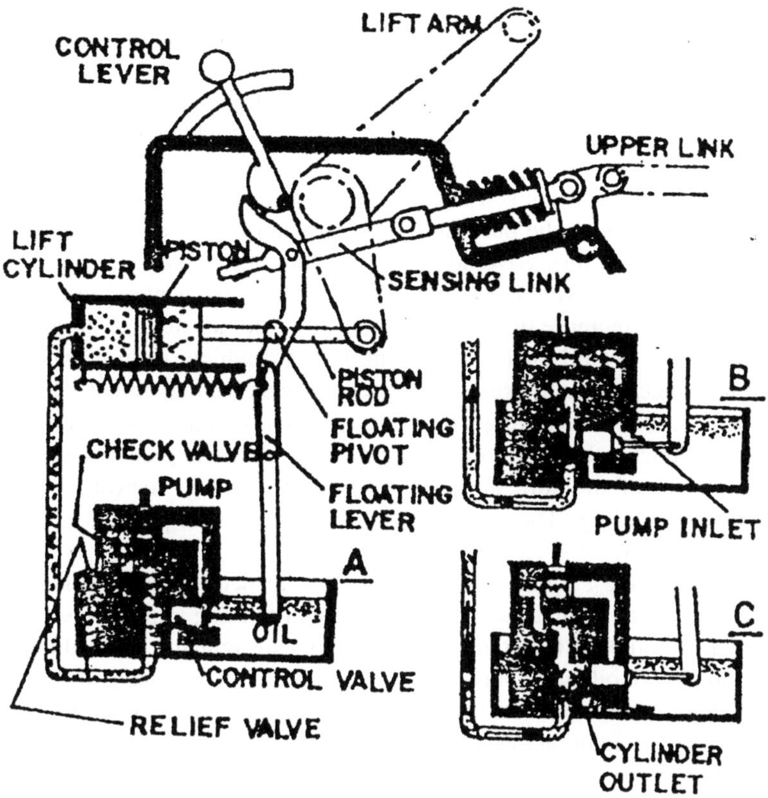

Fig. 3 The original Ferguson draft control hydraulic system

implements. These features were realized with the development of a suitable linkage arrangement between the tractor and the implement, a system of hydraulic controls, and an appropriate tractor. The linkage system, referred to as "the three-point linkage" or "the three-point hitch" was designed with two links at the bottom to pull the implement and one link at the top. It is interesting to note that the original design had two links at the top and one link at the bottom (www.ferguson-museum.co.uk). However, Ferguson quickly realized that this arrangement would not work. The top link served two purposes; namely (a) it served as a rigid restraining brace between the tractor and the implement which prevented the front end of the tractor from rearing backwards off the ground; and (b) as the implement pulled back on the bottom two links (when in the ground), it pushed the top link forward which in turn activated a hydraulic mechanism that tended to exert a lifting force on the implement.

The lifting tendency added weight to the rear wheels and increased traction. It also acted as a counterweight to the front end, shifting front weight to the rear wheels and providing useful traction. Due to the implement reaction forces, the Ferguson tractor could be made lighter and smaller while actually gaining increased stability. This was one of the key features of the Ferguson system. Another important aspect was the method for lifting and lowering the implement and for controlling the depth of the implement in the soil. This was provided by a hydraulic mechanism that was built into the tractor and could be controlled either manually or automatically. Manual control was furnished by a small lever which caused the built-in hydraulic mechanism to lift, lower and adjust the working depth. Automatic control utilized the resistance of the soil as the implement was pulled forward, to activate a control spring through the pushing or compression forces set up in the top link. The control spring in turn governed the hydraulic mechanism, causing it to react automatically, and so govern the working depth of the implement as preset manually. In this manner, an increase in the soil resistance would not stop the progress of the tractor. Rather, it would push the top link forward which, in turn, would cause the hydraulic mechanism to exert a lifting tendency on the implement. This, in turn, would increase the tractive weight on the rear wheels, thereby adding to the pulling ability of the tractor. Furthermore, to the extent that the implement was momentarily lifted out of the soil, allowed the tractor to move forward even under hard-pulling conditions. The implement in the two extreme positions is shown in Fig. 4 (Fraser 1973).

A tractor with the Ferguson system was light, maneuverable, ideally suited for small fields and was applicable to a wide variety of farming implements. The single upper link had improved the draft control but problems remained caused by the continuous pumping of oil under pressure and the tendency to heat and aerate it. In some cases, the problem was so severe that after some 30 min of work, the hydraulic system would airlock. The solution that was proposed by Ferguson was to put the hydraulic control valve on the suction side of the pump. This, in effect, cut off the supply of oil to the pump when no oil under pressure was required to raise the implement. The pump and the control valve had to be submerged in oil. However, Ferguson could foresee no real problems with this arrangement. The control valve, which would be

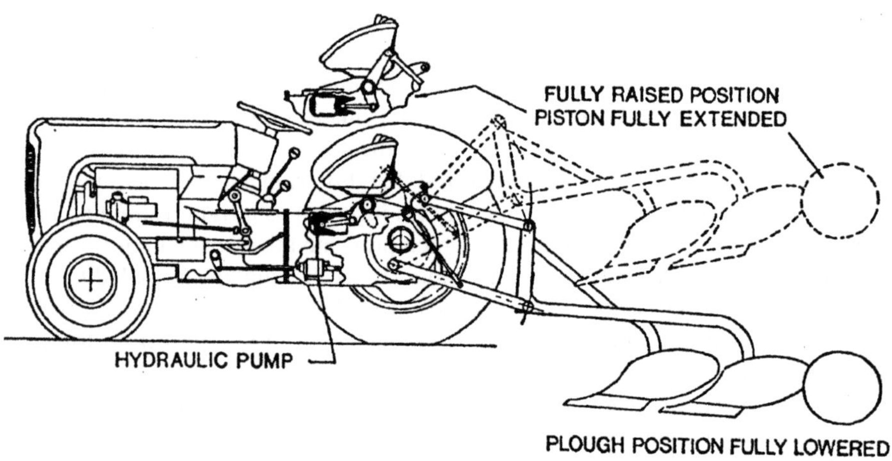

Fig. 4 The implement in the two extreme positions

actuated by the draft forces in the upper link, would be a sleeve running in a housing which was part of the pump. In the housing there would be two ports or openings (an inlet port and an outlet port). When the sleeve valve was centrally located it would block off both ports, thereby retaining the oil in the ram cylinder and holding the lower links at their actual height. If there was an increase in draft as the implement went deeper, the extra compression load in the upper link would slide the valve one way to uncover the inlet port and allow the pump to suck in oil and raise the implement back to the pre-set depth of work. If there was a decrease in draft, the valve would slide the opposite way to uncover the outlet port and allow the oil to escape from the ram cylinder back into the sump, thus lowering the implement.

The automatic mechanism actuated through draft forces in the upper link could, of course, be over-ridden by the lever next to the driver's seat when the driver wanted to raise or lower the implement on the headland or make an adjustment in the working depth. This solution was referred to as suction side control and turned out to be the basic answer to the problem. However, some mild tendency towards bobbing of the implements under certain conditions was to remain for several years. It was finally eliminated by detail changes to the control valve design. Suction side cut-off was the last breakthrough needed to make hydraulic draft control a completely sound and practical proposition. A detailed illustration of the design of the Ferguson system is shown in Fig. 5 (Morling 1979).

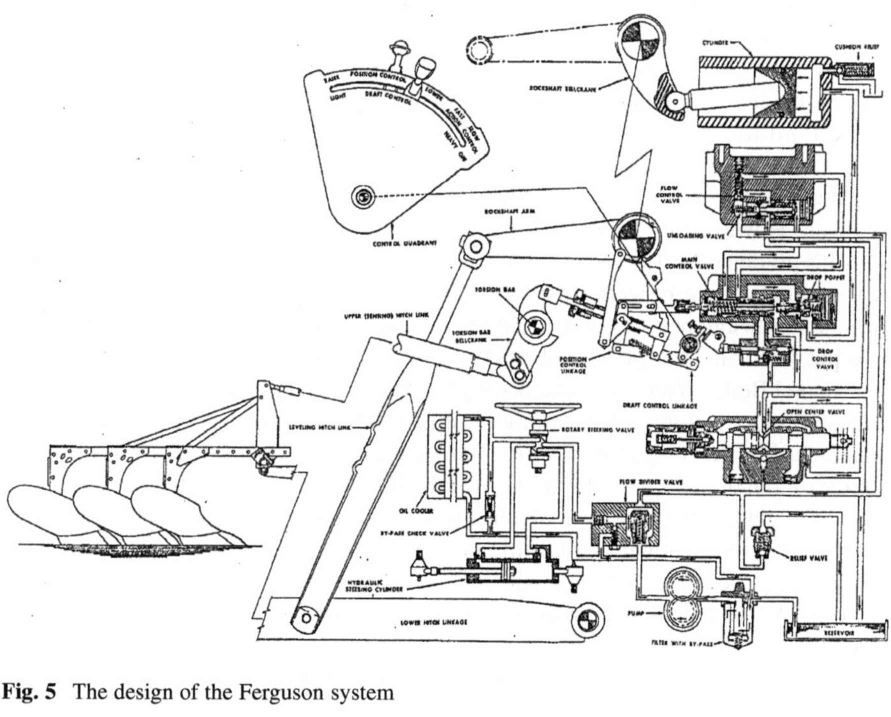

Fig. 5 The design of the Ferguson system

On the Circulation of Works

In 1929, Ferguson claimed to have solved the power farming problem and, in 1930, a demonstration of the tractor system was scheduled for the Northern Ireland government at Andersonstown, on the outskirts of Belfast. However, there were technical details that needed to be resolved and there was the problem of finding the capitol for large-scale production. He asked the Government of Northern Ireland for money to establish his tractor production in Ulster. Lord Craigavon had suggested that financial support would be made available, but nothing happened. The first Ferguson prototype tractor was ready in 1933. The tractor weighed only 16 cwt, compared to the 30 cwt of the lightest tractor then in production. Most of these tractors were built in Belfast and had steel wheels, before tractors had pneumatic tyres. In 1935 after much experiment and engineering development, Ferguson finally claimed to have perfected the Ferguson System. It had taken him almost 16 years to bring the application of hydraulics to mounting implements behind the tractor to what he regarded as a commercially viable platform. This may seem an exceptionally long period of time but it should be remembered that the pumps, control valves, and hydraulic cylinders that he required for the tractor application all had to be designed and developed with no apriori knowledge. The Brown-Ferguson Company was formed about this time and produced the 1354 Model A tractor. The tractors were manufactured in Huddersfield, England, by the David Brown Company. Since the name "Brown" was better known than the name "Ferguson," the tractors were named the Brown-Ferguson tractors and were introduced in 1936.

Then in 1939, Ferguson entered into a partnership with Henry Ford to sell tractors that incorporated the Ferguson System. This led to the Ford-Ferguson tractor (or the Ford tractor – Ferguson system). More than 300,000 of these tractors were manufactured and sold around the world between 1939 and 1947. It is interesting to note that Ford had never considered entering into any form of a partnership with Ferguson. However, he knew if he was to do business with him, he would have to break his golden rule. They never signed a formal contract, their word was their bond, and they sealed their bargain simply by the shaking of hands. Ford was so impressed with the Ferguson system that he told him "This machine is a work of genius…you have put yourself on a plane with such inventors as the Wright Brothers, Marconi, Edison, and Bell… You will make history!" (Wymer 1961). Soon after World War II, Ferguson and Sir John Black of the Standard Motor Company, Coventry, began manufacturing the Ferguson tractor. The Ferguson TE-20 model (TE referred to Tractor England, there was also a TO model where TO referred to Tractor Overseas) a light-weight tractor was one of the models. It was Ferguson's most successful design, commonly known as the Little Grey Fergie (or Wee Fergie). Approximately 500,000 of these tractors were manufactured at the Banner Lane plant in Coventry between 1946 and 1954. Hundreds of thousands of successor models were built at this plant for some 50 years until it closed in 2002. The TE-20 was a small but effective design and is a popular collectors' item for enthusiasts today (Condie 2001).

In 1947, Ford's family (in particular Henry Ford II) reneged on the Ford-Ferguson contract. So in 1948, 9 years after the partnership was formed, Ferguson brought a lawsuit against the Ford Motor Company. The lawsuit was settled in 1952 with Ferguson receiving an estimated 3.3 million pounds sterling (approximately 9.25 million US dollars) in royalties. Ferguson's costs alone were thought to have been in the region of one million pounds sterling. It is important to note that the court refused to renew Ferguson's patent citing its major importance to agriculture and all tractor manufacturers. Surely this was a lasting testament to the engineering ability and ingenuity of Ferguson. In 1953, the Ferguson organization and the Massey-Harris Company, Toronto, amalgamated to form the Massey-Harris-Ferguson Company (later renamed as the Massey-Ferguson Company). The Massey-Harris company was responsible for producing the world's first commercially successful self-propelled combine in 1938. The merger brought together their twin skills in harvesting machinery and tractor design to produce one of the world's most powerful forces in farm equipment. In fact, the merger created a corporation with the second highest sales of farm machinery in the world (Neufeld 1969). However, it was another stormy partnership and in 1954, Ferguson resigned his office in the company. In 1995, the Massey-Ferguson company was purchased by the US-based AGCO Corporation.

Modern Interpretation of the Main Contributions

This section presents a kinematic and dynamic force analysis of the three-point hitch mechanism (Hain 1953, 1955, 1956, 1959, 1973; Hain and Skalweit 1957; Thaer 1956; Phillips 1958; Brock 1954). The method of kinematic coefficients is used to provide geometric insight into the kinematic analysis and the Newton-Euler technique, which also provides physical insight, is used for the dynamic force analysis (Uicker et al. 2003).

(a) Kinematic Analysis. A simple model of the hitch is obtained by assuming that it is symmetrical about the longitudinal axis of the tractor. In this case, each side mechanism (Brock 1954) can be modeled by a planar six-bar mechanism (see Fig. 6). The mechanism is essentially two four-bar linkages coupled by link 3. The input link 2 is actuated by a hydraulic actuator which causes the blade (link 6), in general, and blade point P, in particular, to lift out of and lower into the soil. The blade is attached to the tractor by a lower lift arm (link 4) and an upper link (link 5).

The vectors required for the kinematic analysis are shown in Fig. 6. There are two independent vector loops; namely: the vector loop for links 1, 2, 3 and 4; and the vector loop for links 1, 4, 5 and 6. The first vector loop equation can be written as

$$\overset{\sqrt{}\,\mathrm{I}}{\overline{R}_2} + \overset{\sqrt{}\,?}{\overline{R}_3} - \overset{\sqrt{}\,?}{\overline{R}_4} - \overset{\sqrt{}\,\sqrt{}}{\overline{R}_1} = 0 \tag{1}$$

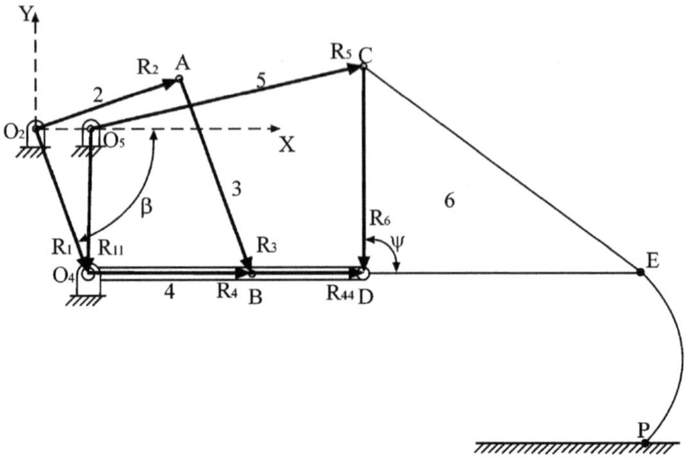

Fig. 6 The three point hitch and the vectors for the mechanism

where the first symbol above each vector indicates the magnitude and the second symbol indicates the direction of the vector with respect to the fixed X-axis. The known quantities are denoted by √, the unknown variables are denoted by ?, and the independent variable (or input θ_2) is denoted by I. For a given position of the input link, the position of links 3 and 4 can be obtained by several techniques, such as trigonometry, or a numerical method, for example, the Newton-Raphson iterative technique (Phillips 1958). Differentiating the X and Y components of Eq. (1) with respect to the input position and writing the resulting equations in matrix form, gives

$$\begin{bmatrix} -R_3 \sin\theta_3 & R_4 \sin\theta_4 \\ R_3 \cos\theta_3 & -R_4 \cos\theta_4 \end{bmatrix} \begin{bmatrix} \theta_3' \\ \theta_4' \end{bmatrix} = \begin{bmatrix} R_2 \sin\theta_2 \\ -R_2 \cos\theta_2 \end{bmatrix} \tag{2}$$

where θ_3' and θ_4' are referred to as the first-order kinematic coefficients of links 3 and 4, respectively (Uicker et al. 2003). Using Cramer's rule, these kinematic coefficients are

$$\theta_3' = \frac{R_2 R_4 \sin(\theta_4 - \theta_2)}{\text{DET}} \text{ and } \theta_4' = \frac{R_2 R_3 \sin(\theta_3 - \theta_2)}{\text{DET}} \tag{3a}$$

where the determinant of the coefficient matrix in Eq. (2) can be written as

$$\text{DET} = R_3 R_4 \sin(\theta_3 - \theta_4) \tag{3b}$$

Differentiating the X and Y components of Eq. (1) twice with respect to the input position and writing the resulting equations in matrix form, gives

$$\begin{bmatrix} -R_3\sin\theta_3 & R_4\sin\theta_4 \\ R_3\cos\theta_3 & -R_4\cos\theta_4 \end{bmatrix} \begin{bmatrix} \theta''_3 \\ \theta''_4 \end{bmatrix} = \begin{bmatrix} R_2\cos\theta_2 + R_3\cos\theta_3\,\theta'^2_3 - R_4\cos\theta_4\,\theta'^2_4 \\ R_2\sin\theta_2 + R_3\sin\theta_3\,\theta'^2_3 - R_4\sin\theta_4\,\theta'^2_4 \end{bmatrix} \tag{4}$$

where θ''_3 and θ''_4 are referred to as the second-order kinematic coefficients of links 3 and 4, respectively. Using Cramer's rule, these kinematic coefficients can be written as

$$\theta''_3 = \frac{-R_2R_4\cos(\theta_2-\theta_4) - R_3R_4\cos(\theta_3-\theta_4)\theta'^2_3 + R_4^2\theta'^2_4}{\text{DET}} \tag{5a}$$

and

$$\theta''_4 = \frac{-R_2R_3\cos(\theta_2-\theta_3) - R_3^2\theta'^2_3 + R_3R_4\cos(\theta_3-\theta_4)\theta'^2_4}{\text{DET}} \tag{5b}$$

where the determinant is given by Eq. (3b).

The second vector loop equation can be written as

$$\vec{R}_5 + \vec{R}_6 - \vec{R}_{44} - \vec{R}_{11} = 0. \tag{6}$$

Differentiating the X and Y components of this equation with respect to the input position gives

$$-R_5\sin\theta_5\,\theta'_5 - R_6\sin\theta_6\,\theta'_6 + R_{44}\sin\theta_4\,\theta'_4 = 0 \tag{7a}$$

and

$$R_5\cos\theta_5\,\theta'_5 + R_6\cos\theta_6\,\theta'_6 - R_{44}\cos\theta_4\,\theta'_4 = 0. \tag{7b}$$

Then writing these two equations in matrix form gives

$$\begin{bmatrix} -R_5\sin\theta_5 & -R_6\sin\theta_6 \\ R_5\cos\theta_5 & R_6\cos\theta_6 \end{bmatrix} \begin{bmatrix} \theta'_5 \\ \theta'_6 \end{bmatrix} = \begin{bmatrix} -R_{44}\sin\theta_4\,\theta'_4 \\ R_{44}\cos\theta_4\,\theta'_4 \end{bmatrix}. \tag{8}$$

Using Cramer's rule, the first-order kinematic coefficients of links 5 and 6 can be written as

$$\theta'_5 = \frac{R_{44}R_6\sin(\theta_6-\theta_4)\theta'_4}{\text{DET}} \quad \text{and} \quad \theta'_6 = \frac{R_{44}R_5\sin(\theta_4-\theta_5)\theta'_4}{\text{DET}} \tag{9}$$

where the determinant of the coefficient matrix in Eq. (8) is

$$\text{DET} = R_5R_6\sin(\theta_6-\theta_5). \tag{10}$$

Differentiating the X and Y components of Eq. (6) twice with respect to the input position and writing the answers in matrix form, gives

$$\begin{bmatrix} -R_5 \sin\theta_5 & -R_6 \sin\theta_6 \\ R_5 \cos\theta_5 & R_6 \cos\theta_6 \end{bmatrix} \begin{bmatrix} \theta_5'' \\ \theta_6'' \end{bmatrix} = \begin{bmatrix} R_5 \cos\theta_5 \theta_5'^2 + R_6 \cos\theta_6 \theta_6'^2 - R_{44} \cos\theta_4 \theta_4'^2 - R_{44} \sin\theta_4 \theta_4'' \\ R_5 \sin\theta_5 \theta_5'^2 + R_6 \sin\theta_6 \theta_6'^2 - R_{44} \sin\theta_4 \theta_4'^2 + R_{44} \cos\theta_4 \theta_4'' \end{bmatrix} \quad (11)$$

Using Cramer's rule, the second-order kinematic coefficients of links 5 and 6 can be written as

$$\theta_5'' = \frac{R_6^2\theta_6'^2 + R_5R_6 \cos(\theta_5 - \theta_6)\theta_5'^2 - R_6R_{44} \cos(\theta_4 - \theta_6)\theta_4'^2 - R_6R_{44} \sin(\theta_4 - \theta_6)\theta_4''}{DET} \quad (12a)$$

and

$$\theta_6'' = \frac{-R_5^2\theta_5'^2 - R_5R_6 \cos(\theta_6 - \theta_5)\theta_6'^2 - R_5R_{44} \cos(\theta_4 - \theta_5)\theta_4'^2 - R_5R_{44} \sin(\theta_4 - \theta_5)\theta_4''}{DET} \quad (12b)$$

where the determinant is given by Eq. (10).

Using the chain rule, the angular velocity and acceleration of link j (= 3, 4, 5 and 6) can be written, respectively, as

$$\omega_j = \theta_j' \, \omega_2 \quad \text{and} \quad \alpha_j = \theta_j'' \, \omega_2^2 + \theta_j' \, \alpha_2. \quad (13)$$

Now that the angular velocities and accelerations of the links are known, the kinematics of the blade point P can be investigated. The vector equation for point P, see Fig. 6, can be written as

$$\overset{??}{\overline{R_P}} = \overset{\surd\surd}{\overline{R_1}} + \overset{\surd\surd}{\overline{R_{44}}} + \overset{\surd\surd}{\overline{R_{66}}} + \overset{\surd\surd}{\overline{R_{67}}} \quad (14)$$

where $\overline{R}_{66}$ is the vector from pin D to point E and $\overline{R}_{67}$ is the vector from point E to point P. Differentiating the X and Y components of this equation with respect to the input position, the first-order kinematic coefficients of point P are

$$X'_P = -R_{44} \sin\theta_{44}\theta_4' - R_{66} \sin(\theta_6 + \psi)\theta'_6 - R_{67} \sin\theta_6\theta_6' \quad (15a)$$

and

$$Y'_P = R_{44} \cos\theta_{44}\theta_4' + R_{66} \cos(\theta_6 + \psi)\theta'_6 + R_{67} \cos\theta_6\theta_6'. \quad (15b)$$

Then differentiating these equations with respect to the input position, the second-order kinematic coefficients of point P are

$$\begin{aligned} X''_P = & -R_{44} \cos\theta_{44}\theta_4'^2 - R_{44} \sin\theta_4\theta_4'' - R_{66} \cos(\theta_6 + \psi)\theta_6'^2 \\ & - R_{66} \sin(\theta_6 + \psi)\theta_6'' - R_{67} \cos\theta_6\theta_6'^2 - R_{67} \sin\theta_6\theta_6'' \end{aligned} \quad (16a)$$

and

$$\begin{aligned} Y''_P = & -R_{44}\sin\theta_{44}\theta_4'^2 + R_{44}\cos\theta_4\theta_4'' - R_{66}\sin(\theta_6+\psi)\theta_6'^2 \\ & + R_{66}\cos(\theta_6+\psi)\theta_6'' - R_{67}\sin\theta_6\theta_6'^2 + R_{67}\cos\theta_6\theta_6'' \end{aligned} \tag{16b}$$

Finally, the velocity and the acceleration of point P can be written, respectively, as

$$\bar{V}_P = (X'_P\,\hat{i} + Y'_P\,\hat{j})\,\omega_2 \tag{17a}$$

and

$$\bar{A}_P = (X''_P\,\hat{i} + Y''_P\,\hat{j})\,\omega_2^2 + (X'_P\,\hat{i} + Y'_P\,\hat{j})\,\alpha_2. \tag{17b}$$

The geometry of the path of point P can also be expressed in terms of kinematic coefficients. The unit tangent and unit normal vectors to the path of point P are defined, respectively, as

$$\hat{u}^t = \frac{X'_P\,\hat{i} + Y'_P\,\hat{j}}{R'_P} \quad \text{and} \quad \hat{u}^n = \hat{k}\times\hat{u}^t = \frac{-Y'_P\,\hat{i} + X'_P\,\hat{j}}{R'_P}, \tag{18a}$$

where

$$R'_P = \pm\sqrt{(X'_P)^2 + (Y'_P)^2}. \tag{18b}$$

The positive sign is chosen if the change in the input position is positive, that is counterclockwise, and the negative sign is chosen if the change in the input position is negative, that is, clockwise.

The radius of the curvature of the path of point P can be written as (Uicker et al. 2003)

$$\rho_P = \frac{R_P'^3}{X'_P Y''_P - Y'_P X''_P} \tag{19}$$

where the first and second-order kinematic coefficients of point P are given by Eqs. (15) and (16), respectively. The X and Y coordinates of the center of the curvature of path of point P can be written, respectively, as

$$X_{CC} = X_P - \rho_P\left[\frac{Y'_P}{R'_P}\right] \quad \text{and} \quad Y_{CC} = Y_P + \rho_P\left[\frac{X'_P}{R'_P}\right]. \tag{20}$$

In the design of a three point hitch, the length of link 2 and the range of motion of this link are predefined by the hitch itself (Ambike and Schmiedeler 2007). The length of link 4 (the rocker) is variable, and the designer can select the length of the coupler to produce the desired path, or orientation, for the implement. The desired performance, however, is not defined by traditional motion generation task

Fig. 7 The free body diagram of link 2

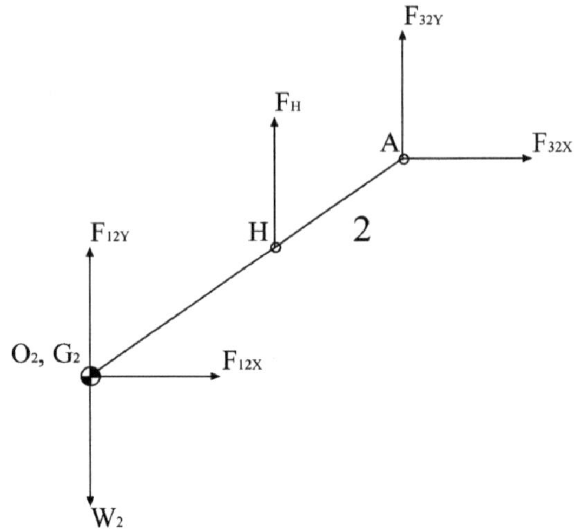

requirements (Kinzel et al. 2006). Rather, the typical design goals are to locate the instant center of the coupler (Uicker et al. 2003) (also referred to as the pitch point) in a desirable location, to provide the necessary ground clearance for the implement when it is not in use and to allow for some elevation of the implement without significant change in its orientation.

(b) Dynamic Force Analysis. For convenience, the centers of mass of links 2, 4, and 5 are assumed coincident with the ground pivots O_2, O_4, and O_5, respectively, and the centers of mass of links 3 and 6 are assumed coincident with points B and E, respectively. The actuating force, denoted as $\overline{F}_H$, is provided by a hydraulic actuator which is assumed to act vertically upward at point H, see the free body diagram of link 2 shown in Fig. 7.

From Newton's second law, the sum of the external forces in the X and Y directions can be written, respectively, as

$$F_{12X} + F_{32X} = m_2 A_{G_{2X}} = 0 \tag{21a}$$

and

$$F_{12Y} + F_{32Y} + F_H - W_2 = m_2 A_{G_{2Y}} = 0, \tag{21b}$$

where F_{12X}, F_{12Y}, F_{32X}, and F_{32Y} are the X and Y components of the internal reaction forces at pins O_2 and A, respectively, and W_2 is the weight of the link. From Euler's equation, the sum of the external moments about the mass center of link 2 can be written as

$$R_{22} \cos\theta_2 F_H + (R_2 \cos\theta_2 F_{32Y} - R_2 \sin\theta_2 F_{32X}) = 0. \tag{22}$$

The free body diagram of link 3 is shown in Fig. 8.

Fig. 8 The free body diagram of link 3

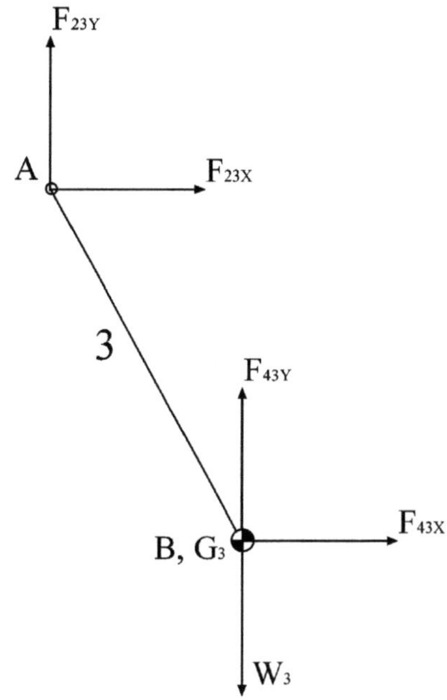

The sum of the external forces in the X and Y directions can be written, respectively, as

$$F_{23X} + F_{43X} = m_3 A_{G_{3X}} \tag{23a}$$

and

$$F_{23Y} + F_{43Y} - W_3 = m_3 A_{G_{3Y}}. \tag{23b}$$

The sum of the external moments about point A can be written as

$$(R_3 \cos\theta_3 F_{43Y} - R_3 \sin\theta_3 F_{43X}) - R_3 \cos\theta_3 W_3 = I_{G_3}\bar{\alpha}_3 + m_3 \bar{R}_3 \times \bar{A}_{G_3}. \tag{24}$$

where the acceleration of the mass center of link 3 (which is coincident with pin B) is obtained from a kinematic analysis. The vector equation for point B can be written as

$$\bar{R}_B = \bar{R}_1 + \bar{R}_4. \tag{25}$$

Differentiating the X and Y components of this equation with respect to the input position, the first and second-order kinematic coefficients of point B, respectively, are

$$X_B^{'} = -R_4 \sin\theta_4 \theta_4^{'} \tag{26a}$$

$$Y'_B = -R_4 \cos\theta_4 \theta'_4 \tag{26b}$$

$$Y''_B = -R_4 \sin\theta_4 \theta''_4 - R_4 \cos\theta_4 \theta'^2_4 \tag{26c}$$

and

$$Y''_B = R_4 \cos\theta_4 \theta''_4 - R_4 \sin\theta_4 \theta'^2_4. \tag{26d}$$

The X and Y components of the acceleration of point B can then be obtained from the relationships

$$A_{BX} = X'_B \alpha_2 + X''_B \omega_2^2 \qquad \text{and} \qquad A_{BY} = Y'_B \alpha_2 + Y''_B \omega_2^2. \tag{27}$$

The free body diagram of link 4 is shown in Fig. 9. The sum of the external forces in the X and Y directions can be written, respectively, as

$$F_{14X} + F_{34X} + F_{64X} = m_4 A_{G_{4X}} = 0 \tag{28a}$$

and

$$F_{14Y} + F_{34Y} + F_{64Y} - W_4 = m_4 A_{G_{4Y}} = 0. \tag{28b}$$

The sum of the external moments about the mass center of link 4 can be written as

$$(R_4 \cos\theta_4 F_{34Y} - R_4 \sin\theta_4 F_{34X}) + (R_{44} \cos\theta_4 F_{64Y} - R_{44} \sin\theta_4 F_{64X}) = I_{G_4} \alpha_4. \tag{29}$$

The free body diagram of link 5 is shown in Fig. 10. The sum of the external forces in the X and Y directions can be written, respectively, as

$$F_{15X} + F_{65X} = m_5 A_{G_{5X}} = 0 \tag{30a}$$

and

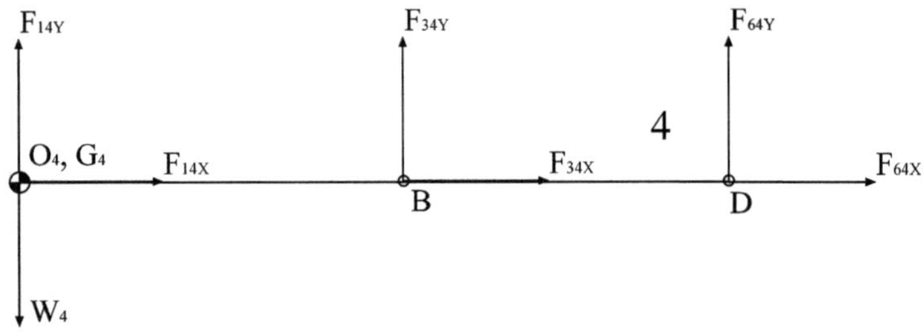

Fig. 9 The free body diagram of link 4

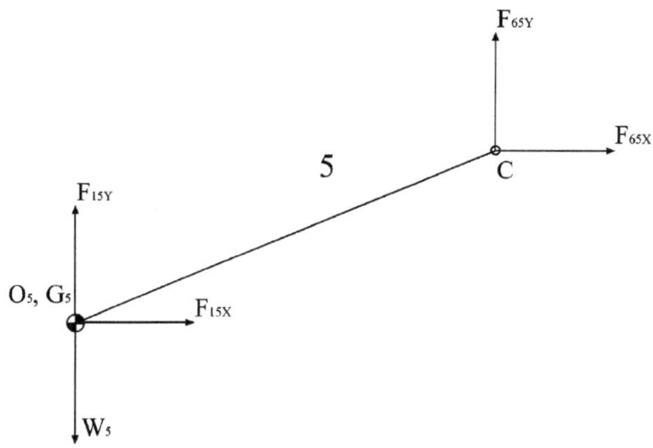

Fig. 10 The free body diagram of link 5

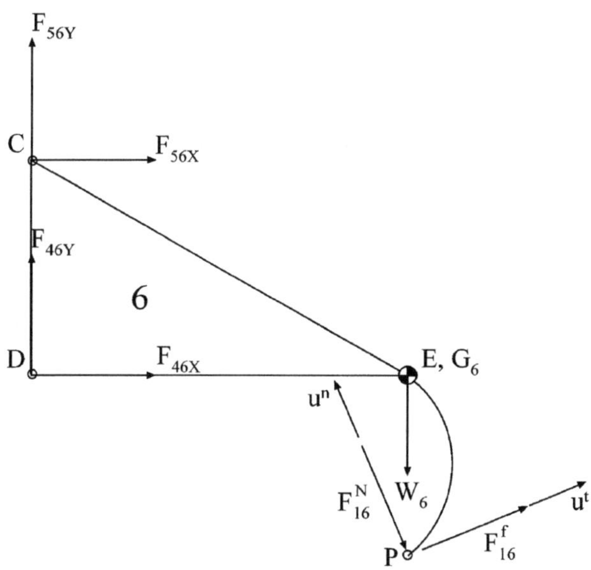

Fig. 11 The free body diagram of link 6

$$F_{15Y} + F_{65Y} - W_5 = m_5 A_{G_{5Y}} = 0. \tag{30b}$$

The sum of the external moments about the mass center of link 5 can be written as

$$R_5 \cos\theta_5 F_{65Y} - R_5 \sin\theta_5 F_{65X} = I_{G_5} \alpha_5. \tag{31}$$

The free body diagram of link 6 is shown in Fig. 11. The contact force between the ground and the blade point P is assumed to be known and the normal force is assumed to be acting downward, that is, in the opposite direction to the unit normal vector.

Also, the friction force opposes the velocity of point P, that is, the friction force is in the opposite direction to the unit tangent vector and can be written as

$$F_{16}^{f} = \mu\, F_{16}^{N} \tag{32}$$

where μ is the coefficient of friction. The sum of the external forces in the X and Y directions can be written, respectively, as

$$F_{56X} + F_{46X} - (F_{16}^{f})_P u^{tx} - (F_{16}^{N})_P u^{nx} = m_6 A_{G_6X} \tag{33a}$$

and

$$F_{56Y} + F_{46Y} - W_6 - (F_{16}^{f})_P u^{ty} - (F_{16}^{N})_P u^{ny} = m_6 A_{G_6Y} \tag{33b}$$

where u^{tx}, u^{ty}, u^{nx}, and u^{ny} are given by Eq. (18a). The sum of the external moments about point D can be written as

$$R_{DCX}F_{56Y} - R_{DCY}F_{56X} - R_{DEX}W_6 + \overline{R}_{DP} \times (-\overline{F}_{16}^{f} \bullet \hat{u}^{t} - \overline{F}_{16}^{N} \bullet \hat{u}^{n}) = I_{G_6}\overline{\alpha}_6 + m_6\, \overline{R}_{DE} \times \overline{A}_{G_6}. \tag{34}$$

The vector equation for the mass center of link 6 (which is coincident with point E) can be written as

$$\overline{\mathrm{R}}_{\mathrm{E}} = \overline{\mathrm{R}}_1 + \overline{\mathrm{R}}_{44} + \overline{\mathrm{R}}_{66}. \tag{35}$$

Differentiating the X and Y components of this equation with respect to the input position, the first and second-order kinematic coefficients of point E are

$$X_E^{'} = -R_4 \sin\theta_4\theta_4^{'} - R_{66}\sin(\theta_6 + 180 - \Psi)\theta_6^{'} \tag{36a}$$

$$Y_4^{'} = R_4 \cos\theta_4\theta_4^{'} + R_{66}\cos(\theta_6 + 180 - \psi)\theta_6^{'} \tag{36b}$$

$$\begin{aligned} X_E^{''} = {} & -R_4 \sin\theta_4\theta_4^{''} - R_4\cos\theta_4\theta_4^{'2} - R_{66}\sin(\theta_6 + 180 - \psi)\theta_6^{''} \\ & - R_{66}\cos(\theta_6 + 180 - \psi)\theta_6^{'2} \end{aligned} \tag{36c}$$

and

$$\begin{aligned} Y_E^{''} = {} & R_4\cos\theta_4\theta_4^{''} - R_4\sin\theta_4\theta_4^{'2} + R_{66}\cos(\theta_6 + 180 - \Psi)\theta_6^{''} \\ & - R_{66}\sin(\theta_6 + 180 - \Psi)\theta_6^{'2}. \end{aligned} \tag{36d}$$

The X and Y components of the acceleration of point E can then be obtained from relationships similar to Eq. (27).

The total number of unknown variables for the dynamic force analysis is 15, and there are 15 equations, see Eqs. (21)–(24), (28)–(31), (33) and (34). The solution can be obtained by writing the equations in matrix form and taking the inverse of the 15 × 15 coefficient matrix. However, to provide insight into the analysis, and as a check of the matrix inversion, the solution can be obtained from an inspection of the 15 equations. A possible rocedure is to solve Eqs. (31) and (34) for the internal reaction forces F_{56X} and F_{56Y}, Eq. (33a) for the internal reaction forces F_{46X}, Eq. (33b) for the internal reaction force F_{46Y}, Eq. (30a) for the internal reaction force F_{15X}, Eq. (30b) for the internal reaction force F_{15Y}, Eqs. (24) and (29) for the internal reaction forces F_{34X} and F_{34Y}, Eq. (28a) for the internal reaction force F_{14X}, Eq. (28b) for the internal reaction force F_{14Y}, Eq. (23a) for the internal reaction force F_{23X}, Eq. (23b) for the internal reaction force F_{23Y}, Eq. (22) for the hydraulic actuator force F_H, Eq. (21a) for the internal reaction force F_{12X} and Eq. (21b) for the internal reaction force F_{12Y}.

To illustrate the numerical procedure, consider $O_2 O_4$ = 34.0 cm, O_2A = 34.0 cm, O_4B = 36.0 cm, AB = 46.0 cm, O_4D = 61.0 cm, O_2O_5 = 11.5 cm, O_5C = 62.5 cm, CD = 46.0 cm, DE = 61.0 cm, EP = 38.0 cm, $\beta = 70^\circ$, $\psi = 90^\circ$ and the line EP is parallel to the line CD. The input link is rotating counter-clockwise from an initial position $\theta_2 = -10^\circ$ to a final position $\theta_2 = +20^\circ$, with an angular velocity and acceleration as shown in Figs. 12a and b, respectively. For the specified range of motion, the X and Y coordinates of the path of point P, obtained from Eq. (14), and the radius of curvature of the path, see Eq. (19), are as shown in Figs. 13a and b, respectively.

The velocity and acceleration of point P versus the input position are as shown in Figs. 14a and b, respectively.

The masses of the links are m_2 = 1.5 kg, m_3 = 2 kg, $m_4\}$ = 4 kg, m_5 = 2.5 kg, and m_6 = 25 kg and the mass moments of inertia of the links, about their respective centers of mass, are I_{G2} = 0.030 N m-s^2, I_{G3} = 0.045 N m-s^2 and I_{G4} = 0.075 N m-s^2 I_{G5} = 0.095 N m-s^2, and I_{G6} = 0.250 N m-s^2. When the blade is in the soil, that is, – 85 cm $\geq Y_P \geq$ – 115 cm, the coefficient of friction and the friction force at point P are assumed to be $\mu = 0.30$ and F_P^f = 100 N respectively.

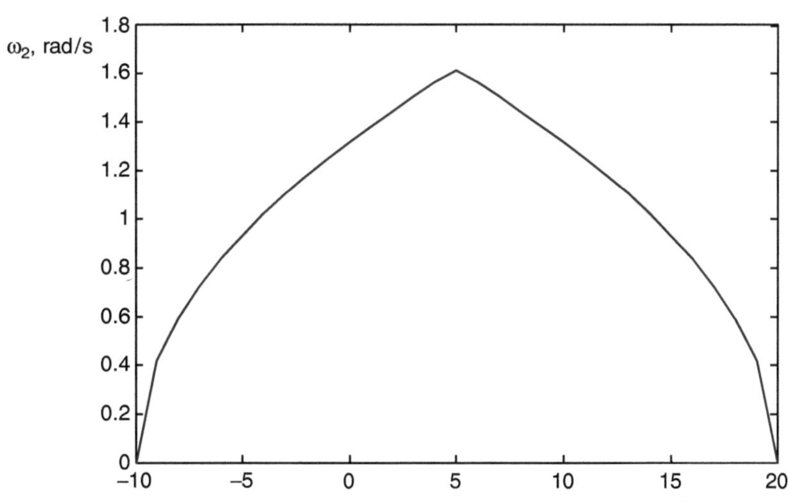

Fig. 12 (**a**) The angular velocity of the input link 2. (**b**) The angular acceleration of the input link 2

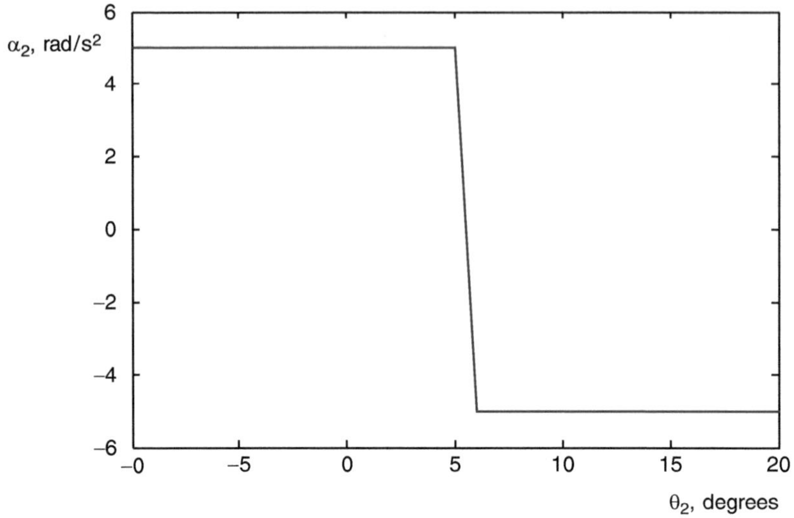

Fig. 12 (continued)

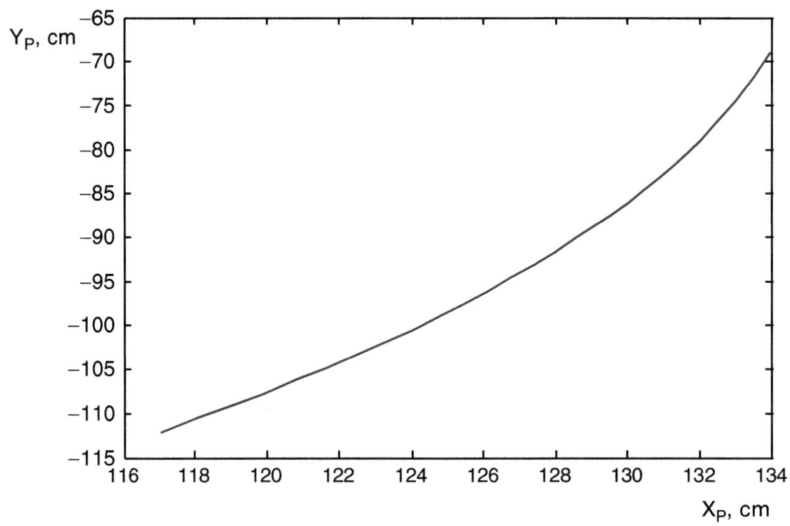

Fig. 13 (**a**) The X and Y coordinates of the path of point P. (**b**) The radius of curvature of the path of point P

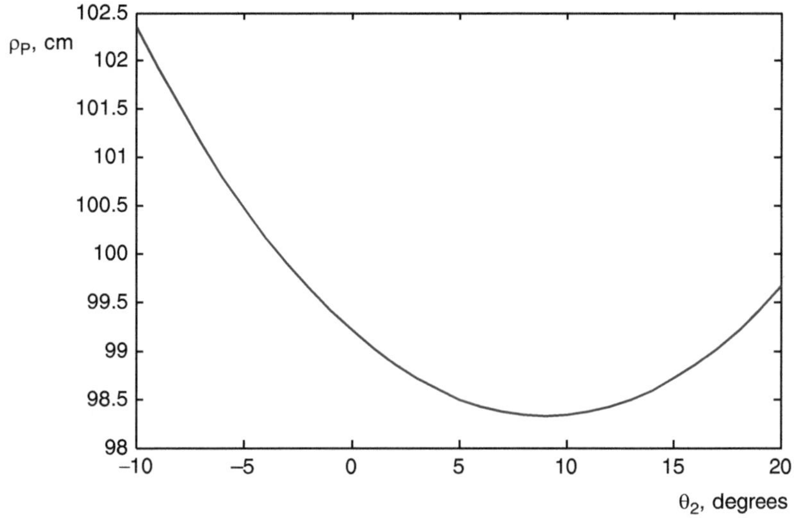

Fig. 13 (continued)

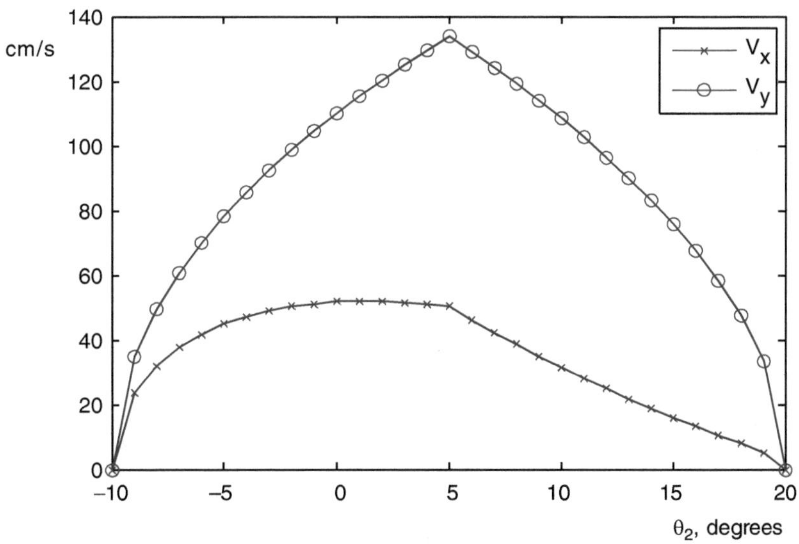

Fig. 14 (**a**) The velocity of point P versus the input position. (**b**) The acceleration of point P versus the input position

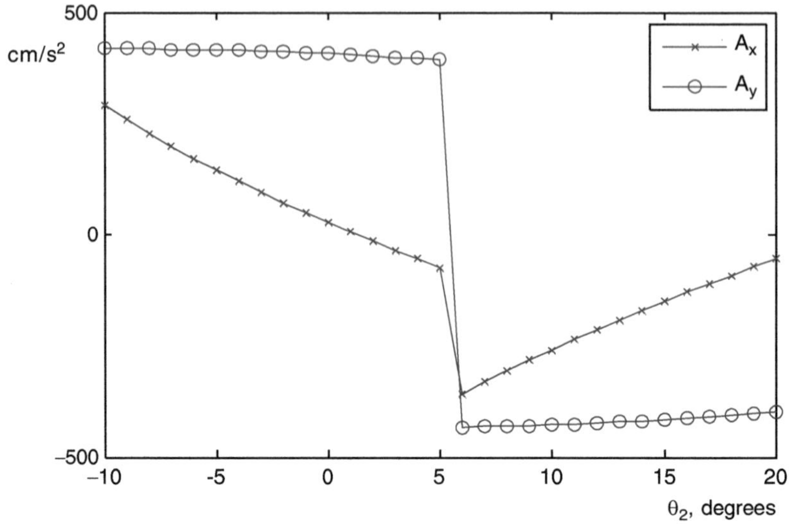

Fig. 14 (continued)

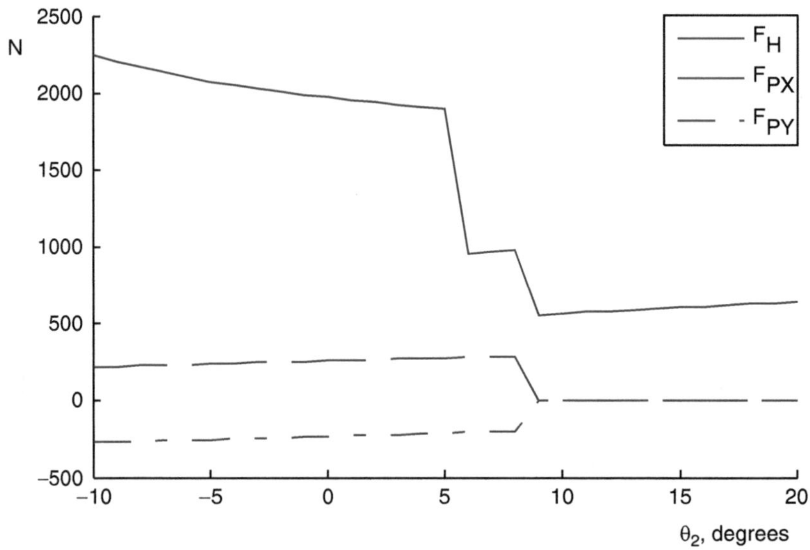

Fig. 15 The X and Y components of F_P and F_H

Due to space considerations, a complete set of numerical results for the dynamic force analysis will not be presented here. However, for illustration purposes, consider the force that is required by the hydraulic actuator to overcome the contact force on the blade at point P. The X and Y components of this force against the position of link 2 are as shown in Fig. 15. From the solution to the dynamic force analysis, a plot of the corresponding hydraulic actuator force against the position of the input link is also shown on the figure. It is worth noting that the numerical results of the force analysis can be used, for example, to demonstrate the influence of the link dimensions on the loads acting on the links of the three-point hitch and on the tractor. The results can also be used by the designer to optimize the link dimensions and to investigate the dimensions proposed by Ferguson in his original design. Also, the hydraulic actuator can be sized based on the maximum contact force that is expected at P.

Concluding Remarks

Henry George (Harry) Ferguson died at Abbotswood, Stow-on-the-Wold, Gloucestershire, in his bath about 8:30 am, Tuesday, October 25, 1960 (10 days before his 76th birthday). He had been suffering from a manic depressive condition and, over the years, this condition had become more severe. The doctor attending Ferguson said that in July, 1959, Ferguson was found in similar circumstances in his bath. On that particular occasion, the doctor had come to the conclusion that Ferguson was suffering from an overdose of barbiturate tablets. The pathologist and the coroner at Ferguson's inquest indicated that Ferguson had died from an overdose of barbiturate tablets but that there was no evidence to show whether the tablets were self-administered or ingested accidentally. Ferguson was cremated 3 days after his death on Friday, October 28.

Even in the last few months of his life, Ferguson was planning a new venture in tractor distribution in his native Northern Ireland. He saw the exploitation of land resources as the way to prosperity and lower prices. He had even argued in pre-World War II days that Nazism might have been contained by intensive cultivation of German soil. During his last few months, all his activities were centered on his car project which he regarded to be of fundamental importance. This project received tremendous publicity but remained shrouded in secrecy. One prominent motoring correspondent spoke of Ferguson as the man behind the car that never was. One of Ferguson's business associates said at the time that the car was a tremendous technical advance on anything seen so far and should be put on the market at the earliest opportunity. A motoring correspondent of a large newspaper said that competent observers had stated that Ferguson's car design secrets did indeed spell of unheard of efficiency.

In spite of his great gifts and rare personal qualities, Ferguson would not have been an easy partner to work with. In fact, he was an extremely difficult man. He had a fanatical attention to detail and insisted, for example, that his employees wear clean boiler suites and display white handkerchiefs. When Harry stumbled upon an idea, he had the vision and the tenacity to pursue it over many decades. He sought out people to pursue his ideas with him and he had the ability to inspire them to make contributions beyond their own talents. In most cases, he won from them uncompromising and long-lasting loyalty. It will never be known, however, how much his own intense individualism frustrated agreement with those in government circles who wanted to collaborate with him. Ferguson held very strange views on economics, believing very passionately in reducing prices so as to widen the market that he continued to do so regardless of the general inflation forces affecting his costs (Neufeld 1969). Negotiating for sales of his assets with Massey-Ferguson, on one occasion in 1953, he agreed to the toss of a coin to settle the difference between $16 million and $17 million. Alas, he called tails and lost the $1 million. Later he was presented with a cigar box with the tossed coin fixed on top and the engraving, "To our friend and partner Harry Ferguson, A gallant sportsman". His inventions, especially the three-point hitch, and his dedication to an affordable tractor, had a great influence on the agricultural economies of the world. His economic philosophy was guided by the belief that the best way to improve the total economy was through cutting the costs of production of agricultural products which control the cost of living. The measure of his success can be gauged by the fact that over 85% of all wheeled farm tractors produced in the 1960s were based on the Ferguson system or derivations of this system (Feilden 1970).

It is unfortunate that the relentless genius of Ferguson did not bring more opportunities for economic growth in the land of his birth. One of his greatest disappointments was the refusal of the Northern Ireland Government, on two occasions, to support his tractor project. Ferguson claimed that his love for Ulster was greater than his love for any other country on earth. He was proud to be an Ulsterman and it is most fitting that there are several plaques and monuments to him in his native land. There is a plaque on the farm house of his birth, in Growell, and another on the site of his original showroom which is now the Ulster Bank building, Lower Donegall Street, Belfast. The Ulster Folk and Transport Museum at Cultra has an early tractor and plough and a full-scale replica of his aeroplane. There is also a granite memorial to his pioneering flight on the North Promenade in Newcastle, County Down. Ferguson received an honorary degree of Doctor of Science from Queen's University, Belfast, in 1948, and an honorary degree of Master of Engineering from Trinity College, Dublin, in 1949. He was offered a knighthood on two occasions, by the Labor and Conservative Governments (Wymer 1961). The British Prime Minister, Sir Winston Churchill, proposed to Ferguson that he should be offered a knighthood and in November 1953 he received a formal letter from 10 Downing Street, London (the home of the Prime Minister). The letter asked if this mark of Her Majesty's favor would be agreeable with Ferguson. However, Ferguson declined the offer on the grounds that this honor should be reserved for people such as servicemen and statesmen whose financial reward for their labor is comparatively small. He believed they should not be given to businessmen or industrialists who have the opportunity to become famous or amass wealth and all that wealth can bring them. In his letter of reply to Churchill he wrote, "I fear I have seen, so often, the harmful effects of an Honors List for industrialists, that I believe I have come to the right conclusion when I ask that the Prime Minister would not submit my name to the Queen in this connection." He, among many others, greatly underestimated his accomplishments and contributions to Britain and the world. A knighthood would have been no less than this man deserved.

It is also unfortunate that the history books will tend to remember Henry George (Harry) Ferguson as a businessman who fought and won a David and Goliath legal action against the Ford Motor Company. This is a grave injustice, for Ferguson was a superb innovator and engineer who built up his business on the basis of very small capitol resources. Harry Ferguson revolutionized mechanized farming in the 1940s. His invention of the three-point linkage system meant that one tractor could be used to pull a variety of implements. Before this invention, farmers had to have a different tractor for different jobs. His system also brought greater safety and stability to tractors. He was the cofounder of the Coventry tractor manufacturing company of Massey-Ferguson. The TE20 tractor was built in Coventry, England, and reached production of more than half a million, with hundreds of thousands of successor models built at the plant for almost 50 years. It was fitting that in 2008, Coventry considered Ferguson for one of the first 10 names for the city's Walk of Fame. In 2006, the Northern Bank Limited, Northern Ireland, issued a 20 pounds sterling banknote with a portrait of Ferguson and the Belfast plough. On August 16, 2008, Growell opened a memorial garden, near the home of his birth, to honor his contributions to society. At the time of writing this article, plans are underway to commemorate the 100th anniversary of his first powered flight in Ireland in that small monoplane on December 31, 1909, at Hillsborough. Ferguson may not have soared very high that particular day, but his mechanical genius was to lift mankind to new heights.

Acknowledgements The author wishes to thank the Belfast Telegraph for providing the newspaper articles concerning the death of Harry Ferguson and the subsequent inquest. Articles were found in the 25th, 26th and 28th of October, 1960, and the 8th and 9th of November, 1960, issues of the newspaper. Thanks are also due to the Belfast City Council, the Public Record Office of Northern Ireland, and the Ferguson Family Museum, Isle of Wight for the information they provided. The author would also like to thank Mr. Michael Kean for the schematic drawings for the kinematic analysis and dynamic force analysis of the three point hitch. Finally, the author would like to acknowledge Mr. Jack Woods for the detailed information he so willing provided in numerous emails on background and historical aspects of the Ferguson family.

References

Ambike SS, Schmiedeler JP (2007) Application of geometric constraint programming to the kinematic design of three point hitches. J Appl Eng Agric (Am Soc Agric and Biol Eng) 23(1):13–21

Brock HL (1954) Tractor hitches and hydraulic systems – a tractor designer's viewpoint. SAE Trans 62:187–195

Condie AT (2001) Great tractor builders: Ferguson. Ian Allan Publishing, Surrey, England

Feilden GBR (1970) Invention, innovation and design, first Harry Ferguson Memorial Lecture. Queen's University of Belfast, Belfast, Northern Ireland

Flackes WD (1960) The man and the legend: strange genius of Harry Ferguson, newspaper article, published in the Belfast Telegraph, October 26. Belfast, Northern Ireland

Fraser C (1972) Harry Ferguson: inventor and pioneer. John Murray (Publishers) Limited, England/London

Fraser C (1973) Tractor pioneer: the life of Harry Ferguson. Oberlin Printing Company, Ohio University Press, Athens, Ohio

Hain K (1953) Anwendung des Gelenkvierecks bei der Konstruktion des Ackerschleppers. Proceedings of the 1st VDI Conference on Mechanisms, pp 127–134
Hain K (1955) Der Entwurf von Gelenkgetrieben für den Ackerschlepper (The design of linkages for the tractor), VDI-Berichte, vol 5. VDI-Verlag, Düsseldorf, pp 11–19
Hain K (1956) Die Dreipunktaufhängung am Ackerschlepper – Bewegung. VDI-Berichte 12:85–91
Hain K (1959) Wirkungen des Gelenkspiels beim Dreipunktanbau des Pfluges an den Ackerschlepper. Konstruktion 11(9):344–348
Hain K (1973) Front-und Hecklader am Ackerschlepper in Gabelstapler-Wirkungsweise, Getriebebeispiel-Atlas. VDI-Verlag, Düsseldorf, p. 10.3.4.
Hain K, Skalweit H (1957) Dreipunktanbau: Kompromiß zwischen Zugfähigkeit des Schleppers und Tiefenhaltung des Pfluges. Landtechn, Forschung 7(5):127–133
Kinzel EC, Schmiedeler JP, Pennock GR (2006) Kinematic synthesis for finitely separated positions using geometric constraint programming. J Mech Design Trans ASME 128(5):1070–1079
Morling RW (1979) Agricultural tractor Hitches – analysis of design requirements, ASAE Distinguished Lecture Series – Tractor Design. American Society of Agricultural Engineers, No. 5, pp 1–28, St. Joseph, Michigan
Neufeld EP (1969) A global corporation: a history of the International Development of Massey-Ferguson Limited. University of Toronto Press, Toronto, Canada
Phillips JR (1958) Some problems of mechanics met in designing the disc plough. Transactions of the Fifth International Congress of Agricultural Engineering, vol 2. Printed by Commission Internationale de Genie Rural, Gembloux, Belgium, pp 1144–1160
Rae JB (1980) Harry Ferguson and Henry Ford, published by the Ulster Historical Foundation. W. & G. Baird, Ltd., Antrim, County Antrim, Northern Ireland
Thaer R (1956) Untersuchungen über die Dreipunktaufhängung der Geräte am Schlepper (Investigations on the three point hitch of agricultural machines behind the tractor). Grundlagen der Landtechnik, vol 7. VDI-Verlag, Düsseldorf, pp 72–88
Uicker JJ Jr, Pennock GR, Shigley JE (2003) Theory of machines and mechanisms, 3rd edn. Oxford University Press, New York
Wymer N (1961) Harry Ferguson, The 'Living Biographies' Series, Phoenix House Limited, Strand, London, England (Phoenix House is now an imprint of the Orion Publishing Group, Orion House, 5 Upper Saint Martin's Lane, London, England)
The Public Record Office of Northern Ireland (P.R.O.N.I.), 66 Balmoral Avenue, Belfast, Northern Ireland. www.proni.gov.uk
http://en.wikipedia.org./wiki/belfast
Statistics for Northern Ireland. www.nirsa.gov.uk
http://en.wikipedia.org./wiki/ferguson
The Ulster Folk and Transport Museum, Cultra, Holywood, County Down, Northern Ireland. http://www.uftm.org.uk
The Ferguson Family Museum, Kings Manor Farm, Copse Lane, Freshwater, The Isle of Wight, PO40 9TL. www.ferguson-museum.co.uk
Patents filed by Ferguson can be located at: http://www.google.com/patents
Patents filed by Ferguson can also be located at the European Patent Office: http://ep.espacenet.com

Juanelo (1501–1585)

Emilio Bautista, José Luis Muñoz, and Javier Echávarri

Abstract Juanelo Turriano was a clockmaker, astronomer, mathematician, and a mechanical and hydraulic engineer. He worked for kings Charles I and Philip II during the Iberian Empire. His most important device was called the "Dancing Machine" and was built in the sixteenth century in Toledo, the capital city of Spain at that time, to raise water from the River Tagus to the city, 90 m above the river, with only the power of the water flow. He is also famous because he is supposed to be co-author of the "The Twenty-One Books of Devices and Machines of Juanelo Turriano" written in the second half of the sixteenth century.

Biographical Notes

Giovanni Torriani, known as Gianello Turriano or Juanelo Turriano, was born in 1501 in Cremona (Italy) to a humble family. During the long nights as a shepherd, he became very interested in astronomy and others sciences (Del Campo y Francés 1997).

When Charles I of Spain became Charles V of Germany, Juanelo's life changed. The governor of Milan, Ferrante Gonzaga, wanted to give Charles V an astronomical clock made in the fourteenth century by Giovanni Dondi, which was one of the most important medieval technological works. Charles V's interest in mechanical engineering, art and especially clocks was common knowledge. Dondi's clock did not work and Juanelo was the only one able to repair it. This made him a really famous clockmaker. And he received orders from all the courts in Europe for his timepieces.

Around 1530 he came to Spain and became clockmaker to Charles V (Valverde Sepúlveda 2001); he even accompanied Charles V when he retired to the Monastery of Yuste in 1557. One year later, Charles V died and his son Philip II became King

E. Bautista (✉), J.L. Muñoz, and J. Echávarri
DIM: Divisi6n de Ingeniería de Máquinas, Universidad Politécnica de Madrid,
José Gutiérrez Abascal 2, P.O. Box 28006 Madrid, Spain
e-mail: ebautista@etsii.upm.es

M. Ceccarelli (ed.), *Distinguished Figures in Mechanism and Machine Science: Their Contributions and Legacies, Part 2*, History of Mechanism and Machine Science 7,
DOI 10.1007/978-90-481-2346-9_5,

of Spain. Philip II was not as interested in clocks as his father, thus Juanelo worked for him in other fields such as mathematics. For this, Philip II called him "The Wise Mathematician". At the king's court, he became friends with several important Spanish architects and engineers (Del Campo y Francés 1997).

A few years later, thanks to his remarkable water machine in Toledo, Juanelo became known as a magician by the Toledo inhabitants. Even though he was very wellknown at that time, he was very poor because no one paid him for his work; what is more, he had to pay for some of his hydraulic works with his own money.

Juanelo died on 13 June 1585 in Toledo and was buried in the church of the Monastery of El Carmen in Toledo. Up until the last moment, he demanded payment from the city of Toledo but he was never paid, and his family continued with the complaints to the city and even to the king.

Legend says that Juanelo was very old in his last years of life, so he used one of his automatons called "The Stick Man" that begged in the streets, and when someone gave him a coin, he bowed (Cervera Vera 1996). This legend provided the name for a street in Toledo: "Calle del Hombre de Palo" ("The street of The Stick Man").

Figure 1 shows a plaque in the street in Toledo dedicated to "The Stick Man", the text reads "Along this street passed the wooden automaton built by Juanelo Turriano, clockmaker for Charles V, to the amazement and perplexity of the crowd". There is another street named after Juanelo in the centre of Madrid.

Although scarcely any of his works have survived, Juanelo was wellknown during the Renaissance. There are a great number of references to him and especially to his works in many books of engineering of the sixteenth and following centuries,

Fig. 1 A plaque in the Street dedicated to "The Stick Man" of Toledo

as well as in the Spanish literature of Góngora, Lope de Vega and Quevedo. As for portraits, we know that Charles V ordered one, but that portrait was lost as well as the one that Juan de Herrera had in his house. The only surviving objects that bear his likeness are a marble bust and a coin.

The bust, shown in Fig. 2a, was made by Leon de Leoni and is kept in the Santa Cruz Museum in Toledo; it represents Juanelo when he was approximately 45 years old. There is a copy, three times larger, of the bust in the Royal Palace of Madrid, shown in Fig. 2b.

The coin, shown in Fig. 3, is kept in the Archaeological Museum of Madrid. It is thought to have served Titian and El Greco and other anonymous artists of the seventeenth century as inspiration to paint some portraits which look like Juanelo, shown in Fig. 3 (Del Campo y Francés 1997).

As the Iberian Empire grew, Spain became the number one world political and economic power, enabling its empire to reach out to the five continents. It was said that, over the dominions of King Philip II, "the Sun never sets".

a b

Fig. 2 (**a**) A marble bust of Juanelo Turriano by Leon de Leoni; (**b**) copy of the bust in the Royal Palace of Madrid

Fig. 3 Portrait of Juanelo Turriano on a coin, and a later Portrait of Juanelo (XVII anonym)

Nevertheless, there was a lack of qualified technicians and scientists in Spain at that time (Lusa Monforte 2004), which is quite evident in a letter of Francés de Álava (Bautista et al. 2007), written in the second half of the sixteenth century: "The persons I know in Spain engaged in the service of His Majesty as engineers, except for Fratin (Fratino) and Antonelli, are Jorge Setara, who lives in Perpignan and Baptista Antonelli, brother of the cited Antonelli, who lives in Peñíscola and Cristóbal Antonelli, his nephew, who nowadays is living in Barcelona, and Tiburcio (Spanocci), the one who His Majesty sent to Fuenterrabía, and Felipe Tercio (Terci) who lives in Lisboa. All of them are foreigners, and I do not know a single Spaniard who knows the half of what they do, although I have racked my brains…". In order to avoid such a negative position, a discussion arose about the convenience of establishing a scientific-technical centre in the court.

Juan de Herrera, Juanelo Turriano, Jerónimo de Ayanz and other outstanding engineers and architects in the court, were supposedly decisive in King Philip II deciding the foundation of a popular "Technical Academy" in Madrid (Esteban Piñeiro 2002–2003). The Academy was founded in 1583 and intended for everyone interested in jobs related to the different arts of all the technical activities practised at that time.

List of Main Works

Juanelo's main activities regarding engineering can be divided into two different groups: on the one hand, the design and construction of mechanical devices and, on the other hand, the elaboration of manuscripts.

Mechanical devices:

- Clocks: the "Astrarium" clock and the tower clocks of the monastery of El Escorial near Madrid.
- Two hydraulic devices for raising water from the river Tagus to the city of Toledo.

Manuscripts:

- Breve discorso alla Majestad de Re Catolico interno la reduttione dell anno et restitutione del Calendario con la dichiaratione deglo instrumenti da eso per mostrarla in atto prattico (Brief speech for His Catholic Majesty about the Calendar...).
- "Trattato dell'acque" (Water Treatise).
- Los Veintiún Libros de los Ingenios y las Máquinas de Juanelo Turriano (The Twenty-One Books of the Devices and Machines of Juanelo Turriano).

Review of the Main Works on Mechanism Design

Although he did not usually write about his works, they are cited in many books of other authors at that time. This is our main source of information because none of his clocks, machines or automatons have been found.

In 1530, Ferrante Gonzaga gave Charles V a medieval astronomical clock called the "Astrarium", which was repaired by Juanelo, shown in Fig. 4a. Not only he did repair it, but he also made another one which also showed the planetary

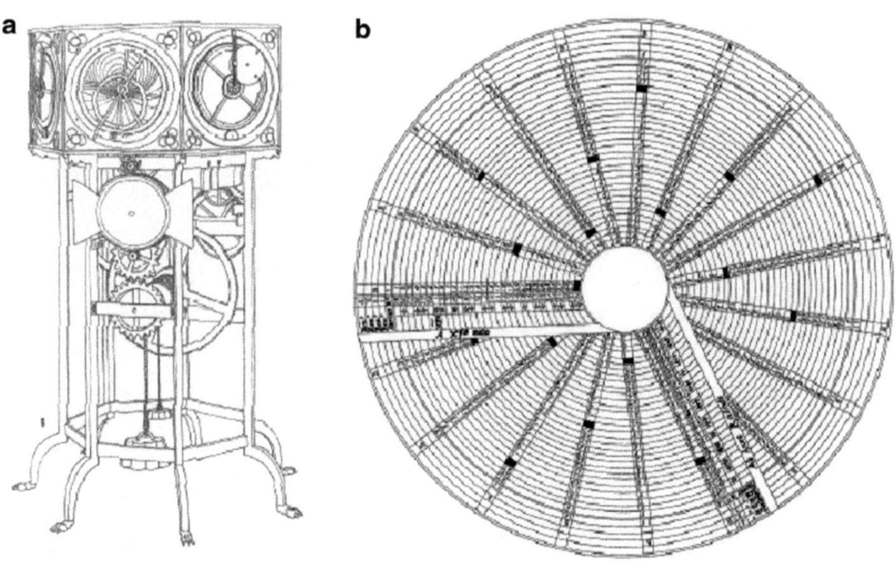

Fig. 4 (**a**) Astronomical clock; (**b**) table obtained from the "Breve discorso alla Majestad de Re Catolico..."

movements. This was called "Crystalline" and it showed the hours, minutes, solar and moon hours, as described planetary movement and it showed the signs of the Zodiac. It was built with more than 1,500 parts, three springs which produced the movement and eight planetary spheres.

When Charles V died and Philip II became king of Spain, Juanelo met Juan de Herrera, one of the most important Spanish architects of the Renaissance and who built the monastery of El Escorial near Madrid (1563–1584). Juanelo took part in the project by helping to build the tower clocks (Del Campo y Francés 1997).

His last job as a clockmaker and astronomer was the solution to the problem of the Catholic calendar. Pope Gregory XIII asked all the kings of Europe to ask the most intelligent people in their courts to give their opinion on this topic. Juanelo not only gave his remarks but also he wrote "Breve discorso alla Majestad de Re Catolico interno la reduttione dell anno et restitutione del Calendario con la dichiaratione deglo instrumenti da eso per mostrarla in atto prattico", where he detailed two solutions for Easter to be always on the same date (Fig. 4b).

Juanelo also invented automatons with human aspects. Apart from the cited "The Stick Man", another famous automaton attributed to Juanelo is "The lady from Vienna", shown in Fig. 5.

His main work is considered to be the hydraulic device built in Toledo between 1565 and 1569, which was ordered by the Marquis of Vasto.

In those days, Toledo suffered from a shortage of water in its reservoir, so they needed a machine which was able to raise water from the River Tagus up to the

Fig. 5 The automaton called "The Lady of Vienna" (García-Diego 1982)

Alcázar, the highest part of Toledo; this meant that the water had to be raised 90 m and it could not be done with a pump because the pipes would not withstand the pressure, as happened with the previous device built by German mining engineers (Bautista et al. 2000).

It had to raise 12,400 L round-the-clock, which meant that the machine had to work non-stop. The current of the Tagus itself served as the driving force as well as supplying the water needed for the city. It was the machine's size that made it very significant as a work of mechanical engineering (Porres Martín-Cleto 1987).

We have no complete drawing of Juanelo's device but there is a manuscript description with a rough diagram, shown in Fig. 6, made by the Precentor of Évora (Portugal) in 1604.

His machine was undoubtedly a great feat of engineering for the time, both in size and complexity. The sizes of its parts and the high forces received were a considerable challenge to the technical know-how of that period. In addition, being a machine with so many moving parts meant the dynamic effects would be significant and its joints subjected to considerable tribological actions (Bautista et al. 2007).

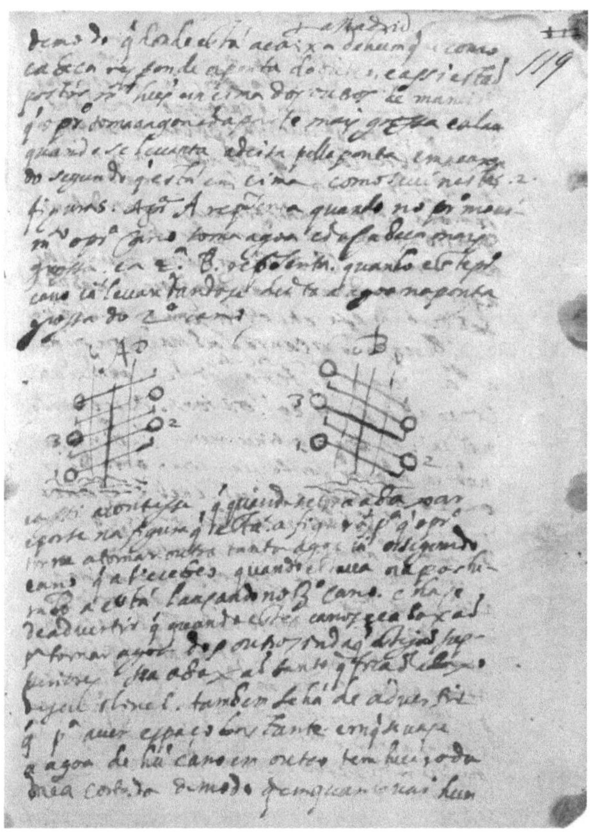

Fig. 6 Manuscript description of the hydraulic device, by the Precentor of Évora (Portugal)

An attempt to explain how this machine worked was published by Luis de Escosura in 1888, though the author himself recognised his doubts about the reconstruction proposed, shown in Fig. 7a. It was based on Valturio's adjustable ladder (1534) shown in Fig. 7b, obtained from the book "De Re Militari" (On the Military Arts).

The different documents of the period and later studies enabled Ladislao Reti to make a fairly close reconstruction of the mechanical device, in 1967, which is shown in Fig. 8.

a b

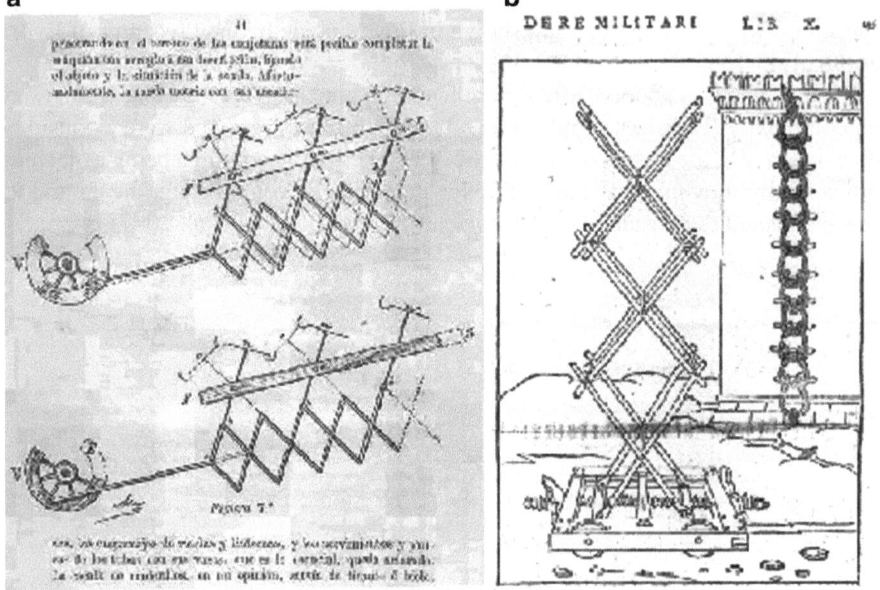

Fig. 7 (**a**) Explanatory sketch of Juanelo's machine, by Luis de Escosura (1888); (**b**) Valturio's adjustable ladder (1534)

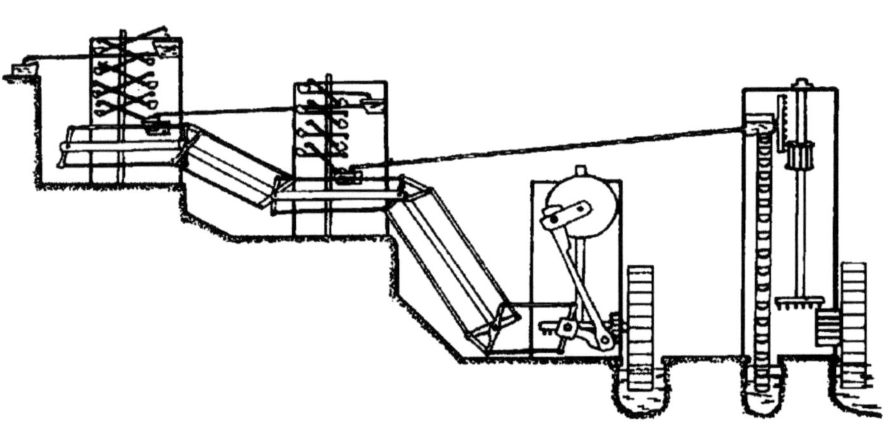

Fig. 8 Diagram of Juanelo's device in Toledo, according to Ladislao Reti's reconstruction

This illustration shows how the river flow operated two water wheels with paddles. The first moved a mechanism with a water wheel that raised the water several metres to a reservoir. The second wheel used another connecting rod-crank-based mechanism to start up the rocking movement of the vertical systems called towers, which enabled the water to be raised from the reservoir under atmospheric pressure.

It was designed to overcome any difference in level, since, as the water ran in contact with the air, no excesses of pressure occurred due to the pumping height, as would happen in a usual pipe system. A greater difference in level only required an increase in the number of sequentially connected towers.

The tower diagram in Fig. 9a shows two working positions for the device, based on a set of pivoted buckets and cups that raised the water in stages, due to the backward and forward motion of the cups. When the water reached the top, it was fed through some pipes to the next tower, and so on until it reached its destination.

Both the device and the way it worked were very curious at the time. These facts contributed to its inventor's fame and the name by which it became popularly known, "The Dancing Machine". A contemporary traveller named Kenelm Digby wrote in respect of the device "…and so the two sides of the machine were like two legs that trod the water in turn".

This gives strength to Nicolás García Tapia's reconstruction (2002) concerning the existence of arches for collecting water from both sides of a tower.

Indeed, Nicolás García Tapia's reconstruction in Fig. 9b of Juan Ramos' diagram shows there were two input and two output arches, situated either side of the tower. When the first cup was introduced into the arch to take water, the opposite one was raised, passing the water on to the next, and each container was full when its

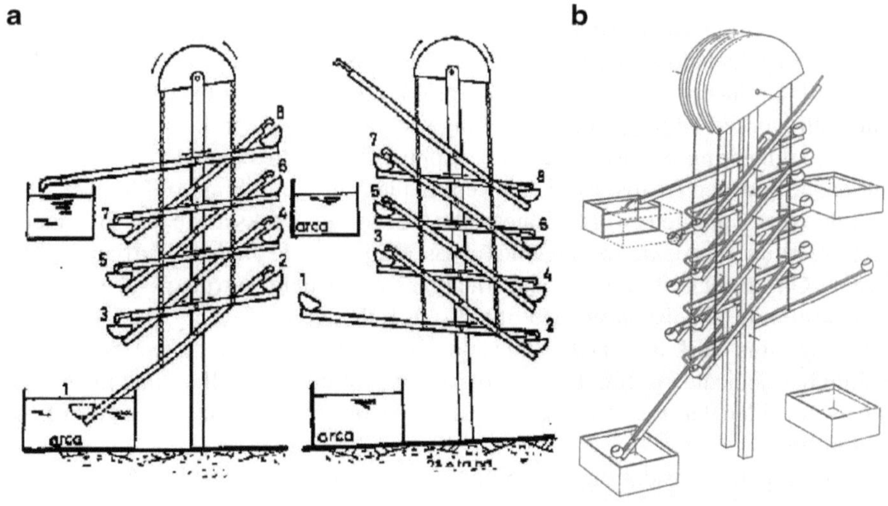

Fig. 9 (**a**) Details of Juanelo's hydraulic device; (**b**) a recent three-dimensional reconstruction of the tower by Nicolás García Tapia

opposite one was empty. The water was thus made to flow continuously with a co-ordinated and precise movement.

Once the device was finished they measured the water volume and it was 50% per cent greater than expected, which meant 17,000 L of water per day.

The first device was a great success, so there was a proposal to build three more devices. However, only the first project came to fruition.

The second device was started in 1575 and finished in 1581 and the water that it supplied was for the town.

When Juanelo died, the only one who knew how to use the device was his grandson who died in 1597.

In 1617, the machine finally stopped working due to lack of maintenance and knowledge of how to use it. The last official document where it is mentioned was an inventory written in 1639.

A machine with a tower quite similar to Juanelo's device can be found in the famous book called "Le diverse et artificiose machine del Capitano Agostino Ramelli" ("The various and ingenious machines of Captain Agostino Ramelli"), a large catalogue of machines finished in 1588, a few years after the death of Juanelo.

Ramelli's book contains 195 superb figures of various kind of machines along with detailed descriptions of each one, including comments on how it was built and how it worked. He endowed his drawings with such clarity that he could be later studied in a large number of fields.

One of the more than 100 water-raising machines described is shown in Fig. 10 (Ceccarelli 2006). It was driven by a water-wheel which moved two gears, the movement being transmitted to a tower with a water-raising system in stages, using a method similar to Juanelo's hydraulic device.

Juanelo shared his knowledge with some of the most important engineers in Spain, during important irrigation works carried out while Philip II was king. The last time that Juanelo was asked for his opinion about a large project was in 1580. The best engineers of that time, Giorgio Fratin, Juan Baptista and Cristobal Antonelli worked on the project of the biggest dam in the world.

One engineering treatise which was supposedly due to Juanelo is the "Trattato dell'acque" (Water Treatise) which is kept in the National Library of Florence. It is considered as a previous work of one of the few written works attributed to Juanelo: "Los Veintiún Libros de los Ingenios y las Máquinas de Juanelo Turriano" (The Twenty-One Books of Devices and Machines of Juanelo Turriano). These books were written in Spanish at the end of the sixteenth century.

Though this work is signed by Juanelo, some experts say that it was not written by him but dedicated to him because of his fame at that time all around the world. Others consider that the reference to Juanelo is used to remark the quality of the manuscript.

Some experts consider the possibility of this manuscript having been written by various different co-authors. In the opinion of Nicolás García Tapia (1987, 1990a, b), one of them could be Pedro Juan de Lastanosa, because the manuscript contains words taken from a Spanish dialect spoken in the area where Lastanosa lived.

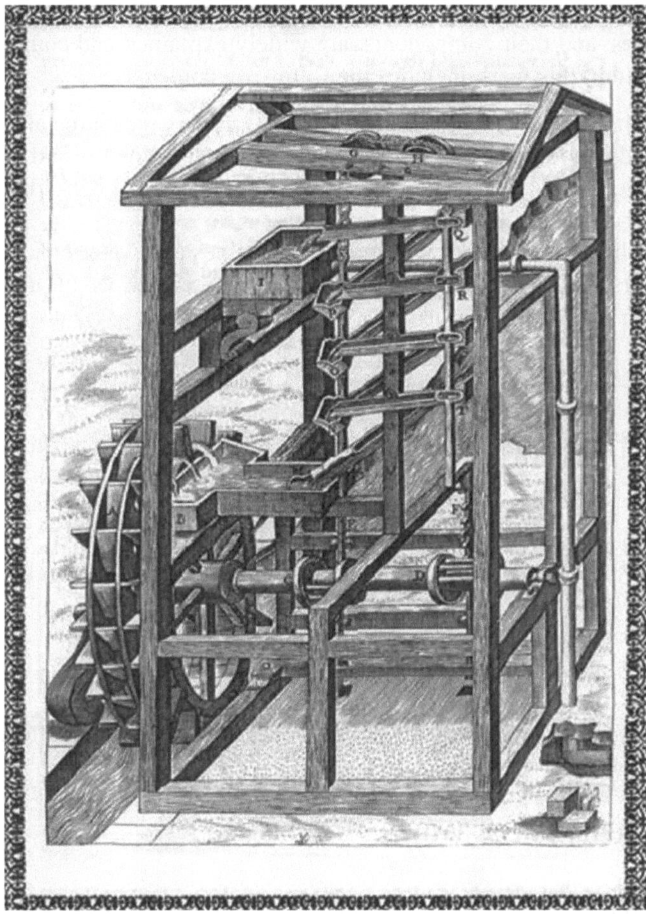

Fig. 10 A water-raising machine included in Ramelli's book

The search for the authors of the Twenty-One Books has promoted the in-depth study of this work.

This collection of books, ordered to be written by "the Catholic King Philip II, King of Spain and the New World", represents an impressive written source for the knowledge of the mechanical arts known and practised in the sixteenth century. Some historians and engineers define it as a Machine Encyclopaedia (Goicolea 2000).

These books, finished in approximately1570, include descriptions of a high number of machines of the time, classified by their function (Instituto Cervantes 1998). They are remarkable in their references to practical devices, such as pumps, mills, elevators, presses and others machines, mainly powered by water, wind, gravity or animal traction.

The front page and some illustrations of machines are shown in Fig. 11. All these machines, and their components, are widely explained and commented on in the text. In detail, this work includes the following contents:

- The first chapter talk about water: its quality, properties, source and effects.
- The second chapter examines the signs that we can follow to find water.
- Chapter 3 teaches how to distinguish between drinkable and non-drinkable water.
- The fourth chapter talks about water levels.
- The fifth chapter teaches how to make water pipes and water conduits.
- In Chapters 6–10 we can read about the different ways of moving water, and storing it.
- Chapter 11 talks about different kinds of mills.
- Chapter 12 teaches how to separate flour from straw.
- Chapter 13 explains the different uses of mills
- Chapter 14 talks about boats that were used to cross rivers when there was no bridge.
- Chapter 15 talks about wooden bridges.
- Chapter 16 talks about materials like wood and stone.
- Chapter 17 includes more precise information on stone: how to treat it and its quality.
- Chapter 18 sets out different ways of building stone bridges.
- Chapter 19 talks about how building in ports should be done.
- Chapter 20 explains how to protect ports from the open sea.
- The last chapter talks about tides, water clocks and the different effects of water.

On the Circulation of Works

Thanks to official letters, literature and some engineering books of that age, we have been able to earn something about Juanelo's devices, mainly regarding the hydraulic devices built in Toledo.

During the Iberian Empire, the interest in mechanical engineering remained patent with the manuscript "Trattato dell'acque" (Water Treatise). The treatise is written in Spanish and was discovered recently in Florence by the researcher María Teresa Cacho. It has 84 pages and approximately 350 images and is supposedly a preliminary work of the most important manuscript attributed to Juanelo: "Los veintiún libros de ingenios y máquinas" (The Twenty-One Books of Devices and Machines), completed in 1570.

The Twenty-One Books kept in the National Library of Madrid (www.bne.es) have 949 pages and 509 images, including most of the illustrations of the previous Water Treatise. In addition, in these Twenty-One Books, the author improves the clarity of the comments and the quality and details of the figures.

The original manuscript was not published at that time due to the information about materials and works that required the Royal Permission of Philip II. Its publication would have allowed a dissemination comparable to the most

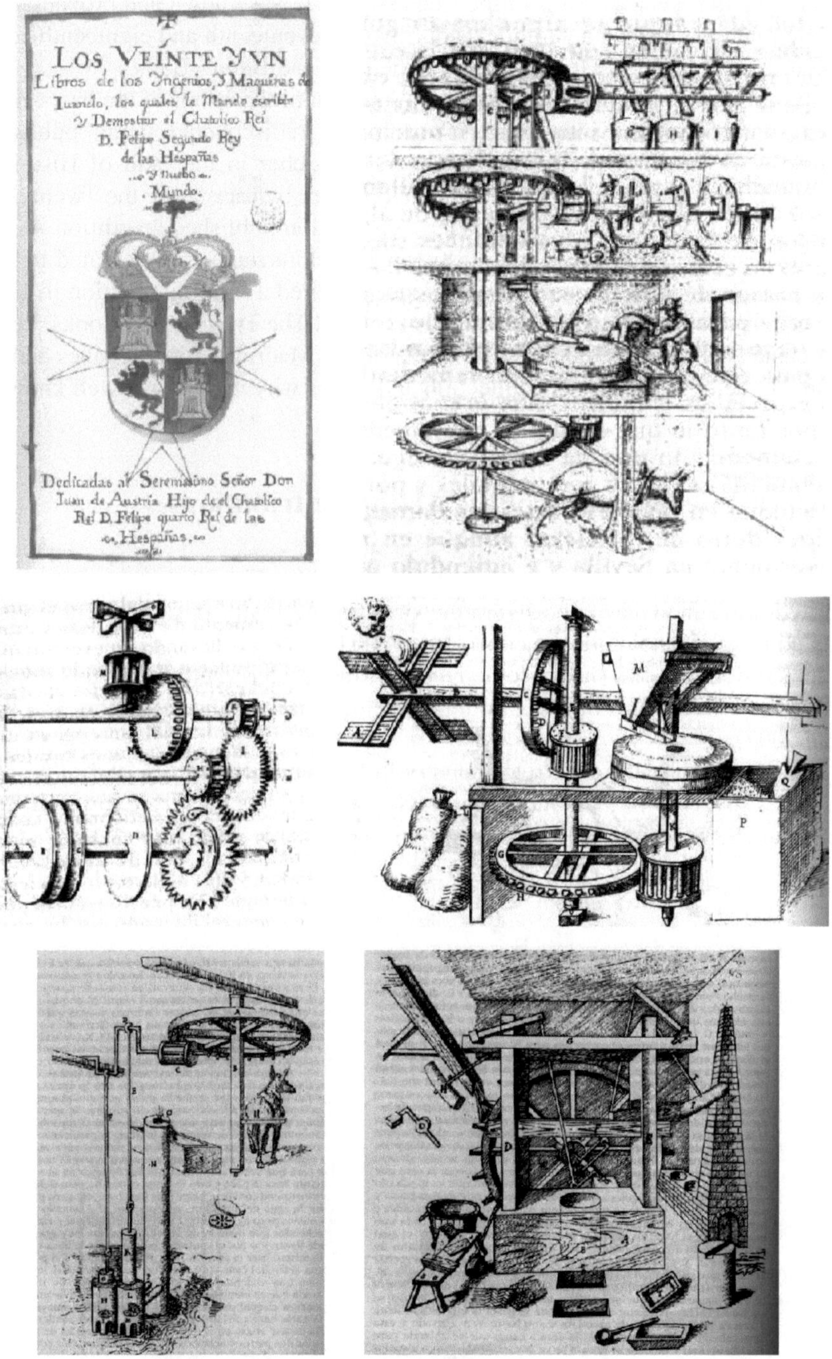

Fig. 11 Front page and some machines in the Twenty-One Books

remarkable European treatises of the Renaissance. It was known and probably used by important engineers and architects during the seventeenth and eighteenth centuries, such as Gómez de Mora, Teodoro Ardemans and Benito Bails.

These Twenty-One Books remained in hiding until the 1960s of the past century when some historians interested in technology rediscovered them, publishing diverse works. Professor Alexander Keller, researcher in the field of History of Science, contributed to the spread of Juanelo's work by translating the Twenty-One Books into English. The Civil Engineers Association published an edition in 1983 and the Juanelo Turriano Foundation (www.juaneloturriano.com), created in 1987 by the Engineer José Antonio García-Diego, prepared a facsimile edition in 1997.

There is another copy of this work, also entitled The Twenty-One Books, kept in Barcelona. This issue is quite similar to the one in Madrid, with 609 pages and 404 figures. It belongs to a private collection and this is why it is not yet well known.

Modern Interpretation of the Main Contributions

Although we do not know exactly what they were like, Fig. 12 shows some modern reduced-scale models of the hydraulic devices originally designed and built by Juanelo (www.juaneloturriano.com) (Jufre 2007).

In addition, Fig. 13 shows some recent virtual reconstructions of the hydraulic devices and the spheres of the astronomical clock made by Juanelo (Bermejo 2005, Jufre 2007).

It is remarkable that the Proto-industry in the sixteenth century, regarding early industries involved in manufacturing goods for trade, has been recently studied by Keller and Silva Suárez (2004), following the contents of The Twenty-One Books of

Fig. 12 Reduced-scale models of the hydraulic devices

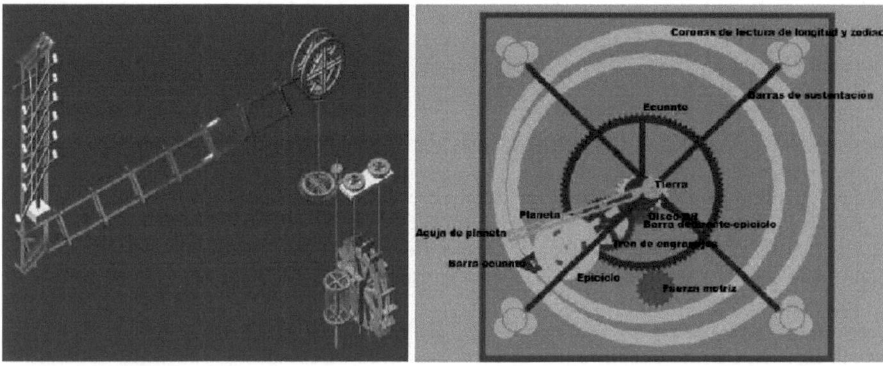

Fig. 13 Virtual models of some famous devices: the hydraulic devices and the spheres of the astronomical clock

Devices and Machines. The detailed analysis of the treatise from this point of view enhances the knowledge on the state of technology at that time concerning different kinds of grinding mills and other devices moved hydraulically, such as oil-presses or fulling hammers. Other Proto-industrial processes are also described, such as the washing of wool, and the production of starch, alum, vitriol and saltpetre.

The discovery of the Twenty-One Books of Devices and Machines, was added to the recent studies developed by García Tapia (1990a, b, 2001) concerning "Royal Privileges for Inventions" (or Spanish Patents) mainly at the end of the sixteenth and the beginning of the seventeenth centuries.

The quantity and quality of the registered inventions, with or without Royal Patent, lets us suppose a rapid increase in the number of scientists and technicians, coming mainly from Italy and other European countries, to serve the Spanish Crown, constituting a genuine community, essential for the technical and scientific growth of the Iberian Empire during its Golden Age.

References

Bautista E, Muñoz JL, Leal P (2000) The dancing machine: the secret of raising the water uphill. In: Ceccarelli M (ed) Proceedings of the International Symposium on History of Machines and Mechanisms HMM 2000, Cassino (Italy). Kluwer, Dordrecht

Bautista E, Bernardos R, Ceccarelli M, Díaz A, Díaz V, Echávarri J, Leal P, Muñoz JL (2007) Breve historia ilustrada de las máquinas. Foundation of the Valencian Internacional University, Spain

Bermejo Herrero M, González Conde L (2005) Reconstrucción virtual del Artificio de Juanelo Turriano para elevar agua del río Tajo a Toledo, XVIIth Ingegraf International Congress, Seville (Spain)

Ceccarelli M (2006) El Renacimiento de las maquinas: primer desarrollo de la ingenieria mecanica moderna, invited lecture, Congreso de Historia del Patrimonio e de la Ingenieria, Las Palmas, pp 3–26

Cervera Vera L (1996) Documentos Biográficos de Juanelo Turriano, Foundation Juanelo Turriano

Del Campo y Francés A (1997) Semblanza iconográfica de Juanelo Turriano, Foundation Juanelo Turriano
Escosura Morrog L (1888) El artificio de Juanelo y el Puente de Julio Cesar "El artificio de Juanelo y el puente de Julio Cesar". Memorias de la Real Academia de Ciencias Exactas, Físicas y Naturales de Madrid
Esteban Piñeiro M (2002–2003) Las academias técnicas en la España del siglo XVI, Cuadernos de Historia de la Ingeniería, Vol. V
García Tapia N (1987) Pedro Juan de Lastanosa y Pseudo Juanelo Turriano, Revista de la Sociedad Española de Historia de las Ciencias y de las Técnicas, Vol. 10, Issue 18–19
García Tapia N (1990a) En defensa de Lastanosa, Revista de Obras Públicas
García Tapia N (1990b) Patentes de invención españolas en el Siglo de Oro. Ministerio de Industria y Energía
García Tapia N (2001) Un inventor navarro. Jerónimo de Ayanz y Beaumont (1553–1613)
García Tapia N, Carrillo Castillo J (2002) Tecnología e Imperio. Ingenios y leyendas del Siglo de Oro. Turriano, Lastanosa, Herrera, Ayanz, Nivola
García-Diego JA (1982) Los relojes y autómatas de Juanelo Turriano, Albatros
Goicolea ZJ (2000) Azudes, molinos y otros aspectos de "Los Veintiún Libros de los Ingenios y Máquinas", Revista de Obras Públicas
Instituto Cervantes (1998) Felipe II: los ingenios y las máquinas: ingeniería y obras públicas en la época de Felipe II
Jufre Garcia X (2007) Nuevo modelo del artificio de Juanelo, Segundo Congreso Internacional de Patrimonio e Historia de la Ingenieria (CIPHI). Las Palmas de Gran Canaria, Spain.
Keller AG, Silva Suárez M (2004) Protoindustria: una perspectiva desde Los Veintiún Libros de los Ingenios y Máquinas, Técnica e ingeniería en España. I el Renacimiento
Lusa Monforte G (2004) La técnica en España durante el renacimiento, Cuadernos de Historia de la Ingeniería, Vol. VI
Porres Martín-Cleto J (1987) Nuevas noticias sobre el artificio de Juanelo, Anales Toledanos
Ramelli A (1588) Le diverse et artificiose machine
Reti L (1967) El artificio de Juanelo en Toledo, Diputación provincial de Toledo
Severim BF, Peregrinaçao de Baltasar de Faria Severim, Chantre de Evora, ao Mosterio de Guadalupe, no anno de 1604, manuscript kept in the National Library of Lisboa
Turriano J (1990) Breve discurso a S.M. el Rey Católico en torno a la reducción del año y reforma del calendario. Introduction by J.A. García-Diego and J.M. González Aboin. Published by Foundation Juanelo Turriano
Turriano J., Trattato dell'acque, 16th century. Original Manuscript kept in The National Library of Florence
Turriano J., Los Veintiún Libros de los Ingenios y las Máquinas, 1570, two original manuscripts kept in Madrid and Barcelona
Turriano J (1997) The twenty one books of engineering and machines, published in Spanish and English, Vols. 1–7. Prologue by Pedro Laín and José Antonio García-Diego, Doce Calles
Valturio, "De Re Militari", published in Paris in 1534
Valverde Sepúlveda J (2001) Juanelo Turriano, el relojero del emperador, Rubiños-1860

Websites

National Library of Madrid: www.bne.es
Foundation Juanelo Turriano: www.juaneloturriano.com

José María Lanz y Zaldívar (1764–1839)

Carlos S. López Cajún

Abstract The Mexican José María Lanz y Zaldívar was one of the authors, together with the Spaniard Agustín de Betancourt, of the first book about industrial kinematics, dealing with mechanisms classification in a systematic fashion. The book, written in French, was published first in Paris in 1908 and later was translated to English and German. Furthermore, at the end of the eighteenth and the first part of the nineteenth centuries, apparently Lanz y Zaldívar played an important role in developing modern engineering in Europe.

Biographical Notes

The mathematician, mechanical and hydraulic engineer Lanz y (and) Zaldívar was born in Campeche, México in 1764. He pursued studies at the *Instituto Campechano* and sometime after he was instructor at the same institution. Successively, he moved to Spain where he studied naval engineering, and obtained the position of Cadiz' Alférez (similar to Lieutenant) of the Marine Guards. He was at the service of the Royal Fleet of Spain with the above-mentioned official rank. From 1782, Lanz, with the degree of ship's lieutenant was in service in the military *fragata* San Fernando at La Habana's offshore. By the end of 1782 he was moved to the fragata Santa Lucía. The latter used to make trips from la Habana (Cuba) to Veracruz (Mexico) as was registered in the so-called *hoja de servicios* (sheet of services) shown in Fig. 1. Then, he was sent to Yucatán by Colonel Borja to work on the *henequen* (hemp) and the *jarcia* (ropes) being fabricated with this fiber. Indeed, in 1783 Lanz was living in Yucatán, where he studied the possibility of using *henequen* to fabricate naval *aparejos* and wrote

C.S. López Cajún (✉)
Facultad de Ingeniería, Universidad Autónoma de Querétaro,
Cerro las campanas s/n, Col. Las campanas, 76010 Querétaro, QRO, México
e-mail: cajun@uaq.mx

M. Ceccarelli (ed.), *Distinguished Figures in Mechanism and Machine Science: Their Contributions and Legacies, Part 2*, History of Mechanism and Machine Science 7,
DOI 10.1007/978-90-481-2346-9_6, © Springer Science+Business Media B.V. 2010

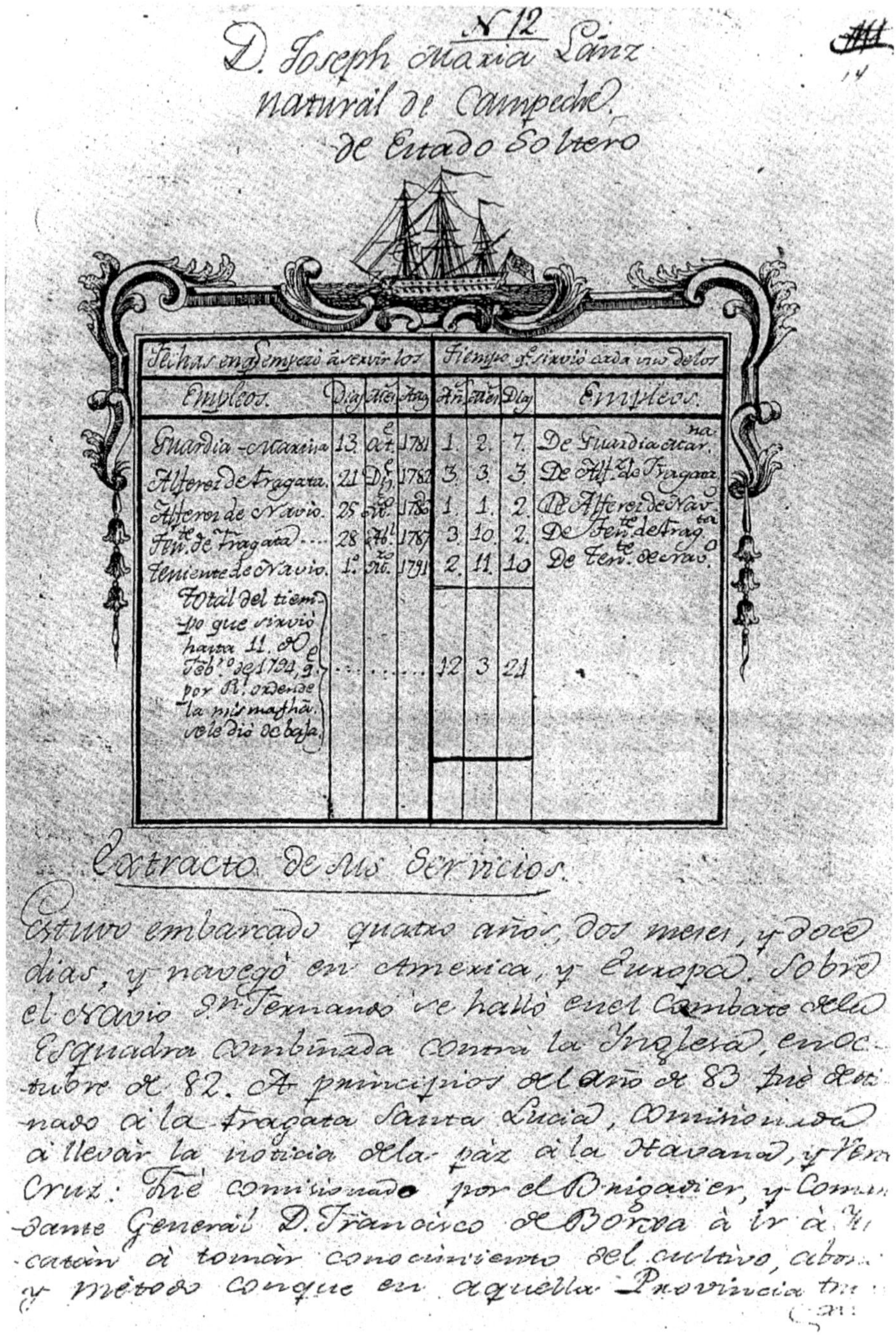

Fig. 1 Lanz's sheet of services (Archivo de la Marina de Guerra de Viso del Marqués)

a report on this task. (This report was published in the *Registro Yucateco*, although, according to Justo Sierra O'Reilly, it was thought that the real author was Policarpo Antonio Echánove y Arzubia. However, Joaquín Lanz Trueba in the brochure *Estudios sobre el henequén*, published in Campeche in 1926, showed that Lanz y Zaldívar was the first to write about this plant.) On the 1st of January of 1784, Lanz returned to La Habana, where for some time he was in service on the *fragata* Santa Dorotea.

Lanz returned to Madrid in 1789 to participate, under Mendoza's (Lanz' advisor) direction, in a trip through France, England, Germany, Sweden, Poland and Russia to collect data, designs and documentation. After the trip through Europe, and back in Spain, he became director of a clock factory.

Because of his outstanding scientific activity, he obtained a scholarship for studying in Paris where he had met Monge and Betancourt by the end of 1789. Lanz y Zaldívar got married in Paris where he decided to stay with his wife. However, from 1802 to 1805 he lived in Madrid and taught at the *Escuela de Caminos y Canales* (School of Roads and Channels). Actually, Lanz lived in Spain in an intensive epoch of intellectual renovation fostered by the egregious figure of Carlos III and his successor Carlos IV.

In 1808, together with Agustín Betancourt, he published the book *Essai sur la composition des machines* in Paris (Lanz and Betancourt, 1808). This is a basic book about industrial kinematics. By the end of the eighteenth and the first part of the nineteenth centuries, Lanz played an important role in developing modern engineering in Europe.

He was expelled from Paris because of his ideological links with the French revolution. Nevertheless, it seems inaccurate that he was receiving Napoleon I's orders, as pointed out by Antonio Rumeu de Armas (1983) in his book, but was actually working on the orders of Napoleon's brother Jose I Bonaparte, King of Spain. Because of that, he went back to Spain and took second-level political positions in the administration. (Rumeu's book has seven chapters, three appendices, and reproduces among other illustrations, the first page of Lanz's service sheet, the cover of the first edition of *Essai sur la composition des machines*, and two of Lanz's letters. One letter is addressed to the Conde Fernán de Núñez and the other one is addressed to the Principe de la Paz, dated 1793 and 1796, respectively.) In 1832, the Board of the *Real Conservatorio de Artes* proposed to the Treasury Minister that Lanz y Zaldívar be appointed (the equivalent of) Full Professor. Lanz y Zaldívar died in Paris in 1839.

According to Rumeu's book, Antonio Gutiérrez, a student of Lanz y Zaldívar, described him, at the age of 67, as kind and good with a reddish face, gross, and short. As a recognition of him after his death, in 1848 a portrait was ordered for Antonio García (Vicuña 1888), and supposedly was exhibited at the *Real Conservatorio de Artes*. So far, unfortunately, the author did not found the portrait.

Essai sur la Composition des Machines

Introduction

Modern activity relating to mechanism design can be considered as starting with the foundation of the *École Polytechnique* in Paris in 1784. This can be also considered the beginning of the Theory of Machines and Mechanisms (TMM), that greatly evolved from the nineteenth century to a topic of modern engineering that today we call Machine and Mechanism Science (MMS). The founding fathers of TMM can be considered to be those who established the *École Polytechnique* and particularly the first teachers of the disciplines that have contributed to the formation of modern industrial engineering. Among those first teachers, also investigators, one can recognize Gaspard Monge, Lazare Carnot, Luis Lagrange, Jean Pierre Nicolas Hachette, Louis Poinsot, Paul Réné Binet, Claude Louis Marie Henri Navier, Dominique François Jean Arago, Jacques Philippe Marie Binet, Jean Victor Poncelet, Jean Marie Constant Duhamel, Félix Savary, Alexix Thérèse Petit, Gaspard Gustave Coriolis, Michel Chasles. Some of them were also *École* students, and later they became professors of mechanics and/or machines at the *École*.

In particular, however, one can recognize in Gaspard Monge (1746–1818) the first teacher of TMM, since he first planned specific classes on mechanism analysis and design at the beginning of the *École Polytechnique*. Monge proposed these classes as the final application of his program on Descriptive Geometry. But only in 1806 the Council of the *École* did approve the teaching for 10 h of mechanism analysis. Meanwhile, Monge had investigated mechanisms by informing other teachers on this topic. His successor was Jean Pierre Nicolas Hachette (1769–1834), who was in fact the teacher who gave the first classes on mechanisms (Hachette 1811) and continued his teaching activities for a long time.

Among those students who attended the first classes on mechanisms which were conducted under the supervision of Monge, one can recognize José María Lanz y Zaldívar and Agustín de Betancourt. Their personalities and activities can be considered very significant and illustrative of early modern industrial engineers, not only for the brilliant academic and technical contributions, but even for the intense engineering activity, and with a wide radius of action. However, Lanz and Betancourt are yet not fully considered in the history of modern TMM, since they are not cited or at the most, mentioned with little information (García-Diego 1985; Rubio and Cuadrado 2000; López-Cajún and Ceccarelli 2005; Ceccarelli et al. 2006).

In this section, details of contributions and activities by Lanz are reported and discussed with the aim of stressing not only the historical significance, but also the pioneering activities in modern engineering. Particular attention has been addressed to the revision of the *Essai sur la composition des machines* that was first published at Paris in 1808, as the text of the early teachings on mechanisms, and in the form that was inspired by Monge, but given by Hachette (1811).

The *Essai sur la composition des machines* was written as related to an Elementary Course on Machines as an application of classes of Descriptive

Geometry given at the *École Polytechnique*. The course was planned in 1795, but only Hachette started it in 1806.

The first draft of the *Essai* was reorganized by Lanz in the years 1805–1807. Lanz and Betancourt reviewed the *Essai* together during the 4 months of coincidence in Paris and submitted it to the *École Polytechnique* in the second semester of 1808. At the same time, Hachette began a course on the Elements of Machines, and he prepared a program of lessons and tables to classify elementary machines. Thus, since Monge and Hachette had knowledge of the book prepared by Lanz and Betancourt, they proposed its publication within the framework of the *École Polytechnique*. Hachette reviewed the manuscript with the idea of adapting it for the above-mentioned program and the *Essai* was published in 1808, and included Hachette's program as shown in Fig. 2. In this first edition, the table of classification

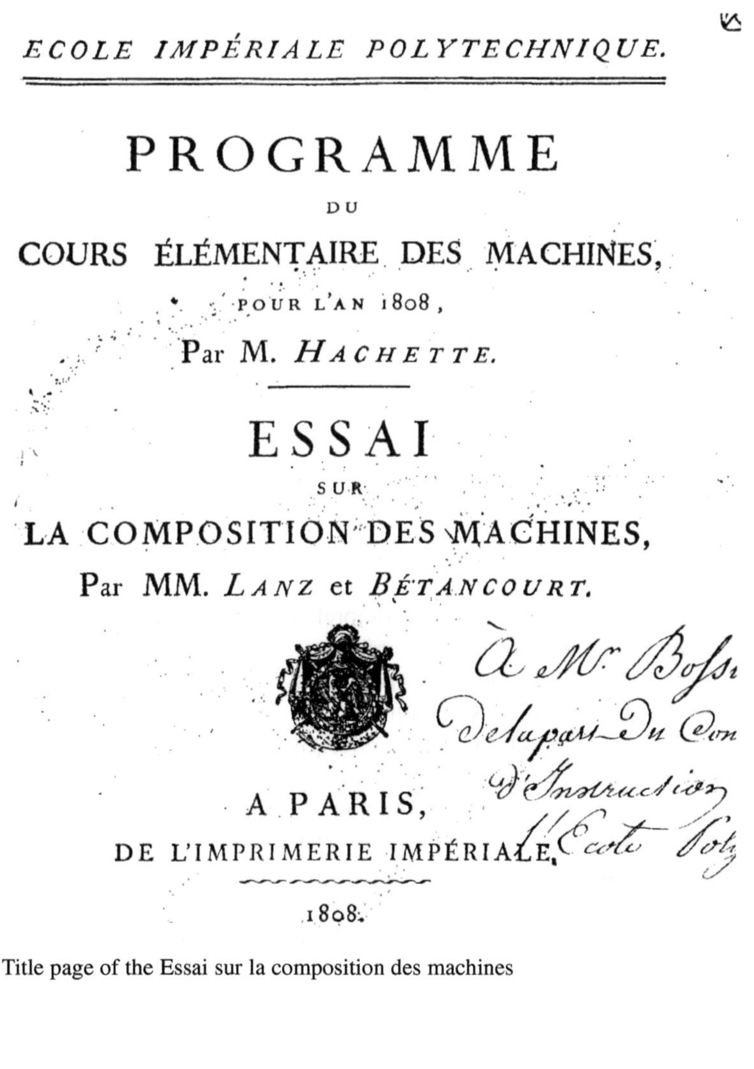

ECOLE IMPÉRIALE POLYTECHNIQUE.

PROGRAMME

DU

COURS ÉLÉMENTAIRE DES MACHINES,

POUR L'AN 1808,

Par M. *HACHETTE.*

ESSAI

SUR

LA COMPOSITION DES MACHINES,

Par MM. *LANZ* et *BÉTANCOURT.*

à Mr Bossu
Delapart Du Con
d'Instruction
l'Ecole Poly

A PARIS,

DE L'IMPRIMERIE IMPÉRIALE,

1808.

Fig. 2 Title page of the Essai sur la composition des machines

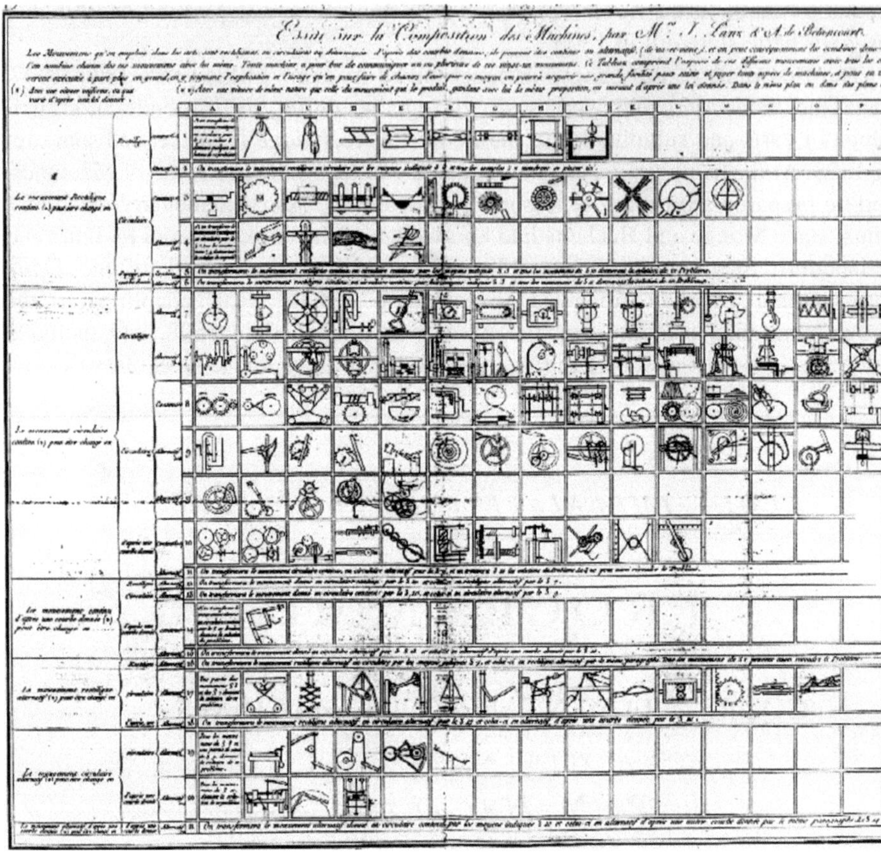

Fig. 3 The table for mechanism classification by Lanz and Betancourt

of Hachette appears at the beginning together with one more complete table by Lanz and Betancourt (Fig. 3).

The content of the book is based on a comprehensive classification of mechanisms that were obtained from considering circular and straight-line motions both for input and output links in continued or alternative modes. Thus, ten classes of mechanisms were identified, as shown in the Table of Fig. 4 and in Fig. 5 by using basically existing well-known mechanism designs and referencing to several technical sources of that time.

The classes of mechanisms are combinations of circular and straight-line input-output motions taking into consideration continuous or alternating motions that could have had practical applications at that time, but still of current interest. Indeed, the treatment of the mechanism that is identified in the table of the classification is described and discussed in detail both for general and for specific purposes. This was done by means of a modern kinematic approach and addressing attention to motion properties and mechanism behavior.

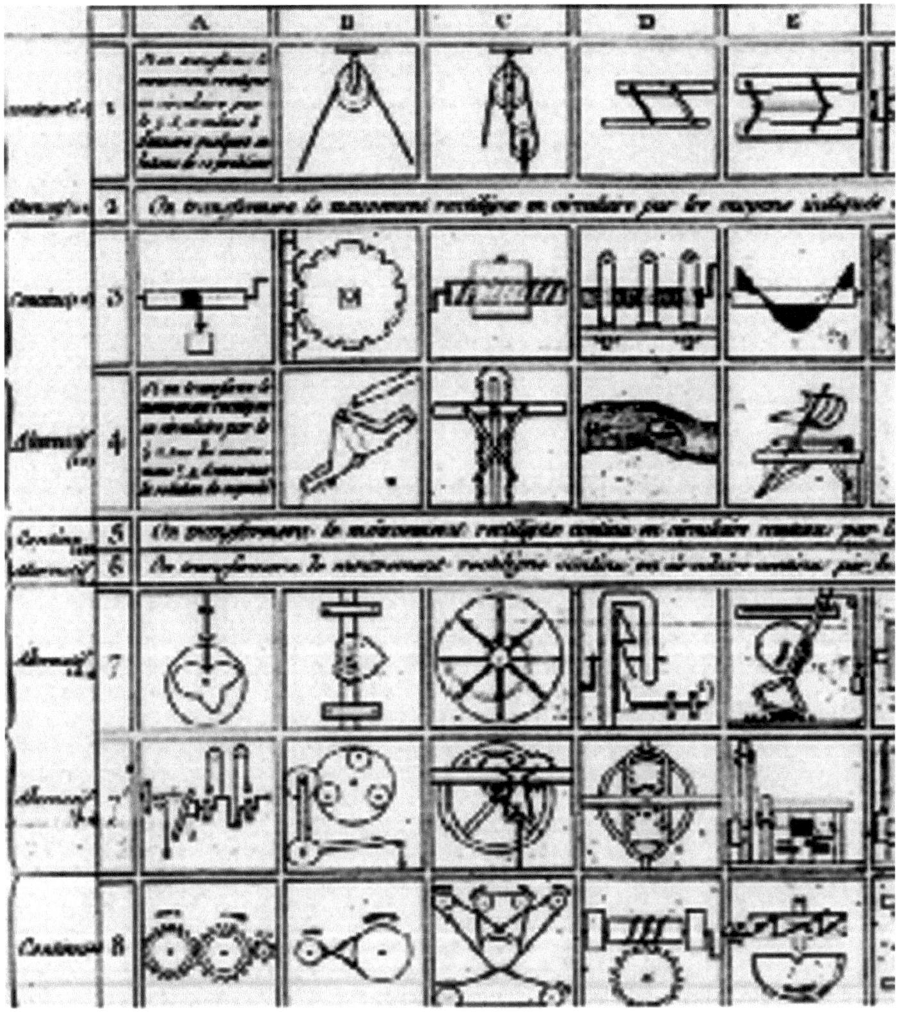

Fig. 4 A section of the table of mechanism classifications

Considerations are also deduced for design purposes. Examples from the *Essai*, which are explained on pages 114–115 and 109–110, respectively, are illustrated in Figs. 5 and 6. In particular, an analysis of the Watt linkage for straight-line guidance is shown in Fig. 5. This was done by using several configurations of the mechanism to emphasize the efficiency of it. The mechanisms shown in Figs. 5 and 6 were analyzed from a practical viewpoint, also showing alternative solutions.

The work in the *Essai* can be considered as a synthesis of the personal experiences of Lanz and Betancourt after maturation within the framework of the rigorous formation at the *École Polytechnique* under the supervision of Monge and Hachette.

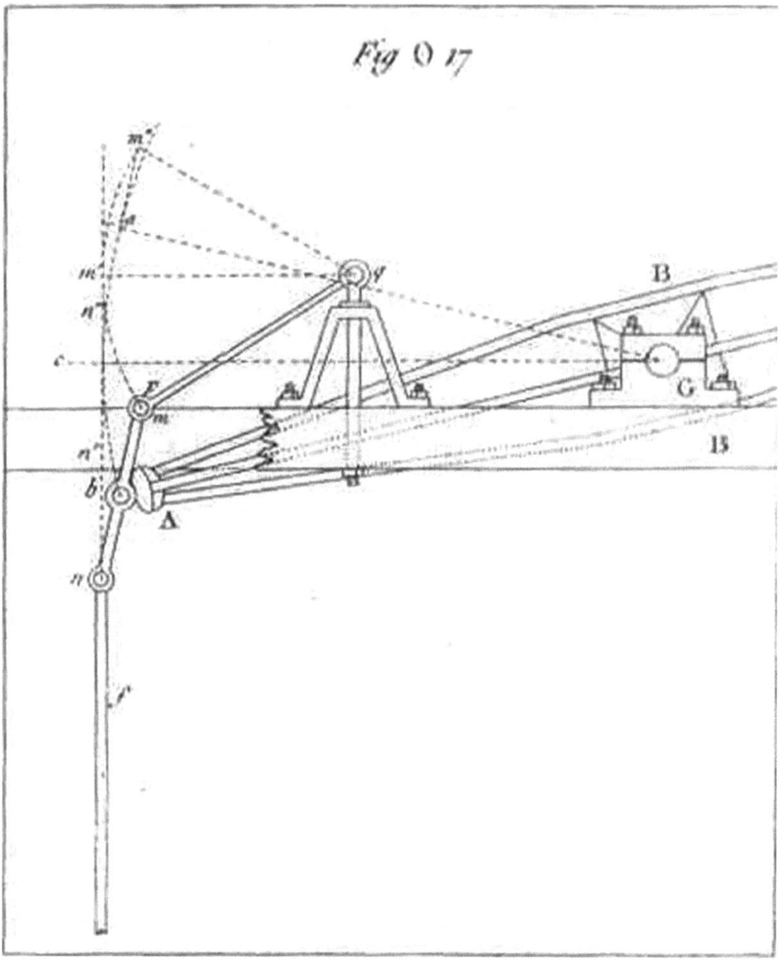

Fig. 5 Example of the Essai: the Watt four-bar linkage

The systematic approach was a successful attempt to treat, with a unique framework the great variety of mechanisms that were designed and used at that time, including an effort for the conception of additional mechanisms. The early kinematic analysis is used in the *Essai* to study and identify basic properties of mechanisms, from a general abstract viewpoint, in order to free the mechanism designs for specific purposes, as used in the previous *Theatrum Machinarum.*

In the *Essai*, in addition to the machine collections, are included the then-existing bibliography, inventors and scope of use of the machines. Indeed, at the end of the eighteenth century, there was already a need for a modern collection of used mechanisms with detailed technical analysis. These were produced much more in depth than in the past handbooks like the *Theatrum Machinarum*, mainly in France, as illustrated by first patents or by new collections like, for example, in the *Enciclopedie.*

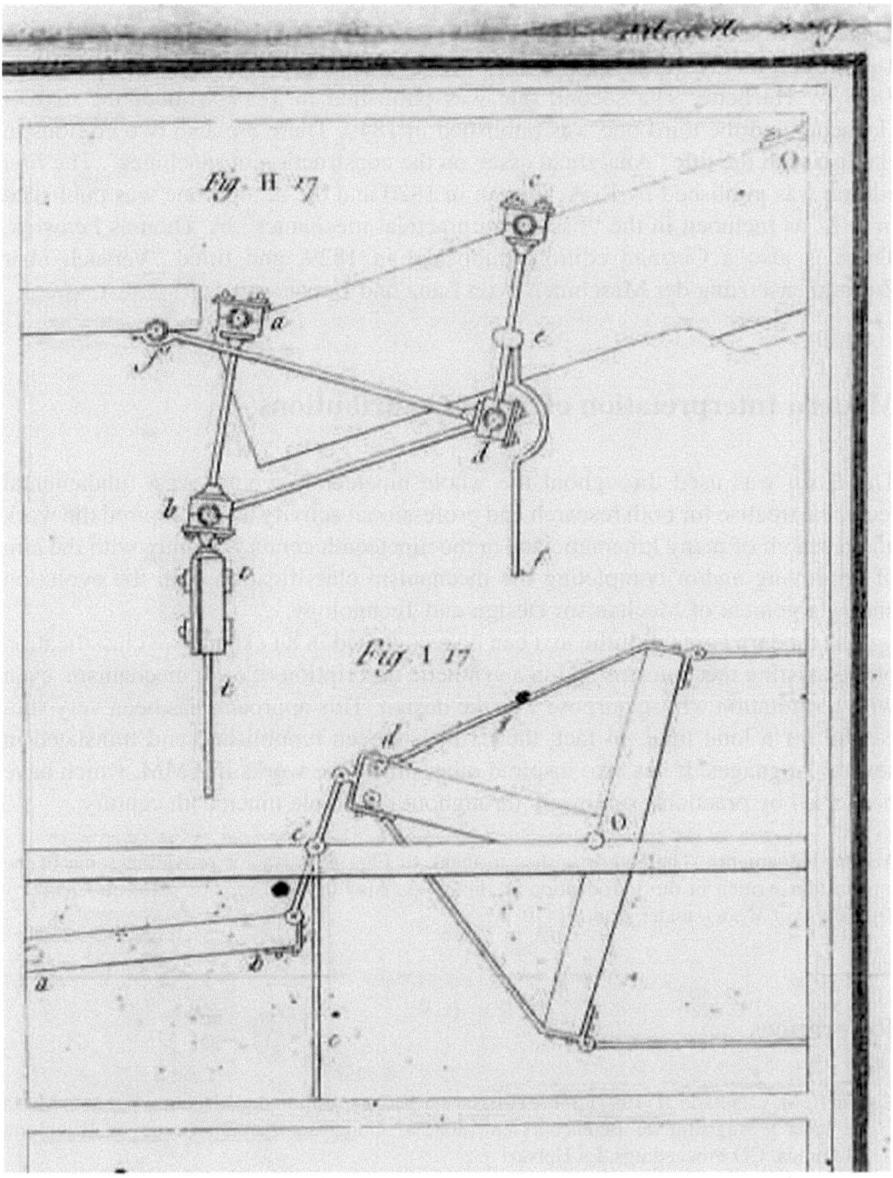

Fig. 6 Example of analysis of linkage mechanisms in the Essai: a study of different practical solutions

On the Circulation of the *ESSAI*

There are three editions in French from the *Essai*. The first one was published in 1808, and it includes the *Programme du cours Élémentaire des Machines pour l'an 1808* by Hachette. The second one was published in 1819 without the text of Hachette, and the third one was published in 1840. There are also two editions in English with the title "Analytical essay on the construction of machines". The first edition was published by R. Ackerman in 1820 and the second one was published in 1822 as included in the "Essays on practical mechanics" by Thomas Fenwick. There is also a German edition, published in 1829, and titled "Versuch über Zusammensetzung der Maschinen" von Lanz und Betancourt.

Modern Interpretation of Main Contributions

The *Essai* was used throughout the whole nineteenth century as a fundamental technical treatise for both research and professional activity and it inspired the work and research of many kinematicians in the nineteenth century, mainly with the aim of improving and/or completing the mechanism classification with the evolution and enlargement of Mechanism Design and Technology.

The modern content of the text can be recognized in an exhaustive classification of the existing mechanisms and in a synthetic description of each mechanism, even with formulation whose purpose was the design. This approach has been very successful for a long time. In fact, the *Essai* has been republished and translated in several languages. It has also inspired other milestone works in TMM, which have been used by practicing engineers throughout the whole nineteenth century.

Acknowledgements The author wishes to thank to Elga Egorova for providing some of the information written in the Introduction of this work. Also the *Consejo Nacional de Ciencia y Tecnología of Mexico* under grant 51410-Y.

References

Ceccarelli M, Cuadrado JI (1997) Sobre el Essai sur la Composition des Machines por Jose Maria de Lanz y Augustin de Betancourt en 1808, 3° Congreso Iberoamericano de Ingeniería Mecánica, CD Proceedings, La Habana

Ceccarelli M, Erogova O, Cuadrado-Iglesias JI, López-Cajún C (2006) V. Pavlov, "Agustin Betancourt: an early modern scientist and engineer in TMM", ASME Paper No. DETC2006-99198

García-Diego JA (1985) En busca de Betancourt y Lanz. Editorial Castalia, Madrid

Hachette JNP (1811) Traitè elementaire des machines, Paris

Lanz JM, Betancourt A (1808) Essai sur la composition des machines, Paris

López-Cajún C, Ceccarelli M (2005) Facts on MMS by Lanz and Betancourt. Proceedings of the Third International Workshop on HMMS, Moscow, Russia, pp 64–69

Rubio FJ, Cuadrado JI (2000) Development of 'Course of Machines' at the Ecole Polytechnique from its origin to the middle of the XIX-th century. In: Ceccarelli M (ed) Proceedings of

HMM2000, International Symposium on History of Machines and Mechanisms. Kluwer, Dordrecht
Rumeu de Armas A (1983) El científico mexicano, José María Lanz, fundador de la cinemática industrial, Madrid
Vicuña G (1888) Lanz y Betancourt: Bosquejo bibliográfico, Madrid

Nicolae Manolescu (1907–1993)

Theodor Ionescu

Abstract Nicolae Manolescu is the founder of the Romanian school of Mechanism and Machine Science (MMS) and one of the IFToMM founders.Through his extensive scientific activity, he approached numerous MMS topics but, above all, he carried out studies on the numerical, structural and kinematic analysis and synthesis of the plane kinematic chains and of various degrees of mobility mechanisms, as well as on the kinetostatic and dynamic analysis of mechanisms. Throughout his entire activity he promoted MMS at national level and essentially contributed to the worldwide recognition of the Romanian school of MMS.

Biographical Notes

Nicolae Manolescu was born on 11 April 1907 in Adjudul Vechi, a small town in the eastern part of Romania as the second son of Ion Manolescu, a railway employee, and Ecaterina, housewife.

He began primary school in 1913 in his home town and continued his education in Focsani, a town with a rich cultural tradition. After leaving the secondary school in 1926, he entered the Polytechnic School in Timisoara at which, in 1931, he took a graduate degree in electromechanical engineering.

Completing military service in 1932, he began his engineering career at the Braila town Municipal Works and, after a time, he moved to the CFR (Romanian State Railways) where he held several technical and managerial positions between 1932 and 1956.

Although he had worked as an assistant lecturer during his last year at Timisoara Polytechnic School and delivered lectures at the Military College, the year 1949 marked the beginning of his long-standing, complex university career.

T. Ionescu (✉)
CFR – Romanian State Passenger Railways, Bd, Dinicu Golescu 38, 010873, Bucuresti, Romania
e-mail: theodor.ionescu@cfr.ro

M. Ceccarelli (ed.), *Distinguished Figures in Mechanism and Machine Science: Their Contributions and Legacies, Part 2*, History of Mechanism and Machine Science 7,
DOI 10.1007/978-90-481-2346-9_7,

Between 1949 and 1959, he was an academic at the Railways Technical Institute of Bucharest and at the Oil and Gas Institute of Bucharest (1950–1958). Combining research with teaching, Nicolae Manolescu climbed the academic ladder through hard work, professional competence and complete dedication.

In 1956, he joined the Polytechnic Institute of Bucharest. His rich technical and scientific culture enabled him to also offer courses in the domains of Strength of Materials, Machine Parts, Dynamics of Railway Vehicles, Thermotechnics, Mechanics, Mathematics.

Professor Manolescu got his Ph.D. degree in MMS in 1969 with a comprehensive work "Contributions to the Numerical, Structural and Kinematic Synthesis of Assur Groups, of Kinematic Chains, of Mechanisms and of Articulated Plane Motor-Mechanisms". In recognition of his overwhelming activity, he was conferred in 1971, with the scientific title of "Doctor Docent" in the field of Mechanism and Machine Science.

Endowed with a prodigious memory and a remarkable working power, he fully engaged himself in guiding younger colleagues and subordinates, conducting research and doctoral theses, contributing to national and international technical journals, lecturing at world famous technical universities, organizing and taking part in symposia, conferences and congresses. He also held the Chair of Mechanism and Machine Theory between 1962 and 1972 and was the Dean of the Transportation Faculty of the Bucharest Polytechnics between 1958 and 1972. He also acted as a technical adviser to the Ministry of Transportation.

As a consulting professor, he continued to carry out academic teaching, doing research and managing the Romanian branch of IFToMM from 1972 until his death in 1993.

Figure 1 presents Professor Nicolae Manolescu addressing the audience at his 80 years anniversary in 1987.

Nicolae Manolescu crowned his lifetime's scientific and technical activity by his election in 1991 as a corresponding member of the Romanian Academy.

A great promoter of MMS in Romania and in the world, he was one of the IFToMM founders and an active member of the IFToMM Technical Committee for Linkages and Cams and of the IFToMM Permanent Commission for Standardization of Terminology. He was the founder of the Romanian branch of IFToMM, e.g. the Romanian Association for the Theory of Mechanisms and Machines (ARoTMM) and held its chair for a long period. It was he who initiated the "SYROM" International Symposium on Theory and Practice of Mechanisms, within IFToMM, which has taken place in Romania every 4 years since 1973, bringing together world MMS personalities and promoting their activities and works.

The results of his studies and research can be found in many courses, academic books and treatises on MMS, in more than 150 articles in technical journals, and more than 120 papers at national and international conferences and congresses.

In his private life, Professor Manolescu had a great tragedy, loosing his first wife and his two children, in the Second World War bombardments in 1944.

He died on 10 October 1993 as a tragic consequence of a car accident.

Fig. 1 Nicolae Manolescu in 1987, addressing the audience at his 80 years anniversary (Author's photo archive)

List of Main Works

Here are some of his main works:

- Books/Courses/Treatises (in Romanian): "Theory of Mechanisms and Machines", 1955; "Theory of Mechanisms and Machines. Kinetostatics and Dynamics", 1958; "Collection of Problems in Mechanism and Machine Theory ", 1963; "Collection of Problems in Mechanism and Machine Theory", 1968; "Theory of Mechanisms and Machines", 1972.
- Doctoral thesis (in Romanian): "Contributions to the numerical, structural and kinematic synthesis of Assur Groups, of kinematic chains, of mechanisms and of articulated plane motor-mechanisms", 1969.
- Papers/articles (in Romanian) : "On the mechanisms made up of different family kinematic chains", 1961 ; "Structural synthesis of plane articulated kinematic chains with 10 elements, 13 kinematic pairs, and degree of freedom $L = 4$", 1969.
- Papers/articles (in English or French): "Sur la détermination du degree de mobilité des mecanismes", 1963; "Une methode unitaire pour la formation des chaines cinematiques et des mecanismes plans articules avec differents degrés de liberté et mobilité", 1964; "Systematisation et classification des mécanismes – moteurs plans articules, a deux degrés de mobilite '*total*', $(M = 2)$ et '*partial*', $(M = 2)$", 1965; "The methods of formation of Assur groups function of the number of loops (Zg) and of the rank of links (J)", 1966; "Structural

synthesis of plane kinematic chains with multiple joints and five degrees of freedom $L_3 = 5$ '*total*', $L_3 = 5$ '*partial*' and $L_3 = 5$ '*fractional*'", 1967; "Structural and kinematic synthesis of plane driving-mechanisms with multiple joints and two degrees of mobility: '*Total*' $M_3 = 2$ and '*Partial*' $M_3 = 2$", 1968; "For a united point of view in the study of the structural analysis of kinematic chains and mechanisms", 1968; "Criteria of forming the plane jointed mechanisms with multiple joints and simple links with e = 11 links and $M_3 = 2$ degrees of mobility '*partial*', '*partial-total*' and '*fractional*'", 1971; "La détermination des variants indépendante des fermes Baranov avec e = 9 élements en utilisant la méthode de graphisation inverse", 1971; "The method of determination of the number of structural kinematic variants of KCsj by joints simplifying", 1971; "A method based on Baranov trusses and using graph theory to find all the planar jointed kinematic chains and mechanisms", 1973; "La comparaison des chaines cinématiques et des mécanismes au point de vue structural a l'aide de la théorie des graphes", 1975; "The unitary method of structural synthesis of all the planar jointed Kinematic chains (KCmjsl)", 1979; "A unified method for the formation of all planar jointed kinematic chains and Baranov trusses", 1979; "La formation des fermes Baranov avec $L_3 = 0$ et 1 a l'aide de la méthode unitaire elaboree pour les chaines cinématiques plans articulees", 1981; "The history of the original methods used in the synthesis of the planar kinematic chains with different degrees of freedom (liberty)", 1988; "L'état actuel du nombre des variants nonisomorphes des chaines cinematiques plans avec articulations multiples et éléments simples (CCpames) et e = 12, $C_5 = 16$ et $L_3 = 4$ degres de liberte", 1989; "Sur la synthese structurale des chaines cinématiques plans avec articulations multiples, $L_3 = 6$ degrés de liberte et e = 8 et 10 elements simples (CCpames)", 1989; "The planar initial jointed kinematic chain", 1993; "L'histoire de l'optimisation de la synthèse des chaines cinématiques plans articulés avec differentes degrés de liberté", 1993.

Review of Main Works

Looking back at the life and work of Manolescu, we have to notice his vast scientific and pedagogical activity concerning mechanisms and machines and, at the same time, his total commitment to engineering. His studies and research were essential for understanding and solving many difficult problems related to the practice of mechanisms and machines in industry and transportation.

As it would be unrealistic and practically impossible to present his lifetime activity within these few pages, we are going to give a concise selection of his most interesting scientific topics.

Concerning the mobility degree of mechanisms, Manolescu extended the original Ozol formula to

$$M_f = \sum_{m=f+1}^{5} (6-m)\,c_m - (6-f)Z, \tag{1}$$

where Z is the number of independent cycles of the mechanism, c_m the number of kinematic pairs of class m, $f = (6 - b)$ – family of the space in which a free element has b degrees of freedom, M the degree of mobility of the mechanism of family f.

Manolescu also established methods for determining the family f of a mechanism by exploring the joining conditions of the kinematic elements and/or by identifying the Assur groups making up the mechanism, thus facilitating the Dobrovolski formula application:

$$M_f = (6-f)n - \sum_{m=f+1}^{5} (m-f)\,c_m \tag{2}$$

where n is the number of kinematic elements.

Figure 2, from Manolescu et al. (1972), presents, for exemplification, the structural scheme of a steam locomotive's distribution mechanism with 14 links and 20 kinematic pairs and, in conformity with relation 2, a degree of mobility $M = 2$. This is also obvious when considering the Assur groups composing this mechanism: dyads (2, 3), (4, 5), (12, 13) and a double triad (6, 7, 8, 9, 10, 11) and the remaining two driving elements : links 1 and 14.

Manolescu considered the problem of structural and numerical synthesis of the plane kinematic chains with simple elements and simple and multiple joints and of the derived mechanisms, as well, which are fundamental for the study of mechanisms structure, kinematics and kinetostatics. Consequently, his studies and research in this domain were developed throughout his entire life, the prominent results and gains being discussed in many papers and communications.

The purpose was to elaborate an optimized methodology for the structural synthesis of planar-jointed kinematic chains (PJKC), capable of delivering all kinematic chains which were able, in their turn, to generate all existing plane mechanisms. A lot of scientists strived to achieve this objective, among whom were Crossley, Kurt Hain, Woo, Kiper, Schian, Mruthyuniaya, Hwang, Rao, but Manolescu played the major role in elaborating it.

These configurations, being of greatest interest from the theoretical point of view, unfortunately may not be found, in their integrality, to be viable and applicable in the real practice of machines and mechanisms.

In his research, Manolescu introduced new concepts and definitions, such as motor mechanism (MM) – a mechanism with a driving element; initial mechanism (IM) – a mechanism made up of fixed and driving elements; Assur group (AG) – a subsystem in the componence of a mechanism, characterized by a degree of mobility $M = 0$; degree of mobility "*total*" – when the movements of all driven elements depend simultaneously on the movements of all driving elements; degree of mobility "*partial*" – when the movements of the driven elements do not depend simultaneously on the movements of the driving elements; degree of mobility "*fractional*" – when

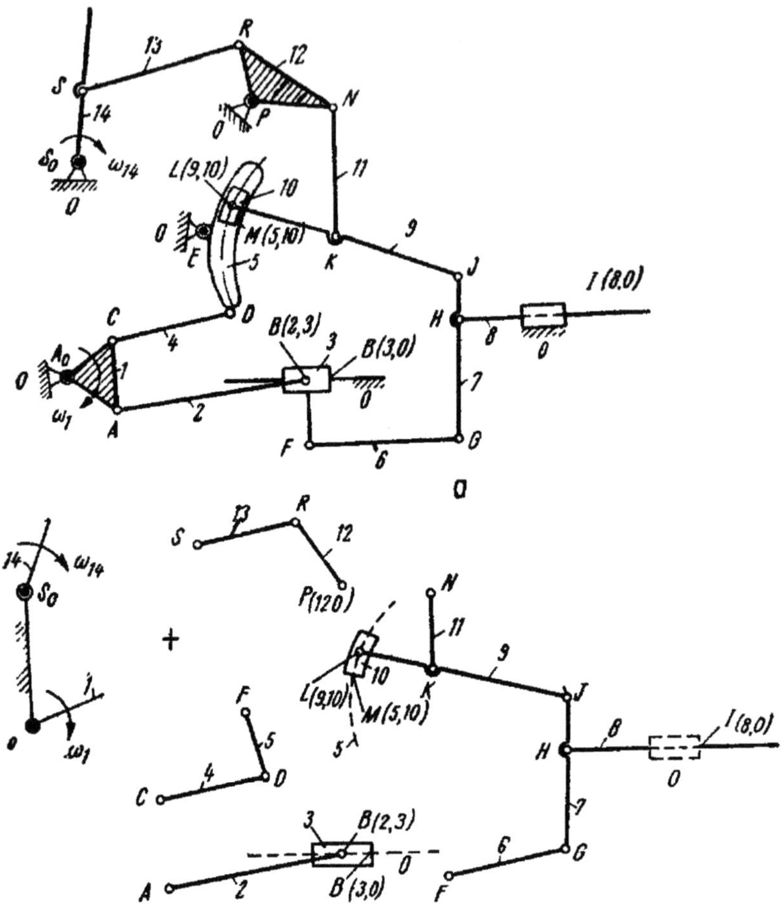

Fig. 2 The steam locomotive distribution mechanism (Manolescu et al. 1972)

there are some driven elements whose movements do not depend on the movements of all driving elements; Baranov truss (BT) – a closed (overconstrained) kinematic chain, which by loosing an element give birth to a corresponding Assur group; graphisizing operation(G) – transforming a structural scheme of a kinematic chain into a graph; inverse graphisizing operation (G) – transforming a graph into a structural scheme of a kinematic chain; joints simplification (JS)- transforming a multiple kinematic pair into several simple kinematic pairs; amplification with *dyads* (AD) – connecting Assur groups of the "*dyad*" type to a kinematic chain.

Figure 3, Manolescu (1964), shows the generation of planar-jointed kinematic chains, using Assur groups and leading to mechanisms with a degree of mobility "*total*" (a), "*partial*" (b) and "*fractional*" (c). These aspects were taken into consideration by many researchers such as Cebisev, Gruebler, Kraus, Crossley and Kurt Hain, who initially proposed the terms " *total*" and "*partial*".

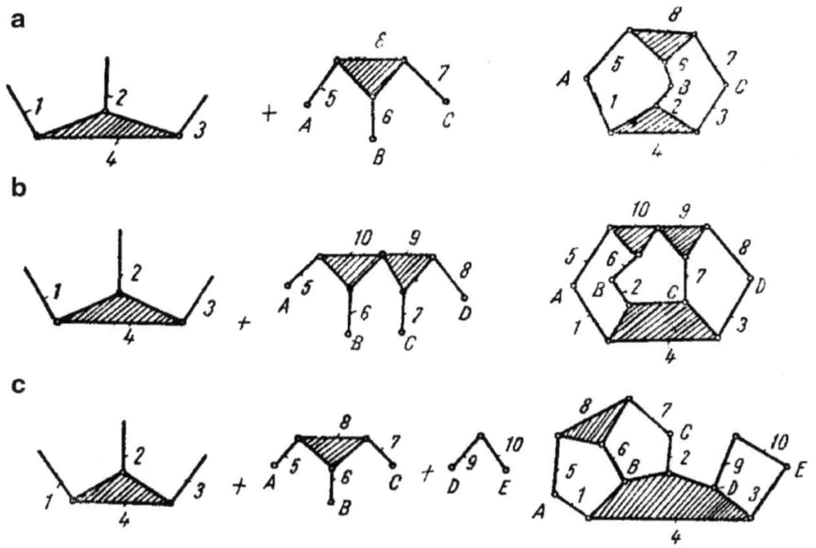

Fig. 3 Generation of planar jointed kinematic chains with a degree of freedom "total", "partial" and "fractional" (Manolescu 1964)

All these concepts of Manolescu are widely used now by the MMS Romanian school and also by many other scientists and researchers in the MMS field.

Over time, Prof. Manolescu developed and experimented several methodologies for the structural and numerical synthesis of kinematic chains and mechanisms. He used them comparatively on sample cases, even combining them and, eventually, arriving at his own conclusions upon their particularities and advantages/disadvantages. He was permanently struggling to improve them, with the declared intention of discovering all the (possible) existing configurations of planar jointed kinematic chains.

A first methodology is based on Assur groups. The plane kinematic chains of a given degree of freedom are obtained by successively amplifying an initial chain (of the same degree of freedom) with Assur groups, until the achievement of the considered number of kinematic elements and kinematic pairs. Finally, the solutions are selected from pre-established criteria and only the distinctly nonisomorphic configurations are retained.

With regard to mechanisms, Fig. 4. in Manolescu, Environment and Planning (1979), presents Manolescu's vision of the two methodologies used for the formation of planar driving mechanisms (PDM) : the methodology based on Assur groups – dyad, triad, tetrad (columns 1–3) and the classical methodology (Reuleax 1876), by defining the fixed link of a planar-jointed kinematic chain and then the driving link (columns 5, 4, 3).

A second methodology developed by Prof. Manolescu is derived from Baranov trusses amplified by nonassuric groups. This was used mostly as a comparative and verifying methodology and gave way to a third and more elaborated methodology,

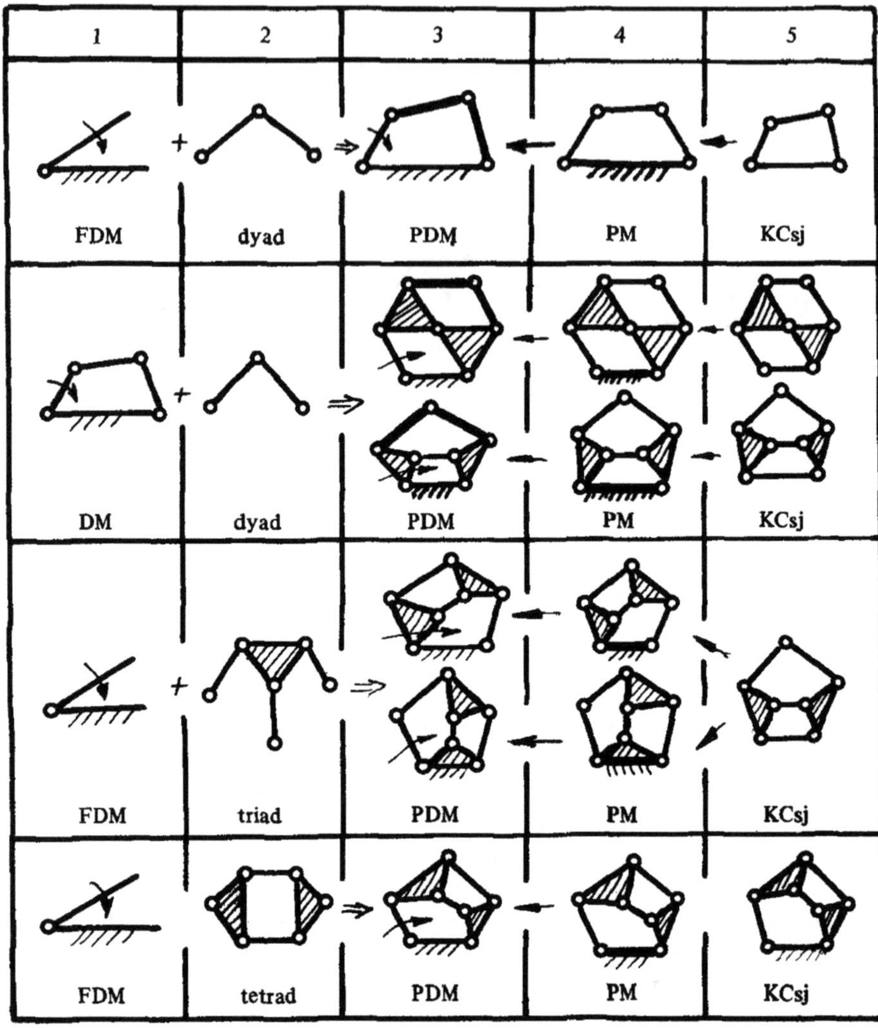

Fig. 4 Two methods of formation of planar-driving mechanisms (Manolescu, Environment and Planning B 1979)

based on graph theory. The graph theory takes into consideration the associated graph of a planar jointed kinematic chain (PJKC). This associated graph has knots corresponding to the kinematic chain elements and, simultaneously, the lines corresponding to the kinematic pairs.

An example for obtaining "Stephenson" and "Watt" planar-jointed kinematic chains from a Baranov truss with five elements by the successive operations of graphisizing (G) and joints simplification (JS), is shown in Fig. 5, by Grecu (2007).

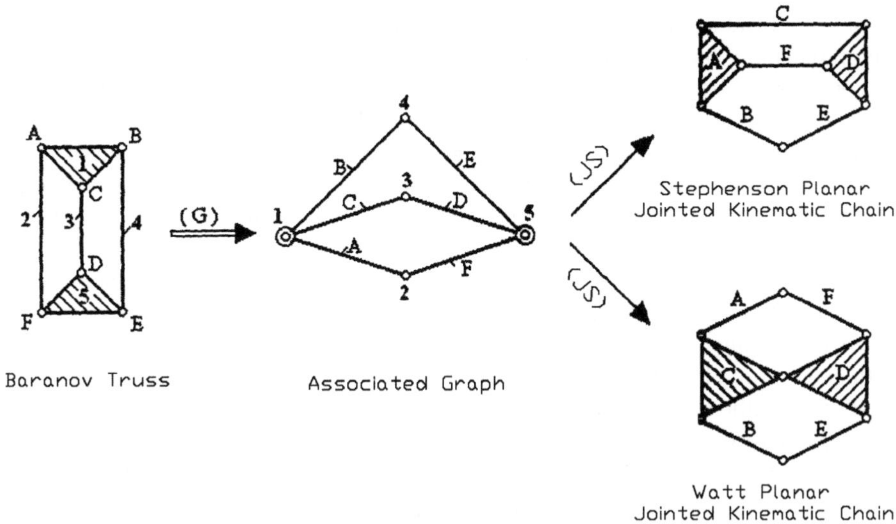

Fig. 5 Generation of "Stephenson" and "Watt" planar jointed kinematic chains from the Baranov truss by (G) + (JS) operations (Grecu 2007)

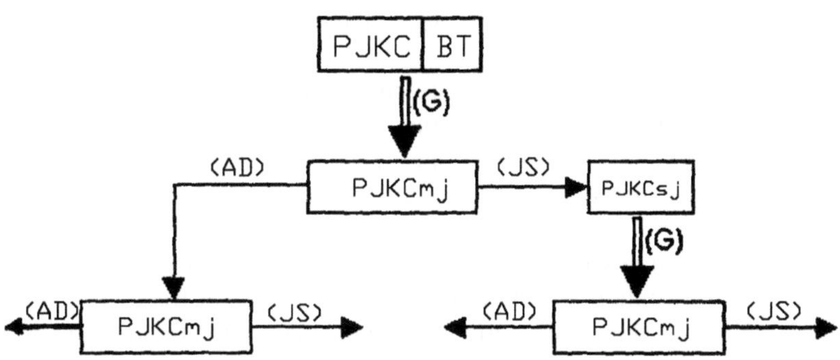

Fig. 6 The basic algorithm of Manolescu's structural synthesis Methodology (**a**) Manolescu et al. (1972); (**b**) Grecu (2007)

By noticing that the associated graph of a planar-jointed kinematic chain (or Baranov truss) can itself be considered a planar-jointed kinematic chain with multiple joints (PJKCmj) (see Fig. 5.), Manolescu delivered his ultimate structural synthesis methodology based on the combination and multiple use of operations of graphisizing (G), joint simplification (JS) and amplification with *dyads* (AD). The basic algorithm of this methodology is presented in Fig. 6. by Manolescu et al. (1972) and by Grecu (2007).

Starting from a planar-jointed kinematic chain (PJKC) or Baranov truss (BT), by the graphisizing operation one comes to its associated graph, which in its turn can be considered a planar jointed kinematic chain with multiple joints (PJKCmj). By applying the operation of joints simplification (JS), we come to another planar-jointed kinematic chain with simple joints (PJKCsj), of the same degree of freedom; or, by applying the operation of amplification with *dyads* (AD), we come to another different planar-jointed kinematic chain with multiple joints (PJKCmj) with more elements but with the same degree of freedom.

The process can be continued by repeatedly applying, in an arborescent manner, these operations of graphisizing (G), joints simplification (JS), amplification with *dyads* (AD), thus getting several new solutions of planar-jointed kinematic chains with even more elements.

By this methodology, Manolescu was able to verify the computer-aided solutions obtained by researchers in India and Taiwan and to achieve his goal of discovering all the existing configurations. The methodology can be also extended to the study of kinematic chains with superior kinematic pairs, cams, cylindrical gears, etc.

In his last studies, Manolescu introduced the concepts of "planar initial jointed kinematic chain" (PIJKC) and "planar initial jointed kinematic chain with simple links" (PIJKCsl).

The following tables are relevant for these concepts and express some of Manolescu's last published achievements.

Figure 7, from Manolescu (1993) and also Grecu (2007) presents the nonisomorph variants of planar initial jointed kinematic chain with simple links-PIJKCsl.

In Figure 8, of Manolescu (1993), are shown the generation of planar-jointed kinematic chains with simple links, multiple joints and degree of freedom $L = 4$.

In Figure 9, of Manolescu (1993), the generation of planar-jointed kinematic chains with simple links, multiple joints and degree of freedom $L = 5$ is explained.

Figure 10, of Manolescu (1993), shows the generation of planar-jointed kinematic chains with simple links, multiple joints and degree of freedom $L = 6$.

Professor Nicolae Manolescu definitely left his mark upon the Romanian school of mechanisms and machines. His studies and findings are fundamental to the further development of MMS in Romania and abroad. For hundreds of polytechnics graduates, he will be remembered as a great teacher, an innovative, sympathetic and creative one. His life and dedicated work are, above all, an inspiration for all of us who knew him and have read his works. Figure 11 shows Professor Nicolae Manolescu with some of his pupils and collaborators at his 1992 anniversary at Bucharest Polytechnic Instuitute.

Degree of Freedom	Structural Characteristics		Structural schema of the PIJKC sl
	Number of links	Number of Multiple Joints	
$L_3 = -1$	$e = 1$	$C_5^1 = 2$	
$L_3 = 0$	$e = 2$	$C_5^1 = 3$	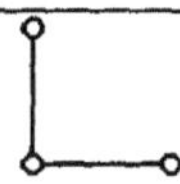
		$C_5^1 = 1; C_5^2 = 1$	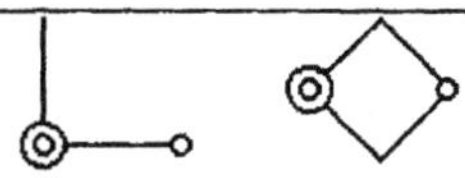
		$C_5^3 = 1$	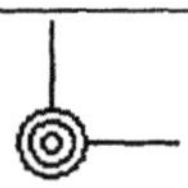
$L_3 = 1$	$e = 1$	$C_5^1 = 1$	
$L_3 = 2$	$e = 2$	$C_5^1 = 2$	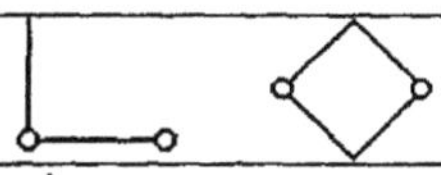
		$C_5^2 = 1$	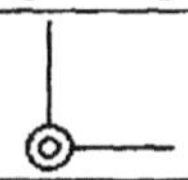
$L_3 = 3$	$e = 1$	$C_5^1 = 0$	
$L_3 = 4$	$e = 2$	$C_5^1 = 1$	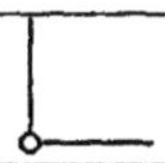
$L_3 = 5$	$e = 3$	$C_5^1 = 2$	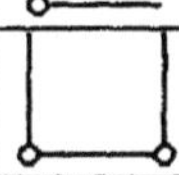
		$C_5^2 = 1$	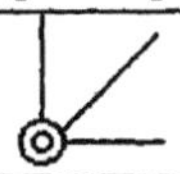
C_5^1 = Simple joint; 	C_5^2 = Double joint; 	C_5^3 = Triple joint 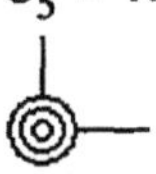	

Fig. 7 The nonisomorph variants of planar initial jointed kinematic chain with simple links (**a**) Manolescu (1993); (**b**) Grecu (2007)

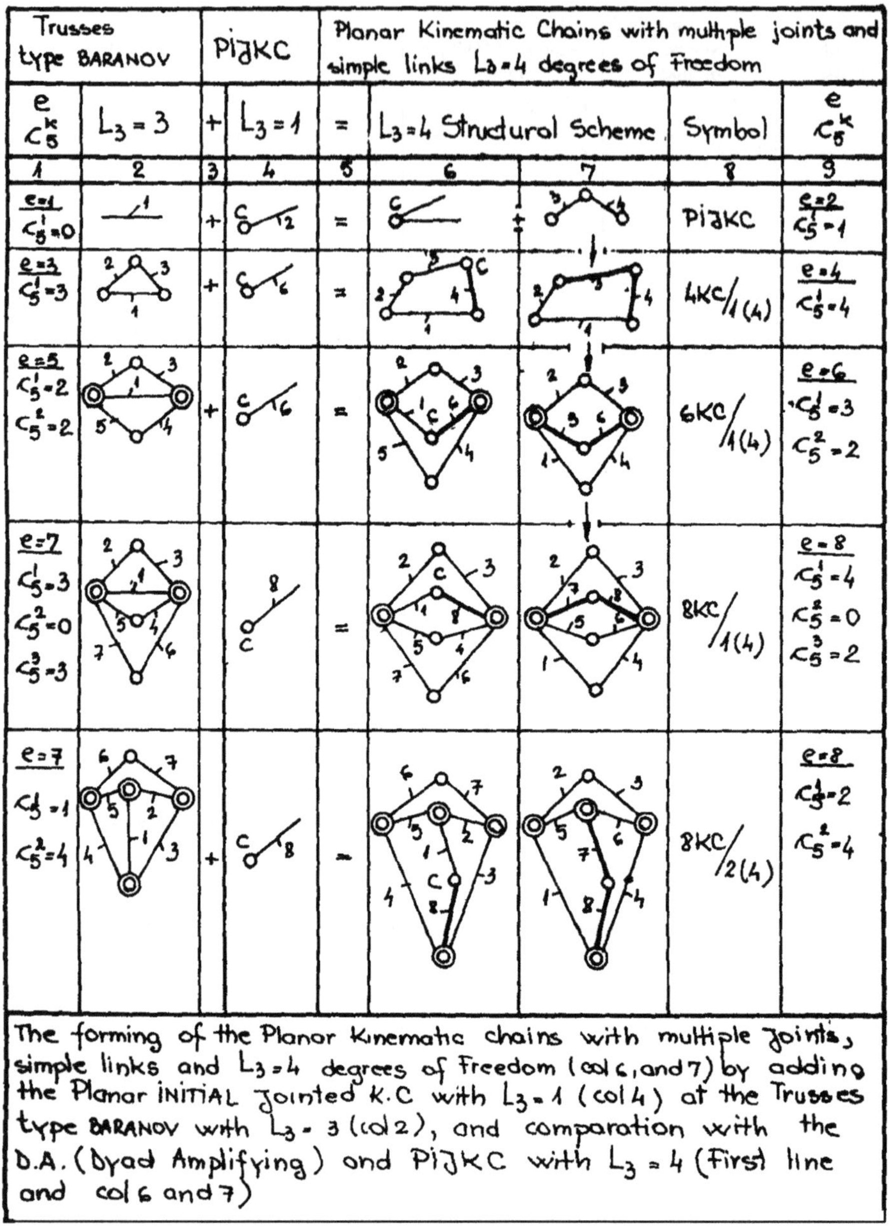

Fig. 8 The generation of planar jointed kinematic chains with simple links, multiple joints and degree of freedom L = 4 (Manolescu 1993)

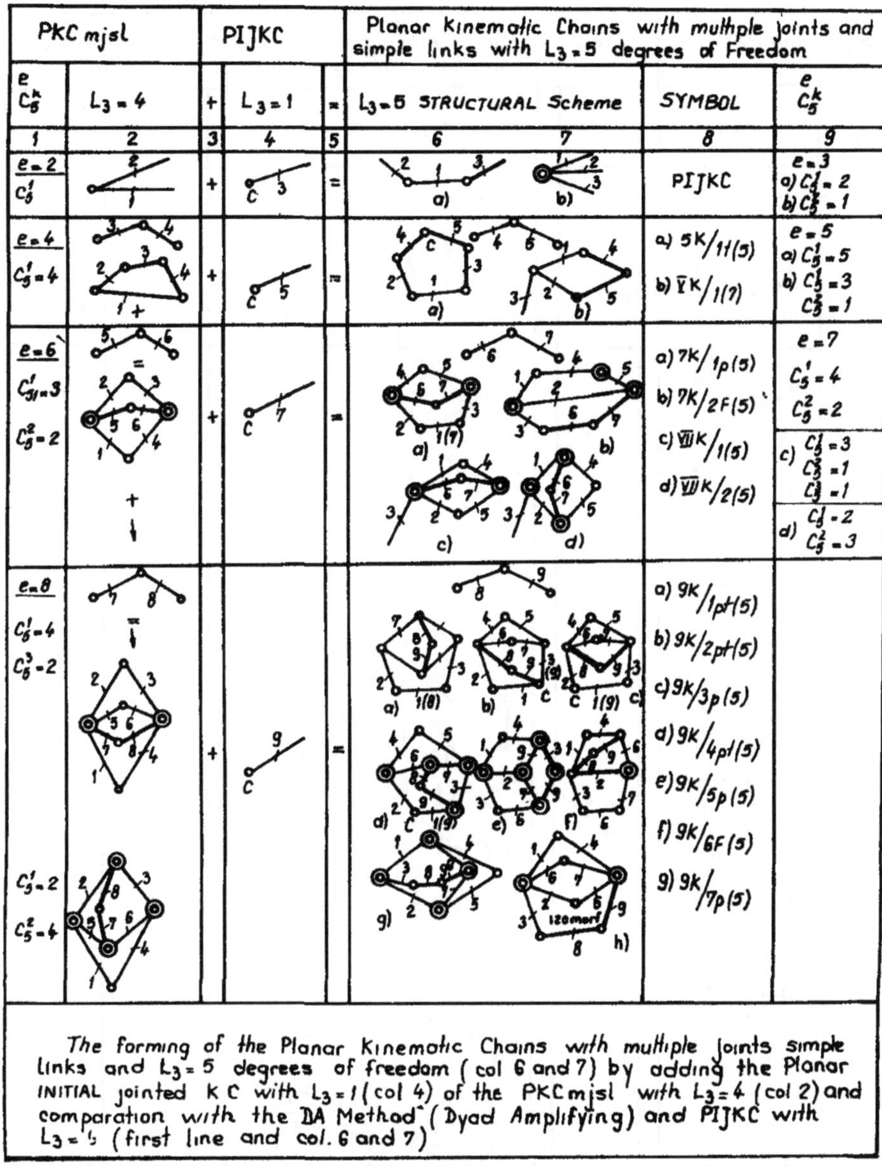

Fig. 9 The generation of planar jointed kinematic chains with simple links, multiple joints and degree of freedom L = 5 (Manolescu 1993)

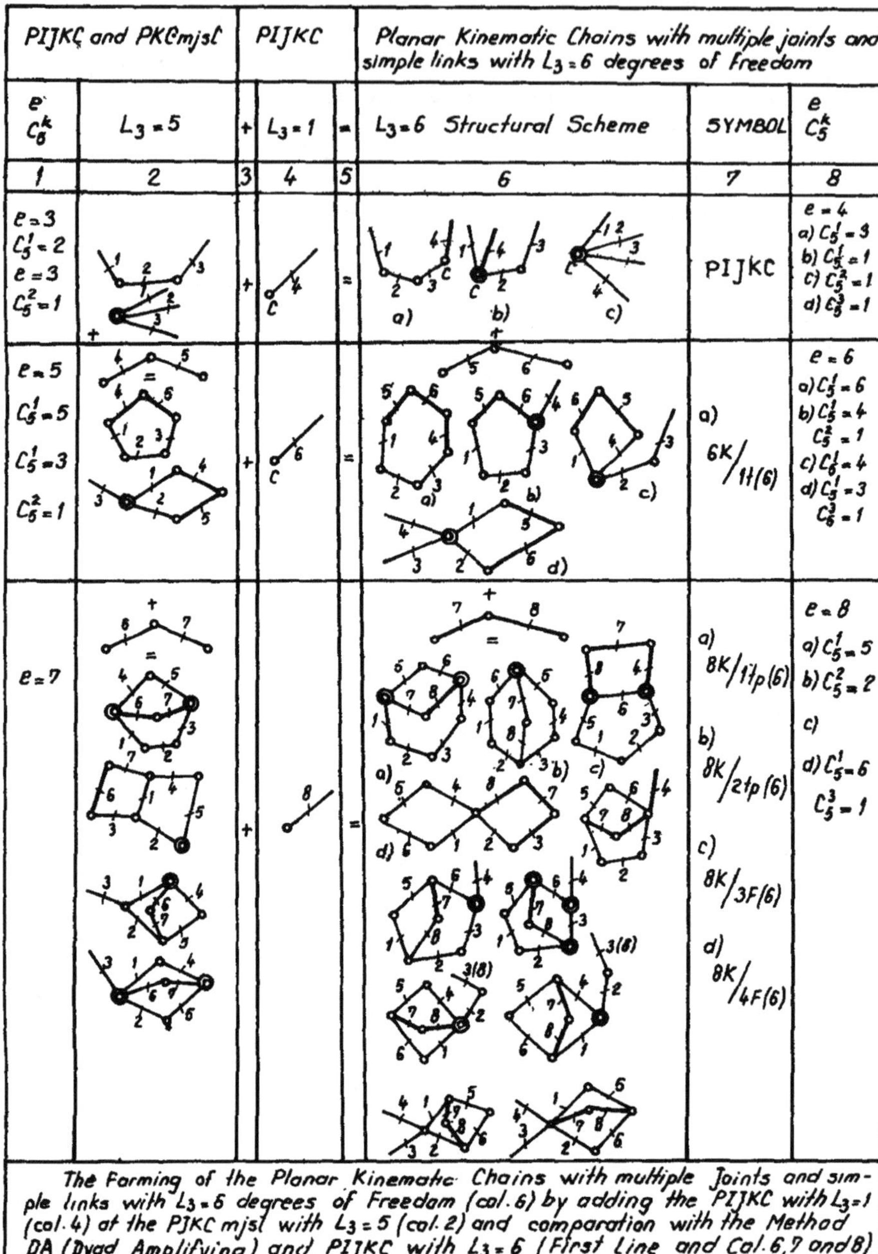

Fig. 10 The generation of planar jointed kinematic chains with simple links, multiple joints and degree of freedom L = 6 (Manolescu 1993)

Fig. 11 Nicolae Manolescu at his 1992 anniversary at Bucharest Polytechnic Institute, chair of Mechanisms and Machines (flanked, from left to right, by S. Cononovici, A. Lascu, F. Duditza and R. Bogdan, T. Ionescu, P. Alexandru, B. Grecu, F. Nitzu, I. Tempea) (Author's photo archive)

References

Manolescu N (1955–1956) Theory of mechanisms and machines, vol 1–4. Litografia I.C.F., Bucuresti (in Romanian)

Manolescu N, Maros D (1958) Theory of mechanisms and machines. Kinetostatics and dynamics. Editura Tehnica, Bucuresti (in Romanian)

Manolescu N et al (1963) Collection of problems in mechanisms and machines theory, vol 1. Editura Didactica si Pedagogica, Bucuresti (in Romanian)

Manolescu N et al (1968) Collection of problems in mechanisms and machines theory, vol 2. Editura Didactica si Pedagogica, Bucuresti (in Romanian)

Manolescu N, Kovacs F, Oranescu A (1972) Theory of mechanisms and machines. Editura Didactica si Pedagogica, Bucuresti (in Romanian)

Manolescu N (1961) On the mechanisms made up of kinematic chains of different families. Studii si Cercetari de Mecanica Aplicata, XII, nr. 3. Bucuresti, pp 589–615 (in Romanian)

Manolescu N, Manafu V (1963) Sur la détermination du degré de mobilité des mécanismes. Buletin Institut Politehnic Bucuresti, XXV, nr. 5, Bucuresti, pp 45–66

Manolescu N (1964) Une Methode Unitaire pour la Formation des Chaines Cinématiques et des Mécanismes Plans Articules avec differentes Degrés de Liberté et Mobilité. Revue Roumaine Science et Technique, serie de Mecanique Appliquee, IX, 6, Bucuresti, pp 1263–1300

Manolescu N (1965) Sistematisation et Classification des Mecanismes-Moteurs Plan Articules a deux Degrés de Mobilite "total" ($M_3 = 2$), et "partial" ($M_3 = 2$). Revue Roumaine Science et Technique, Serie de Mecanique Appliquee, X, 4, Bucuresti, pp 999–1042

Manolescu N, Antonescu P, Erceanu I (1966) The methods of formation of Assur Groups function of the number of loops (Zg) and of the rank of links (J). Buletin Institut Politehnic Bucuresti, XXVIII, nr. 3, Bucuresti, pp 59–83

Manolescu N, Tempea I (1967) Structural synthesis of plane kinematic chains with multiple joints and five degrees of freedom $L_3 = 5$ "total", $L_3 = 5$ "partial" and $L_3 = 5$ "fractional". Revue Roumaine Science et Technique, Serie de Mecanique Appliquee, XII, 5, Bucuresti

Manolescu N (1968) Structural and kinematic synthesis of plane driving mechanisms with multiple joints and two degrees of mobility: "total" $M_3 = 2$ and "partial" $M_3 = 2$. Revue Roumaine Science et Technique, Serie de Mecanique Appliquee, XIII, 1, Bucuresti

Manolescu N (1968b) For a united point of view in the study of the structural analysis of kinematic chains and mechanisms. J Mech 3(3):149–169

Manolescu N (1969) Contributions to the numerical and structural synthesis of Assur groups, of kinematic chains, of mechanisms and plane jointed motor mechanisms. Doctoral dissertation, Institutut Politehnic Bucuresti (in Romanian)

Manolescu N, Tempea I (1969) Structural synthesis of plane jointed kinematic chains with e = 10 Elements, c5 = 13 Joints and Degree of Freedom $L_3 = 4$. Buletin Institut Politehnic Bucuresti, XXXI, nr. 4, Bucuresti (in Romanian)

Manolescu N (1971) Criteria of forming the plane jointed mechanisms with multiple joints and simple links, with e = 11 Links and $M_3 = 2$ Degrees of Mobility "partial", "partial-total " and "fractional". Proceedings of the 3rd World Congress on TMM, vol D, paper D-11, Kupari-Dubrovnik, pp 161–176

Manolescu N, Ardeleanu T (1971) La Détermination des Variantes Indépendentes des Fermes Baranov avec e = 9 Eléments en utilisant la Méthode de Graphisation Inverse. Proceedings of the 3rd World Congress on TMM, vol D, paper D-12, Kupari-Dubrovnik, pp 177–188

Manolescu N, Tempea I (1971) The method of the determination of the number of structural kinematic variants of KCsj by joints simplifying. Proceedings of the 3rd World Congress on TMM, vol C, paper C-12, Kupari-Dubrovnik, pp 145–162

Manolescu N (1973) A method based on Baranov Trusses and using graph theory to find all the planar jointed kinematic chains and mechanisms. Mech Mach Theory 8(8):3–22

Manolescu N, Dranga M (1975) La Comparaison des Chaines Cinématiques et des Mécanismes au point de vue Structurel a l'aide de la Théorie des Graphes. Buletin Institut Politehnic Bucuresti, XXXVII, nr.1, Bucuresti, pp 41–46

Manolescu N (1979) The unitary method of structural synthesis of all the planar jointed kinematic chains (KCmjsl). Proceedings of the 5th World Congress on TMM, vol I, Montreal, pp 514–518

Manolescu N (1979) A unified method for the formation of all planar jointed kinematic chains and baranov trusses. Environ Plann B (London) 6:447–454

Manolescu N, Martineac A (1981) La Formation des Fermes Baranov avec $L_3 = 0$ et 1 a l'aide de la Méthode Unitaire élaboreée pour les Chaines Cinématiques Planes Articulées. Proceedings of the 3rd International Symposium on Theory and Practice of Mechanisms, SYROM '81, paper 15, Bucharest, pp 137–158

Manolescu N (1988) The history of the original methods used in the synthesis of the planar kinematic chains with different degrees of freedom (liberty). Proceedings of the Conference on Mechanism Construction, Liberec, pp 145–154

Manolescu N (1989) L'etat actuel du Nombre des Variantes Nonisomorphes des Chaines Cinématiques Planes avec Articulations multiples et Eléments simples (CCpames) et e = 12, c5 = 16 et $L_3 = 4$ Degrés de Liberté. Proceedings of the 5th International Symposium on Theory and Practice of Mechanisms, SYROM ' 89, vol I, Bucharest, 1989, pp 531–548

Manolescu N (1989) Sur la Synthese Structurale des Chaines Cinematiques Planes avec Articulations Multiples, $L_3 = 6$ Degres de Liberte et e = 8 et 10 Elements Simples (CCpames). Proceedings of the 5th International Symposium on Theory and Practice of Mechanisms, SYROM '89,vol IV, Bucharest, pp 113–123

Manolescu N (1989) The planar initial jointed kinematic chain. Proceedings of the 5th International Symposium on Theory and Practice of Mechanisms, SYROM ' 89, vol I/2, Bucharest, pp 549–556

Manolescu N (1993) L'histoire de l'optimisation de la Synthèse des Chaines Cinématiques Planes Articulées avec differentes Degrés de Liberté. Proceedings of the 6th International Symposium on Theory and Practice of Mechanisms, SYROM '93, vol 1, paper 22, Bucharest, pp 169–192

Crossley F (1965) The permutation of kinematic chains of eight members or less from the graph theoretic viewpoint. In: Developments in theoretical and applied mechanics, vol 2. Pergamons, Oxford, pp 467–486

Dijksman EA (1975) Kempes (focal) linkages generalized, particularly in connection with Hart's second straight-line mechanism. Mech Mach Theory 10(3):445–460

Grecu B (2007) Structural synthesis of mechanisms – main objective of scientific papers of Prof. N. Manolescu, Homage paper on the occasion of N. Manolescu 100th birth anniversary, Polytechnic Institute of Bucharest, April 2007 (in Romanian)

Reuleaux F (1976) The kinematics of machinery: outlines of a theory of machines. Translated and edited by A.B.W. Kennedy. Macmillan, London

Mruthyuniaya T (1979) Structural synthesis by transformation of binary chains. Mech Mach Theory 14:221–231

Kojevnikov C (1979) Osnovnaia Structurnovo Sinteza Mehanizmov. Naukova Dumka, Kiev (in Russian)

Kiper G, Schian D (1975) Die 12 gliedrigen Grublerschen kinematischen Ketten, VDI-Z, No. 6, Maerz (II), pp 283–288 (in German)

Dijksman EA, Timmermans EA (1994) Look-out for prime-chains with a prescribed number of mobility degrees of freedom. Mech Mach Theory 29(5):653–672

Mruthyuniaya T, Raghayan M (1984) Computer aided analysis of the structure of kinematic chains. Mech Mach Theory 19(3):357–368

Mruthyuniaya T (1984) A computerized methodology for structural synthesis of kinematic chains. Part II. Application to several fully or partially known cases. Mech Mach Theory (Pergamon Press) 19(6):497–505

Mruthyuniaya T (1984) A computerized methodology for structural synthesis of kinematic chains. Part III: Application to the new case of 10 link, three freedom chaises. Mech Mach Theory (Pergamon Press) 19(6):507–530

Wen Miin Hwang, Yii Wen Hwang (1992) Computer aided structural synthesis of planar kinematic chains with simple joints. Mech Mach Theory (Pergamon Press) 27(2):189–199

Rao AC, Rao CN (1993) Loop based pseudo-hamming values, testing isomorphism and rating kinematic chains. Part I; Inversions, prefered frames and actuators. Part II. Mech Mach 28(1):113–143

Gogu G (2005) Mobility of mechanisms: a critical review. Mech Mach 40:1068–1097

Baranov G (1948) Klassificatia Stroenie, Kinematica i Kinetostatica ploskih Mehanizmov s Parami pervovo Roda, Trudi seminara po TMM, AN SSSR, T.V, vol 20 (in Russian)

Francesco Masi (1852–1944)

Marco Ceccarelli

Abstract In this paper, the figure and work of Francesco Masi is presented as an influent professor in the field of TMM in Italy. His main contribution, still of present-day significance, can be recognized in the rigorous systematization of Reuleaux's classification of mechanisms together with an analytical approach for analysis and synthesis of mechanisms. He was also a scientist and engineer in approaching design problems and teaching developments in mechanical engineering at large, and particularly in TMM, lubrication and machine drawing.

Introduction

The second half of the nineteenth century can be considered as the Golden Age for TMM (Theory of Machines and Mechanisms) since it has been developed to high levels of knowledge and practical applications. This happened mainly all around Europe during the Industrial Revolution. But in a very rich literature, some relevant works and personalities are forgotten in today's memory and consideration for the History of TMM. This is the case of most of the Italian literature on kinematics, and the Italian personalities are rediscovered only now, together with their contributions regarding TMM success, both in formation of engineers and research developments with practical applications in new machinery, as pointed out in Ceccarelli (2000).

The Italian tradition in the field of mechanics goes back to the first studies in the twelfth century and continues over the centuries with considerable contributions.

In particular, mechanics received a great boost by the work of Galileo Galilei (Galilei 1600) and Guidobaldo Del Monte (Del Monte 1577) in the sixteenth century and their influence was considerable throughout the eighteenth century too, as

M. Ceccarelli (✉)
LARM: Laboratory of Robotics and Mechatronics, DiMSAT- University of Cassino,
Via di Biasio 43, 03043, Cassino (Fr), Italy
e-mail: ceccarelli@unicas.it

M. Ceccarelli (ed.), *Distinguished Figures in Mechanism and Machine Science: Their Contributions and Legacies, Part 2*, History of Mechanism and Machine Science 7,
DOI 10.1007/978-90-481-2346-9_8, © Springer Science+Business Media B.V. 2010

pointed out in Ceccarelli (1998, 2000). With this cultural background, new achievements were obtained in mechanics so that in the eighteenth century mechanics and mechanical engineering were addressed as separate mature disciplines apart from mathematics. Thus, specific books were written and academic courses were established in several Italian Universities.

These former developments produced a very promising environment for modern enhancements. Italian personalities are recognised and well known from the seventeenth and eighteenth centuries as, for example, Paolo Branca (Branca 1629), Evangelista Torricelli (Torricelli 1919), Guido Grandi (Grandi 1739), Paolo Frisi (Frisi 1765, 1777), Ruggero Boscovich (Boscovich 1763), only to cite some. But the cultural influence of Italian universities was reduced so that in the nineteenth century they were not considered centres of excellence as they had been in the previous century. Nevertheless, exiting activity was carried out in the field of mechanical engineering mainly with the aim of establishing a modern view of the field. Moreover, specific developments were obtained and original contributions were proposed. However, these works circulated with difficulty, even in Italy, probably since at the beginning of the century, there was a considerable number of small kingdoms and later, the main aim was the attempt to obtain a unified culture and organization when the Italian nation was achieved. Another reason for forgetting about Italian works can be the fact that the nineteenth century can be considered as a recent past in Italian culture and because of this, it is not yet considered worthy of attention. In fact, in the Italian libraries, the books of the past century are not classified in ancient funds but neither are they catalogued in modern databases. Most the time they are just stored somewhere. In addition, although memory of the personalities still exists, their works are sometimes considered as out-of-date, without taking into account that they have been useful as an important basis for modern developments.

In the second half of the nineteenth century, schools of engineering were established in most of the Italian universities, but it was only after the Decree of 3 July 1879 that the course "Kinematics Applied to Machinery" was introduced on a regular basis in the curricula of industrial engineers (Pugno 1959).

Nevertheless, only in the oldest Italian universities there was a well-established tradition and in-depth works were carried out in TMM. In fact, we have found textbooks (sometimes handwritten) for regular academic courses on TMM or kinematics given in Turin, Milan, Padua, Pisa, Bologna, Rome, and Naples.

In Turin, following the work by Lagrange and Papacino, courses on TMM were given by Carlo Ignazio Giulio (Giulio 1846), Galileo Ferraris, Domenico Tessari (Tessari 1890), Scipione Cappa (Cappa 1890). Other contributions were given by Elia Ovazza (who seems to be the first female engineer in Italy) (1890), and M. Panetti (1913).

In Milan, after Ruggero Boscovich, relevant Academic personalities can be recognised in Ernesto Cavalli (Cavalli 1882, 1889, 1908), and Gian Antonio Maggi, (Maggi 1896, 1914). Secondary works were developed by Ugo Ancona (Ancona 1896), Giusto Bellavitis (1852), and Ernesto Padova (Padova 1884).

In Bologna, after Gaetano Giorgini (1836), Domenico Chelini (Chelini 1862), and Francesco Masi (Masi 1883, 1897a) made significant contributions to TMM.

In Rome, works on TMM were written by Ugo Cerruti (Cerruti 1898) and Carlo Saviotti (Saviotti 1890). A temporally presence was Lorenzo Allievi, who published a masterpiece (Allievi 1895) on planar kinematics after giving lectures for only 1 year (Ceccarelli and Koetsier 2008). A late successor of Saviotti was Anastasio Anastasi (1908).

In Naples, an important personality on mechanics at large was Giovanni Battaglini (Battaglini 1870, 1873), and in the field of TMM Dino Padelletti (1884), had a good reputation.

In addition, professors were involved in promoting the academy and specifically TMM in more than one university. An emblematic example is Ernesto Cavalli who taught in Milan, Livorno, Pisa and Naples, as reported in his textbooks (Cavalli 1882, 1889, 1908).

The figure and work of Francesco Masi in Bologna University can be considered representative of the Italian fecundity in the field of kinematics of mechanisms, beside the relevance of his work and activity in mechanical engineering generally.

This paper is an attempt to pay due tribute to the personality of Francesco Masi (1852–1944) by surveying his teaching activities and his works that were published and circulated, unfortunately, only in the newly established Italian nation.

Biographical Notes

Francesco Masi (Fig. 1) was born on 28 February 1852 in Guastalla (Reggio Emilia). He obtained his engineer's degree in 1875 at the Royal School of Engineering in Turin. His first employment in 1875 was as professor of Physics and Mechanics at

Fig. 1 A portrait of Francesco Masi (1852–1944)

the Technical College in Cagliari. In 1877, he returned to Bologna to the Royal School of Engineering as assistant professor in the field of Mechanics of Machinery. In 1891, he was appointed full professor at Bologna University where he remained until his death on 30 November 1944 in Bologna. He married Teresa, the daughter of the famous Italian poet Giosuè Carducci, but they never had children.

Unfortunately, no detailed biographical information can be found to better characterize the wide interests that Francesco Masi had during his long life. Masi focused his attention on TMM but also on the fields of hydraulics, technical drawing, mechanics of agricultural machines by developing teaching activities in Technical Colleges and at Bologna University.

List of Main Works

The following publications can be considered as the main works os Francesco Masi. (All are written in Italian and were published in Italy only) (Masi 1927):

1. The bridge on Panaro river in Vignola, 1875.
2. The Rectory Factory for baskets in Bologna, 1880.
3. On the couplings for spherical four-bar linkages, 1880.
4. The irrigation in Modena lands, 1882.
5. Handbook for Applied Kinematics – New classification of mechanisms, 1883 (Masi 1883).
6. Determination of clear spans of bridges, 1885.
7. Methods for corrosion defense, 1885.
8. Properties of Watt curve, 1890.
9. Lectures on Hydraulics at Royal School of Engineering in Bologna, 1893.
10. Experimental methods in the teaching at Technical Professional Colleges, 1890.
11. Treatise on Civil and Industrial Constructions as co-author, published by F. Vallardi, 1890 and successive editions.
12. Drawing of mechanical parts for Technical Colleges, 1891 (Masi 1891).
13. Study of vein obtained by means of shining projections, 1895.
14. Theory of Mechanisms, 1897 (Masi 1897a).
15. New views on theoretical and experimental investigations on friction, 1897 (Masi 1897b).
16. Experiences on friction, 1897 (Masi 1897c).
17. Kinematics of mechanisms and Kinetics of machines, 1920.
18. Plough for hills, Patent, 1920.
19. Lecture Notes of Mechanics for Agriculture at the Royal Agriculture Institute in Bologna, 1926.

The variety of subjects in the above-mentioned publications shows a large variety of interests on which Francesco Masi continuously worked throughout his long academic career. In fact, he was well reputed and his memory is still persistent in the Italian Academy, not only in the field of TMM but even for his contributions in machine drawing and tribology.

Teaching Activity

Francesco Masi paid much attention to his teaching activity in the framework of both University formation of engineers and Technical Colleges for technicians. He started his career just after he obtained his engineer's degree by giving classes at a technical college in Cagliari. This experience was so fruitful for him that he continued to teach at technical colleges, even when he held the position of full professor at the University of Bologna. In particular, he devoted his efforts mainly to the subjects of technical drawing and machine design in the technical college Aldini-Valeriani in Bologna until 1906. During this period, he prepared and published the textbooks (10 and 12 on the list) (Masi 1927). In particular, the textbook on technical drawing (Masi 1891) (Fig. 2) was particularly appreciated and indeed was well used even at the level of University teaching for a long period as a significant development of the Italian tradition following the results of Quintino Sella, as pointed out in Ceccarelli and Cigola (2007).

In this textbook, Masi collected 60 drawings of the main, basic components of machines that are represented in great details for their construction, by using projections and many sections with further developments for technical views and drawing. As a conclusive example, he reported the technical drawing of a locomotive. He also included discussions and expressions both with theoretical and practical formulation for the design of mechanical parts within each drawing table. In addition, for each drawing, he explained the generalities of the technical drawing and specific particularities for the representation of details as a further development of the work by Quintino Sella, who got inspiration from Monge's Descriptive Geometry. Masi emphasized the need of writing the sizes within the figures of the drawings by quoting all the dimensions, although he recommended the use of a proper scale to indicate the real proportions. He developed technical drawing with more technical emphasis by strongly reducing the artistic features of shadows and chiaro-scuro according to Sella's approach too. Nevertheless, he recommended and used to represent through limited signs, the objects as lightened by rays at 45° from the left. He used broken-line segments as a conventional indication for different materials. Today, this textbook on Technical Drawing receives much interest from the historical viewpoint, since its completeness and clarity make it a fundamental source for understanding the historical development of technical drawing and its standardization mainly but not only in an Italian framework.

The textbook No.19 in the list (Masi 1927) on agriculture mechanics was written by Francesco Masi as a result of his teaching at the Royal Agriculture Institute in Bologna that he started in 1903.

Masi collected most of his research experiences and novel contributions in the textbooks for TMM (Masi 1883, 1897a), that are Nos. 5 and 14 in the list

Fig. 2 Cover page of the textbook on mechanical drawing by Francesco Masi, published in 1891

(Masi 1927), Fig. 3. They can be considered his main works. Both books on TMM were a great success, since they were used for a long time, even in other universities in Italy. In fact, they can still be found in the libraries of many Italian schools of engineering. Successively, he revised and completed the content for more wider TMM teaching in the textbook No.17 in (Masi 1927) that was published in 1920 but it was not successful as the other two.

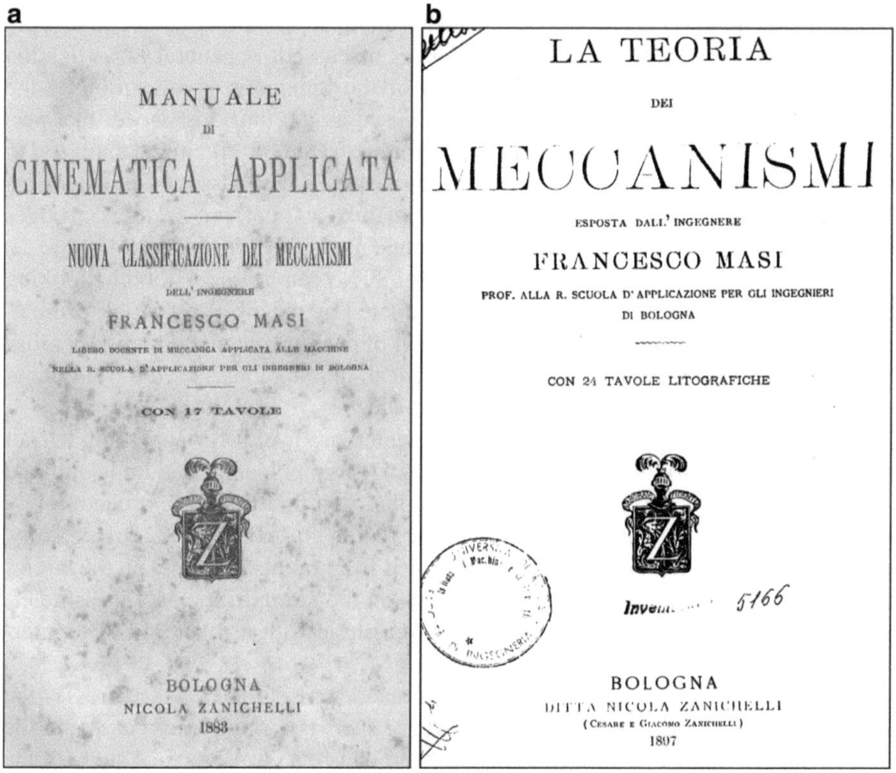

a

MANUALE
DI
CINEMATICA APPLICATA
NUOVA CLASSIFICAZIONE DEI MECCANISMI
DELL' INGEGNERE
FRANCESCO MASI
LIBERO DOCENTE DI MECCANICA APPLICATA ALLE MACCHINE
NELLA R. SCUOLA D'APPLICAZIONE PER GLI INGEGNERI DI BOLOGNA
CON 17 TAVOLE
BOLOGNA
NICOLA ZANICHELLI
1883

b

LA TEORIA
DEI
MECCANISMI
ESPOSTA DALL' INGEGNERE
FRANCESCO MASI
PROF. ALLA R. SCUOLA D'APPLICAZIONE PER GLI INGEGNERI
DI BOLOGNA
CON 24 TAVOLE LITOGRAFICHE
Inven. 5166
BOLOGNA
DITTA NICOLA ZANICHELLI
(CESARE E GIACOMO ZANICHELLI)
1897

Fig. 3 Cover pages of works by Francesco Masi on TMM and mechanism design: (**a**) Handbook of applied kinematics in 1883. (**b**) Theory of mechanisms in 1897

The book Masi (1897a) has a clear teaching purpose while the book Masi (1883) is much more a treatise for a new classification of mechanisms. Indeed, the book Masi (1897a) includes the content of the book Masi (1883) in a specific chapter on mechanism classification and together with Masi (1920), they complete a set that Masi prepared for the Course of Mechanics of Machinery that he taught continuously throughout his long career.

In Masi (1897a), the Theory of Mechanisms is treated both in theoretical aspects and practical applications. The first three chapters deal with the fundamentals of theoretical kinematics by explaining the kinematics of relative motion, determination of centrodes, curvature analysis of trajectory, determination of inflection circle and cycloid curves by using a graphical approach and an analytical formulation.

Chapter 4 is devoted to the study of the motion of a body on a plane by looking at the number of contact points from a kinematic viewpoint. Chapter 5 attaches the problem of instantaneous motion for the determination of the centrodes and axoides of the motion. Chapter 6 is a detailed analysis of gears with theoretical formulation and practical design rules. Chapter 7 summarizes the classification procedure that Masi elaborated in Masi (1883) as an improvement of Reuleaux's work (Reuleaux 1875).

The symbolic notation is introduced together with the notion of kinematic inversion. Successively, the general synthesis problem is discussed. A general classification of mechanisms is introduced as a function of the kinematic composition of the kinematic chain of mechanisms. This is proposed as the unifying aspect that permits us to treat all the mechanisms within a unique frame for kinematic analysis. In fact, the remaining Chapters from 8 to 15 describe all the classes of mechanisms that are included in the classification, namely linkages, gearing systems, wedges and cams, spur gears, epicyclic trains, ratchet gears, and belt transmissions.

The successful teaching approach in Masi (1897a) can be understood by looking at the clarity of the explanations that are completed by means of essential formulation and exhaustive schemes. In the discussion of each mechanism, theory is properly combined with a discussion for practical applications. In the following, examples are discussed to show Masi's approach.

In Fig. 4, kinematic schemes that are presented are useful for analyzing a Watt parallelogram. In particular, in Masi's Figure 257, a graphical procedure is outlined to draw a Watt curve of a point O of the coupler with the following procedure: by using the circles with centers A and D whose radius are AB and DC, respectively, a general point P of the curve can be determined from points M and N on the circles through the segment MP that is long like OB on the straight-line MN. N is determined by using the CB length as the radius of a circle with its center in M to intersect the circle with its center in D.

Thus, once the procedure for analysis formulation is outlined, Masi stresses the main kinematic characteristics of the Watt curve such as a sixth-order curve and the consequent order of contact of the curve with straight-line yy. A practical study of the curve is further suggested by using the variable mechanism design of Figure 258 to investigate differences in the curve as a function of the position of the tracing point O on the coupler. Results are shown in the drawn Figures from 259 to 264 that have been obtained experimentally, as reported in Fig. 5. Then, Figure 265 is proposed to outline a synthesis procedure and evaluation of the maximum deviation of the curve from the straight-line yy, through graphical determination. The sequence of the graphical construction for the synthesis is represented as a function of the data CD and angle 2α. By using angles β and λ (which are between OO_1 and yy), the maximum deviation ε can be computed as $OO_1 \sin \lambda$, whose simplification is obtained from geometric reasoning on Figure 266 to obtain $\varepsilon < AB / 2{,}592$ as a practical expression.

In Fig. 6, the study for constraining the planar motion of a body is discussed to explain the thesis that four contact points are necessary and sufficient to avoid relative translation of the body but they are only necessary to avoid relative rotation. The discussion is illustrated by looking at the intersection points of the normal lines at the contacts to determine the triangles abc and cde in Figure 55: if abc is inside cde, the relative rotation is impossible.

The extension to a general case with any number of contacts is discussed by referring No. 56 in Fig. 6 in which the basic polygon is determined by the normal lines from the contacts. Then, the conditions for possible relative rotation depend on the topology of the polygon considering the orientation of the sides: when the polygon is convex with the inside not overlapping circuits, the relative rotation is not possible.

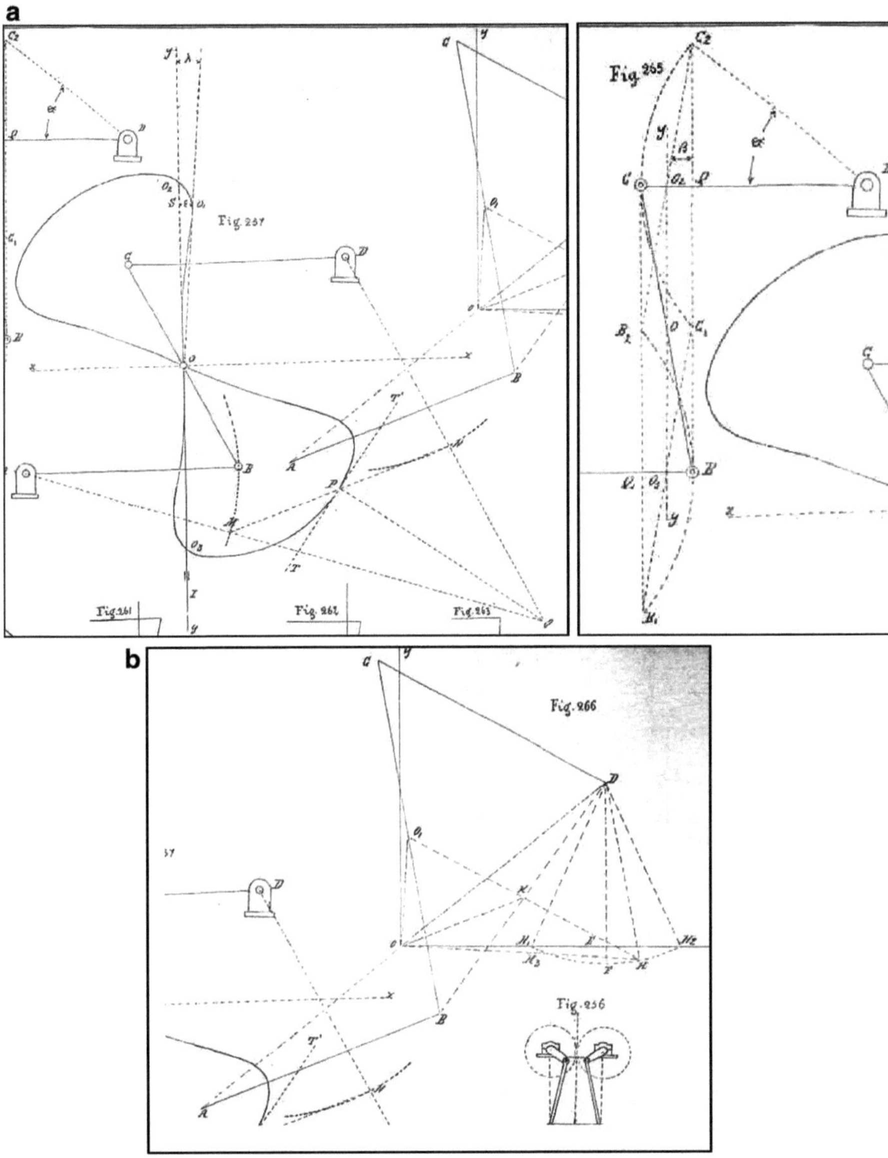

Fig. 4 Kinematic schemes for Watt parallelogram in Masi (1897a): (**a**) Figs. 257 and 265; (**b**) Fig. 266

In Fig. 7a, the case of hypoid gears is discussed by using the scheme of Figure 148 to give practical indications for a successful manufacturing and efficient use (equal normal pitches, small transmission forces, limited skew distance) and to compute the angular velocity ratio as ω_2/ω_1 after some algebraic manipulations of

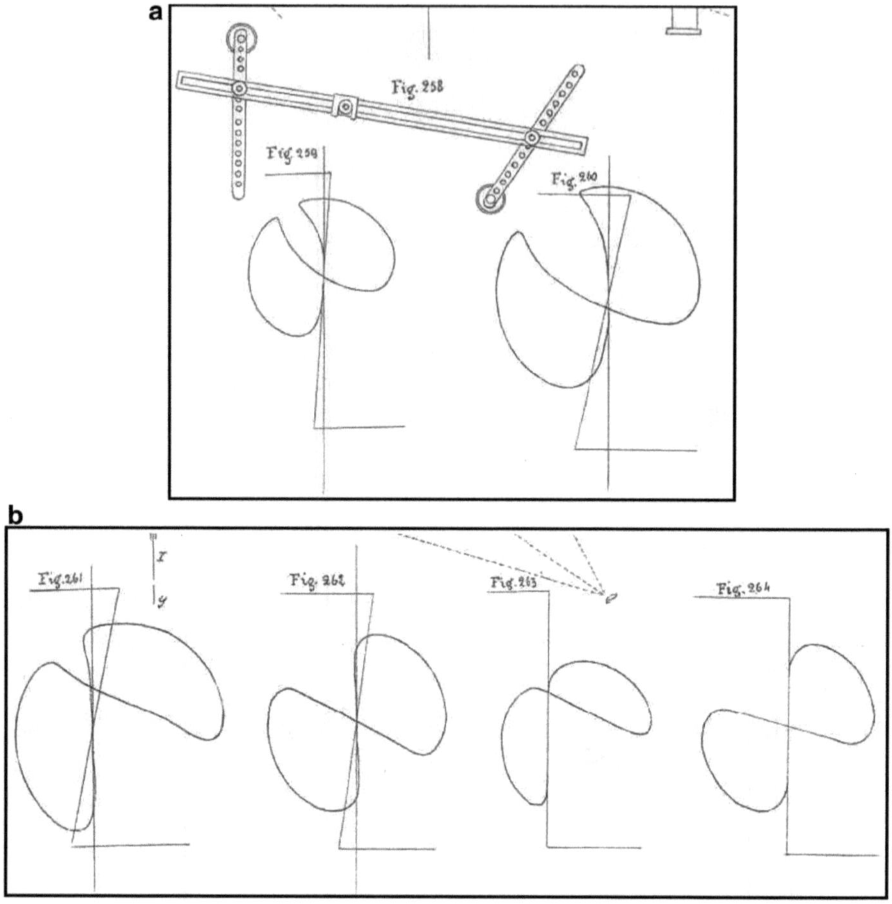

Fig. 5 A parametric study on coupler curves of the Watt parallelogram in Masi (1897a): (**a**) Figs. 258–260; (**b**) Figs. 261–264

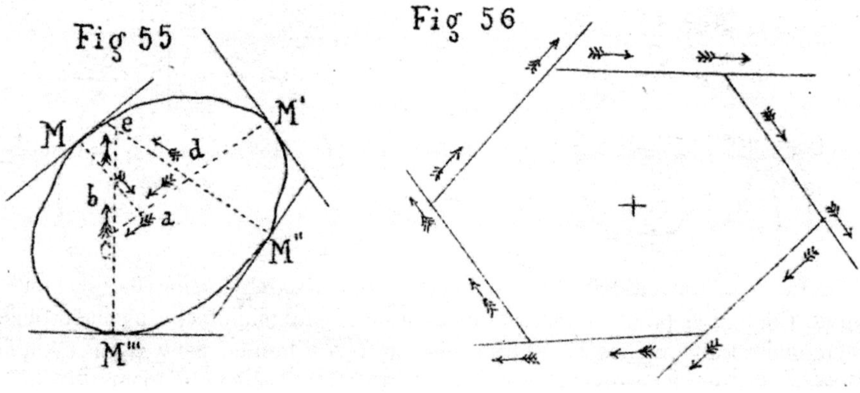

Fig. 6 Models for static analysis of multiple contacts in Masi (1897)

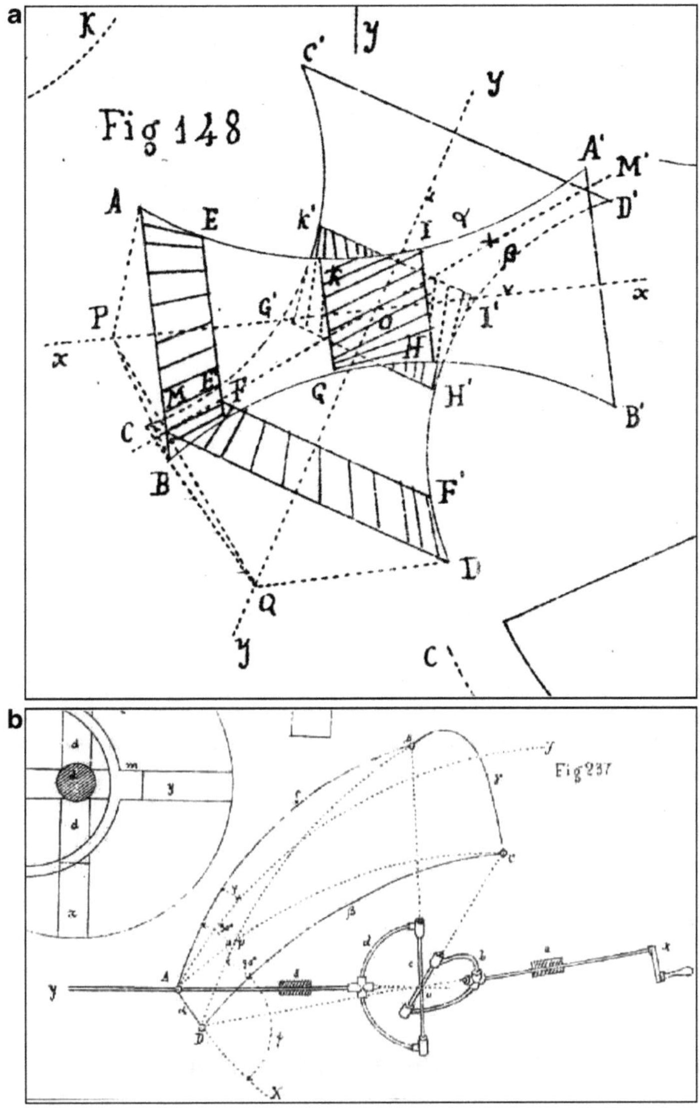

Fig. 7 Kinematic schemes in Masi (1897a) for: (**a**) hypoid gears; (**b**) Hooke's joint

the fundamental design parameters as n_1/n_2, with n_1 and n_2 as the number of teeth in gears 1 and 2, respectively.

In Fig. 7b, similarly, the Hooke joint (mainly known in Italy as the Cardan joint) is studied to characterize its kinematic operation as a function of its design parameters with formulation derivation and information for successful uses like in tower clocks, windmills, and land agriculture machinery. In particular, by referring to the scheme No. 237 in Fig. 7b, Masi deduced the well-known ration input-output in

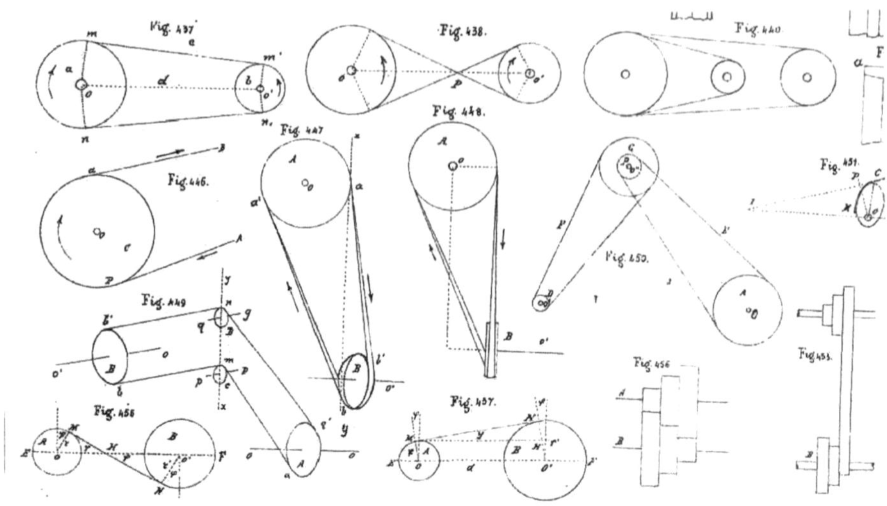

Fig. 8 Schemes and design solutions for belt transmissions in Masi (1897a)

term of cos $\alpha/(1-\sin^2\alpha \sin^2\varphi)$ in a straightforward way by using the time derivative of geometrical expressions for the angle α among points A, O, and D and angle φ among points B, A, and Y in the structure and operation of the mechanism.

In Fig. 8, there are schemes that Masi used to describe the functionality and variety of belt systems for power transmissions. Figure 8 is an example summarizing the approach that Masi used to treat mechanisms giving theory (even trough drawings), computations, design details, and information of applications for each case. The large variety of studied mechanisms can be appreciated by the 515 schemes that he drew in his book (Masi 1897a).

It is worth noting that most of the treatments of reported mechanisms can be still considered today to be of both of theoretical/computational validity and practical interest.

Modern Interpretation of the Main Contributions

In the field of TMM, Francesco Masi made relevant contributions in a systematic approach for teaching TMM, as an historical evolution of the work by C. Giulio (Giulio 1846) in Italy and as a modern organization by following the works of R. Willis (Willis 1841) and F. Reuleaux (Reuleaux 1875). In addition, he has contributed to the scientific developments of topics regarding mechanism classification and mechanism design. The brilliant organization of the teaching on mechanisms and related theories can be appreciated in the organization of his textbooks, as outlined in previous section.

Relevant is the enhancement that Masi elaborated in Masi (1883) for the mechanism classification by using Reuleaux's approach and formalism in Reuleaux

(1875). Masi revised and completed the mechanism notation by expressing the following formalism: type symbols for general design characteristics of bodies, like for example, C is for cylinder, P is for prismatic box, R_d is for geared wheel; shape symbols for the specific geometry of the body, like for example, C^+ is for bulk cylinder (pivot), C^- is for hole cylinder (bearing), P^+ is for bulk prismatic box (slider), P^- is for prismatic guide; relationship symbols for describing the relations between two elements of a kinematic chain, like for example, || is for parallel axes, x is for crossing lines, < is for inclined lines. Therefore, any mechanism can be represented by a formula, when additional signs are used for fixed bodies (segment with dashed lines), elastic bodies (wave segment), and moving bodies (straight segment) like, for example, the cases shown in Fig. 9.

a

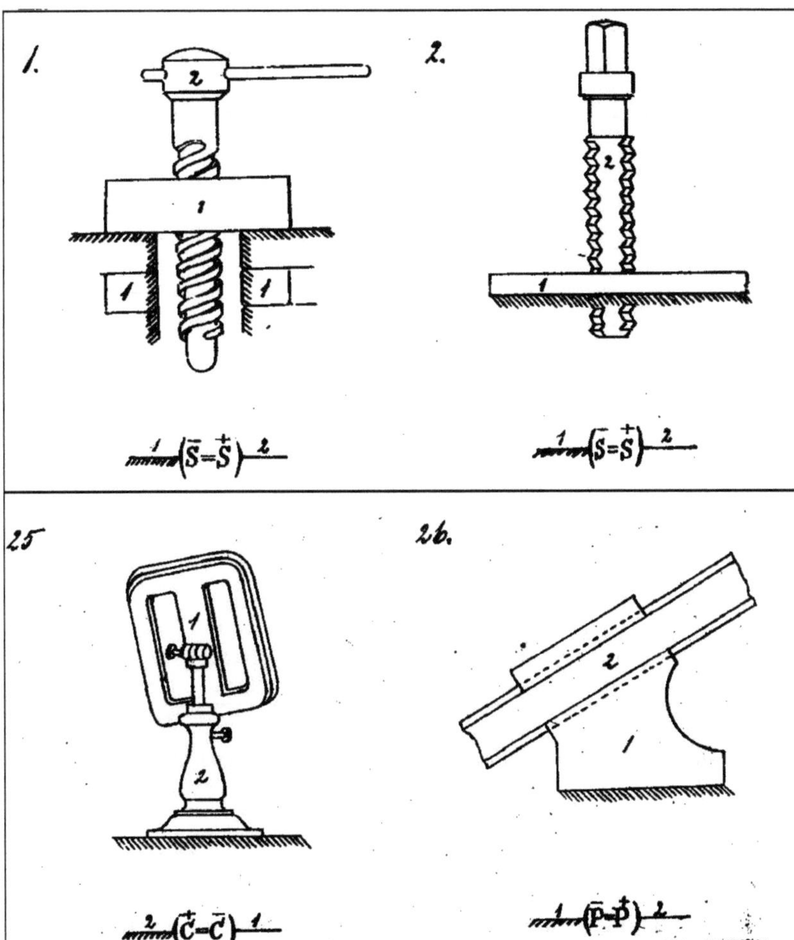

Fig. 9 Examples of notation and formalism for analytical representation of mechanisms in Masi (1883): (**a**) basic pairs; (**b**) cam mechanisms

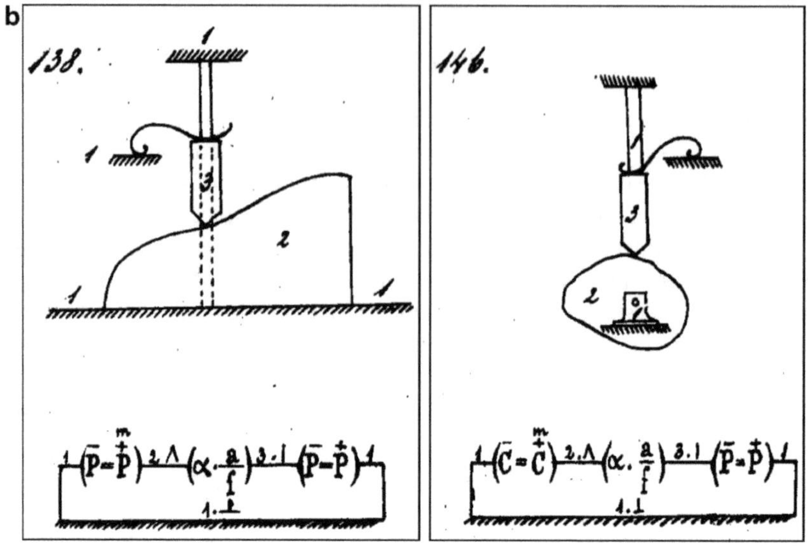

Fig. 9 (continued)

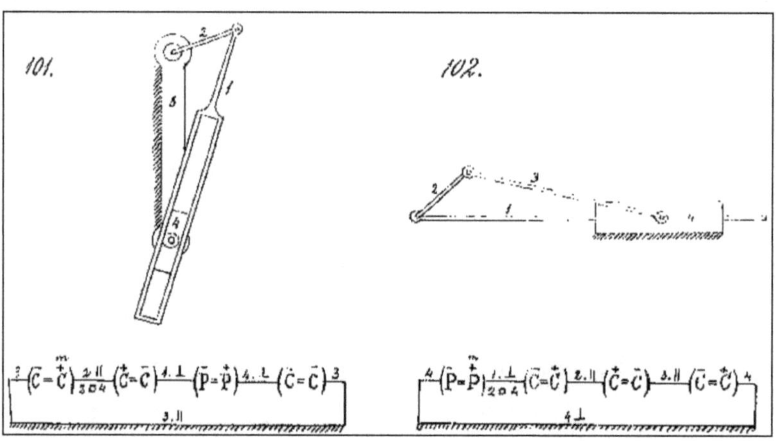

Fig. 10 Notations and schemes of mechanisms from the slider-crank chain in Masi (1883)

The notation was an attempt to develop a more analytical representation of mechanisms, avoiding encumbering mechanism drawings, even with the aim of using it for analytical treatment of the basic kinematic properties. But, as we know, it was unsuccessful because of the complexity both in writing and understanding, as the examples in Fig. 10 illustrate for the case of slider-clank mechanisms and inversions.

Some practical interest of this notation can be recognized in the definition of mechanisms by means of kinematic inversion, like the example in Fig. 10 illustrates when one considers permutation of the order of the mechanism components in choosing the frame link for the mechanism. The notation is attractive because it also can be very synthetic in showing the variety of possibilities of a kinematic chain like in the example of Fig. 11.

In addition, the formalism of the proposed notation can be considered as part of the classification rules, when one considers that the characters/rules of the classification can be expressed as functions of the introduced symbols for the fundamental kinematic properties of kinematic design of mechanisms.

Following Reuleaux's work, Masi proposed his view and formulation of mechanism classification. His view also takes into account the previous attempts of an exhaustive classification of existing mechanisms that were elaborated during the nineteenth century (Ceccarelli 2004). The basic rules of Masi's classification of mechanisms can be recognized in the concepts of simple and compound mechanisms, the kinematic chain and its inversion, categories and classes of mechanisms (Masi 1883).

Simple mechanisms are considered to be composed of one circuit of linked bodies only; compound mechanisms are those with more circuits of links with a different nature. The concept of a kinematic chain, which is the kinematic structure of mechanisms when any link is considered as a fixed frame, is useful to consider the kinematic architecture of mechanisms as mechanism topology. The kinematic inversion consists in considering any link of the mechanism as a fixed frame for different specific mechanisms as introduced by Reuleaux (1875).

A category of simple mechanisms is identified by the number of pairs of different types that are in the kinematic chain. Thus, the first category for so-called homogenous mechanisms is made of all the pairs with components of the same nature, even repeated in the chain like, for example, revolute joints or prismatic

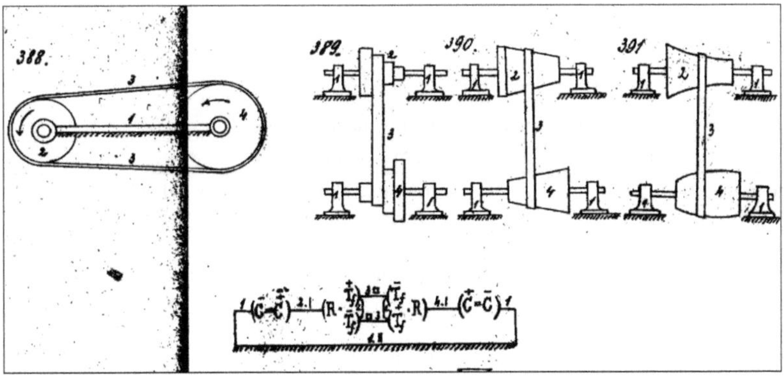

Fig. 11 Notation and schemes for a variety of belt transmissions as represented by one formula in Masi (1883)

joints, and so on. The second category for binary mechanisms is composed of two pairs of different natures, even repeated in the chain like, for example, revolute joint with prismatic guide or revolute joint with screw, and so on. Although there were a wide range of theoretical possibilities, Masi identified eight categories only, those of practical engineering interest that are made of articulation joint ($C^+ = C^-$ or $P^+ = P^-$), point with a line ($\alpha.\alpha$), two lines ($a.\alpha$), two screws ($S^+ = S^-$), two wheels (R.R), two gears ($R_d.R_d$), one gear with a pawl ($R_d.a$); compression and traction elements (with a complex notation!).

In each category, there are several possible mechanisms that can be grouped in classes with the same kinematic design. Thus, for the first category, the first class can be identified as grouping all the linkages that are composed of revolute joints (that Masi advised as the only practical ones in the category). The second class is composed of cams and wedges; the third class includes sockets; screws are in the fourth class; wheel connections are the fifth class; gears are the sixth class; the seventh class is composed of ratchets; and the eighth class is given by elements working by compression or traction.

Compound mechanisms can be classified similarly to simple mechanisms by categories and classes depending on the number of kinematic chains and their nature, respectively.

By using the above-mentioned concepts and rules, Masi classified the mechanisms according to Tables 1 and 2, by starting from the eight practical simple classes only. He advised that many of the counted mechanisms in the classes can

Table 1 Classification of simple mechanisms in (Masi 1883)

Category	Number of classes	Description of classes
Homogeneous	8	Linkages
Binary	28	Wedges and cams, sockets, screws, wheels, gears, ratchets, mechanism by traction and compression
Ternary	56	Gears with sockets, ratchets with gears
Quaternary	70	
5th	56	
6th	28	
7th	8	
8th	1	

Table 2 Classification of compound mechanisms in Masi (1883)

Category	Number of classes	Description of classes
Homogeneous	225	Compound linkages, screws, wheels, gears, ratchets, mechanisms by traction and compression
Binary	25,200	Gears and linkages, sockets and traction elements, ratchets and gears, ratchets and linkages, linkages and compression systems
Ternary	1,873,200	Same as binary
Quaternary	10,362,600	Same as binary
205th	… 1	Same as binary

be determined only theoretically (with the help of the notation!) and therefore they should be invented. Likewise, extending the consideration to more simple categories with other basic rules/structures, any other mechanisms can be invented but practical applications should be thought to be specifically part of the invention.

Thus, he applied his theory for mechanism classification by giving a unifying view of mechanisms that are also the basis of the systematic organization for teaching mechanism design. He also gave a computation of the design possibility, as reported in Table 2, in which he calculated up to 25,200 binary mechanisms and 10,362,600 quaternary mechanisms.

Regarding mechanism design, Masi clearly expressed the basic characteristics through kinematic schemes throughout his textbooks (Masi 1883, 1897a), with the aim of facilitating understanding the kinematics and the potentiality of mechanisms in practical applications.

Those comprehensive views, whose examples are illustrated in Figs. 4–8, were achieved by Masi through his encyclopedic knowledge of mechanisms that he elaborated, thanks also to his classification procedure.

Circulation of Masi's Work

Although Masi's works were very relevant and successful during his time, they were rarely considered after the Second World War. This is due mainly to two facts, namely the graphical approach and orientation of Italian users, by whom the works were written. The graphical approach was developed and greatly developed in the second half of the nineteenth century and Masi strongly used it when completed, together with some analytical formulation. But the real limitation for the circulation of Masi's work can be considered in Italian writings and orientation to the teaching in the Italian schools of engineering. This feature rapidly circulated Masi's work in Italy, mainly at the time of his career in the nineteenth century. The main orientation to Italian users can be understood when one considers that the Italian nation was only unified in 1860 and, since then, great efforts have been made to produce a common national frame for university formation.

We can consider three main topics in which Masi made relevant contributions that circulated successfully the memory of which is persistent in Italian schools of engineering, namely technical drawing for machinery, friction and early tribology, TMM and mechanism design.

Technical drawing for machinery was of great interest for technical colleges and professional activity for many years. Masi's work was considered relevant for University formation. Today, this work is addresse mainly for historical investigations on the developments of drawing technique and representations of machine elements, as indicated in Ceccarelli and Cigola (2007).

Masi's works on friction can be considered to be a significant early development of modern tribology and because of that character, they were much considered at the beginning of the Italian community on tribology. Masi's works on tribology

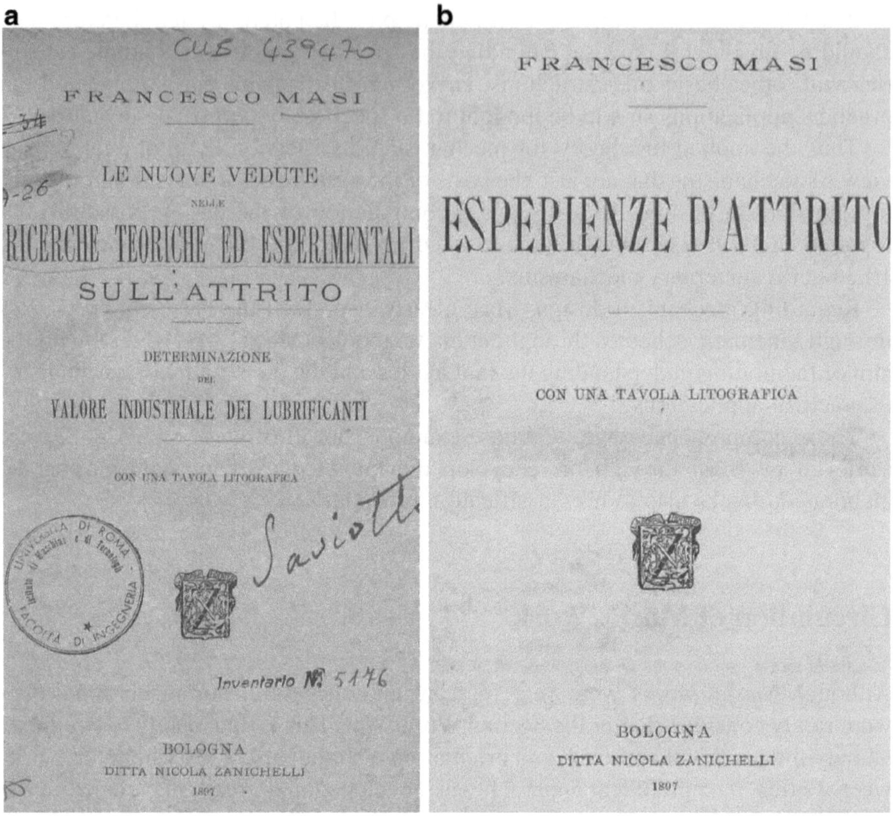
a

FRANCESCO MASI

LE NUOVE VEDUTE
NELLE
RICERCHE TEORICHE ED ESPERIMENTALI
SULL'ATTRITO

DETERMINAZIONE
DEL
VALORE INDUSTRIALE DEI LUBRIFICANTI

CON UNA TAVOLA LITOGRAFICA

BOLOGNA
DITTA NICOLA ZANICHELLI
1897

b

FRANCESCO MASI

ESPERIENZE D'ATTRITO

CON UNA TAVOLA LITOGRAFICA

BOLOGNA
DITTA NICOLA ZANICHELLI
1897

Fig. 12 Cover pages of works by Francesco Masi: (**a**) new perspectives on theoretical and experimental researches on friction in 1897b; (**b**) experiences on friction in 1897c

(Masi 1897b, c) are No.15 and 16 in the list (Masi 1927), Fig. 12, can be remembered mainly as reports of investigations.

His treatise (Masi 1897b), Fig. 12a, is a clear discussion of the state-of-arts in the field of friction evaluation that is oriented to machinery applications referring to lubrication properties. The most important theories of the time (by Petroff, Hirn, Thurstin, Kirchweger, Poiseuille) are also overviewed by referring to Italian experiences in order to propose further developments that Masi applied to practical cases for evaluation and choice of lubricant oils. An example of those results is shown in Fig. 13 also indicating the details and modernity of Masis' studies.

The experimental approach of the investigation led Masi to conceive a suitable new procedure and equipment, like the test-bed in Fig. 14 that he used to develop further knowledge of the topic of friction and lubrication, and to obtain useful results for practical engineering with determination of friction and lubrication properties as a function of environment and operation variables. Both works (Masi 1897b, c) were used even as practical handbooks on industrial engineering.

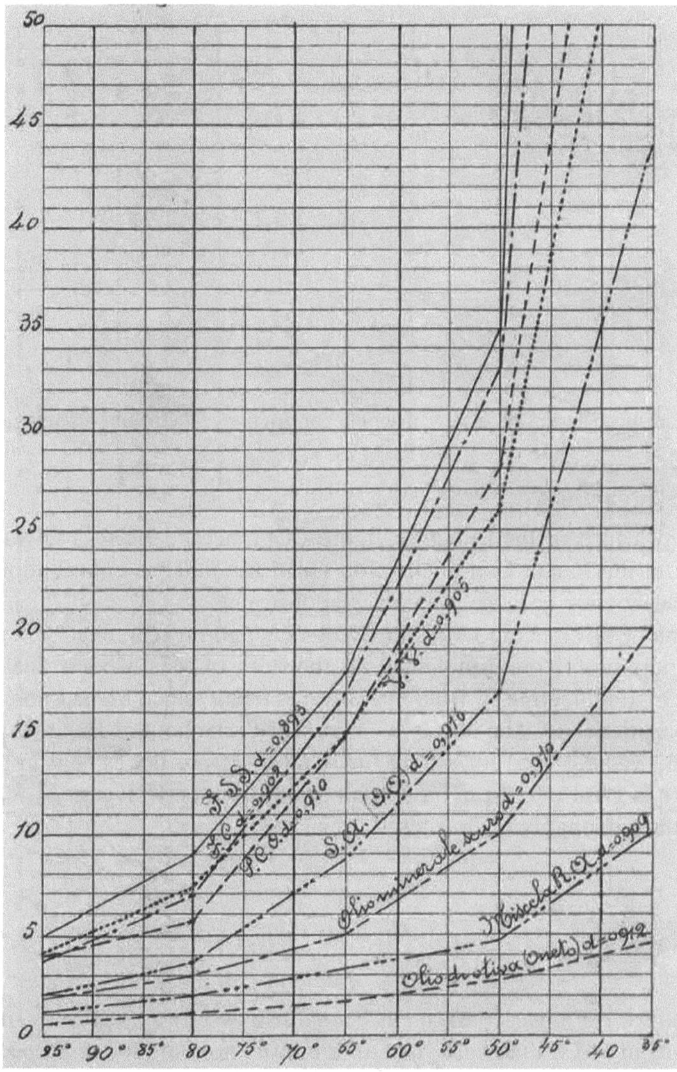

Fig. 13 Experimental results for viscosity characteristics of several lubricant oils in Masi (1897b)

However, the most successful works by Masi, also in terms of circulation, can be considered to be the books Masi (1883, 1897a) on TMM and Mechanism Design. They were used for long periods and even when they were successed by new textbooks, they were always considered as a sources of inspiration both for technical contents and conceptual explanations of mechanisms and their kinematic proprieties.

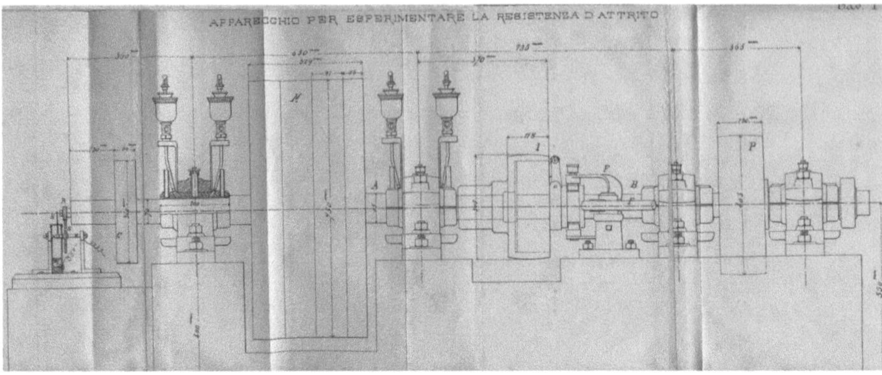

Fig. 14 Mechanical design of a new machine for friction experiments as reported in Masi (1897c)

The significance of the scientific figure of Francesco Masi is nowadays well recognized in the Italian Community for TMM but still his contributions are not know abroad, mainly because of language constraints.

The memory of Francesco Masi in Bologna is still present, even at level of current University professors. Emblematic is the story of Masi's house that he left to the School of Engineering in Bologna after his death, since he had no children to leave his proprieties to. The house, located near the School of Engineering with a nice garden, was not used for a long time and therefore it was sold in the 1950s. Today, there is great interest in having it back within the University's fold, also to recognize the personality of Francesco Masi.

Conclusions

The personality of Francesco Masi has been illustrated by emphasizing his relevant contributions in TMM that were and still are influent in the Italian Academy. Masi always developed his research activity with a vision to teaching activity and by looking at practical engineering applications, even when he encountered problems with very theoretical formulations. His work has been an inspiration for many generations in Italy and is still considered as an interesting source. Unfortunately, Masi's work is not known abroad and this paper is an attempt to refresh the memory and to point out modernity of his publications in TMM.

Acknowledgements The author wishes to thank the library of the Department of Mechanics and Aeronautics of "La Sapienza" University of Rome (Italy), the library of the Department of Mathematics of Politecnico di Milano (Italy), the library of Politecnico di Torino (Italy), the University library of Bologna (Italy), the University library of Padua (Italy), the University library of Pisa (Italy), the University library "Boaga" of Roma (Italy), the University library of Torino (Italy), the Apostolic Library in Vatican, and Library of Montecassino Abbey for help in the search for original material.

References

Allievi L (1895) Cinematica della biella piana, Regia Tipografia Francesco Giannini & Figli, Napoli
Anastasi A (February 1908) Studio di alcuni meccanismi per limatrici, Annali della Società degli Ingegneri e degli Architetti Italiani, n.3
Ancona U (1896) Teoria generale delle macchine, Milano
Battaglini G (1873) Trattato elementare sulla meccanica razionale, B. Pellegrino, Napoli
Battaglini G (1870) Sul movimento geometrico infinitesimo di un sistema rigido, Rendiconto dell'Accademia delle Scienze, Napoli, Vol. IX, pp 89–100
Bellavitis G (1852) Applicazione della Cinematica alla curvatura di tutte le traiettorie descritte dai punti di un sistema piano invariabile, Estratto dal Vol. IV, serie III, Regio Istituto veneto di scienze, lettere ed arti
Boscovich R (1763) Theoria philosophiae naturalis, Venetiae
Branca G (1629) Le machine, Roma
Cappa S (1890) Meccanica applicata alle macchine. Tipografia Giorgis, Torino
Cavalli E (1882) Elementi di cinematica teorica – ad uso delle scuole di applicazione per gli ingegneri. Hoepli, Milano
Cavalli E (1889) Lezioni di Meccanica applicata alle macchine. Regia Scuola d'applicazione per gli Ingegneri, Roma
Cavalli E (1908) Elementi di Meccanica applicata alle macchine. Tipografia Trani, Napoli
Ceccarelli M (1998) Mechanism schemes in teaching: a historical overview. ASME J Mech Des 120:533–541
Ceccarelli M (2000) Italian kinematic studies in XIXth century. Proceedings of IFToMM International Symposium on History of Machines and Mechanisms HMM2000. Kluwer, Dordrecht, pp 197–206
Ceccarelli M (2004) Classifications of mechanism over time. Proceedings of IFToMM International Symposium on History of Machines and Mechanisms HMM2004. Kluwer, Dordrecht, pp 285–302
Ceccarelli M, Cigola M (2007) Gaspard Monge e la sua influenza in Italia tra Geometria Descrittiva e Teoria dei Meccanismi, L'Ingegneria e la sua storia, Di Leo (Editor), Marlin, Salerno, pp 108–117
Ceccarelli M, Koetsier T (2008) Burmester and Allievi: A Theory and Its Application for Mechanism Design at the End of 19th Century, Journal of Mechanical Design, Vol 130, July 2008, pp. 072301-1:16, http://link.aip.org/link/?JMD/130/072301, 2009, (DOI: 10.1115/1.2918911)
Cerruti U (1898) Cinematica (Dispense litografate1897–98), Università di Roma, Roma
Chelini D (1862) Dei moti geometrici e loro leggi nello spostamento di una figura di forma invariabile. Tipografia Gamberini e Parmeggiani, Bologna
Del Monte G (1577) Mechanicorum liber, Pesaro
Frisi P (1765) Cosmographiae Physicae et Mathematicae, Ex Tipografia Marelli, Mediolanum
Frisi P (1777) Instituzioni di Meccanica, d'Idrostatica, d'Idrometria e dell'Architettura Statica, e Idraulica, Galeazzi Regio Stampatore, Milano
Galilei G (1600) Le meccaniche, 1600, in: "Opere di Galileo", edited by A. Favaro, Firenze, 1964–66.
Giorgini G (1836) Intorno alle proprietà geometriche dei movimenti di un sistema di punti di forma invariabile, Memorie di Matematica e Fisica della Società Italiana delle Scienze, Tipografia Camerale, Modena, Tomo XXI, pp 1–54
Giulio CI (1846) Sunti delle Lezioni di Meccanica applicata alle arti. Tipografia Pomba, Torino
Grandi G (1739) Instituzioni Meccaniche, Firenze
Maggi GA (1896) Principi della teoria matematica del movimento dei corpi. Bernardini, Milano
Maggi GA (1914) Lezioni di Cinematica. Spoerri, Pisa
Masi F (1883) Manuale di cinematica applicata: Nuova classificazione dei meccanismi. Zanichelli, Bologna
Masi F (1891) Corso di Disegno: Organi delle Macchine, Vallardi Editore

Masi F (1897a) La Teoria dei Meccanismi. Zanichelli, Bologna
Masi F (1897b) New perspectives on theoretical and experimental researches on friction. Zanichelli, Bologna
Masi F (1897c) Experiences on friction. Zanichelli, Bologna
Masi F (1927) Scientific and academic curriculum, Bologna (in Italian)
Ovazza E (1890) Il Poligono funicolare in Cinematica. Clausen Libraio, Torino
Padelletti D (1884) Sopra un'estensione del concetto di polo e caratteristica in Cinematica, Rendiconto della Regia Accademia delle Scienze di Napoli, Fascicolo 4° e 5°, maggio, pp 1–2
Padova E (1884) Sulla teoria dei moti relativi, Annali di Matematica pura ed applicata, Serie II, Tomo XII, Milano, pp 265–282
Panetti M (1913) Corso di Meccanica applicata alle macchine. Gnocchi Tipografia, Torino
Pugno GM (1959) Storia del Politecnico di Torino, Torino
Reuleaux F (1875) Theoretische kinematic, Braunschweig
Saviotti C (1890) Teoria dei meccanismi, Roma
Tessari D (1890) La cinematica applicata alle macchine. Loescher, Torino
Torricelli E (1919) in: Opere Torricelli E, Loria G and Vassura G (eds), Stamperia Montanari, Faenza.
Willis R (1841) Principle of mechanisms. Longmans Green & Co, Londonp

Adam Morecki (1929–2001)

Teresa Zielinska and Krzysztof Kedzior

Abstract Adam Morecki contributed significantly to the advancements in several fields of mechanical science. He worked on the classification of mechanisms, developed the model of skeletal muscles co-operation, stimulated great progress in modeling and design of elastic manipulators and walking machines. The paper presents Morecki's life, the history of his fundamental works, and the facts proving his contribution towards the development of mechanics.

Biographical Notes

Professor Adam Morecki was born on 5 September 1929 in Kraków (Cracow), Poland. His youth was associated with the terrifying period of World War II. The experience of that time affected the whole of his life. He lived very intensely, exploiting every minute, as if he was living not only for himself but also for those who were not so lucky and did not manage to survive (Fig. 1).

He obtained an M.Sc. degree in 1951 from the Faculty of Electronics at the Academy of Mining and Metallurgy (now: AGH University of Science and Technology) in Cracow. In 1949, he started to work as an assistant in the Chair of Mechanics and Strength of Materials at the Academy. In 1952, he was selected from many candidates for studying at the Institute of Mining in Moscow, where in 1955 he received his Ph.D. degree for the dissertation on "*Investigation into safe braking process and rational selection of the brake type*".

After returning to Poland, he was appointed by the Polish Ministry of Higher Education to work at Warsaw University of Technology (WUT). Together with Professor Jan Oderfeld, he was assigned the important task of founding the chair of the theory of machines and mechanisms.

T. Zielinska (✉) and K. Kedzior
Institute of Aeronautics and Applied Mechanics,
Warsaw University of Technology, 00-665, Warsaw, ul.Nowowiejska 24
e-mail: teresaz@meil.pw.edu.pl

M. Ceccarelli (ed.), *Distinguished Figures in Mechanism and Machine Science: Their Contributions and Legacies, Part 2*, History of Mechanism and Machine Science 7,
DOI 10.1007/978-90-481-2346-9_9, © Springer Science+Business Media B.V. 2010

Fig. 1 Portrait of Adam Morecki

He was motivated all his life by a sense of that task, and his academic career developed as follows:

1955–1958	— Assistant professor
1958–1968	— Associate professor
1968–1974	— Professor and
1974–2001	— Full professor at the Faculty of Power and Aeronautical Engineering, WUT

In September 1969, Adam Morecki organized the second World Congress on the Theory of Machines and Mechanisms. During that event, held in Zakopane, Poland, the International Federation for the Theory of Machines and Mechanisms (IFToMM) was founded. Professor Morecki, together with other members of the founding committee, signed the IFToMM foundation act.

He served as the head of the Dynamic Measurements Chair (1961–1970), head of the of Theory of Machines and Mechanisms Chair (1978–1984), vice-dean of the faculty (1961–1964), head of the Research Team of Robotics and Biomechanical Engineering (1964–2000). Being a very active person, he also worked as a part-time professor at the Institute of Fundamental Research Problems of the Polish Academy of Sciences in Warsaw (1955–1965) and at the Institute of Mechanics and Vibroacoustics at the Academy of Mining and Metallurgy in Cracow (1986–1997).

For many years, he was a general secretary of the IFToMM. Then, he was the chairman of IFToMM Robotics Technical Committee between 1982 and

1989. In the years 1992–1995, Professor Morecki held the position of IFToMM president. He was a member of the Executive Council of the International Federation of Robotics (1989–1991, 1996–1997) and a member of the Executive Council of the International Centre for Mechanical Sciences (CISM) from 1986.

He remained extremely active, also running domestic affairs; for example, he was the president of the Polish Society of Biomechanics (1988–1992), vice-president of the Committee for Automatic Control and Robotics of the Polish Academy of Sciences, a member of the executive board of the Committee for Machine Building (PAS), president of the Polish Committee of TMM affiliated to the Committee for Machine Building PAS, a member of the Committee for Biocybernetics at the Institute of Biocybernetics and Biomedical Engineering PAS, a member of the Academy of Engineering in Poland and a member of the Warsaw Scientific Society.

In the years 1991–1995, serving as the general secretary of IFToMM, he organized over 70 congresses and meetings of technical committees and other boards in Europe, the USA and Canada. When he was a co-chairman of the Program Committee of CISM-IFToMM, he organized 12 symposia on Robots and Manipulator Systems (the symposia were referred to as "RoManSy") in the years 1976–1998. With this initiative, Morecki contributed significantly to the promotion of biomechanics. The first RoManSy meetings were marked by verification of some novel results obtained in the study of walking robots. Active orthoses to replace the human amputated limbs were also the subject of research and investigations. During those times, it was a very novel and revoluntary research area. In 1994, he said that "RoManSy 10 opens the 3rd period of activity. Of course it is not easy to predict the future, but in my opinion the theory and practice of robots and manipulators will continue to be of great importance for future development of science and technology". He edited or co-edited the symposia proceedings issued in the years 1973–2000 (Fig. 2). From the formal point of view, those events have demonstrated a long-lasting partnership between the two international institutions; that is CISM and IFToMM. Moreover, he contributed towards the organization of six schools of biomechanics at the International Centre for Biocybernetics of the Polish Academy of Sciences.

His remarkable achievements in science, education and organization were awarded by the Academy of Mining and Metallurgy in Cracow with an Honorary Doctorate. He was also received many awards from various Polish and international organizations; for example, he received the prestigious Engelberger Award for Education in Robotics in Singapore (1995).

Morecki was one of the architects of the Polish School of Theory of Machines and Mechanisms and the founder of the Polish Research School in Biomechanics and Robotics. He supervised 24 doctoral students.

He pursued intense scientific and organizational activities until the last moments of his life. He died on 20 May 2001.

Fig. 2 Volumes of the proceedings of robots and manipulator systems symposia

Course of the Research and Fundamental Monographs

In 1962, together with his colleague Kazimierz Fidelus, Morecki created the research team and launched the investigations into measurements in mechanics, biomechanics and bio-cybernetics, taking into account both the theory and practice. The group was involved in theoretical research as well as in the development of the devices helping them in experiments (Fig. 3).

The first studies were focused on the application of electromyography (EMG) to control a human arm-like manipulator with seven degrees of freedom driven by artificial pneumatic muscles. The first papers on the subject were published as early as in 1964. Also in 1964, the regular "Friday Seminars" were launched. Those meetings soon became well known and attracted many scientists and students from other centers.

In 1971, a monograph presenting the mechanical principles of the biological system's locomotion was published (Morecki et al. 1971). The scope of the monograph covered a variety of biomechanical problems, such as structure of bio-mechanisms, analysis of bio-kinematical chain driven by muscles, problem of muscle force distribution, models of human extremity control, synthesis of bio-manipulators and

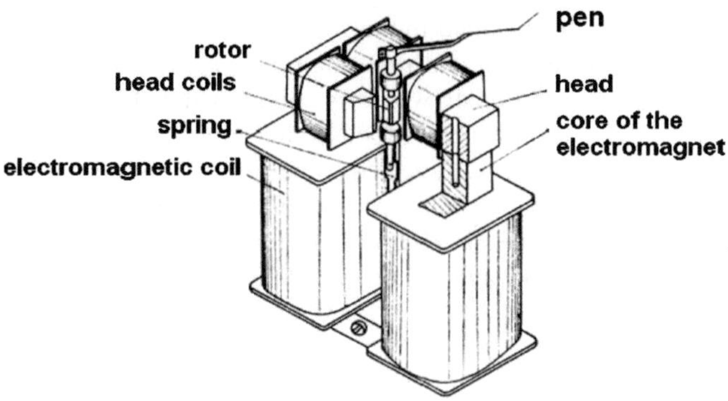

Fig. 3 Scheme of the electromagnetic registrator

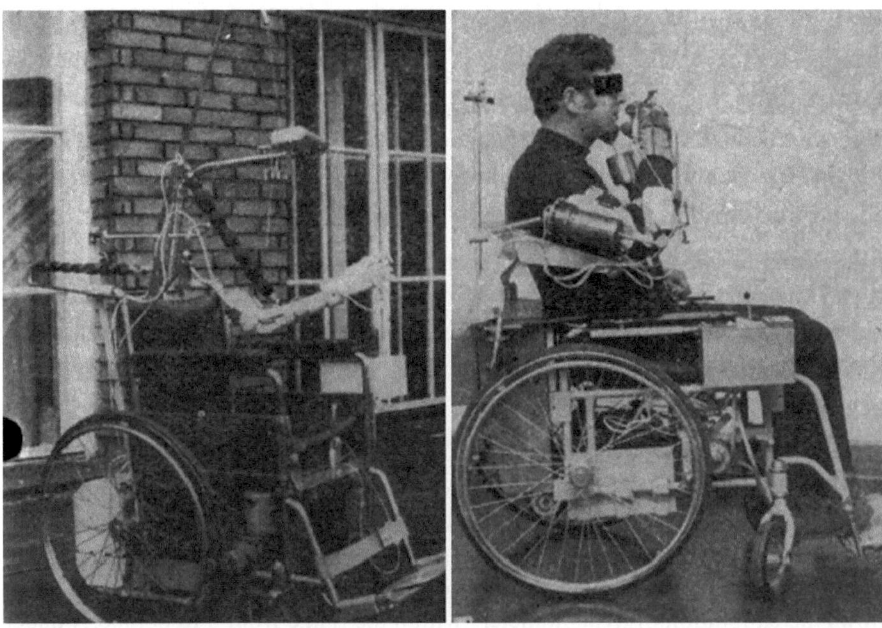

Fig. 4 Orthoses mounted on a wheelchair – 1974

bio-prostheses. The authors presented the state of the art, the results they obtained and guidelines for future research. This fundamental book attracted great interest and encouraged researchers to undertake the study in biomechanics using the interdisciplinary approach.

In the following years, the research conducted by the team tended towards applications. The implanted stimulators of human limb muscles, two types of EMG controlled orthoses (Fig. 4) and EMG controlled prosthesis with pneumatic drives were developed and clinically tested.

This experimental work resulted in the second monograph by the same authors on cybernetic models of animal and robot motion systems (Morecki et al. 1979). The book, first published in Polish, was very soon translated into English.

Since the late 1970s, Prof. Morecki focused his attention on robotics. However, the research he conducted always revealed some sort of "biomechanical flavour".

The spine-type and elephant trunk-type manipulators were designed and built by the research team he supervised. The results created a basis for the monograph published in Polish (Morecki and Knapczyk 1993) and in English (Morecki and Knapczyk 1999).

Contribution Towards the Development of Mechanics

The scientific activity of Adam Morecki covered a wide range of issues concerning applied mechanics and biomechanics. He contributed to the study of mechanism kinematics and dynamics, theory of gears, application of electrical methods to measurements of mechanical parameters, bionics of motion, biomechanical engineering and rehabilitation. He initiated the research and formulated basic concepts of the theory of anthropomorphic robots and manipulators. In view of the research subject his scientific activity can be divided into four periods.

The first period, leading to his Ph.D. degree, was devoted to classical mechanisms and the research into the methods for their improvement.

The second period started when he took the job at Warsaw University of Technology and devoted his attention to the measurements of mechanical quantities in view of both the theory and practice. For the first time in Poland, he used strain gauges in solving both the industrial and research problems. He also initiated the production of tensometric amplifiers and mercury commutators. Among his other significant achievements, one should mention here the fact that he inspired the design and supervised the team working on the construction and implementation of a special machine for dynamometer calibration. The main designer of those machines was W.Narkiewicz. The design was patented. The machines were used in Poland and Germany. They are still in use in Poland (Fig. 5). The bigger machine allows the calibration of dynamometers within the range from 0 to 60,000 N under automatic control with an accuracy of 0.01%. The smaller machine has the range up to 5,000 N. The automatic control system ensures a fast and stable change of the loading mass. For example, the change by 100 kg takes only 2 s. Over 40 years experience of the machine exploitation has proved their good quality and reliability. The first prototype was ready in 1961 and during the next 10 years, the design was gradually improved. As compared to other similar devices, the machines offer much faster and more stable changes of the load and higher accuracy. The Polish machine works in the Central Office of Measures and is used as a calibration tool for the force templates in all Polish laboratories.

The third period of Professor Morecki's scientific life began around 1962 with his fascination for biomechanics and bio-cybernetics (Fig. 6). It brought about his most significant achievements, including the pneumatics muscles (Fig. 7) and the

Fig. 5 Machines used for calibration of force patterns, on the left – a machine with the range 0–60,000 N, on the right – a machine with the range 0–5,000 N, the picture was taken in 2008

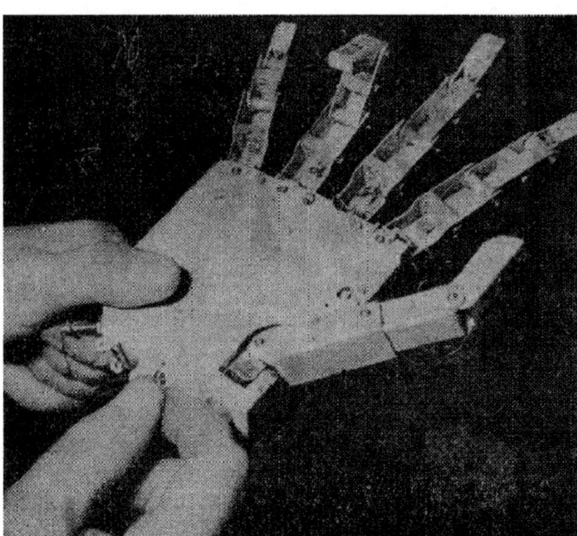

Fig. 6 Artificial hand – 1970

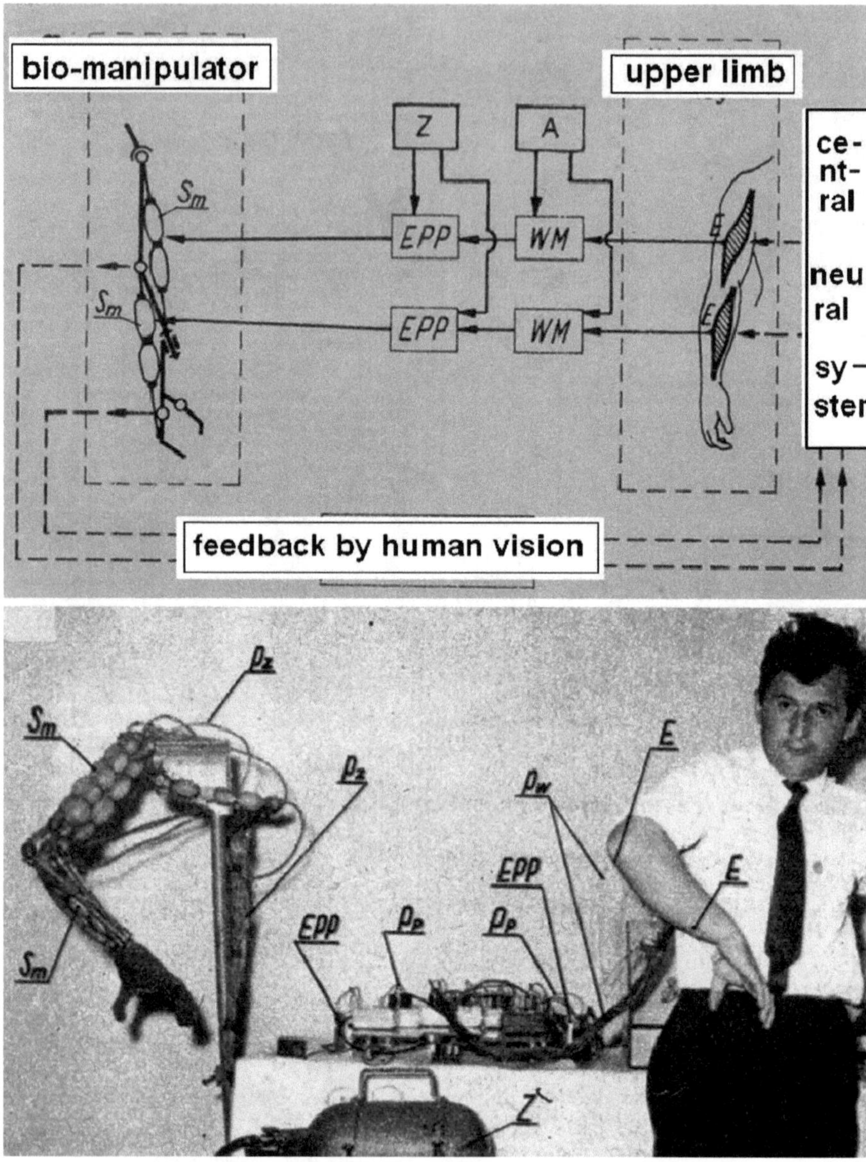

Fig. 7 The concept of pneumatic prothesis control introduced by Morecki, in which EMG signals were employed, and the experimental rig

prosthesis activated by EMG signals (Fig. 8). Roughly speaking, the period lasted until his death. It should be noted that during those years, Professor Morecki not only launched the pioneering research aimed at solving new problems but also stimulated the improvement of the devices already available and necessary for

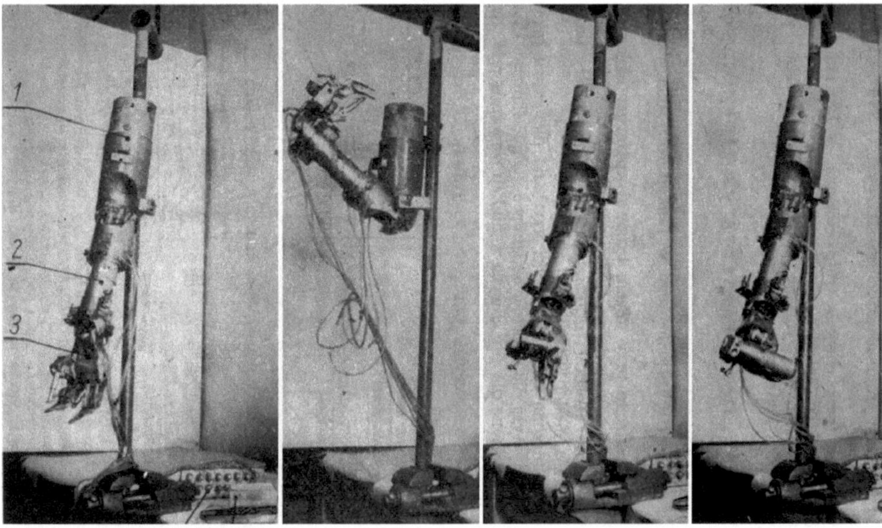

Fig. 8 Orthoses controlled by EMG signals with pneaumatic actuators – 1975

experimental purposes. One can mention here the eight-channel recorded for registering the electric signals monitored by sensors. The measurements were recorded on a paper tape as functions of time, the monitored signal range was 0–4 V with frequency up to 100 Hz, the measurement accuracy was 5%. The recorder was used by the research team supervised by Professor Morecki in the investigations into electro-stimulation.

Professor Morecki co-operated with the Academy of Physical Education (now the Józef Piłsudski University of Physical Education) in Warsaw. Within the framework of this collaboration, a variety of devices for monitoring the efforts of sportsmen were created by the team supervised by him; for example, dynamometric platforms measuring the ground reaction forces during walking, strain gauge insoles and devices measuring the linear and angular moments of a ball.

During the fourth part of his scientific activity, he concentrated his research on nature-inspired robots; that is, elastic manipulators and walking machines (Morecki et al. 2002). Many ideas of Adam Morecki were embodied in working prototypes. For example, it is worthwhile to mention here the elastic manipulator, the design properties of which were inspired by the elephant trunk, and the other one which referred to the structure of the spinal cord (Fig. 9). To develop a manipulator of the elephant trunk type, the animal trunk was studied beforehand together with its muscles. The ranges and positions during a particular manipulation were recorded and analyzed. The real muscle structure of the elephant trunk was studied to design manipulator driving systems.

Professor Morecki had a great gift for creative transformation of biological patterns into advanced mechanical devices. Before starting the design, an in-depth

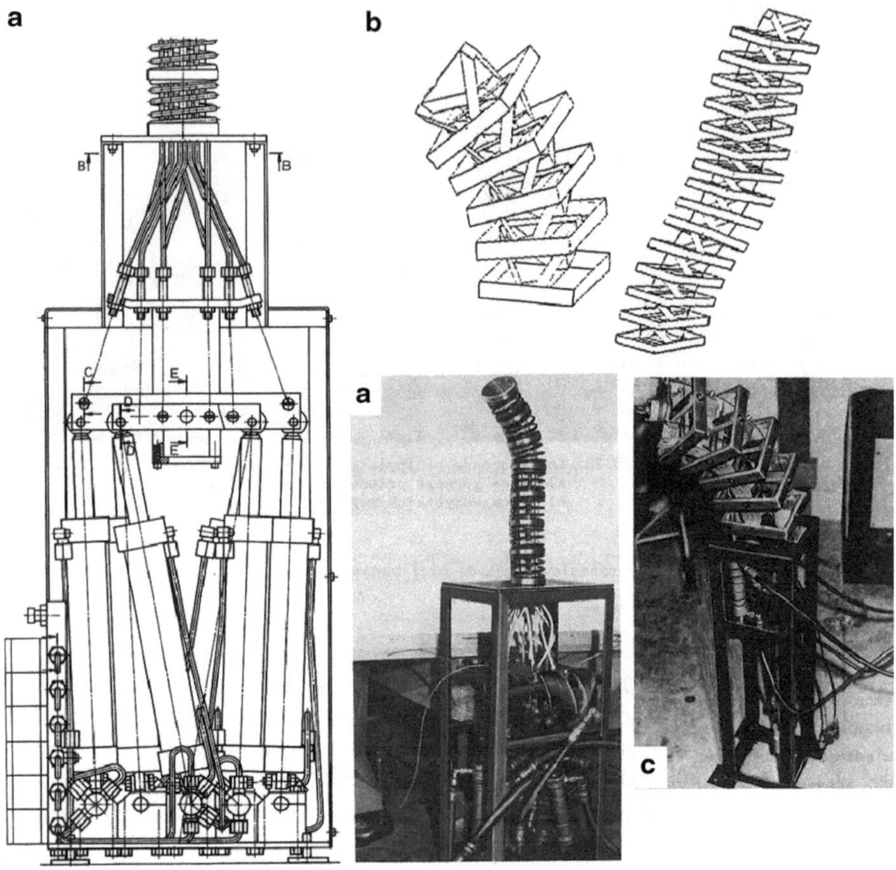

Fig. 9 Elastic manipulators, their structures and prototypes (**a**) elephant trunk type, (**b**) spinal cord type

study and brainstorming were conducted, having in mind the performances of biological systems (Fig. 10). That profited in the development of several unique prototypes.

Throughout the whole of his scientific activity, he was convinced that scientists dealing with mechanics should look for inspiration in nature (Morecki et al. 1988). That idea, among many other achievements, brought about the design of a four-legged walking machine similar to animal quadrupeds, produced by the research team in the years 1973–1976. The leg structures of a four-legged walking machine were similar to those of horse legs. A. Morecki showed that carefully investigated and suitably generalized properties of different types of motion observed in nature, can be successfully applied to motion synthesis of some mechanical systems; such as, anthropomorphic manipulators and walking machines (Morecki and Zielinska 1992, 1993).

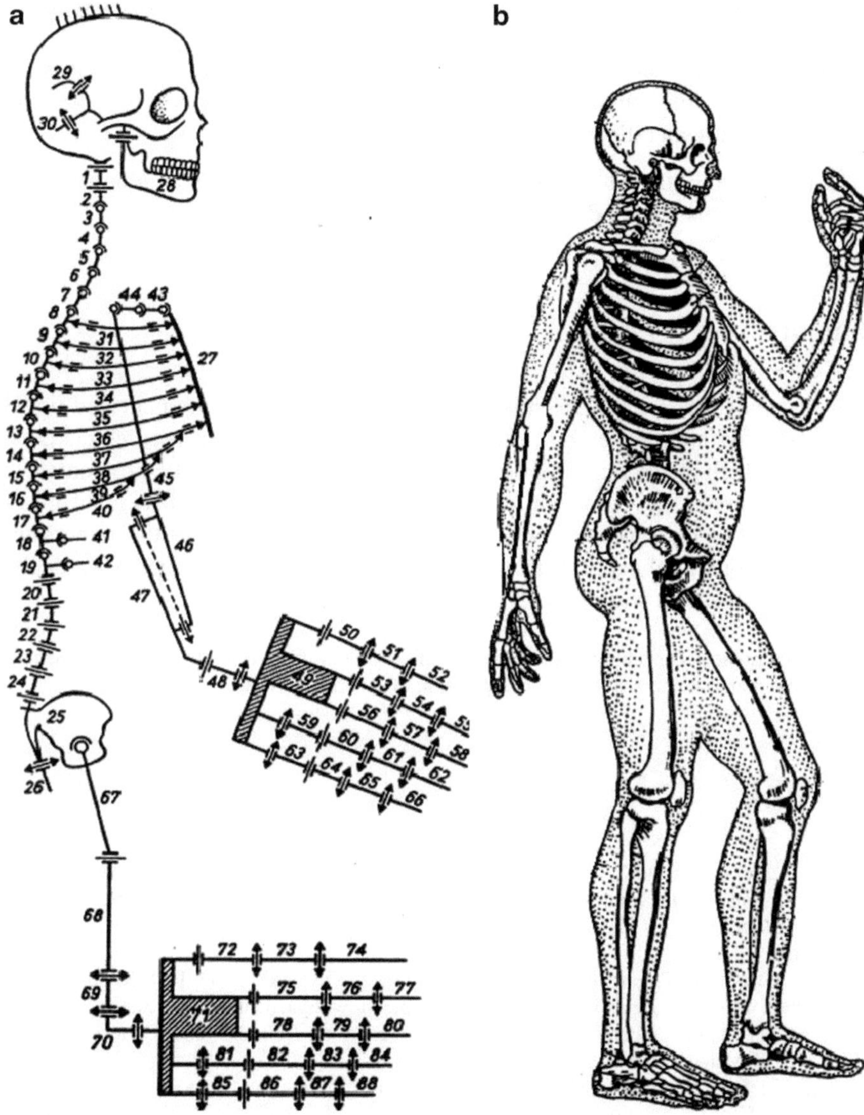

Fig. 10 Human body: (**a**) kinemarical structure proposed by Morecki, (**b**) view of the skeleton

Below is a list of the successfully accomplished tasks that Professor Morecki considered to be his most important scientific achievements:

- Devising a new method for mechanism classification
- Development of a model representing force distributions in skeletal muscles

- Putting forward a concept of stimulator activation schemes for human extremity control
- Putting forward some modeling and design methods for elastic manipulators
- Investigations into human and walking machine locomotion properties that brought about general models of locomotory behavior

The results he obtained were published in 324 papers and 38 monographs. He obtained 17 patents for his designs.

Summary

Adam Morecki was a great scientist in view of both the commitment and novel problem-solving approaches he fostered. He had great intuition besides his deep knowledge which resulted in finding solutions to important scientific problems. He stimulated scientific development in his students and other researchers concentrating their attention on issues crucial for progress in the field. Both of the authors are his followers and highly appreciate his support and humanistic attitude towards the people he co-operated with.

Acknowledgements The authors express their thanks to Dr. Cezary Rzymkowski for his help in collecting the material necessary for the preparation of this manuscript.

References

Morecki A, Ekiel J, Fidelus K (1971) Bionics of motion. Fundamentals of external control of biomechanism and human limbs. Polish Scientific Editor – PWN, Warsaw (in Polish)

Morecki A, Ekiel J, Fidelus K (1979) Cybernetics motion systems of animals and robots. Warsaw (in Polish)

Morecki A, Jaworek K, Pogorzelski W, Zielinska T, Fraczek J, Malczyk G (1988) Robots system – elephant trunk type elastic manipulator combined with a quadruped walking machine. Proceedings of the Second International Conference of Robotics and Factories of the Future'87. Springer, pp 649–656

Morecki A, Zielinska T (1992) Indirect control method of the motion of walking machine legs. 8th CISM-IFToMM Symposium on Theory and Practice of Robots and Manipulators Systems. Warsaw University of Technology Publication, Warsaw, pp 330–339

Morecki A, Zielinska T (1993) Quadruped walking machine: creation of the model of motion. Robots and biological systems. Towards a New Bionics NATO ASI Series F, vol. 102, pp 206–222

Morecki A, Knapczyk J (eds) (1993) Basics of robotics. WNT, Warsaw (in Polish)

Morecki A, Knapczyk J (eds) (1999) Basics of robotics. CISM Courses and Lecture Notes, Springer (in English)

Morecki A, Knapczyk J, Kedzior K (2002) Theory of mechanisms and manipulators. Fundamentals and application examples. WNT, Warsaw (in Polish)

Feodor Orlov 1843–1892

Alexander Golovin, Valentin Tarabarin, and Zinaida Tarabarina

Abstract Professor Feodor Orlov was one of the founders of the Moscow School of Applied Mechanics. His activities from the year 1872 until his death in 1892 were connected with the two biggest academies: the Moscow University and the Imperial Moscow Technical School (IMTS). He had graduated at the Moscow University and then worked there as an invited professor at the chair of applied mechanics. The chair of Applied Mechanics of IMTS (now TMM) has existed as an independent educational department since 1872, in other words, since the time when Orlov began to teach his course. In 1868, the Imperial Moscow Technical Secondary School (IMTS) was by law confirmed as the Supreme Educational School. It had the following divisions: Mechanical, Chemical and Mechanical-Building. The academy status required an important rise in the scientific level of all courses. Orlov's main course was that of applied mechanics, which he, according to the Great Russian mathematician and mechanician Professor N. Zhukovsky, was the best in Russia. Moreover, Orlov prepared and read courses about thermodynamics, hydraulics, steam engines and the resistance of materials. The creating of the mechanic collection in the Moscow University and the IMTS was Orlov's greatest service. In the IMTS (today it is called the Bauman Moscow State Technical University), the main part of this collection, gathered by Orlov, is housed. Orlov dreamt about going in for abstract mathematics. But his financial position made him spend most of his time working with applied disciplines. However, he had written a number of unique mathematical articles. He was one of the founders of the Polytechnic Society and its vice-president.

A. Golovin (✉), V. Tarabarin, and Z. Tarabarina
Department of Theory of Mechanisms and Machines, Bauman Moscow State Technical University, 2nd Baumanskja Str. 5, 105 005, Moscow, Russia
e-mail: aalgol@mail.ru waltar44@mail.ru zintar@mail.ru

M. Ceccarelli (ed.), *Distinguished Figures in Mechanism and Machine Science: Their Contributions and Legacies, Part 2*, History of Mechanism and Machine Science 7, DOI 10.1007/978-90-481-2346-9_10,

Biographical Notes

Feodor Orlov (Fig. 1) was born in 1843 in a village, Velikoe, of the Novgorod province (Yershov 1854; Orlov 1885/86; Zernov 1895/96; Zhoukovsky 1898/99, 1921; Scientific School of Bauman Moscow State Technical University 1995; Volcnkevich 2000). He was the son of the medical officer who graduated at Moscow University. After his father's death in 1851, he left for Moscow. Because of his difficult financial position, Feodor Orlov had been admitted to the Yaroslavl gymnasium at public expense. In Yaroslavl, he lived under the guardianship of his uncle, who was a professor of Yaroslavl's seminary. The deep religious upbringing of Orlov can be explained by the moral influence of his uncle. Orlov graduated from the gymnasium with a gold medal in 1859 and then entered Moscow University, where he graduated in 1863. Because of the poor financial status of his family Orlov, had to give private lessons. In the University, he was one of the elite students who came to the professor of the mechanics. Brashman took their meetings very seriously and he organized them every weekend. After graduating at the University, Orlov did not leave it and continued working at the Chair of Abstract Mathematics to get ready for receiving his master's decree. In 1867, he published his first scientific work "The proof of Euler's theory". In 1869, he supported his dissertation, which was called "About reciprocity of differential equations", for the degree of master. After the death of Professor A. Yershov, the chair of Applied Mathematics in the Moscow University became free. The Council of the University decided that this place must be occupied by a person who had a very good grounding in mathematics, which would guarantee good knowledge of Mechanics, and recommended Orlov for this position. At the same time, the Council interceded with the Ministry for Orlov to be sent abroad in preparation for his future job. Orlov agreed to move abroad not without doubts because he had his own scientific plans in the field of Mathematic. Though he understood that this offer

Fig. 1 Feodor Orlov (1843–1892)

could help him get a well-paid job and become the provider for his family. In September 1869, he left for St. Petersburg where he met two great Russian scientists: P. Chebyshev and J. Vyshnegradsky. They gave Orlov some practical advice about living abroad and handed him letters of recommendation to European scientists. Chebyshev told Orlov about a new leading mechanism which let Chebyshev create a unique construction of a steam engine. During the period between 1869 and 1870, Orlov attended the Abstract and Applied Mathematics lectures in Zurich in Switzerland. He also translated Chebyshev's article about parallelograms for the "Civilingeneur" magazine. In 1870, he was elected to be the head of the Practical Mechanics and Thermodynamics' chair in IMTS. In the summer of that year he also attended the lectures of professors F. Reuleaux and Christoffel at the Gewerbe-Academie and Weierstrass and Kronecker in the Berlin University. Orlov was mostly influenced by the director of Gewerbe-Academie F. Reuleaux. Orlov listened to the course of kinematics, read by F. Reuleaux with great interest. Many of Reuleaux's ideas, which were not very popular in Russia those days, Orlov used in his educational activities later. He wrote: "Reuleaux gives very good experience of setting forth kinematics with another point of view. The formula with which he shows the structure of mechanism and methods, opened by himself, lead him to the most unexpected results". The demonstrations of machine and mechanism models, by Reuleaux had, made an impression on Orlov. It is necessary to say that of the problems delivered by the rector of IMTS, V. Della-Vos, to Orlov was to study the possibility of getting new models for the chair of Mathematics, established by Professor A. Yershov. Then Orlov attended the mathematics lectures of Professor Katalan at Liege University until the end of 1871. Katalan met the young Russian scientist and became interested in his dissertation for the degree of master and then asked Orlov to translate it into French. This translation was presented to the Belgian scientific academy. Katalan offered Orlov the chance to take part in the competition for the Belgium Academy Prize the works in the field of the theory of surfaces. But Orlov refused to do that as he thought that it would be a denial of his main duties. In his diary he wrote: "I am beginning to suffer from remorse because for the last days I have left applied mathematics and become enthusiastic about my past love – abstract mathematics." Later, Orlov removed to France. He lived in Paris until the autumn of 1872, where he attended the lectures in the Ecole Centrale, Ecole Polytechnique, Sorbonne and the College de France. In the autumn of 1872, he returned to Russia and from that year until his death in 1892, Orlov followed practical mechanics courses at the Moscow University and the IMTS. His inaugural speech, which was read in the Moscow University and the IMTS, was the lecture "About machines". There the development of Reuleaux's ideas about kinematics and kinematic pair was stated for the first time. The professors of the IMTS lived in flats, which were allocated to them by the university administration, not far from the University. Orlov settled in one of them with his family, which consisted of his mother and sister. At the same house, where Orlov's flat was, lived Professor Zhukovsky, who became a friend of Orlov. On the basis of his university preparation and the materials which had been collected abroad, Orlov gave several lecture courses. Firstly, we should discuss his main course – "Applied Mechanics". Moreover, he prepared and read thermodynamics, hydraulics, steam engines and resistance of materials lectures. All the courses reflected not only the modern state of science at that time, but were also methodically worked on. The teaching of the applied mechanics at the Moscow University and the IMTS was raised by Orlov to a high level like at no other University in Russia. Professor Zhukovsky stated that those were the best courses in Russia.

Orlov was curator of the applied mechanics room at the Moscow University and the IMTS. The main part of his collection was acquired from Hustav Voight, who had been given the right to make and sell some of the models from the Reuleaux collection in the 1980s. Besides, Voight's models, models of Alexander Clear from Paris, Joseph Shröder and many others were bought.

The models were bought only after thorough selection as the financing had been very limited. Since Orlov's death, many people noticed that the room of mechanisms was set in an ideal situation. Besides that room, Orlov had organized very good sketching and projection classes. He spent much time working individually with his students. He helped them choose a theme for writing their candidate's work, and lent books from his own library. About half of all candidate's works in the Moscow University were written on applied Mechanics.

In 1874, the Scientific Council of the IMTS despatched Orlov, with scientific aims, to Moscow, Nizhegorodsk and Vladimir province. He examined the manufacturing and factory industries there. In 1876, he was assigned to the Philadelphia exhibition to study the state of techniques in the USA. In 1878 he was sent to the World exhibition in Paris. From all these places, Orlov brought a vast amount of stuff which, unfortunately, was not elaborated upon and printed.

In 1875, a group of students made an offer to the rector of the IMTS, V.K. Della-Vos, to organize a society, which would have such a purpose as "intellectual and moral unity and pecuniary mutual aid to the graduated students". The necessity of such a society appeared because most of the graduated students, especially in the first years after graduating at the IMTS, could find a job with merited payment. The offer was supported by the honorary trustee, prince Sergei Alexandrovich Obolonsky. On 14 May 1877, the Emperor enjoined (1) to organize the Polytechnic Society in the IMTS (2) to give 3,000 rubles to the main fund of the Society from the reserve capital of the IMTS.

The first meeting of the Polytechnic Society took place on 4 January 1878. Orlov actively participated, not only in the organization, but also in the work of most of the departments of the Society. He was one of the authors of the charter and in 1886 was elected to be the vice-president. The society consisted of four departments: the Engineer-Mechanical, Engineer-Technological, Editorial and Inquiry. The Editorial department was headed by Orlov and was occupied with edition the works of the Society and scientific-technical magazines such as "Transactions of the Polytechnic Society" (1879), "Proceedings of the Polytechnic Society" (1882), "Bulletin of the Polytechnic Society" (1888), etc. The Inquiry department was occupied with looking at work for different Russian factories and manufactures for the Mechanic Engineers and processing engineers who had graduated at the IMTS. Facts were listed about the quantity of places and quantity of students who had found a work in the reports of the Society. Some graduates were helped materially.

At the suggestion of Orlov in 1887, a new department of the Society was set up – the Department of Technical Education.

Orlov had more than once addressed meetings of the Society. His lectures were accompanied by demonstrations of new models from the mechanics room of the IMTS. Thus on 18 November 1879, Orlov gave a lecture about the leading mechanism of Hart a lecture about the double detector of Sheffer and Budenberg on 15 January 1881. On 21 November 1891, Orlov delivered his last lecture "About the astatic regulators". On 20 January 1892, Orlov died in the hands of his sister. He is buried in the Alekseevsky cloister next to his mother. In January 1892 after Orlov's death, members of the Polytechnic Society collected funds and established grant in Orlov's name for students.

List of Main Works

1. Orlov F.E. The proof of Euler's theory. Mathematic manual, Volume 2, University Printing Office (Katkov & Co), Moscow, 1867 (in Russian)
2. Orlov F.E. The dissertation about correlation of differential equations. Dissertation was written for receiving the degree of master of abstract mathematics. University Printing Office of A. Mamonov, Bolshaya Dmitrovka, 7, Moscow, 1868 (in Russian)
3. Orlov F.E. Lecture about machines. Inaugural lecture, read at the Moscow University on the 21st of October, 1872, Moscow, Moscow Mathematical Society that belongs to the Moscow University, 1873, reprint from Mathematical Manual published in 1873, Volume 6, 17 pages (in Russian)
4. Lectures about Applied Mathematics that were read by Prof. F.E. Orlov and assembled by students of the 2nd special class, the Practical Academy of Commercial Science, Moscow, 1873, 462 pages, lithographic edition
5. Orlov F.E. Economical significance of machines. The speech of the Imperial Technical School's professor Orlov F.E. The report of Imperial Moscow Technical School for the year 1878, the second publication, Moscow, University Printing Office, 1879
6. Orlov F.E. The theory of resistance of materials. Notes made from lectures of Prof. Orlov. The first course of special class. 1880, 340 pages
7. Hydraulic engines. Lectures of Prof. F.E. Orlov, 1881. The first special class of IMTS – Moscow; ITMS – 286 pages, illustrations, lithographic edition
8. The theory of steam engines. Lectures of Prof. F.E. Orlov. From the books of Prof. A.P. Gavrilenko. 1881/82, Moscow; IMTS – 246 pages, illustrations (lithographic edition)
9. The theory of steam engines. Lectures of Prof. F.E Orlov, 1882/83, Moscow, IMTS, 286 pages, illustrations, lithographic edition
10. Hydraulics. Lectures of Prof. F.E Orlov. 1882, Moscow, IMTS, 238 pages, illustrations, lithographic edition

11. Orlov F.E. The theory of roulette. Odessa, Printing Office of P.A Zelenskii. Extract from the Notes of mathematical department of Novorossiysk Society of Natural Sciences, 1884, vol. 5
12. Orlov F.E. About the quadrature of roulette. Lecture was read on the 15th of November, 1883, Moscow, Moscow Mathematical Society which belongs to the Moscow university, extract from the Mathematical Manual, 1884, vol. 11, 60 pages
13. Orlov F.E. Applied Mechanics, Moscow, the Printing Office of Zubarev, 1885, 453 pages, lithographic edition
14. Lectures about the theories of steam-engines which were read by Professor F.E. Orlov in 1886/87 academic year and then written down by the student of the third year in the Moscow University A. Sidorov, Moscow, MU, 249 pages, illustrations, lithographic edition
15. General theory of machines. Lectures of Prof. F.E. Orlov, the fourth term, Moscow, of Blagushin, 1887, 148 pages, lithographic edition
16. Steam-engines. Lectures of Prof. F.E. Orlov, the second and third courses of St. Petersburg IMTS,1890, Moscow, IMTS, 276 pages, illustrations, lithographic edition
17. Applied mechanics. The lectures of Prof. F.E. Orlov, 1891/92, Moscow, parts 2 and 3, Moscow, the Printing Office of Bogomolov, 1892, printed with a duplicating machine
18. Thermodynamics. The lectures of Prof. F.E. Orlov, 1891/92, Moscow, IMTS, 246 pages, illustrations, lithographic edition
19. Steam-engines. The lectures of Prof. F.E Orlov, a course of the second special class of engineer-mechanical section, Moscow, IMTS, 1892, 357 pages, printed with a duplication machine
20. Theory of machines. Lectures which were read by Prof. F.E. Orlov in IMTS in 1892, edited in 1893 and allowed by Prof. D.S. Zernovoi, lithographic edition
21. Orlov F.E. Diary of the foreign trip 1869–1872, Moscow, Printing Office of G. Lissner and A. Geshel, 1898, 346 pages

Review of Main Works on Mechanism Design

The main directions of Prof. F. E. Orlov's activity were:

- Establishing the applied mechanics course involving theory of mechanism and machines.
- Creating offices of mechanisms to practically support the applied mechanics course.
- Establishing courses "Theory of Resistance of Materials and Theory of Elasticity", "Theory of Steam Engines", "Steam Engines", "Hydraulics".
- Enthusiasm for problems of abstract mathematics.
- Social and educational activity. Orlov's educational work can be clearly represented by his opinion of establishing the applied mechanics course. Other ones can be described briefly. Five intravital lithographic courses of applied mechanics (1887/88, 1889, 1890/91, 1891, 1892), and one posthumous course of the theory

of machines (1893), are stored in the IMTS library. Orlov gave the applied mechanics course in Moscow University (MU) and Imperial Moscow Technical School (IMTS). The former course was obviously more theoretical while the latter was practical. The persevering educational policy of Orlov was to combine analysis of theoretical problems with a demonstration of their practical solution. Orlov improved his course during the 20 years of his educational activity. One of his last courses is presented below.

Applied Mechanics

Part 1 – Theory of Mechanisms (224 pages, 232 illustrations); Part 1 consisted of introduction, general theory of mechanisms and analysis of three classes of mechanisms classified according to connection type of their elements. Introduction. Classification of kinematic pairs according to their DOF. Lower and higher pairs. Mechanism is a system with DOF = 1. Like Professor A. Yershov, Orlov used a two-level classification. The first level was classification based on the type of connection of mechanisms elements. The second level was Willis's classification based on a type of moving transformation. Professor Orlov changed the sequence of relation of materials and theoretical level, as opposed to professor A. Yershov (Fig. 2).

Fig. 2 Foedor Orlov's "Applied Mechanics" (part 1)

General theory of mechanisms. Mechanisms with lower kinematic pairs. Mechanisms with prismatic kinematic pairs. Wedge press, grip. Mechanisms with revolute pairs. Four-bar linkage, parallelogram, antiparallelogram. Crank-slider mechanism. Ellipsograph, Oldhem's clutch, sinus-mechanism, etc., Hooke's joint and his analogues, control mechanism for railway semaphore. Mechanisms with screw pairs. Mechanism with three screw pairs. Practical application. Transformation of crank-slider mechanism. Examples of the mechanisms listed here can be seen in Table 1. A further three classes of mechanism classified according to the connection type of mechanisms elements (Willis's classification) are considered.

Class 1. Moving transmission by direct contact. Willis's momentary gear ratio theorem. Teeth-wheels (constant and variable gear ratio with constant indicium). The methods for profile tracing (Poncelet, Camus, Reuleaux). External and internal involute gearing. Cycloidal gearing. Pin gearing. Teeth profiling by two points. Teeth profiling by arcs of a circle (Willis's method). Hooke's wheels. Rotation axes meet, bevel gearing. Skew rotation axes, hyperboloid wheels, screw and worm gearings. Unicycle elliptical wheel. Cam mechanisms (variable gear ratio with variable indicium). Examples of cams with rectilinear and rotary movements. Types of cam mechanisms. Example of practical application.

Table 1 Examples of mechanisms with lower kinematics pairs

Examples of mechanisms with lower kinematics pairs			
1. Mechanisms with prismatic kinematics pairs			3. Mechanisms with screw pairs
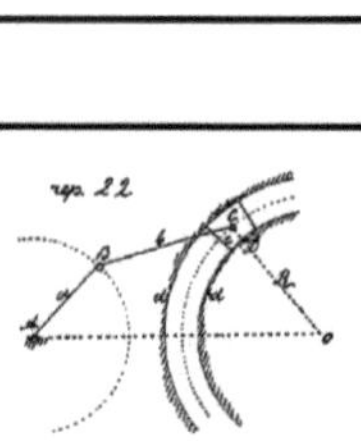 Wedge press	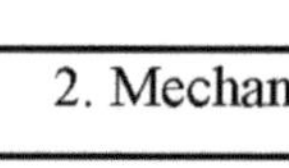 Grip	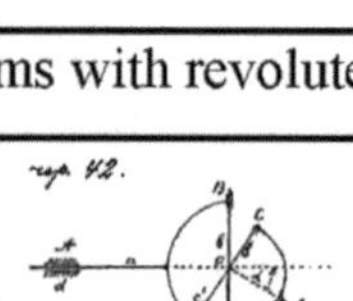	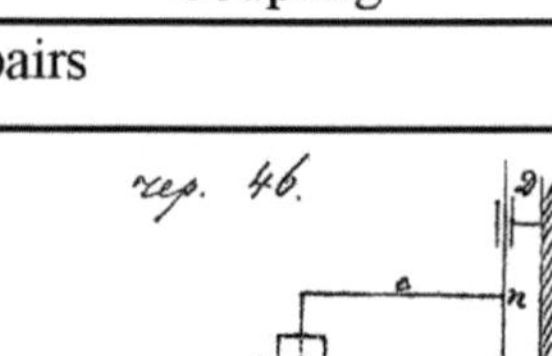 Coupling
2. Mechanisms with revolute pairs			
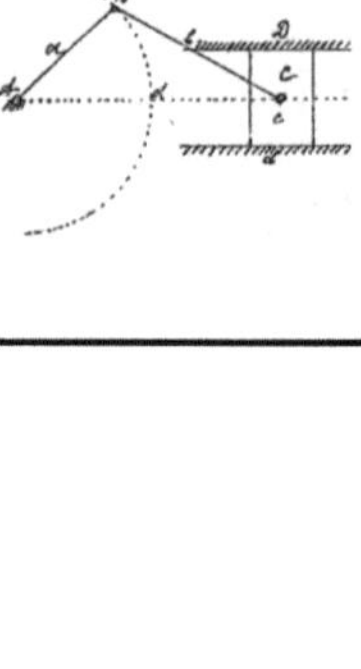		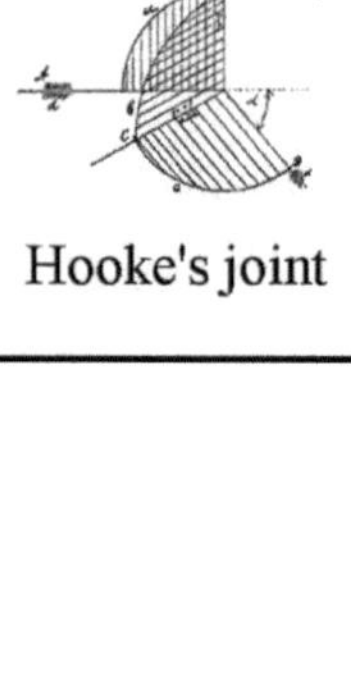 Hooke's joint	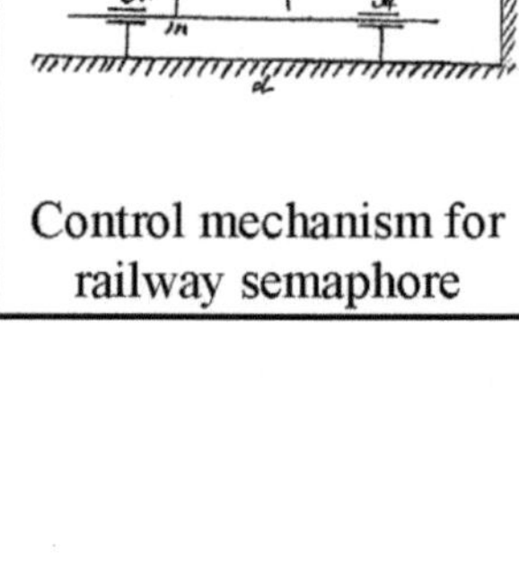 Control mechanism for railway semaphore

Mechanisms with a constant gear ratio with variable indicium. Examples. Ratchet mechanisms, including clockworks.

Examples of the mechanisms listed here can be seen in Tables 2 and 3. A transmission designed by Orlov and manufactured at the IMTS workrooms is one of them. Unfortunately, it was impossible to find this model. It should be mentioned that Orlov gathered a considerable collection of mechanisms made in Germany and France.

Class 2. Moving transmission by coupler. Mechanisms with parallel axes. Four-bar linkage (instantaneous center of zero-velocity, possible variants of moving transformation, parallelogram, antiparallelogram, Grassgoff's theorem). Crank-slider mechanism, sine-mechanism, ellipsograf. Crank – shaft and eccentric. Hooke's joint.

Class 3. Moving transmission by flexible elements or liquid medium. Belts and ropes.

Part 2. Theory of machines (344 pp., 247 ill.) This part accords to the second stage of "Applied mechanics" course (Golovin and Danilenko 2000), namely, the problems of friction and dynamic theory of machines (Table 4), theory of motors (Tables 5 and 6). Dynamic theory of machines. Machine and mechanism. The foundation of machine dynamics. Kinetic energy and dissipation. Speed control of machine and reason for using fly-wheel: gain in mass of fly-wheel or gain in mass of coupler, for example? About regulation: Watt's inertia governor, parabolic and pseudoparabolic governor, etc. Problems of friction. General problems. Some

Table 2 Examples of moving transmission by direct contact (gearing)

Examples of moving transmission by direct contact (gearing)			
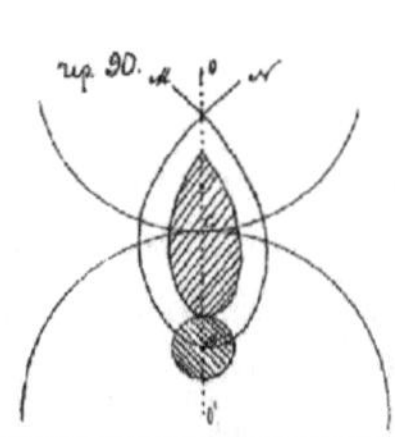	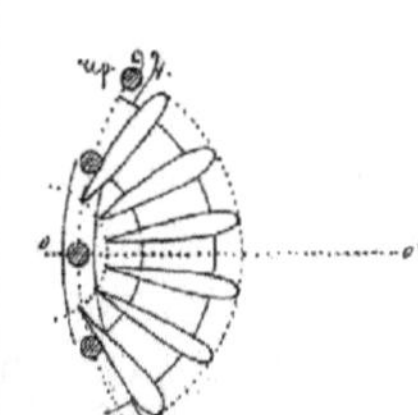	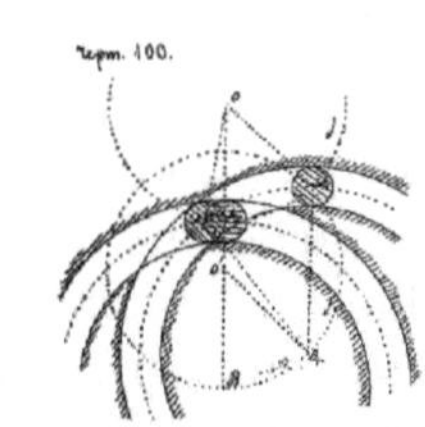	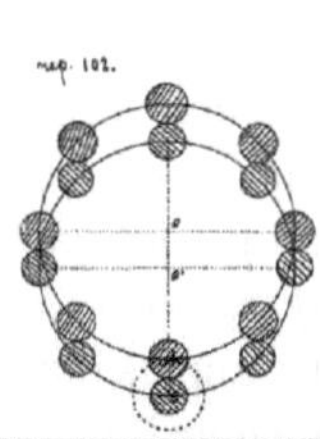
Gears of Reaulaux		Orlov's gearing, fabricate in EMTS	
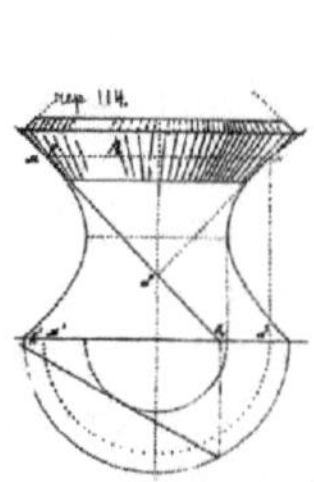	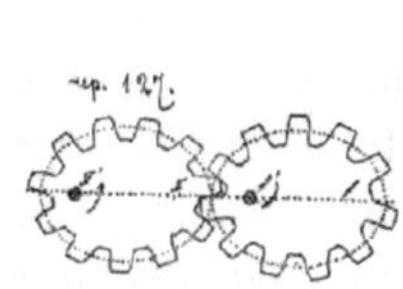	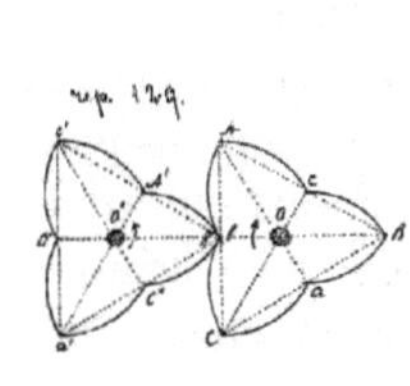	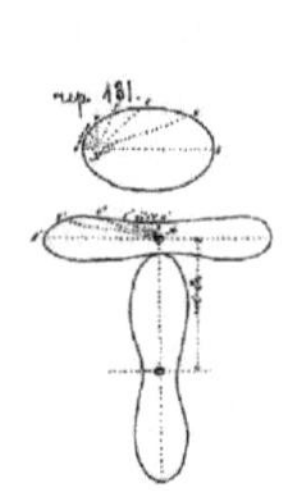
Hyperboloid wheels		Different gearings	

Table 3 Examples of moving transmission by direct contact (cams and other mechanisms)

Examples of moving transmission by direct contact (cams and other mechanisms)		
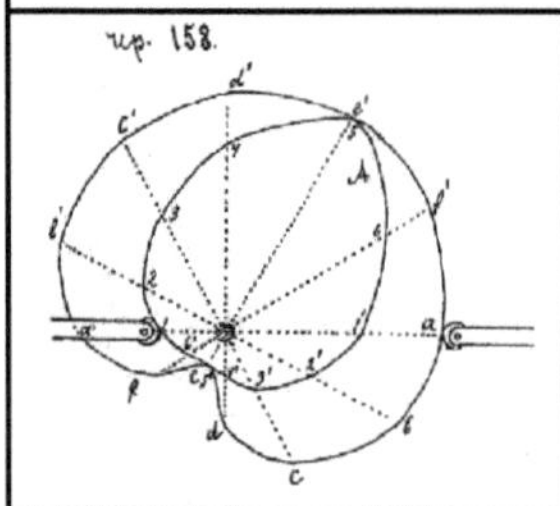	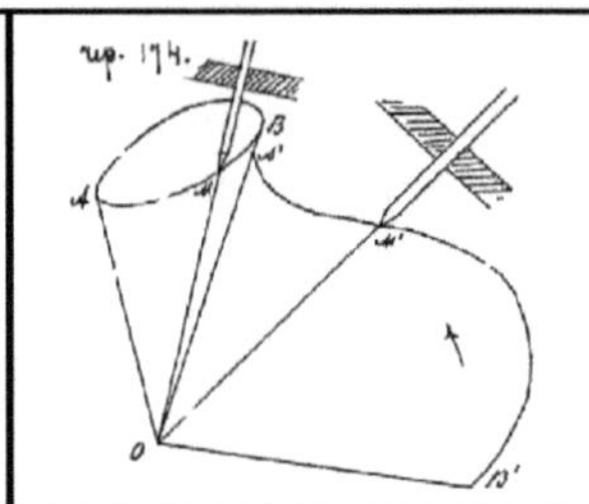	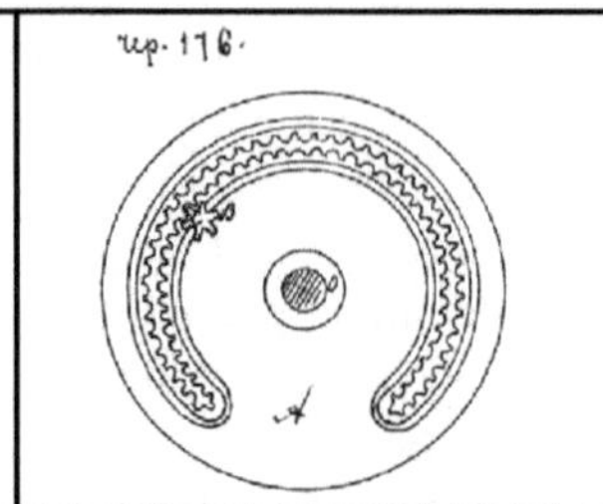
Cams of Morin	3-D cams	Ratchet mechanism
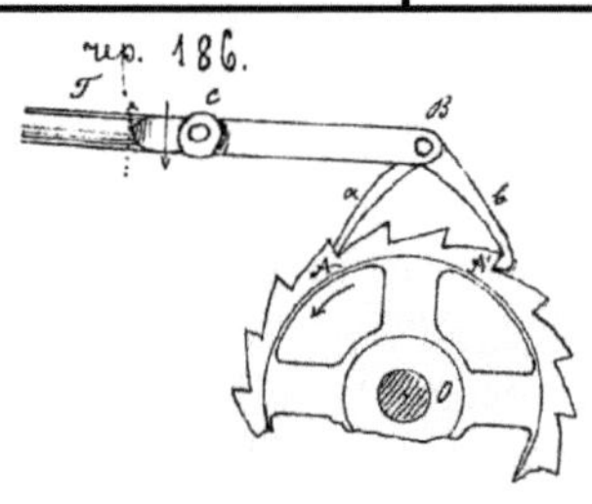	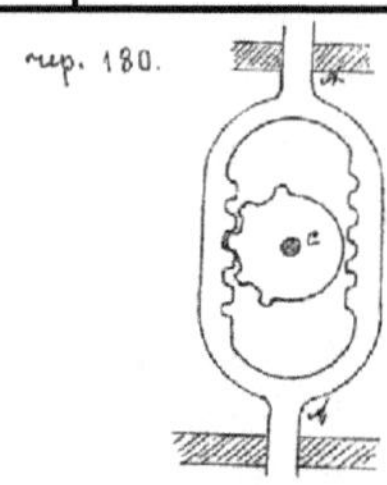	
Other mechanisms		

devices that use the friction effect. Determination of friction coefficient experiment. Prismatic pair: wedge, horizontal and vertical slider. Revolute pair. Screw pair. Teeth-wheels. Transmission of work by friction: friction wheels, friction couple etc. Rolling friction and its application. Friction of flexible bodies. Transmission of work by endless belt. Brakes. Dynamometer of Navier, Imre, Hachette, Watt, etc.

Examples illustrating the problems of friction and dynamic theory of machines can be seen in Table 4. Theory of motors. Living motors. Hydraulics. Hydrostatics. Mains pressure. Equilibrium of body submerged in liquid. Equilibrium of floating body. Determine of mains pressure ponderable liquid. Hydrodynamics. Flowing pressure. Liquid streaming from filler. Liquid moving in pipes. Liquid flow in rivers and canals. Water as motor, dams. Hydraulics machines. Water wheels. Turbines (eight types). Air as motor. Examples considering this part of the course are shown in Table 5.

Thermodynamics. General questions. About fallowes. Caloric machines. Work measuring of steam engine. Steam engines. Papin – Savary steam engines. Installation of Papin's steam engine in Peter the Great's garden in St. Petersburg in 1718. Newcomen and Kawlay. Watt's steam engine. Systems of steam engines and devices. Steam boxes. Furnace, steam boilers. Outfits of boilers (accessories). Manometers, float-gauges, float level gauges, safety valves, feed of boiler. Refrigerators. Examples considered this part of the course are shown in Table 6.

Table 4 Some examples from theory of machines

Some examples from Theory of Machines		
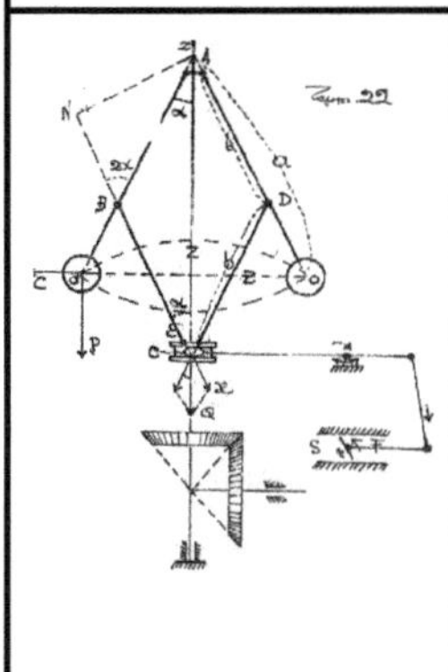	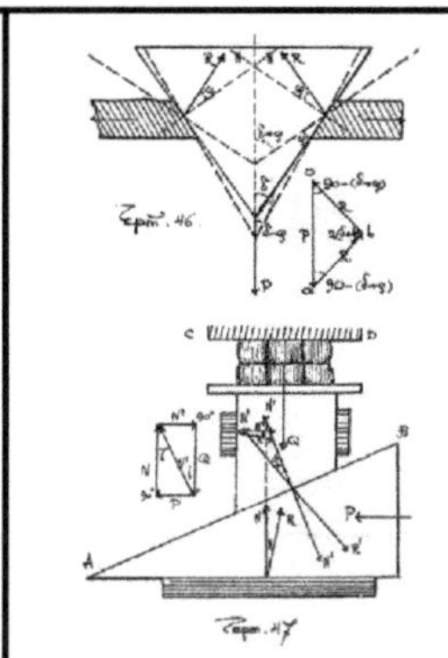	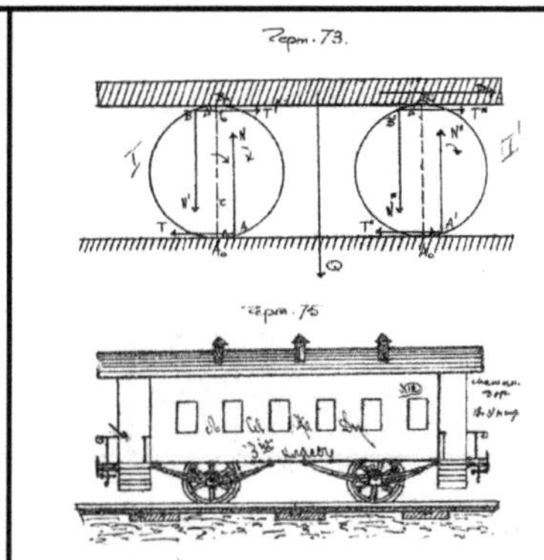
Watt's inertia governor	Horizontal and vertical slider	Rolling friction
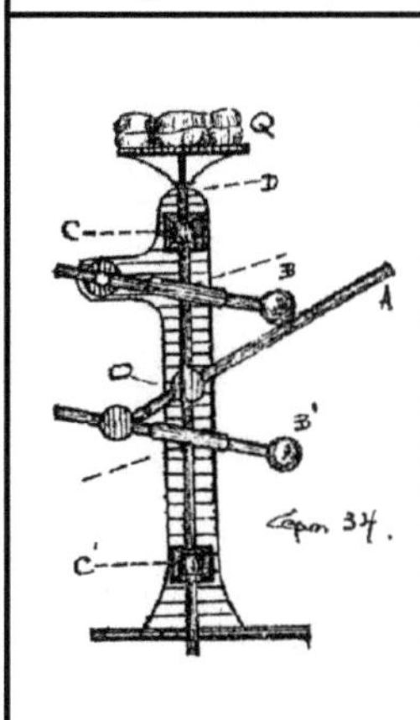	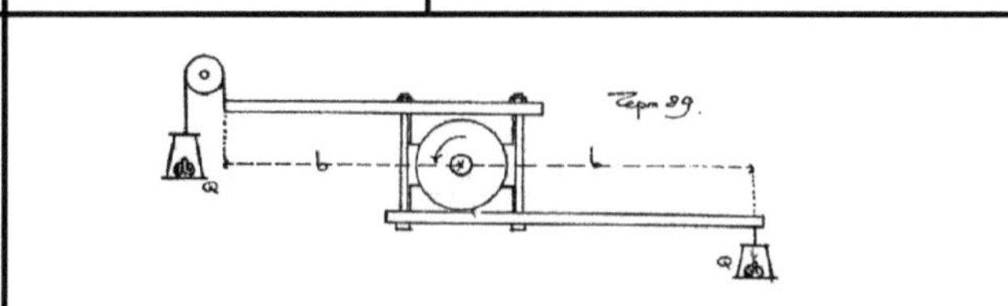	
	Prony's wheel brake	
	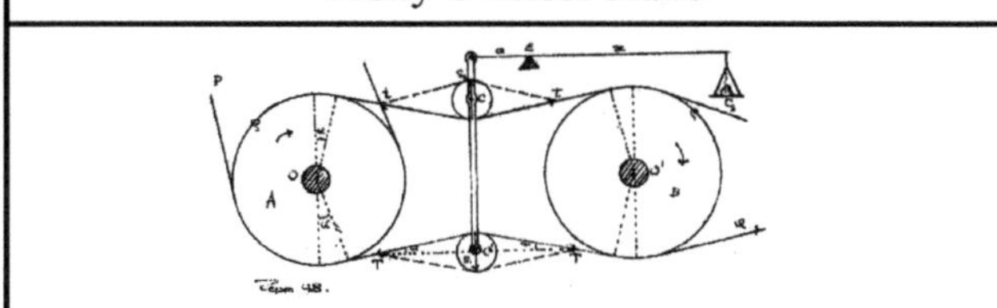	
Lifting jack	The belt dynamometer of Briggs	

Theory of Resistance of Materials (Part 1) and Theory of Elasticity (Part 2)

Title page of the book published in 1880 reads as follows: "Excerpts from Prof. F. E. Orlov's lectures; Course of 1st Special class" (Fig. 3).

Contents of Part 1 (300 pages) corresponds to the classical contents of such courses. The following sections (in modern terms) are analysed consecutively:

Stress and strains, Hooke's law. Geometric properties of sections. Tension and compression. Pure bending. Tangential stress. Eccentric compression. Equation of deflection curve (now referred to as Krylov's method). All cases of two- and multi-support beams.

Table 5 Some examples from theory of machines (motors). Hydraulics

Some examples from Theory of Machines (Motors). Hydraulics		
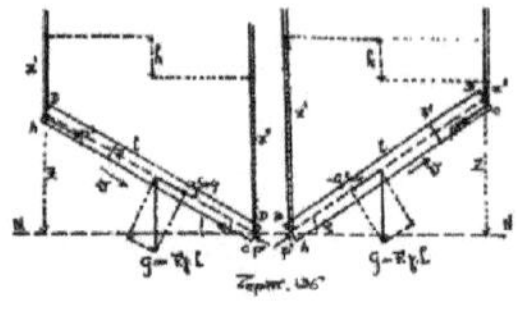	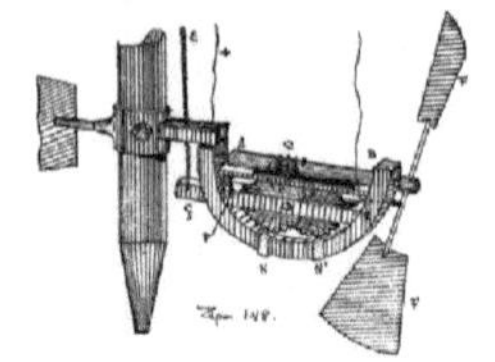	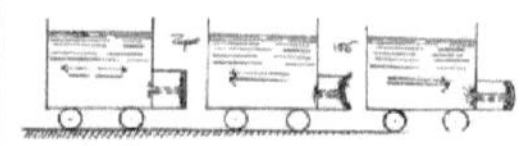
Liquid moving in pipes	Voltman's propeller flowmeter	Water as motor
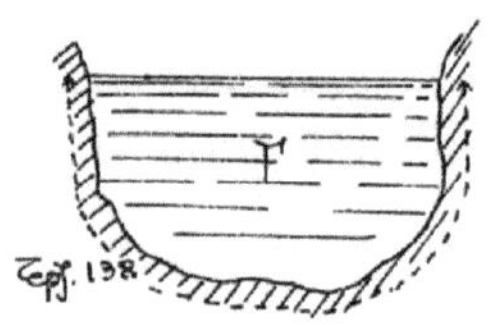	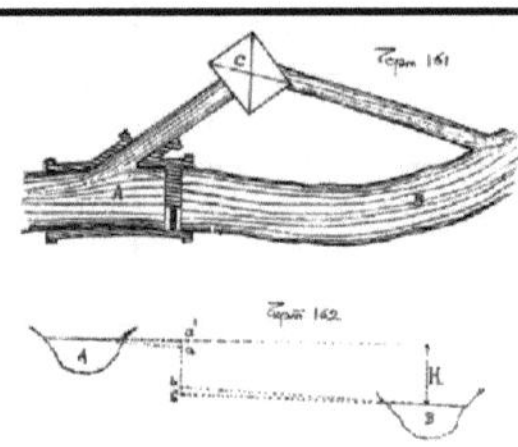	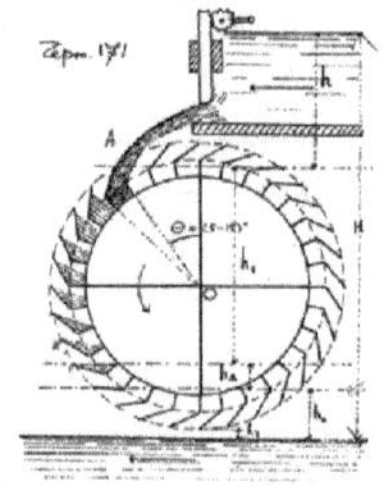
Liquid flow in rivers and canals	Dam, dyke	Overshot water wheels
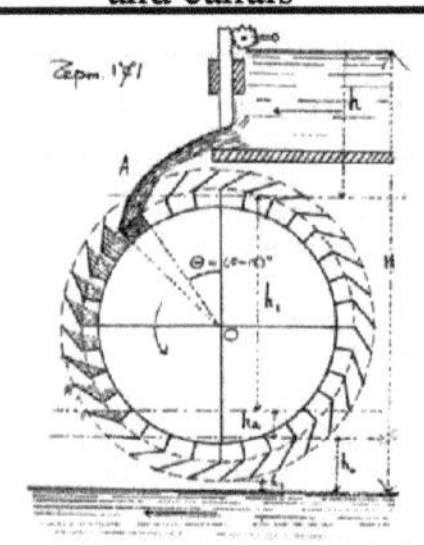	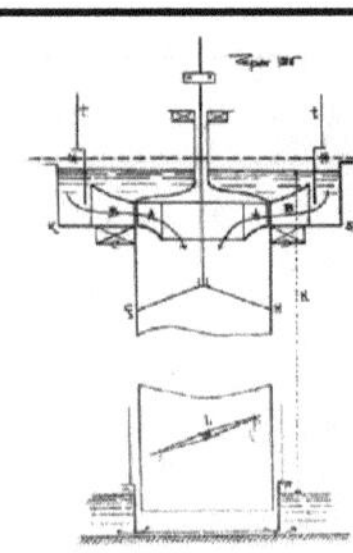	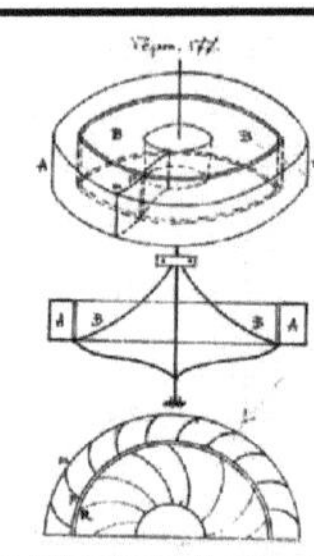
Lovershot water wheels	Turbines	

Stability of compressed beam. Flexure of curvilinear beam. Torsion. Momentless theory of shells.

Theory of elasticity – Part 2 (40 pages) is presented briefly. Actually (in modern understanding, too) the paper presents components of stress and strain tensors, Cauchy's surface, and Lamé's ellipsoid. Correspondence between theory of elasticity and resistance of materials is shown.

Theory of Steam Engines

Title page of the book published in 1881–1882 reads: "Prof. F. E. Orlov's lectures" (Fig. 4). In fact it is an extended course of thermodynamics presented in the

Table 6 Some examples from theory of machines (motors)

<table>
<tr><th colspan="3">Some examples from Theory of Machines (Motors)</th></tr>
<tr><td>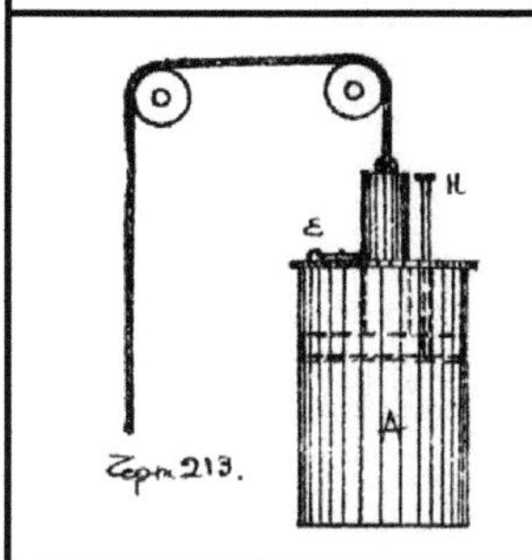</td><td>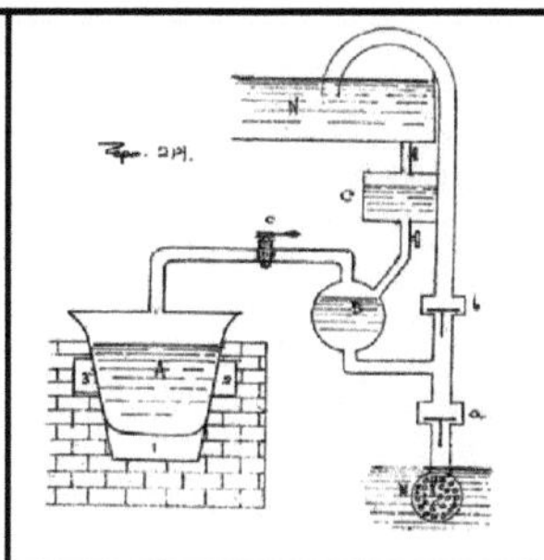</td><td>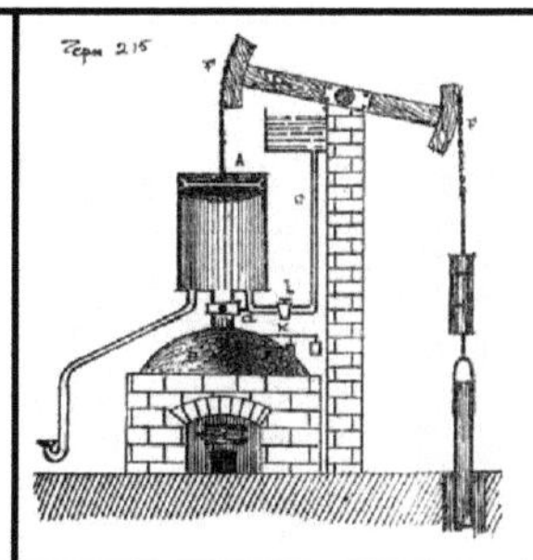</td></tr>
<tr><td>Steam engine of Papin (1690)</td><td>Steam engine of Savary (1698).</td><td>Steam engine of Newcomen (1705)</td></tr>
<tr><td colspan="3">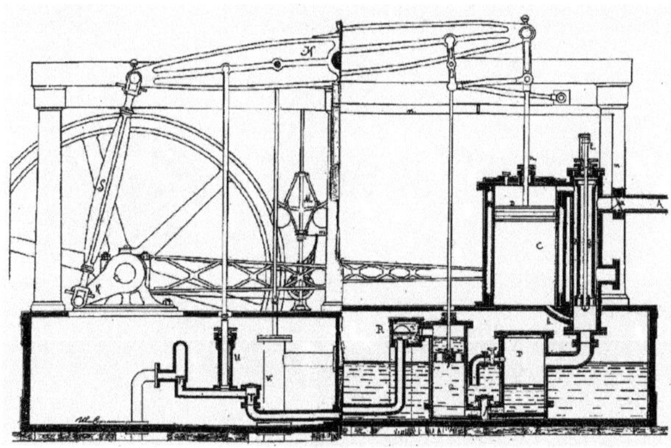</td></tr>
<tr><td colspan="3">Steam engine of Watt</td></tr>
<tr><td>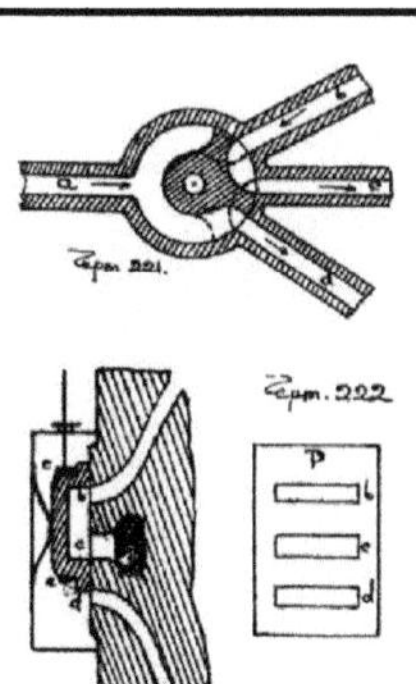</td><td>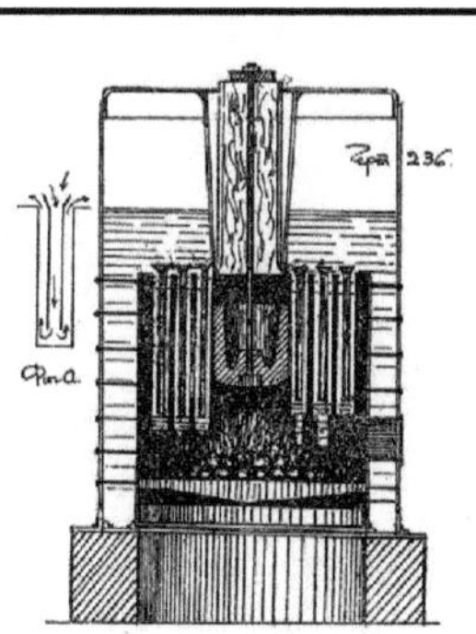</td><td>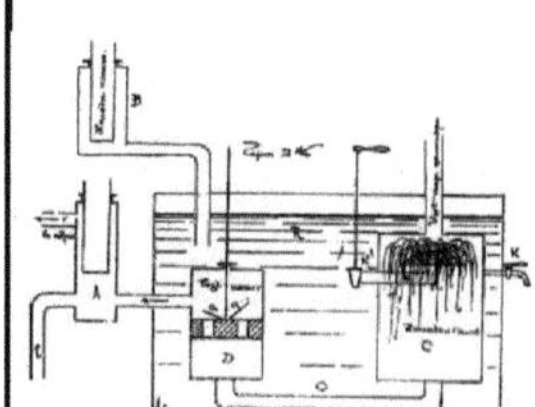</td></tr>
<tr><td>Steam box systems</td><td>Boiler of Fild</td><td>Refrigerator</td></tr>
</table>

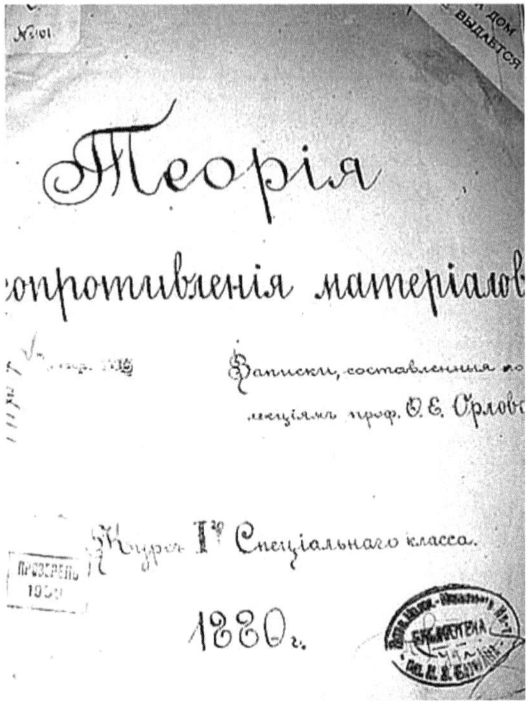

Fig. 3 Orlov's book of 1880

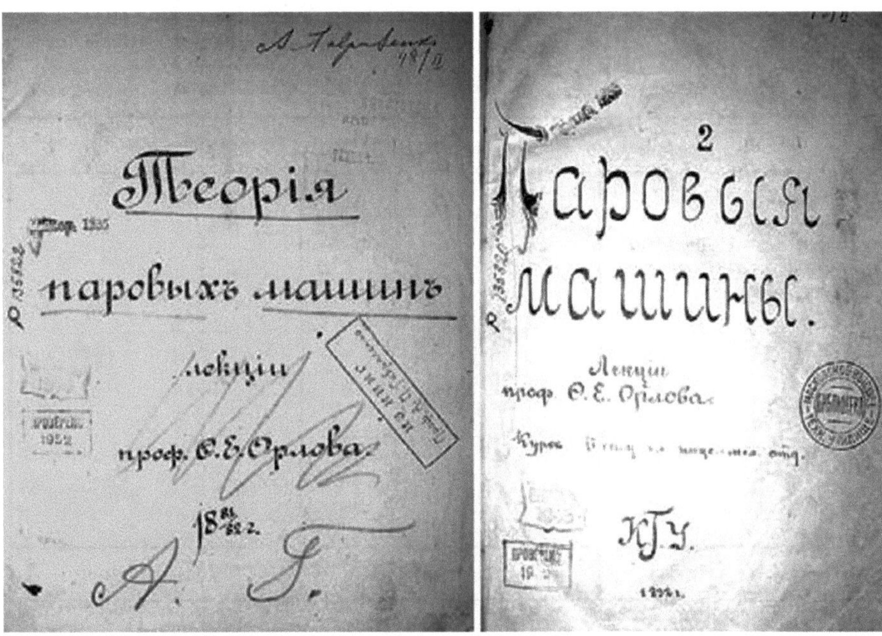

Fig. 4 F. Orlov "Theory of Steam Engines" and "Steam Engines"

course "Applied Mechanics, Part 2 – Theory of Machines". The course is presented as 23 lectures on 197 pages. Part 1 (Thermodynamics, 91 pages) presents the fundamentals of thermodynamics, atmospheric and thermal machines. In the second section (on steam, 42 pages) steam types are considered. All then existing kinds of steam engines are analysed in the third section (theory of steam engines, 68 pages). History of steam engines: Hero of Alexandria, Solomon de Coos, D. Papin, T. Savery, T. Newcomen, J. Watt. Steam engines of Leupold and I. Polzunov are not considered.

Steam Engines. Title page of the book published in 1892 reads: "Prof. F. E. Orlov's lectures, course for Mechanical Engineering Department" (Fig. 4). Book comprises 364 pages. Types of steam engines, indicators and their peculiarities (also defects), are analysed. Theory of double-acting steam engine with one cylinder of a complex steam engine, is presented. Analyses for all units of a steam engine are performed. Dynamic problems (travel adjustment, flywheels, maximum number of turns) are investigated. Railway engines: design, locomotive resistance (friction, air resistance, shocks on splice joints, etc.) steamships.

Hydraulics. Title page of the book published in 1882 reads: "Prof. F. E. Orlov's lectures" (Fig. 5). In fact it is an extended course of hydraulics presented in the course "Applied Mechanics, Part 2 – Theory of Machines". Book consists of 238 pages and two sections: hydrostatics and hydrodynamics. Problems of practical hydraulics: fluid flow in tube and channel, complex pipeline and manifolds, dams and embankments, methods to measure parameters of hydraulic systems, are solved.

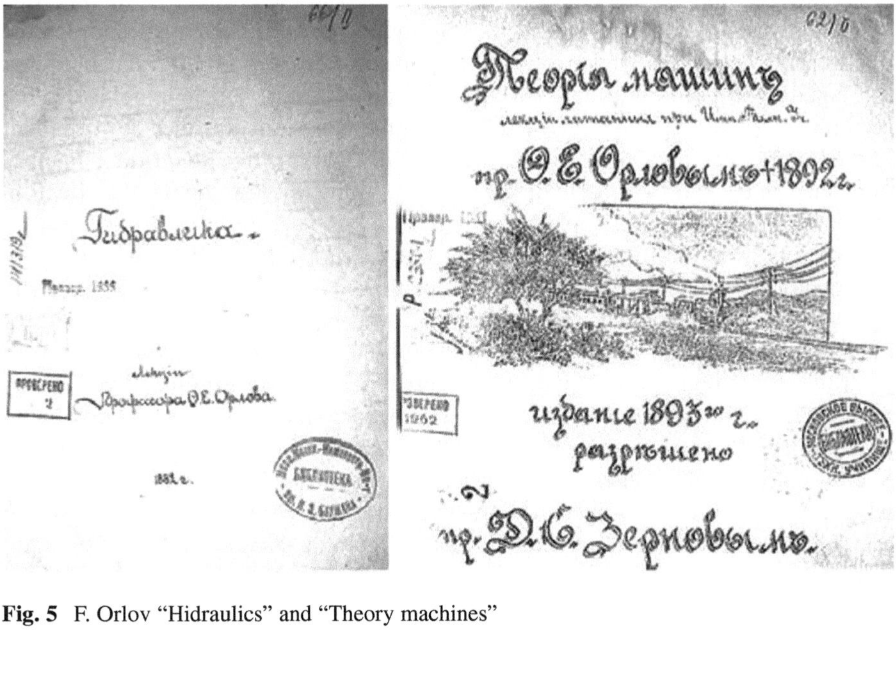

Fig. 5 F. Orlov "Hidraulics" and "Theory machines"

Theory of machines. The book was written in 1892 and published posthumously in 1893 (Fig. 5). It comprises 196 pages. Actually it is an arranged edition of Part 2 of Applied Mechanics.

Certainly, Orlov's main works were coherent with his teaching activity. However, he wrote several original works on abstract mathematics. Orlov the mathematician can be represented by two papers:

Proof of Euler's theorem (1867). The derivation is proposed for condition given by Euler for integrability of differential expression

$$f(y, x, y', y'', \ldots y^n)\,dx,$$

Theorem defines necessary and sufficient conditions for the function integrability. Orlov proposed a more simple and elegant proof. His derivation may be extended to both direct and converse Euler theorems. On Reciprocity of Differential Equations (1868). This was his thesis on the Master Degree in mathematics. Along with the given autonomous differential equation

$$F(x, y, y') = 0,$$

Orlov proposed one more equation which he refers to as one reciprocal to the given one

$$f(\omega', p\omega' - \omega p) = 0,$$

where ω is a straight line segment cut off by the tangent on the axis of ordinates, argument – slope of tangent to axis. It is proved in the paper that the integrals of these two equations are identical. The author gives integration examples for differential equations of various kinds, and notices the connection between the equation's reciprocity and the last multiplier theory. In the conclusion, the reciprocity idea is extended to partial derivative equations. Further development of this work was presented in Orlov's article "Finding the Integrating Factor" published in the fourth volume of the Moscow University Mathematic Collection in 1869.

Excerpts from the Roulette Theory. At the VII Congress of Russian naturalists and doctors, which took place in 1888 in the city of Odessa, Orlov gave the "Excerpts from the Roulette Theory" report. It was published in the Proceedings of the Novorossiysk Naturalists' Society. The problem of analysing movable and fixed centroids for a body whose one point describes a straight line, is solved in the report. This solution involves the theorem on acceleration centers for a body being in a uniform rotary motion, while one of this body's point moves in a straight line. Such a body has all its even order acceleration centres on the said straight line, while odd order acceleration centers are on another line perpendicular to the former one (Fig. 6). Orlov derived the second-order differential equation whose integration can lead to roulettes if the dependence between the radii of the curvature for the centroids is known.

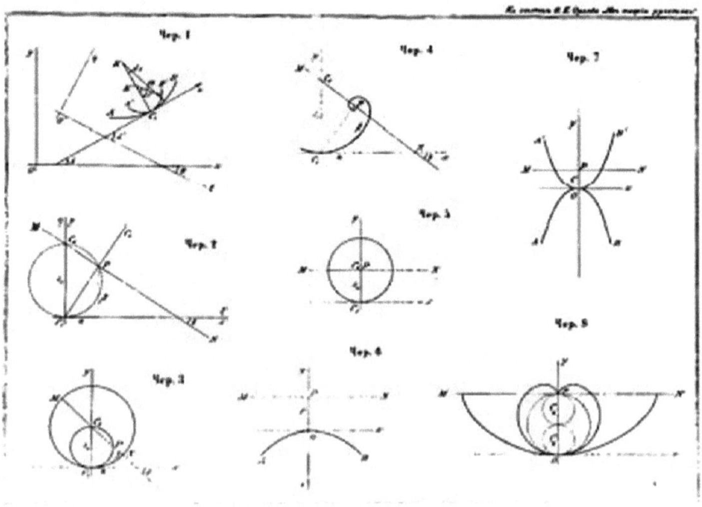

Fig. 6 Pictures from Orlov's article

The solution is illustrated by examples. In particular, on contacting the conjugate convex curves, both cetroids are parabolas and the point describing the straight line is the focus of the movable centroid. A wooden model of Cheby-shev's antiparallel link mechanism (Fig. 7) was fabricated in the IMTS workshop. In this model, the crank point moves along a line close to the straight one. The model also has a movable centroid linked to the crank, and a fixed one linked to the column. This model is supposed to be made by Orlov to visualize the theory presented in the report. This work was continued in the "On the Roulette Quadrature" article. This paper deals with interesting but little-known equations on the roulette quadrature derived in a more general form with the analytical methods being developments of Darboux's method. The said approach allows to define the area between some point's roulette, two perpendiculars constructed from its ends to a fixed centroid, and its contour. Orlov found that all points of the movable figure's plane whose roulettes have the same area, are on concentric circles. Orlov refers to the centre of these circles as the centre of roulette's area. With this point, the roulette areas can be represented by where r – distance from a point to the roulettes' area centre, E_0 – constant, ϑ – rotation angle of the figure under analysis.

Economic Role of Machines (1879). While abroad, Orlov took an interest in societies that supported workers. He visited meetings of these societies in Liège, Zürich and Berlin. The objectives of such societies were treated as follows: "…they are aiming at spreading education to the working class that is the first step towards socialism and the only way to solve the workers' problems". Impressed by these ideas, Orlov came to the conclusion that a scientist can help the working class by improving small thermal machines, and methods of long-distance power transfer from big machines. In his speech, he exploded a theory that the progress of technology and engineering worsens the workers' situation. On the contrary, he was sure

Fig. 7 Model of Chebyshev's "Excerpts from the Roulette theory" antiparallel link mechanism

that the machines would facilitate manual labour. Orlov presented these ideas in his inaugural speech in Moscow University on 21 October 1872. In the conclusion of the speech Orlov told students: "... they would not forget that their high calling is to facilitate human labour by natural force, and that power over this force is given by knowledge, and that knowledge is power".

Cabinet of Applied Mechanics in the IMTS and Moscow University

Orlov supervised offices of practical mechanics in Moscow University and the IMTS. Going abroad on business trip (Fig. 8), and satisfying requests from director of the IMTS V. Della-Vos, Orlov took an interest in collections of mechanisms in different European universities, and he acquired some models. During his trips, Orlov familiarised himself with collections of mechanisms, equipment of lecture halls, drawing offices, and workshops in universities and the technical schools of Zurich, Frankfurt, Berlin, Aachen, Margeburg, Karlsruhe, Liege, Paris, etc. This allowed him to form a clear picture of possible equipment for a mechanical office. Orlov wrote to Relaud "...about the mechanical laboratory organised at the Technical School aiming to be a necessary element of every special technical institution to acquaint students with methods and technique of experimental research".

Fig. 8 The show-window cupboards made in the nineteenth century at the beginning of the formation of the collection

Professor A. Yershov laid the foundation of mechanism collection at the IMTS. Orlov sufficiently updated and replenished the applied mechanics offices. Since the funds to acquire the models were modest, Orlov was thorough in selecting them to resupply the collections. In the 1880s, Hustav Voight in Berlin, Joseph Sröder in Darmstadt, Pierre and Alexandre Claire in Paris, started to manufacture models of mechanisms. Models by Reuleaux and Voight were the most perfect and thus they were preferably bought both for the IMTS and for the university. Some original models were elaborated by Orlov himself and manufactured in the IMTS workshops. Unfortunately, these models are now lost. But during Orlov's lifetime, the mechanism office was in perfect tune.

Toward the middle of the nineteenth century in Russia and abroad, engineering departments were formed at universities, and many higher technical schools and institutes were established (Rubio and Cuadrado 2000). Applied mechanics was one of the basic courses in their curricula. During that time, model offices were created in St. Petersburg (M.F. Okatov), Kiev (I.I. Rakhmaninov), Kharkiv (V.G. Ishmenitsky) universities and many technical institutes. Orlov had a great influence both on the contents of applied mechanics courses in these universities and on model selection for the mechanism offices. At the Polytechnic Society meeting devoted to the memory of Orlov, many spokesmen emphasised his merits of equipping applied mechanics offices. Orlov "...loved the Technical School and took great care to replenishing the main library with the best books in applied mechanics, and of acquiring valuable machines and instruments for the mechanical office under his supervision" (I.V. Aristov, IMTS director in 1883–1902) (Zhoukovsky 1893).

A.K. Eshliman: "Orlov inherited the mechanical office in an unenviable state. Despite the niggardly yearly allowance for office renewal, he succeeded in turning it into a highly valuable manipulative item, following his strict selection criteria. He acquired the collection of guides, models of recent steam distributors and inertia governors. A novice teacher would find strict order and rich content in the mechanical office" (Zhoukovsky 1893).

At the Polytechnic Society meeting on 23 November 1891, Orlov made his last report "On Astatic Governors" with a demonstration of new models from the mechanical office of the IMTS (Zhoukovsky 1893). Pictures of these governors are shown in Fig. 9.

Course of Lectures and Contemporaries

Together with Professors A. Yershov, N. Zhukovsky, D. Zernov, Orlov is by rights one of the founders of the Moscow School of Science of Machines. The great Russian engineer V. Shukhov, who had matriculated to the IMTS in 1876, listened to Orlov's course. Successors of Orlov were Professors L. Smirnov and N. Mertsalov whose pupil is the famous scientist academician I. Artobolevsky.

Courses established by Orlov, primarily of applied mechanics, and also courses of hydraulics, thermodynamics, and steam engines, corresponded to the modern state of engineering and scientific thought in Russia and abroad. Lithographic notes of Orlov's courses are extant. In fact these were the first Russian lectures designed for higher education. They laid the foundations of such disciplines as thermodynamics, steam engines, hydraulics and hydraulic machines, resistance of materials, theory of elasticity, theory of machines and mechanisms.

Fig. 9 The models of centrifugal governors, which were purchased by Orlov in Germany

Modern Interpretation of Main Contributions to Mechanism Design

In 1868, the Moscow Industrial Engineering School (MIES) was transformed into the Imperial Moscow Technical School (IMTS) whose administration established and developed the chairs on the basis of the MIES departments. First, the chairs of general mechanics (1868, Prof. F. Korolev), machine building (1868, Prof. D. Lebedev), general mechanics and machine building (1869, Prof. D. Lebedev; 1872, Prof. F. Orlov), were founded. In 1878, the general mechanics and machine building chairs were transformed into the applied mechanics chair headed by Orlov who is thus an acknowledged founder of the Applied Mechanics chair of the IMTS. In the same year, the theoretical mechanics chair headed by Professor N. Zhukovsky was organized. In the BMSTU, the official date of the applied mechanics chair foundation is regarded as 1873 when Orlov began to deliver a systemised course of applied mechanics. Friendship between Zhukovsky and Orlov greatly influenced the making of the applied mechanics course. The high level of the theoretical mechanics course allowed the improvement the theoretical level of the applied mechanics course. This level was maintained by Orlov's successors, Profs. D. Zernov and N. Mertsalov until the October revolution of 1917. Later, the level dropped due to the change in the student body.

The first systemised course of the theory of mechanisms in Russia is known to be Prof. A. Yershov's work "Foundation of Kinematics or Elementary Theory about Mechanisms of Machines Especially" published in the Moscow University printing house in 1854 (Yershov 1854). It was a kind of encyclopaedia of mechanisms organised according to Willis's classification (Golovin and Mkrtychyan 2007). Orlov's course was designed to a sufficiently advanced basic level and was replete with high theory. The course was perfected by Orlov's successors Profs. N. Mertsalov and D. Zernov whose textbook had several impressions over more than 30 years (Golovin and Tarabarin 2004). In the Theory of Machines and Mechanisms courses, especially since the 1940s of the twentieth century, much attention has been paid to various analyses. Mechanism properties lectures were essentially less attended. However, the very mechanism properties were thoroughly investigated in Orlov's course. For example, the quality analysis for components of the equation of motion is greatly influenced by the mathematical part of Orlov's talent. This tradition was continued in courses by D. Zernov, N. Mertsalov (also a graduate of the mechanical-mathematical department of the Moscow University), and others (Golovin and Tarabarin 2004).

Orlov's activity in the Polytechnic Society not only made a great contribution to the establishment of the Russian, later Soviet, engineering school, but also helped graduates of the IMTS and other technical schools to solve their problems. The society's activities (taking into account the time) can be compared to the modern activity of the ASME.

The collection of mechanisms is a special part of Orlov's work. The foundation of this collection was begun by Prof. A. Yershov who not only bought mechanism models from abroad, but also manufactured some models in the MIES (later IMTS) workshops (Golovin and Mkrtychyan 2007). Orlov continued this tradition, although funds to replenish the collection were limited. Orlov not only extended the collection, but also gave the problemoriented feature to it by taking into account the course needs. This work was continued by D. Zernov, N. Mertsalov, L. Smirnov, V. Gavrilenko (Golovin and Tarabarin 2004, 2005; Tarabarin et al. 2006). Actually, the collection consists of more than 600 exhibits, and is therefore one of the most comprehensive mechanism collections in the world. Moreover, it can be treated as an historical monument of science and technology. Besides, the collection is used as educational media. For example, the model of Watt's steam engine and its parts can demonstrate the common principles of machine design (multiple alternatives of engineering decisions), and of linkage design; fragments of a worn gear set serves to help understand the mechanism of the wearing of a kinematic pair of higher degree, etc. (Borisov et al. 2004).

Acknowledgements The authors wish to thank the employees of the BMSTU Museum and its director Galina Bazanchuk, employees of the BMSTU Library and especially its director Tamara Ageyeva, for documents provided and help in the work on the present paper, and also Tatiana Danilenko, Constantin Danilenko, and Anastasia Golovina for their assistance in translating the manuscript.

References

Scientific School of Bauman Moscow State Technical University (1995). In: Fiodorov I, Kolesnikov K (eds) History of development. BMSTU Publisher, Moscow, pp 424, ill (in Russian)
Volcnkevich I (2000) Outlines of BMSTU history. Mashinostroenie, Moscow, 240 pp (in Russian)
Yershov A (1854) Foundation of kinematics, or elementary theory about motion in general especially and about mechanisms of machines. Edition of Moscow University, Moscow, 266 pp, 227 ill (in Russian)
Orlov F (1885/1886) Application mechanics/Part 1 – Theory of mechanisms (224 pp, 232 ill); Part 2 – Theory of machines (344 pp, 247 ill). Lithographic edition of lectures, Moscow (in Russian)
Zernov D (1895/1896) Application mechanics/Part 1 – Theory of mechanisms (432 pp, ill); Part 2 – Theory of machines (232 pp, ill). Lithographic edition of lectures/Edition of EMTS, Moscow (in Russian)
Zhoukovsky N (1898/1899) Analytical mechanics. Lithographic edition of lectures/Edition of EMTS, Moscow (in Russian)
Zhoukovsky N (1921) Theoretical mechanics. Edition of lectures from 1874 in EMTS, Moscow (in Russian)
Rubio FJ, Cuadrado JJ (2000) Development of "Course of Machines" at the Ecole Polytechnique from its origin to middle of the XIX century. In: Ceccarelli M (ed) Proceeding of International Symposium on History of Machines and Mechanisms. Kluwer, Dordrecht, pp 271–279
Zhoukovsky N (1893) Obituary and outline of Professor's F. Orlov scientific activity. From report of Emperor Moscow University for 1893, 16 p (in Russian)
Golovin A, Tarabarin V (2005) Some models of Russian scientists from mechanisms collection of TMM department. In: Ceccarelli M, Golovin A (eds) Proceeding of Third International Workshop on History of Machines and Mechanisms. BMSTU, Moscow, pp 158–168
Golovin A, Danilenko C (2000) Evolution of theory of mechanisms and machines from G. Monge to present day and modern problems of TMM training in technical university. In: Ceccarelli M

(ed) Proceeding of International Symposium on History of Machines and Mechanisms. Kluwer, Dordrecht, pp 263–270

Golovin A, Tarabarin V (2004) A few pages from Russian TMM in 19th century. In: Ceccarelli M, Kerle H, Mauersberger K (eds) CD-Proceeding of Second International Workshop on History of Machines and Mechanisms, Dresden

Tarabarin V, Golovin A, Tarabarina Z (2006) The collection models of machines and mechanisms of TMM's department of the Bauman Moscow State Technical University. History of Machines and Mechanisms 2006. IFToMM Workshop-Lectures. Cornell University, College of Engineering, Ithaca, New York, 9–10 September 2006

Golovin A, Mkrtychyan D (2007) Alexander Yershov (1818–1867). In: Ceccarelli M (ed) Distinguished figures in mechanism and machine science. Springer, The Netherlands, pp 135–149

Borisov A, Golovin A, Ermakova A (2004) Some example from the history of machinery in teaching TMM. In: Ceccarelli M (ed) Proceeding of International Symposium on History of Machines and Mechanisms. Kluwer, Dordrecht, pp 107–118

Todor Pantelić (1923–1999)

Aleksandar Veg

Abstract By transforming fundamental theoretical solutions into functional mechanisms, Todor Lazar Pantelić became a pioneer of innovative structural synthesis. Most of his work evolved to the level of patents and industrial products. Professor Pantelić's specialty was the four-bar chain in innumerable varieties such as walking, imprinting, and shutter mechanisms. The crown of his innovative work was the helical conveyor, theoretically established, patented, designed, and prototyped. Loved and admired by his students and colleagues, he made a point of introducing engineering talent and creativity into the science of machines and mechanisms. He was one of the founders of the International Federation for the Theory of Machines and Mechanisms (IFToMM) and the first president of its Commission for Collaboration of Science and Industry.

Biographical Notes

So much of who we are – we owe to our teachers. In the constellation of all the great masters I have met, shines brightly the giant star of Professor Todor Pantelić.

Todor Lazara Pantelić (Fig. 1) was born on 2nd July 1923 in Belgrade (then the capital of the Kingdom of Yugoslavia) to a family of a small merchant. He had a carefree childhood in a Belgrade suburb, spending every waking hour in his brother's workshop for mechanical calculators. His regular schooling at King Alexander's Lyceum was interrupted before he was 18 by the bombing of Belgrade on 6 April 1941, which was the beginning of the Second World War in Yugoslavia. Resourceful and restless, he soon took part in the antifascist resistance movement in occupied Belgrade. When the war finally ended in 1945, he was demobilised and became a student of the Belgrade Polytechnics.

A. Veg (✉)
Machine Mechanics Institute, Faculty of Mechanical Engineering, University of Belgrade, Kraljice Marije 16, 11000, Belgrade, Serbia
e-mail: aveg@mas.bg.ac.yu

M. Ceccarelli (ed.), *Distinguished Figures in Mechanism and Machine Science: Their Contributions and Legacies, Part 2*, History of Mechanism and Machine Science 7,
DOI 10.1007/978-90-481-2346-9_11,

Fig. 1 Prof. Todor Pantelić at the international symposium machines and mechanisms, Belgrade, September 1997

His academic career started right after graduation: he became Junior Engineer in 1950, Teaching Assistant in 1954, Assistant Professor in 1960, Associate Professor in 1966 and, finally, Full Professor at the Faculty of Mechanical Engineering, University of Belgrade in 1971 (Evaluation report...).

He has published extensively at home and abroad (180 works and 4 books) and was the author or co-author of 23 national and five international patents. Marked by books, students, travels, patents, friends and family – his life was indeed an amazing one.

Prof. Pantelić was one of the founders of The International Federation for the Theory of Machines and Mechanisms (IFToMM). The decision was made in Varna (Bulgaria) in 1965, during the first Congress in that field, which assembled representatives of 16 countries. The committee including Prof. I. I. Artobolevskii (И. И. АртоболевскИй), Prof. F. R. E. Crossley, Prof. M. S. Konstantinov, Prof. W. Meyer zur Capellen, Prof. G. Bianchi, and Prof. Todor Pantelić, among others, organised the founding conference of IFToMM in Zakopane (Poland) in 1969. The next one was held in 1971 in Kupari (Yugoslavia), hosted by Prof. Pantelić (Fig. 2).

That grandiose gathering, with over 800 participants, has been considered a model of a successfully organised international scientific conference coloured by national charm. The exhibition of applied mechanisms was particularly successful: members of the Executive Board of IFToMM, very impressed, later pronounced Prof. Pantelić as chairman of the Commission for Collaboration of Science and Industry.

The Committee's seat therefore moved to Belgrade, to the Faculty of Mechanical Engineering. Innovative and brilliant, Prof. Pantelić established the Machine Mechanics Institute, and developed it into an incubator where scientists and innovators came for consultations and help with their evolving solutions and prototypes.

Many years later, in 1995, during the Ninth World Congress of IFToMM in Milan, I met Prof. J. R. Phillips who had travelled all the way from Australia to Belgrade in order to work with Prof. Pantelić on a model of mechanism analysed

Fig. 2 Kupari, Third IFToMM World Congress 1971. *Left* to *right*: (1) Ž. Nikolić [Yu], (2) I. I. Artobolevskii [USSR], (3) unknown, (4) P. Genova [Bul], (5) G. Boegelsack [GDR], 6) W. Roessner [GFR], 7) and 8) unknown, 9) N. I. Manolescu [Rom], 10) R. Unterberger [GFR]

in a doctoral thesis. At that time, the Machine Mechanics Institute was already renowned for application of the theory of machines and mechanisms.

The walls of the Institute still bear marks of the creative atmosphere of that time: a photograph (Fig. 3) taken at the Leonardo da Vinci Museum during the Congress in Milan shows the Yugoslav team led by Prof. Pantelić and accompanied by Prof. A. Morecki from Poland (the previous president of the IFToMM), Prof. B. Roth from USA, and Prof. A. P. Bessonov (А. П. Бессонов) from Russia. As I realised only recently, a model of Leonardo's helicoid stands out in the background of the picture, adding a touch of magic: in the preceding years, a helical conveyor had become the crown of Prof. Pantelić's innovative work.

Clearly, it was the Professor's idiosyncrasies that led to this patent conceived apparently contrary to the fundamental principles of mechanics. Namely, the only way to make a helicoid belt revolve around a central post without friction was to convert it into a solid body; for the belt as a floppy form, the task would be impossible. Many were sceptic, but Prof. Pantelić succeeded.

Recognition of creative powers and already achieved results of Belgrade's IFToMM centre came in 1974 with the Symposium on Application of the Theory of Machines and Mechanisms in Dublin (Ireland). Prof. Duffy from the Liverpool Polytechnic was chairing the Organising Committee, and Sir Crossley, president of IFToMM, personally entrusted Prof. Pantelić with chairing the event.

Fig. 3 Milan, Ninth IFToMM World Congress 1995. At the Leonardo da Vinci Museum (with a model of a hellicoid in the background), left to right: Lj. Miladinović [Serbia], A. Sekulić [Ser], A. Veg [Ser], N. Pavlović [Ser], M. Vukobratović [Ser], B. Roth [USA], A. Morecki [Pol], A. P. Bessonov [Rus], unknown, L. Cvetičanin [Ser], M. Pantelić (Prof. Pantelić's wife) [Ser], T. Pantelić [Ser]

In 1997, 26 professors from all parts of the world came to Belgrade Symposium on the Theory of Machines and Mechanisms, organised in his honour: Prof. M. Ceccarelli from Italy, Prof. T. Leinonen from Finland, Prof. E. J. Hahn from Australia, Prof. F. G. Dedini from Brazil, Prof. S. Kato from Japan, and Prof. G. Bőgelsack from Germany.

In 1999, while bombs were falling on Belgrade (again), Prof. Pantelić received an invitation to chair the Tenth World Congress of IFToMM in Oulu (Finland). He thanked for the honour but stayed at home.

Later that year, when he suddenly and unexpectedly died, his wife and three daughters received condolences from all over the world.

Main Works on Mechanism Design

Prof. Pantelić's main works include:

- A four-bar mechanism applied to operations on pieces on a conveying line (Pantelić 1974, 1977b). It represented a new type of mechanism that replaces intermittent mechanism, and it was used for manufacturing 120 pastry production lines (*Minel* Company, Pančevo, Serbia).

- Settlement of motion equivalence between four-bar and slider-crank mechanisms (Pantelić 1976, 1981b), which enabled synthesis of a family of extended slider-crank mechanisms with a member performing approximate curvilinear translation (Pantelić 1975a, 1975b , 1976, 1977b, 1980, 1981a, Stoimenov and Pantelić 1976). It was used for highly efficient pastry production lines, in which the translatory member carries the tools for executing individual operations. One of its inversions was applied to the model of the *walker* mechanism (Pantelić 1980, 1981a, 1981d, 1983a, Stoimenov and Pantelić 1981).
- Synthesis of six-bar linkage with three dwells in the motion of its driven bar (Martinović R and Pantelić 1975a, 1975b). It was the first practical solution of the problem since 1932. It was used to double or treble the number of rows in conveying lines, with the respective dwells in the motion phased by π, or $2/3\pi$.
- An advanced numerical method of the inertial forces harmonic analysis (Četić D and Pantelić and Sekulić 1971; Pantelić, 1981c) applying Bessel functions for balancing inertial forces of basic linkages. The results of this work were applied to balancing the new family of slider-crank mechanisms used for Prof. Pantelić's *walker.*
- Synthesis of four-bar with additional dyad (Pantelić 1977a, 1980, 1981). It generated a mechanism for a deep drawing machine press with very fast approach, relatively slow imprint, and fast return. Compared to the conventional machine presses, productivity of the new one was doubled.
- Research of kinematic chains with zero degrees of freedom, mainly for steel and chain conveyor belts, which contributed to optimisation of their drive and control (Četić D and Pantelić 1980).
- Helical conveyor and overcoming Euler friction (Bukumirović and Pantelić 1975a, 1975b), as the most serious challenge of Prof. Pantelić's career.

Neither a groundbreaking theorem nor a revolutionary new theory is what Prof. Pantelić is to be praised for. He did something else: he made a point of introducing engineering talent and creativity into the science of machines and mechanisms. His art was *simple and efficient transformation of an idea into a product.*

And to that vision he dedicated his whole career, his life too. Working with him, generations of students discovered the beauty of engineering creation, learned to love their profession and to enjoy every day of creative work and every achievement they made.

At the peak of his career, Prof. Pantelić found himself in a unique position to connect creative work of academic communities of all 17 members of the IFToMM within the Commission for Collaboration of Science and Industry, the so-called Commission C.

He invested all his talent and enthusiasm into realisation of the IFToMM *Programme on Collaboration of Science and Industry in the Theory of Machines and Mechanisms.*

Establishing of the Commission C was proposed in 1971 in Düsseldorf (Germany) at the meeting of the IFToMM Executive Council. An action plan in 12 steps was formulated, and Prof. Pantelić (nominated by Academician Artobolevskii) was elected President of the new commission (Fig. 4).

Jugoslovenski nacionalni komitet „IFToMM"
Br. 018
1 74 g.
BEOGRAD, ul. 27. Marta 80
Tel. 29-212 ili 29-241

IFTOMM

INTERNATIONAL SYMPOSIUM - MACHINES AND MECHANISMS, UNIVERSITY RESEARCH WORK AND ITS APPLICATION TO INDUSTRY.

ACKNOWLEDGEMENTS

The Chairman of the Organising Committee wishes to express gratitude to the Convention Bureau of Ireland and to Professors P Leahy and J R McCraith of University College, Dublin for their assistance and cooperation in organising the Symposium.

PROFESSOR T P PANTELIC
FACULTY OF MECHANICAL ENGINEERING
UNIVERSITY OF BELGRADE, YUGOSLAVIA
CHAIRMAN, IFTOMM COMMISSION "C".

Chairman of Organising Committee:
Dr. J Duffy
Reader in Mechanical Engineering
Department of Mechanical, Marine and Production Engineering
Liverpool Polytechnic, U.K.

Fig. 4 Proceedings Dublin 1974, front page

The first meeting of Commission C was held in 1971 in Kupari (Yugoslavia), the second one in 1972 in Miskolc (Hungary), the third one in 1973 in Bucharest (Romania). The pioneering job of harmonising creative work of university researchers and industry designers was soon in full swing, as can be seen from the proceedings of the 1974 IFToMM Symposium in Dublin.

Obviously, Prof. Pantelić was not leading such a creative army only because of his brilliant managerial skills; it was also because he had a gift for closing up the golden engineering triangle *idea→concept→product.*

An eye for harmony and beauty helped him create elegance without effort. "A beautiful construction is certainly a functional one, that's for sure," he used to say.

Design without effort came from the synergy of his exceptional visualisation, thorough understanding of fundamental sciences, and fascination with mechanisms.

Naturally, it resulted in axiomatic design. Although the following examples (chosen from among numerous successful constructions of Prof. Pantelić) seemed technically incomparable, he still found their common point and used it as a basis for conceptual design, which then led to two outstanding solutions (Pantelić 1965, 1983).

The first one is applied to an automatic bun production line (Fig. 5), and the second – to an orthopaedic device.

It all started with a mechanism for imprinting buns on a conveying line. A standard industrial solution widely applied in similar technologies was available: a configuration with coupled motions – horizontal, which follows a bun, and vertical, which imprints it in the process. The challenge was to simplify the configuration by implementing a single member with the resulting path.

The first attempt was synthesis of a four-bar chain where the coupler generates the desired path. But the search went on.

The same operation can be performed by adding new links to a kinematic chain, which follows the same path (Fig. 6). In that case, a link's endpoint in two inverted slider-crank mechanisms performs curvilinear translation, that is follows exactly the desired path. The solution was modified into an eight-link mechanism by omitting one slider and adding one link. The search for more efficient solutions continued.

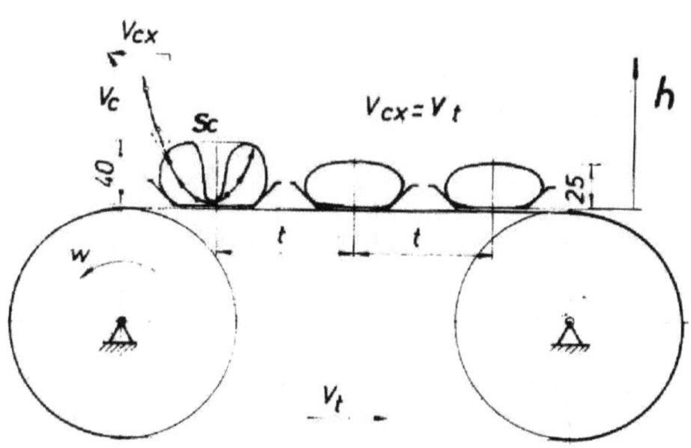

Fig. 5 Path analysis of the bun imprinting tool (Pantelić 1975b)

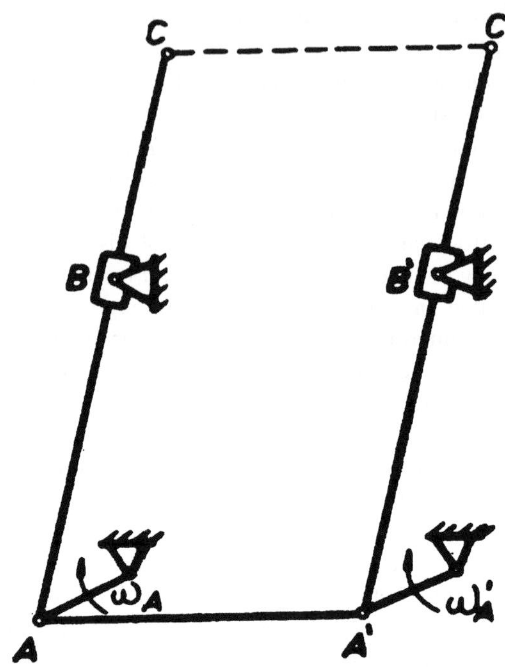

Fig. 6 Inverted slider-crank (Pantelić 1975b)

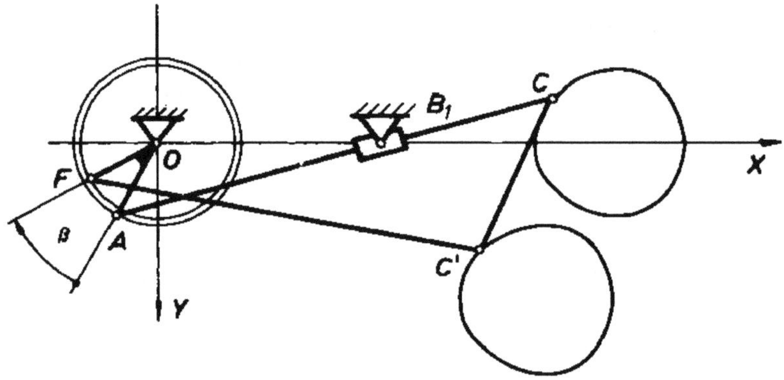

Fig. 7 Slider-crank mechanism with additional dyad – curvilinear translation (Pantelić 1975b)

Prof. Pantelić came up with a new task: to create a six-link inverted slider-crank mechanism with additional dyad, in which coupler CC' performs curvilinear translation with the given path. It took quite a while to finalise the solution of this problem, which had not been solved up to that time.

In the inverted slider-crank mechanism with additional dyad (Fig. 7), translatory motion of the link CC' is due to points C and C' following approximately the same path. Such a simplified configuration was far more suitable for application to production lines.

Once established, synthesis of the extended inverted slider-crank mechanism with a member that performs approximately curvilinear translation, enabled a whole generation of innovative modifications.

The breakthrough occurred when Prof. Pantelić noticed hidden similarities in two apparently very different processes: bun imprinting and – human walking! In operational cycle of the imprinting mechanism (Fig. 8/1) there is a fragment with

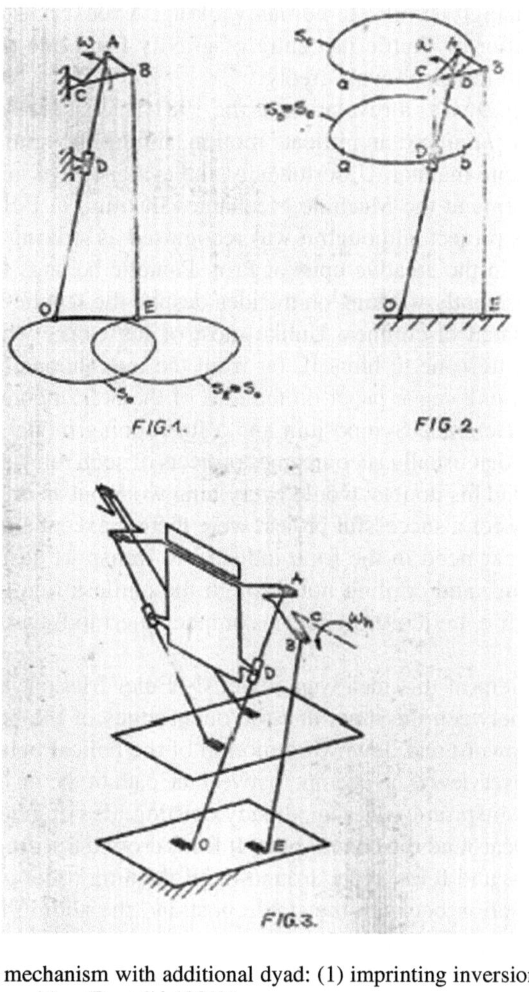

Fig. 8 Slider-crank mechanism with additional dyad: (1) imprinting inversion, (2) walker inversion, (3) walker composition (Pantelić 1981b)

no relative motion between the link CC' and the bun; similarly, from time to time, there is no relative motion between the foot and the floor.

Looking at the new inversion, the so-called *walker* mechanism (Fig. 8/2), one sees that its coupler CD performs curvilinear translation, which enables the *walker* to make a step forward relative to the floor. The basic structure of this mechanism is a four-bar OABE, extended with a coupler CD and a slider D. The transition from the raw concept to the actual "walking model" was long and winding, but Prof. Pantelić had a clear guiding vision.

Two identical walker mechanisms were placed in parallel planes, with cranks shifted for half a cycle and embedded in the point C (Fig. 8/3). The grounding of each walker acts as a foot; when one of them is firmly on the ground, the other is in the air and, due to rotation of the crank, performs a step forward. Alternating and synchronised lifting and dropping of the feet, and their curvilinear translation forward appears amazingly similar to human walking. A rocker also had to be added in the model in order to shuttle its centre of gravity from one plane to the other, enabling the foot in the air to walk freely.

The Walker (Fig. 9) was a revelation of the 1981 Bucharest Symposium *Theory and Practice of Mechanisms*: an intricate motion, similar the human walk, was realised in a very simple manner. Unfortunately, the avalanche of new, easily marketable ideas and patents at the Machine Mechanics Institute in Belgrade diminished the interest for this project, although it was recognised as striking and successful.

A special place in the creative opus of Prof. Pantelić belongs to his helical conveyor. He was persistently working on the idea despite the sceptic attitude expressed in some of the related disciplines. Unlike most of his works which were broadly published, he kept this one to himself, far from the international scene. Except for four master theses, one vague paper on the idea of the helical conveyor, one review paper at the 1997 Belgrade Symposium and a functioning model, there is no other technical evidence that usually accompanies projects of such relevance. Prof. Pantelić himself probably had his doubts: would everything work out as in his vision? All the elements necessary for a successful project were there – except a market boom.

There was a clear need in the food industry to transport pastry in the process of warming, baking, and cooling not through the endless line tunnels (30–70 m long!) but through a helicoid, inside a compact thermodynamically controlled space.

The main problem of this idea was obviously Euler friction; there was also the speed difference between the inner and the outer edges of the belt, as well as the inconformity of upward and downward motion of the helical belt.

The following overview is an attempt to reveal the path of the invention's evolution.

Prof. Pantelić's inspiration was an already existing idea of a helical transporting belt which ascends around the central post. It folds around the dragging roller at the top, then descends, and folds again around the tightening roller at the bottom. The problem is that friction between the stable post and the sliding belt, according to Euler's formula, grows exponentially.

There was also a version with a rotating post that drags the belt upwards; then, after a short linear part, the belt descends the second rotating post and proceeds linearly to the starting point.

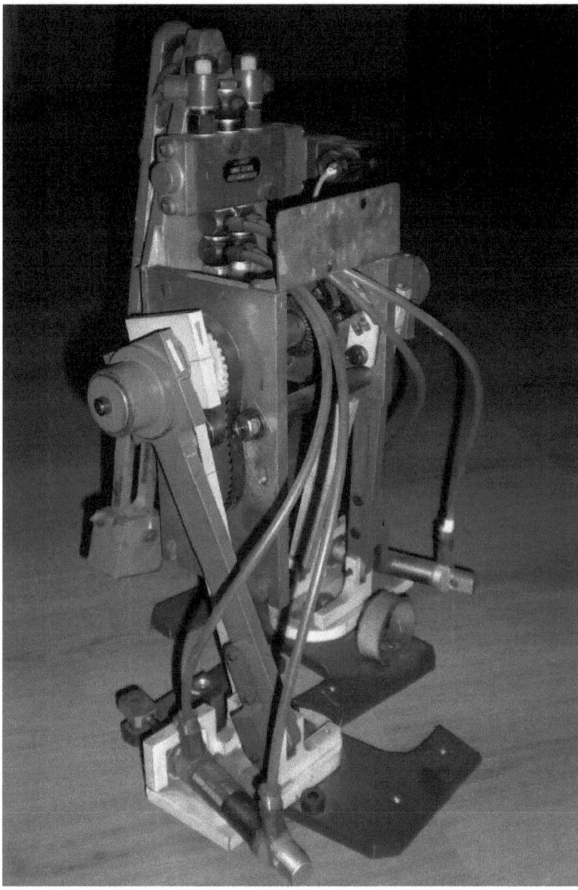

Fig. 9 Mechanical Walker (Photo A. Veg, 2008)

A serious disadvantage of the first solution was the enormous dragging force needed to overcome enormous friction. In the second one, combination of curved and linear paths of the belt, driven in free contact with rotating posts, caused various functional difficulties.

The creative illumination came when Prof. Pantelić wrote down all the necessary conditions and limitations for his helical conveyor:

- The helical belt should not touch the central post.
- Within the helical guides, belt chain members must loose all degrees of freedom, so the belt could rotate as a rigid helicoid (in such a way the dragging force of the helical conveyor becomes approximately equal to the dragging force of the correspondent line conveyor, (Miljković M. and Pantelić 1997)).
- The inner diameter of the solid helicoid must be significantly greater than the diameter of the central guiding cylinder.

– Coming out of helical guides, and also engaging with the driving rollers, belt chain members regain necessary degrees of freedom, so the belt could behave as a solid body.

This was the basis for Prof. Pantelić's new solution. In a consistent and focused developmental phase (Fig. 10), he defined the geometry of the belt; defined the

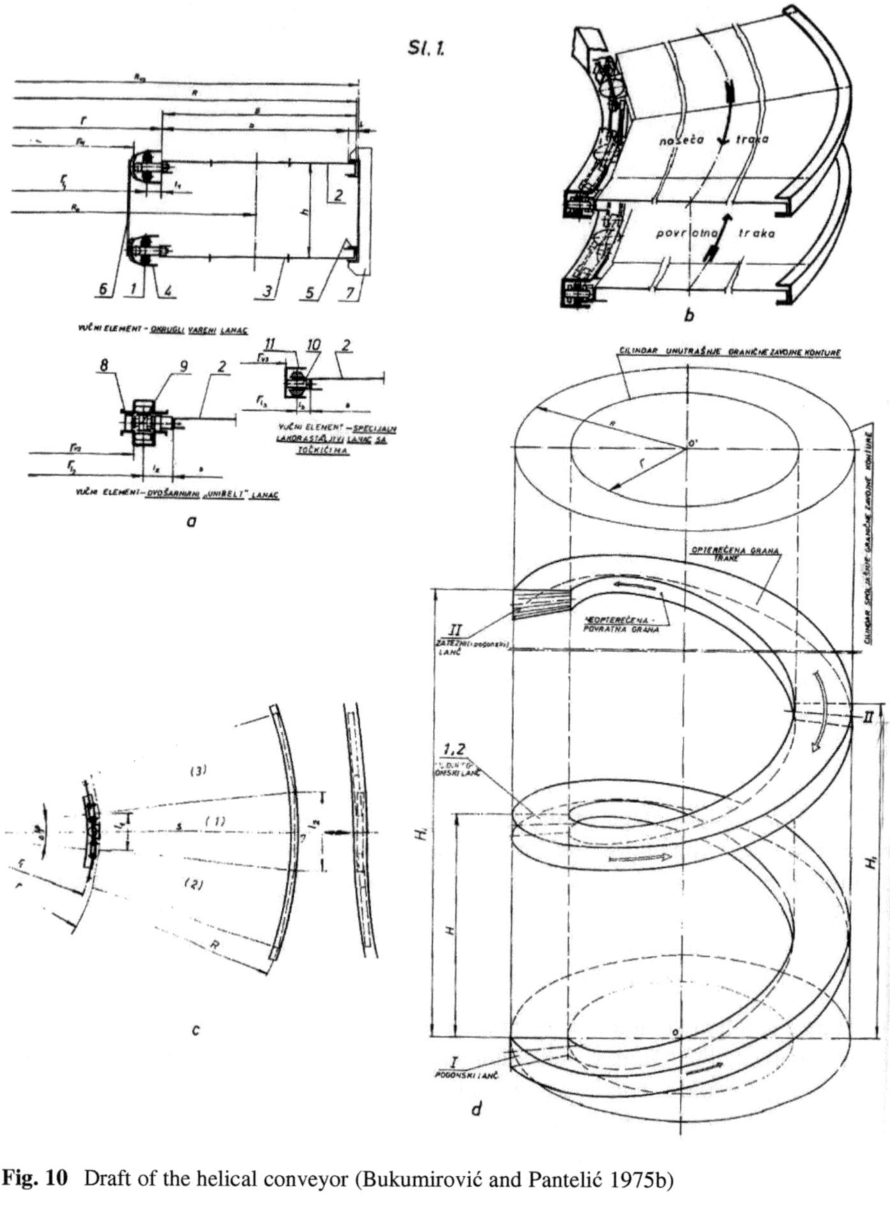

Fig. 10 Draft of the helical conveyor (Bukumirović and Pantelić 1975b)

upward and downward trajectories; constructed the helical guides; calculated contact radii of the driving roller; optimised speeds of the belt chains, that is developed a prototype (Fig. 11). Two models of the helical conveyor were made and they both fulfilled all the technical requirements.

Following the development of this project through documents, I concluded that it was the most serious challenge of Prof. Pantelić's career. Like a sport champion after achieving a record that will obviously stand for many years, he lost any further wish to compete.

The moment prototype of the helical conveyor came to life, all doubts of his colleagues disappeared and, even more importantly, his personal hidden doubts too. I think it was the last and the greatest creative flash of Prof. Todor Pantelić.

From then on, resigning the post of a great inventor, he focused on his teaching mission, leaving space for the upcoming younger colleagues.

Fig. 11 Helicoid prototype (Photo A. Veg 2008)

On the Circulation of Works

There are several aspects of Prof. Pantelić's conspicuously dynamic career.

Firstly, he did all he could to disseminate the theory of machines and mechanisms in the territory of former Yugoslavia. He had postgraduate and doctoral students from all the emerging Yugoslav centres – Niš (Serbia), Skopje (Macedonia), Mostar and Sarajevo (Bosnia and Herzegovina), Podgorica (Montenegro) – as well as from the centres which already had scientific reputation, only not in the field of mechanisms – Ljubljana (Slovenia) and Zagreb (Croatia). Most of those postgraduate and doctoral students became professors at their home universities, broadening the scientific community dedicated to the theory of machines and mechanisms and related fields.

He travelled and lectured extensively and organised symposia on the theory of machines and mechanisms in all those academic centres.

Another significant aspect of Prof. Pantelić's career was theoretical elaboration of insufficiently studied issues in the theory of machines and mechanisms which were needed for his technical innovations.

Application of those newly created mechanisms represents the third aspect of his career – undoubtedly its crown, since it has confirmed all his creative visions. An army of young enthusiastic engineers pursuing their dreams were involved in realisation of his projects. Prof. Pantelić offered a chance to all of them – 22 doctoral, 47 postgraduate, and more than 400 graduate students. They all worked towards their degrees in the innovative atmosphere of the Machine Mechanics Institute in Belgrade.

Prof. Pantelić was the intellectual and spiritual leader of his academic community, powerful, friendly, and informal. Working sessions at the Machine Mechanics Institute would start with his brief announcement: "Gentlemen, I signed a new contract, now let's make it happen." Ideas would come pouring in – what needs to be solved, done, connected, and in what way – from students, assistants, and researchers alike, who were all in a creative fever. At the end of a session, Professor would say: "If anything is still unclear or controversial, make the travel arrangements through the secretariat and go wherever they worked on it before, but come up with a solution in 20 days." And then there would be 20 days of hectic racing, sleepless nights, and excitement so much like the first romantic ones. Such was the creative magic of Prof. Pantelić.

But he did so much more.

He showed us there were no unconquerable fortresses; he showed us how to use a hook, how to catch small and big fish; what to wear to a business meeting; where to find the best men's shoes in the world; where to drink the finest espresso; that Volvo launched its Golden Series – small things like that, and so much more: how to enjoy every moment of life; how to sing in the company of friends; he even used to play for us if an accordion was somewhere in sight.

Finally, he directed his patent rights (1968) to the *Lazar Pantelić Foundation* (named after his father) in order to provide scholarships for excellent but poor students of machine engineering. It was indeed an honour to win them, as some 30 engineers who managed to do so still proudly point out (Contract on patents...).

Acknowledgement This text would have been impossible to write had it not been for the years I have spent learning, working, and forming scientific visions under the guidance of my teacher and mentor Prof. Todor Pantelić.

I wish to thank Prof. Pantelić's close friends and colleagues who helped me reconstruct his creative work and ideas (in random order): Prof. Gerhard Boegelsack (TU Illmenau, Germany), Prof. Života Živković (MEF Niš, Serbia), and Prof. Aleksandar Sekulić (MEF Belgrade, Serbia).

The documents I relied upon were mainly from the library of the Machine Mechanics Institute, as well as from the library and the archives of the Faculty of Mechanical Engineering, University of Belgrade. I am grateful to all the librarians and archivists who kindly helped me find my way around while preparing this text.

References

Bukumirović M, Pantelić T (1975a) Transport u helikoidi. Zbornik radova sa simpozijuma *Mašine i mehanizmi*, JNK IFToMM, Zagreb, tom II, str 53–68 [Helicoidal transport. In: Machines and mechanisms, vol. II. Symposium Proceedings, under the auspices of IFToMM, Zagreb, pp 53–68]

Bukumirović M, Pantelić T (1975b) Helicoidal transport. In: Proceedings of the second international Brazilian congress of mechanical engineering, Rio de Janeiro, p 165

Četić D, Pantelić T (1980) Pokretljivost kinemati čkog lanca sa jednim elasti čnim članom i nula stepeni slobode kretanja. Zbornik radova sa III jugoslovenskog simpozijuma *Mašine i mehanizmi: Univerzitetska istraživanja i njihova primena u industriji*, Mostar, tom II, str 365 [Mobility of kinematic chain with a flexible member and zero degrees of freedom. In: Machines and mechanisms: University research work and its application to industry, vol. II. Proceedings of the Third Yugoslav Symposium, Mostar, p 365]

Contract on patent rights transfer (30 December 1968) to the benefit of the Lazar Pantelić Foundation, Faculty of Mechanical Engineering Archive

Evaluation report on the occasion of Prof. Pantelić's promotion (Faculty of Mechanical Engineering, Belgrade, 1983), Faculty of Mechanical Engineering Archive

Gobeljić A, Stoimenov M, Miladinović Lj, Pantelić T (1981) Problems of balancing the mechanical walker. In: Theory and practice of mechanisms. Proceedings of the Third International Symposium, IFToMM SyROM 81, Bucharest, Romania, pp 107–116

Martinović R, Pantelić T (1975a) Sinteza šesto članog zglobnog mehanizma sa tri pauze u kretanju vođenog člana. Zbornik radova sa simpozijuma. *Mašine i mehanizmi*, JNK IFToMM, Zagreb, str 207–221 [Synthesis of six-bar linkage with three dwells in the movement of its driven bar. In: Machines and mechanisms. Symposium Proceedings, under the auspices of IFToMM, Zagreb, pp 207–221]

Martinović R, Pantelić T (1975b) Synthesis of six-bar linkage with three dwells in the movement of its driven bar. In: Proceedings of the fourth world congress on the theory of machines and mechanisms, Newcastle, p 102

Miladinović Lj, Gobeljić A, Ostojić M, Pantelić T (1980) Problemi uravnotežavanja mehani čkog hoda ča. Zbornik radova sa jugoslovenskog simpozijuma *Mašine i mehanizmi: Buka, vibracije i uravnotežavanje rotora i mašina*, Beograd, str 185 [Problems of the mechanical walker balancing. In: Machines and mechanisms: Noise, vibrations, and balancing of machines and rotors. Proceedings of the Yugoslav Symposium, Belgrade, p 185]

Miljković M, Pantelić T (1997) Principle of achievement of helical conveyors at which drag force does not grow progressively at the change of number of helical strokes. In: Machines and mechanisms. Proceedings of the International Symposium, Belgrade, pp 221–224

Pantelić T (1974) Anwendung von Viergelenkgetriebe zur ausführung des Verfahrens an bewegenden Werkstucken. In: Machines and mechanisms: university research work and its application to industry. Proceedings of the IFToMM International Symposium, Dublin, paper no. D4

Pantelić T (1975a) Primena krivajno-kliznog mehanizma za obavljanje operacija na radnim predmetima koji se kreću. Zbornik radova sa simpozijuma *Mašine i mehanizmi*, JNK IFToMM, Zagreb, str 245–252 [Application of slider-crank mechanism to shape moving items in line production. In: Machines and mechanisms. Symposium Proceedings, under the auspices of IFToMM, Zagreb, pp 245–252]

Pantelić T (1975b) Possible application of slider-crank mechanisms for performing operations on moving bodies. In: Proceedings of the fourth world congress on the theory of machines and mechanisms, Newcastle, p 4

Pantelić T (1976) Zglobni četvorougaonik sa specifi čnim parametrima – ekvivalentan mehanizam krivajno-kliznom. Prezentacija istraživanja, Mašinski fakultet, Beograd [Specific configuration of four-bar mechanism equivalent to slider-crank. Research presentation, Mechanical Engineering Faculty, Belgrade]

Pantelić T (1977a) Primena zglobnog četvorougaonika proširenog dodatnom dijadom kod prese za duboko izvla čenje. Zbornik radova sa simpozijuma *Mašine i mehanizmi: Univerzitetska istraživanja i njihova primena u industriji*, JNK IFToMM, Beograd, str 1–13 [Application of four-bar mechanism with additional dyad for the deep drawing machine press. In: Machines and mechanisms: university research work and its application to industry. Symposium Proceedings, under the auspices of IFToMM, Belgrade, pp 1–13]

Pantelić T (1977b) Primena zglobnog četvorougaonika za obavljanje operacija na radnim predmetima koji se kreću. Zbornik radova sa simpozijuma *Mašine i mehanizmi: Univerzitetska istraživanja i njihova primena u industriji*, JNK IFToMM, Beograd, str 231–240 [Application of four-bar mechanism to shape moving items in line production. In: Machines and mechanisms: University research work and its application to industry. Symposium Proceedings, under the auspices of IFToMM, Belgrade, pp 231–240]

Pantelić T (1980) Jedan primer ostvarenja usaglašenosti potreba tehnološkog procesa i mogućnosti sredstava rada. Zbornik radova sa III jugoslovenskog simpozijuma *Mašine i mehanizmi: Univerzitetska istraživanja i njihova primena u industriji*, Mostar, tom I, str 189 [An example of harmonization between technological needs and technical abilities of machinery. In: Machines and mechanisms: University research work and its application to industry, vol. I. Proceedings of the Third Yugoslav Symposium, Mostar, p 189]

Pantelić T (1981a) Sintez shestizvennogo ricijannogo mehanizma s tseliu obespechenia neizmennoi skorosti na opredelenom intervale. In: Theory and practice of mechanisms. Proceedings of the Third International Symposium, IFToMM SyROM 81, Bucharest, Romania, pp 87–92

Pantelić T (1981b) Zglobni četvorougaonik posebnih parametara, ekvivalentni mehanizam krivajno-kliznom. Zbornik radova sa III jugoslovenskog simpozijuma *Kinematika i dinamika mašina i mehanizama*, IFToMM (JUToMM), Ljubljana, str 67–72 [Four-bar mechanism equivalent to slider-crank. In: Machine kinematics and dynamics. Proceedings of the Third Yugoslav Symposium, under the auspices of IFToMM, Ljubljana, pp 67–72]

Pantelić T (1981c) A contribution to the control of the system with eight input variables. In: Proceedings of the sixth international Brazilian congress of mechanical engineering. COBEM 81, Rio de Janeiro, pp 98–104

Pantelić T (1981d) Application d'un mécanisme a collisse élargi par une dyade sur un model de marcheur mécanique. In: Proceedings of the sixth international Brazilian congress of mechanical engineering. COBEM 81, Rio de Janeiro

Pantelić T (1983a) Application d'un mécanisme a collisse élargi par une dyade sur un model de marcheur mécanique. In: Proceedings of the sixth IFToMM congress on the theory of machines and mechanisms, New Delhi, India, pp 106–111

Pantelić T (1983) Universal method of synthesizing a four-bar linkage as a rigid-body-guidance mechanism and function-and-path generator for up to five precise position. In: Proceedings of the sixth IFToMM congress on the theory of machines and mechanisms, New Delhi, India

Pantelić T (November 1965) Određivanje kinemati čkih veli čina bregastih ravnih mehanizama na bazi ekvivalentnih polužnih mehanizama. *Mašinstvo*, Beograd, str 18–24 [Definition of cam parameters based on equivalent linkages. *Mašinstvo* J (Belgrade):18–24]
Pantelić T, Sekulić A (1971) Primena Besselovih funkcija uravnotežavanja inercijalnih sila ravanskih polužnih mehanizama. Proceedings of the Third World Congress on the Theory of Machines and Mechanisms, IFToMM, Kupari, tom B, str 127–140 [Application of Bessel functions for balancing inertial forces of basic linkages. In: Proceedings of the fourth world congress on the theory of machines and mechanisms, vol. B. IFToMM, Kupari, pp 127–140]
Stoimenov M, Pantelić T (1981) Primena krivajno-kliznog mehanizma proširenog dodatnom dijadom na model mehani čkog hoda ča. Zbornik radova sa III jugoslovenskog simpozijuma *Kinematika i dinamika mašina i mehanizama*, IFToMM (JUToMM), Ljubljana, str 523–528 [Application of slider-crank mechanism extended with additional dyad to a model of mechanical walker. In: Machine kinematics and dynamics. Proceedings of the Third Yugoslav Symposium, under the auspices of IFToMM, Ljubljana, pp 523–528]
Stoimenov M, Pantelić T (June 1976) Mogućnost ostvarivanja krivolinijske translacije člana proširenog kinemati čkog lanca krivajno-kliznog mehanizma. *Tehnika*, Beograd XXXI, str 994–997 [Curvilinear translation of additional member in an extended slider-crank kinematic chain. *Tehnika* J (Belgrade) XXXI:994–997]

Ferdinand Jakob Redtenbacher (1809–1863)

Jörg Wauer, Klaus Mauersberger, and Francis C. Moon

Abstract The rapid development of technology in the twentieth century, especially in the automotive and aircraft industries as well as machine tool and manufacturing engineering, is causally connected to the developments of the scientific principles of mechanical engineering in the nineteenth century. The transformation of mechanical technology from a mainly workshop-based profession into an engineering science, particularly in the area of design of machines and mechanisms, is closely related with the name of Ferdinand Redtenbacher. He was of eminent historical importance for this evolution of mechanical engineering to a technical science in the nineteenth century. Redtenbacher's role as the originator of mechanical engineering as a science in Germany is fundamental and of general interest since the introduction of science into engineering is a "sine qua none" for the development of a powerful high tech industry. As a professor and the director of the Polytechnic School in Karlsruhe, Germany – today the University of Karlsruhe (TH) –, Redtenbacher's influence therefore helped establish the German manufacturing systems engineering even today.

Biographical Notes

At the beginning of the nineteenth century, Germany and Austria were increasingly anxious to establish industries to counter the English manufacturing dominance. This beginning development of capitalist industrialization was accompanied by the

J. Wauer (✉)
Institut für Technische Mechanik, Universität Karlsruhe, 76128 Karlsruhe, Germany
e-mail: wauer@itm.uni-karlsruhe.de

K. Mauersberger
Kustodie, TU Dresden, 01062 Dresden, Germany
e-mail: klaus.mauersberger@tu-dresden.de

F.C. Moon
Cornell University, Ithaca, NY 14853, USA
e-mail: fcm3@cornell.edu

M. Ceccarelli (ed.), *Distinguished Figures in Mechanism and Machine Science: Their Contributions and Legacies, Part 2*, History of Mechanism and Machine Science 7, DOI 10.1007/978-90-481-2346-9_12, © Springer Science+Business Media B.V. 2010

Fig. 1 Lithograph of Ferdinand Redtenbacher (1809–1863)

institution of technical educational establishments modeled after the founding of the Ecole Polytechnique (1794) in Paris. Among the first polytechnic schools in middle Europe were the Vienna Polytechnic Institute (1815) in Austria and the Polytechnic School of Karlsruhe (1825) in Germany.

Early Years

Redtenbacher's childhood and early teens in Austria fell in the time of transition into the industrial age. The hustle of the iron and steel industry of his hometown, later known as the "Austrian Birmingham", were parts of the surrounding scenery for Redtenbacher, in which he was born into. But it was not the primary factor that became decisive for his later lifework. The ideals of education put forward by Vienna Polytechnic Institute and its director Johann Joseph Chevalier von Prechtl, influenced the young generation in Austria including Redtenbacher.

Ferdinand Jakob Redtenbacher (1809–1863) (Fig. 1) was born on July 25, 1809 in Steyr, Upper Austria. His parents' home was closely connected with the systems of acquisition and economy that had barely changed during the centuries. His father, Alois Vincent Redtenbacher, a scion of an old-established reputed dynasty of business magnates from Upper Austria, had inherited Voith's hardware shop in Steyr. Due to his economic wealth, Alois Redtenbacher belonged to the dignitaries of the small city. The father was described as a bright and likeable person, witty in his views and with affinities to literature and languages to pass them on to his children Alois and Ferdinand. Growing up in this contemplative atmosphere of the Biedermeier period, Ferdinand Redtenbacher had only sparse contact with engineering and technology in his young childhood.

The father wanted Ferdinand to become a merchant. At the age of 10 years he was put into a retail apprenticeship at his uncle's general store in Steyr. For 5 long years, Redtenbacher had to hold out and at the end of 1824 he had finished this first education step. It is difficult to say what these unloved years mattered for the adult intellect. However, it can be assumed that the young Redtenbacher could draw such advantages from this severe time as work discipline and practical understanding.

Besides his job, he took time to learn the basics of technical drawing and mathematics through private study, and this may certainly be interpreted as the first independent act to escape the retail career after the apprenticeship. The skills were definitely helpful and necessary when, after the end of the apprenticeship from January 1825 through September of the same year, Redtenbacher was employed by the building authority of Linz, Austria to draw construction plans and to help take geometrical measurements. Also during this time in Linz, Redtenbacher tried through private learning to come closer to becoming a capable engineer. In the fall of 1825, the free-thinking father let the single-minded son go to take up his studies at the Vienna Polytechnic Institute.

Education in Vienna

During Redtenbacher's time of studying in Vienna, reform of technical education began, but it was far from the result realized by Redtenbacher later at the Polytechnic School of Karlsruhe. However, the schools at that time cannot be compared with the technical universities of today. The Vienna school had an entrance age of 13 years for pupils. The preparatory junior school with its two annual classes, taught the subjects of religion, exercises in good reading, German grammar and style, element and (basics of) advanced mathematics, geography, history, drawing, good writing, Italian, French and the possibility to learn English, Czech and Latin. Only after taking these subjects would the technical section start with further instruction, depending on the career aspirations, in subjects today taught in the departments of mathematics, physics, chemistry, architecture, civil engineering or mechanical engineering. All this was certainly less thoroughly practiced than today but in its great variety and large extent, it was no less exhausting than the study of contemporary engineering.

Redtenbacher was studied not exactly a prodigy at the age of 16 years but, with his deficient elementary education, he faced up to the demands of the school. After successfully passing the junior school he studied technical drawing and design (with Arzberger), surveying and mapping, planimetry, drawing theory of maps, leveling and altimetry, drawing of geometrical layouts and maps, practical measuring (with Simon Stampfer), as well as road construction, hydrology, civil engineering economy, drawing of architecture layouts (with Josef Purkinje). He passed all exams with first-class grades and preference when he finished his engineering study. It is important to state that in the early nineteenth century, a mechanical engineering curriculum did not exist and engineering training was associated with what is called today civil engineering.

A special feature of Redtenbacher's studies in Vienna was that from the first school year, he participated, at his own option, in additional so-called "Tentamina", ceremonial public exams where the professors recognized a small selection of their outstanding students and the results of their work. Parallel to his study in the technical section of the Polytechnic Institute he enrolled at the University of Vienna for courses in advanced mathematics (with Andreas von Ettingshausen) and astronomy (with Karl-Ludwig von Littrow), passing the corresponding exams with excellent results.

The appreciation of bright students by his professors offered Redtenbacher, in 1829, an assistantship in mechanics and machine design. After studying with Arzberger for 2 years, his engagement was renewed in 1831 for a further 2 years. His natural teaching ability exhibited in his tutorials on the basics of machine design, became clearer simultaneously with his methodical intellectual power. Furthermore, he had to teach technical drawing and to sketch machines that were manufactured in the workshop of the Polytechnic Institute for the growing model cabinet. No doubt, the easy-going years of his assistantship represented the real student time in which Redtenbacher broadened his mind, through inquisitiveness and diligence.

In Vienna, Redtenbacher also stated for the first time that there is no real knowledge and no certainty excepting such that could be mathematically founded. In this period, Redtenbacher was also for the first time the co-author of a scientific contribution on a fire engine in a Technical Encyclopedia by Prechtl. From this article, it cannot be seen what portion of it was written by his professor and what part by him. Anyway, it is clearly and coherently written and deals with the technical matter in a thorough manner. These 4 years were accompanied by vacations of hiking trips in the vicinity of his homeland where he also inspected machine technology on route (e.g. the grand pumping station of Berchtesgaden, which delivered the brine over the mountain into the brewery of Reichenhall). He also trained himself in freehand drawing, composed outdoor sketches and started to paint, a hobby which he enjoyed for the rest of his life.

Zurich Appointment

The assistantship at the Polytechnic Institute was limited by imperial statute to 4 years and ran out irrevocably in September 1833. In the spring of 1834, his cousin, Dr. Joseph Redtenbacher, accidentally read of a job posting at the Higher Industrial School of Zurich. Dated April 26, 1834, the 25 year old Redtenbacher was appointed as a lecturer of mathematics and descriptive geometry at the Senior Industrial School, initially for 1 year, then finally as a professor.

The technical equipment and experience vital for Redtenbacher's new position was arranged by the Zurich company Escher, Wyss Cie. which started its business in 1805 as a spinnery, later going on to make manufacturing machines for spinning works in its workshop and, since 1836, changed over to the construction of steamboats and locomotives. The head of the company, Hans Caspar Escher, gave Redtenbacher the room to do multi-purposed and comprehensive tests to collect and to separate such data important for the practical daily work that became the starting point of all his subsequent publication activities. His results relating to mechanical engineering were so successful in the Zurich period that Redtenbacher purposed his first publication. However, the emergence of turbines instead of classical water wheels halted his work. Redtenbacher therefore first turned intensively to this new type of power engine, which then became the topic of his first book (in Karlsruhe) before his later work on water wheels and the general principles of mechanical engineering.

Redtenbacher's acuteness of mind grasped many helpful suggestions and practical experiences and amazing theoretical approaches for problems in machine technology. The factory of Escher-Wyss, under the influence of superior English technology with the leading engineers and technicians coming from England, guarded their know-how, which acted as a challenge for Redtenbacher. He wanted to strip off this knowledge secrecy and to place mechanical engineering as the foundation of new scientific principles. However, his main activities were focused on the school. In Zurich, Redtenbacher started thinking about the nature of the engineering school system which constituted the essential part of his lifework along with his scientific publications. Thus, it seems that the nearly 8 year-long restless occupation in Zurich constituted an important waystation for Redtenbacher's career. It let him mature and broaden his mind beyond mechanical engineering. The high reputation which he acquired very quickly brought him membership of the local association of Natural Science in 1834 and spread his reputation beyond Zurich. A token of increasing appreciation was, for example, the corresponding membership in the Lower Austrian association of industry in 1841. In Zurich, he also set the course of his private life: In 1836 Redtenbacher became engaged to his cousin Marie Redtenbacher and married her 1 year later. They had two children, a daughter Marie and a son Rudolf, to whom posterity owes a very informative biography (Redtenbacher 1879) of his father.

Appointment at Karlsruhe

In 1840, Redtenbacher received a call from the Polytechnic School of Karlsruhe. The basic concept of the new institution at Karlsruhe, the first of its kind in Germany, was to integrate the scientific and mathematical basics of engineering competence as an organic unity as part of the Polytechnic schools. The reform in 1832 enforced by Karl Friedrich Nebenius, merged the Civil Engineering School founded by Tulla, the Architectural School tracing back to Weinbrenner, the Industrial School initially domiciled in Freiburg, the School of Forestry, the Trading School and, above all, the Post School. Under Nebenius, the unity of these technical schools, on a unified basis with the preparatory had been much more elaborate than what was first formed when the Polytechnic School was founded in 1825. He had created a type of school somewhere between a grammar school and a university also spatially separated, possessing senate constitution, annular director election by the professors and featured an organization close to a department system of current universities.

Karlsruhe, the capital of the state Baden, with its 25,000 inhabitants, became the permanent home for Redtenbacher. By ministerial resolution of October 13, 1841, i.e. at the beginning of new school year, Redtenbacher was appointed as professor within the Mechanical Engineering Department. It was obvious that Redtenbacher was keen to play a part in Karlsruhe but, at the same time, to get flexibility and space for development to advance the school by pursuing the vision of his progressive concepts.

The subject of Redtenbacher's concern was the senior industrial school. When he started there, it was the technical section that was least developed of all and encompassed the same course of instruction in both engineering chemistry and mechanical engineering. There was no question that it was not mechanical engineering on a scientific basis. Everybody who was interested in becoming a mechanical engineer at that time, normally started with practical training. The industrial school taught theoretical knowledge quite apart from practical experience. One was content to become acquainted with machine elements, studied descriptions of machines and models of them. Dimensions were expected to be given by experience or were modified by empirical trial-and-error methods in the workshop. Strength calculations based on ideas of stress and strain were practically unknown.

During the 1830s of the nineteenth century, and also at the industrial section of the Polytechnic School of Karlsruhe, instruction primarily took place in the workshop in a handicraft-like fashion. Within a few years, Redtenbacher created a fundamental change in this type of education.

Considering the situation of industry in the German state of Baden at that time, the endeavor described in such a manner, at the beginning of Redtenbacher's tenure in Karlsruhe was definitely challenging for him. It is to Redtenbacher's credit that he introduced scientific methods in his lecture as a young 30-year old professor. The scientific treatment of practical engineering problems seemed to Redtenbacher to be not only more economic but also less erroneous than an investigation based on experience. At the same time, Redtenbacher knew that technical education was not sufficient with pure sciences alone, without attending to practical successes.

Herein, Redtenbacher differed from most Paris professors at Ecole Polytechnique. His lecturing was eventually directed to practice. What he had learned rudimentarily as a young student at the Vienna Polytechnic Institute, and what he had taken on as a successful link between science and application from the scientific work of the French polytechnician Jean Victor Poncelet (1826/1838) for his studies on water wheels and turbines during the Zurich time, was now combined by Redtenbacher into a far-ranging, coherent concept. Thus, the bottom line was to bring together both the education carried on by scientific study and practice. Therefore, Redtenbacher in 1846, began to separate the mechanical and the chemical program at the Industrial school into two single programs and in addition to the existing chemical laboratory, he installed a separate machine laboratory. Only after this step was mechanical engineering constituted as an independent department. This was the starting point for mechanical engineering under Redtenbacher and applied chemistry under Carl Weltzien for development into two autonomous and strong departments of international standing.

The fame of the mechanical engineering school of Karlsruhe propagated very fast. From everywhere, the prospective students came for an education at the Karlsruhe Polytechnic School in the mechanical engineering section under Redtenbacher. The number of students increased and at the zenith, Redtenbacher had 359 attendees, where 227 of them were regular students. After a while, Redtenbacher's work governed the scientific standard of his school. Accordingly, in 1857 he was elected as director. Under his directorship, the Karlsruhe Polytechnic School achieved international standing and was a model for the formation of other schools or re-organizations at home and abroad. As noted before, he received several calls from other institutions but he rejected them all. At the beginning of 1850, he received a call from the Austrian Department of Commerce with an offer of the directorship of a company which was the leading establishment of the Austrian railroad. About the same time, Redtenbacher received a prestigious offer for a professorship at the Royal Industrial Institute of Berlin. There were negotiations by letter over a long time, but on October 3, 1850 he decided to remain in Karlsruhe. On August 20, 1854, Redtenbacher also received an exceptionally prestigious call from the president of the Swiss education council, Dr. Kern, for a new professorship of mechanical engineering at the Swiss Polytechnic School of Zurich. It was planned that, under a generous general framework, Redtenbacher should not only undertake the direction of the mechanical engineering section but would participate in the organization of the whole school including the hiring of professors. Working conditions for Redtenbacher in Karlsruhe were already good before these recruiting calls, and the convenient life for him and his family in Germany (the economic situation obviously was further improved and he became privy counsellor so that in the future, some unpleasentnesses he could avoid) again caused him to stay. It must have given him the strongest satisfaction that for the internal organization of the Zurich school, Karlsruhe was taken as the prototype.

According to Redtenbacher's suggestions, the pre-school and the first annular mathematical course and also the Trade School, the Post School and the School of Forestry were abandoned. Instead, a Mining school was founded and philosophy, history, national economy and business, political science and law were included in the curriculum, besides literature and science which had been incorporated long before. The so-called "Education to Industry" already attested in the sense of Friedrich List by Karlsruhe's historian Franz Schnabel, was strongly promoted by these measures.

After the relocation of mechanical engineering in the new building was carried out in 1860, Redtenbacher was at the zenith of his productive life. By establishing humanities, a deeply felt desire of Redtenbacher had come true. Teaching the subjects of humanities at the Polytechnic School meant neither decoration nor luxury, they were a part of the technical studies themselves. What Redtenbacher demanded, or at least expected from his students, with respect to general education essentials beside the usual technical study, he did not excuse himself from either. He did not give up his artistic intentions in painting, music, theatre, poetry until his last years. In connection with his own research, he was inspired by the classical French mathematicians who were seen far above all their contemporarities, especially the English who focused more on practice. The logic and the methodologies of French science were the closest to his heart. From Justus von Liebig he learned to grasp beyond chemistry into adjoining areas. Kuno Fischer who was a recognized philosopher in Jena, gave him an understanding of the newer philosophy, from which he studied Kant and Hegel as well as the much noticed main opus of Hermann Lotze, "Micro Cosmos", and the writings on aesthetics by Friedrich Theodor Vischer. In addition to all the time-consuming teaching and adminstration work and his own literary activities, he even got round to the reading of contemporary historians.

To his admirable self-discipline that he mustered during his studies, during his later years there was added an authoritarian rigorousness towards particular lecturers who became obstinate and were not inclined to follow his plans. He was not always a congenial colleague. He never made a secret of his convictions and he could mercilessly pack a punch in a faculty meeting when mediocrity and small-mindness emerged and when people clung to out-of-date and bureaucratic regulations instead of pursuing the objective requirements. All the time, however, he was backed up by the state authority which was needed to push through his reforms with both meticulousness and impatience against restrictions and obstructions of all kinds, sometimes surely with a good deal of obstinacy as well. Self-willed traits emerged when, for the school year 1861/62, there was simply no re-election of the director – since 1857 year after year, the faculty meeting had voted for him – but he directly pursued his official duties because, to his mind, it would be in the interest of the school to eliminate the previous annular director election by the lecturers of the school by a director appointment for life. The ministry had this opinion of Redtenbacher to prolong his directorship for life which was not explicitly confirmed but condoned and did not urge a re-election. Though some colleagues grumbled, they hesitated to undertake actions against this infringement of their perceived rights, especially because Redtenbacher was at this time starting to be troubled with a stomach illness which developed little by little into a deadly disease. Despite his malady, Redtenbacher still continued his classes during the whole school year 1861/62. In the fall of 1862, after a recuperation journey which should bring recovery, the illness exacerbated such that he had to cease lecturing at Christmas 1862 and also resign the directorate. On April 16, 1863 he silently departed this life blessed by restless work and great success.

These biographical notes, which are a shortened version of Wauer et al. (2008) substantially follow Mauersberger (1989a) and Fuchs (1959) with regard to personal biographical details by Ferdinand Redtenbacher himself (Bader 1858), as well by historical appraisals in Mauersberger (1980, 1985) and Schnabel (1925 and 1938).

List of Works

Theorie und Bau der Turbinen und Ventilatoren, 1844
Theorie und Bau der Wasserräder, 1846
Resultate für den Maschinenbau, 1848
Prinzipien der Mechanik und des Maschinenbaus, 1852
Die Luftexpansionsmaschine, 1852
Die calorische Maschine, 1853
Die Gesetze des Lokomotivbaues, 1855
Die Bewegungsmechanismen, 1857
Das Dynamiden-System. Grundzüge einer Mechanischen Physik, 1857
Die Bewegungsmechanismen, neue Folge, 1861
Die ursprüngliche und jetzige Temperatur der Weltkörper, 1861
Der Maschinenbau, 3 Bände, 1862–1865

Review of Main Works on Machine Science

Within the limits of this contribution, it is hardly possible to completely do justice to Redtenbacher's groundbreaking scientific work; it is too voluminous, too manifold in content and so broad in its scope. Lasting results concern the institutional establishment of mechanical engineering as a science and the formulation of trend-setting methodical design principles. They represent the German way to find a synthesis between the workshop-oriented English technical education close to practice and the highly theoretical machine science of French provenance. Above all, there was the final goal to combine theoretical knowledge, especially of mechanics, with extended practical experience in an appropriate manner. Within this approach, the intention was perceived to extend the brilliant successes of analytical mechanics, in particular to predict the planetary motions. Bearing in mind the fact that machine motion is dictated by terrestrial effects, Redtenbacher aspired to incorporate friction, impact restitution forces and other "noise terms" in a scientific method for predicting the behavior of machines. This could be accomplished not only through reliable knowledge of the mathematical-scientific fundamentals but also through firm workshop experience know-how.

As much as he planned to apply scientific methodology ideals, he did not forget the empirical basis of technology. Intuitively, he developed his knowledge stepwise and started both with the exploration of determinant phenomena and with the examination of different types of machines. In his opinion, every science should begin with induction and end with deduction. His debut publication was in 1844, a book on "Theory and Construction of Turbines and Ventilators" (Redtenbacher 1844), second edition 1860, followed 2 years later by "Theory and Construction of Water Wheels" (Redtenbacher 1846), second edition 1858. This unexpected order had essentially no technical reason. In a letter from his publisher Bassermann of

Mannheim (August 30, 1844), we learn that the "Water Wheels" at that time were ready to print but there was a time lag due to the time consuming production of 30 lithographic tables. Much of the material for both treatises can be traced back to his Zurich time and the collaboration with the company Escher-Wyss, but it was also advanced by study trips 1842 to the industrialized Alsace, 1843 down the Rhine till Cologne, in February 1844 to the Baden Black Forest and in August 1844 to Amsterdam and Haarlem via Cologne, back via Belgium (Figs. 2 and 3).

In the creation of new machines, conventional technical knowledge and the inherited wealth of workshop experience often failed. Therefore, analyses as exact as possible had to be found, in particular for further advancement of such machine innovations. In exemplary form, Redtenbacher tackled the problem. Even though, in the case of turbines, for example the constructions of Fourneyron, Jonval or Schott, seemed to be designed in all details but with respect to limited water resources, there was a need to look for an increase of power or a reduction of energy losses. Additionally, there was the demand for specific dimensioning rules depending upon the application and to harmonize the power requirements with the driven machinery. By his detailed calculation rules based on fundamental hydrodynamic balance equations, Redtenbacher helped to overcome tentativeness in launching such driving aggregates and presented lasting, acceptable construction standards for them.

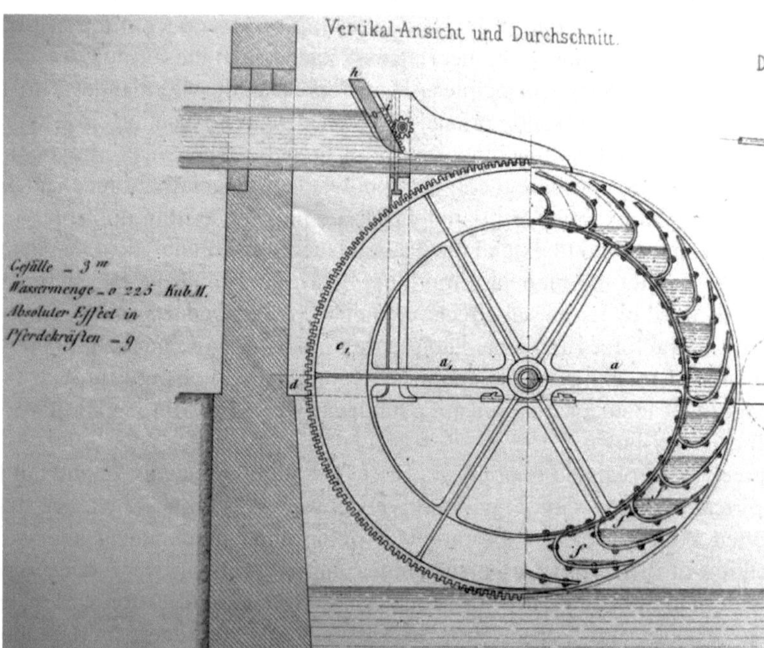

Fig. 2 Water wheel design of Redtenbacher

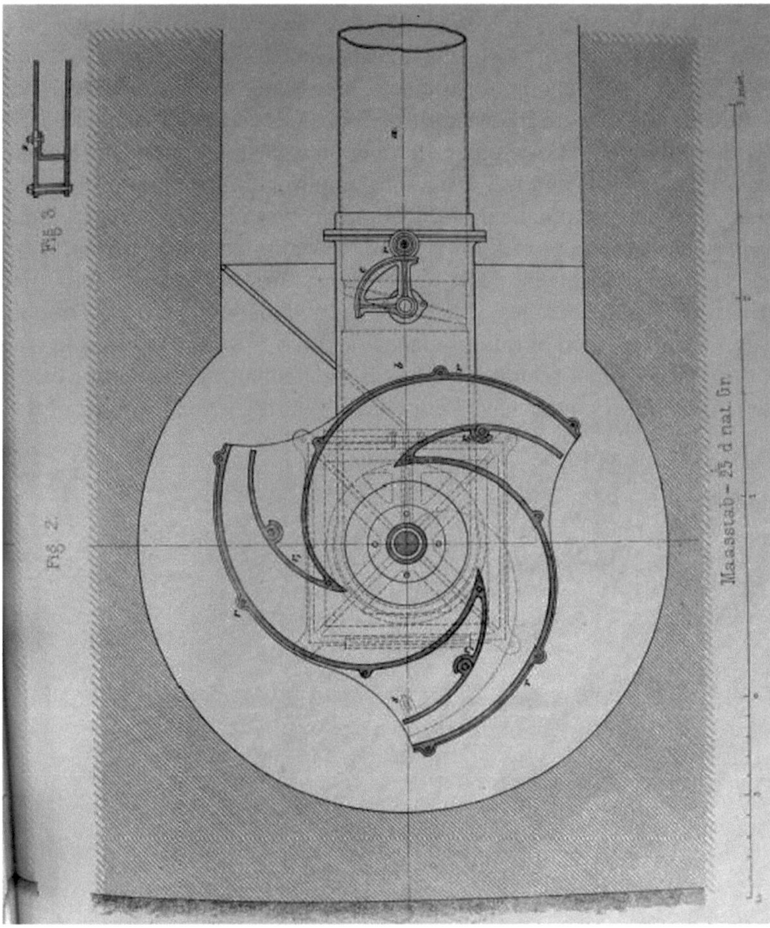

Fig. 3 Water turbine design of Redtenbacher

Moreover, he showed, by comparing issues regarding energy aspects, the advantages and drawbacks of the different designs.

In a report strongly structured according to basic scientific principles, he succeeded in coming up with such effects and to take into account such economic-technical determinants not yet exactly computable in the sense of confirmed scientific knowledge. In this way, Redtenbacher's understanding of sufficiently accurate approximation results was formed to bring them through further research experiments, step by step to a higher level of generality and closeness. Of course, the analytical machine theory vis a vis the French deductive method, was still in its infancy, and there were still too many unknowns in the knowledge basis of the young engineering sciences. For the time being, only use of well-proven technology helped. Recourse to a reliable reference book was among the popular instruments in engineering during that time, as is still true today.

The completion of a design reference book was realized by Redtenbacher in 1848 in his most popular work "Results for Mechanical Engineering" (Redtenbacher 1848) (see Fig. 4) a real manual for students, practically working engineers, designers and industrial managers likewise. It had four German editions and one French edition in his lifetime and two other German ones posthumously. In this work, he presented all those tools and rules in a half-empirical manner, multiply tested and well-established for the practical applications to be indepensible for a beneficial design. He referred therein to accepted standards to dimension all essential machine elements such as gearwheels, pulleys, chains, screws, cables and belts, journals, shafts, rivets and bolts, linkages, stuffing boxes and pistons and transferred them to other scales. This "method of ratios", once borrowed from architecture and, already applied during the time of James Watt, was an elementary method of calculation characterizing scientific mechanical engineering during the early stage of the field.

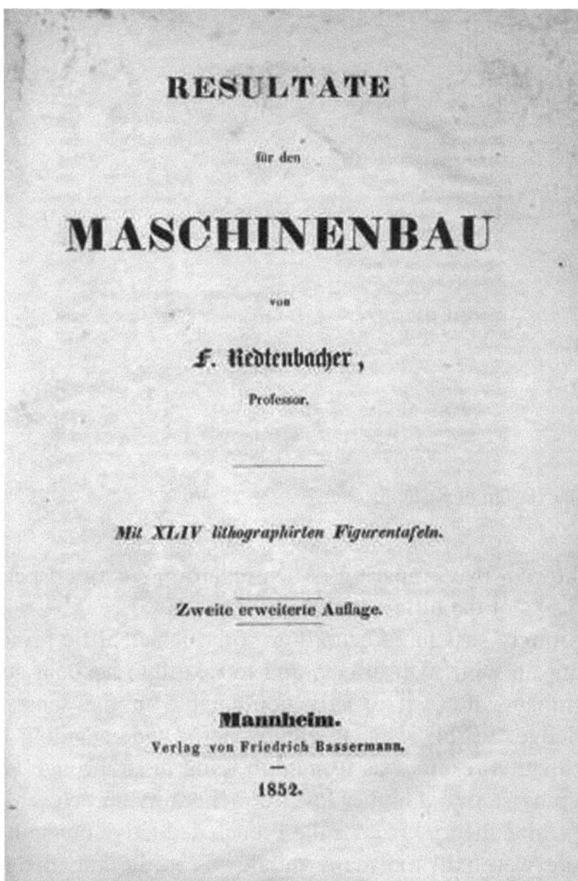
RESULTATE

für den

MASCHINENBAU

von

F. Redtenbacher,

Professor.

Mit XLIV lithographirten Figurentafeln.

Zweite erweiterte Auflage.

Mannheim.

Verlag von Friedrich Bassermann.

1852.

Fig. 4 (**a**) Book facing page "Results for Mechanical Engineering". (**b**) Cycloid mathematical curves from "Results". (**c**) Kinematic mechanism design from "Results"

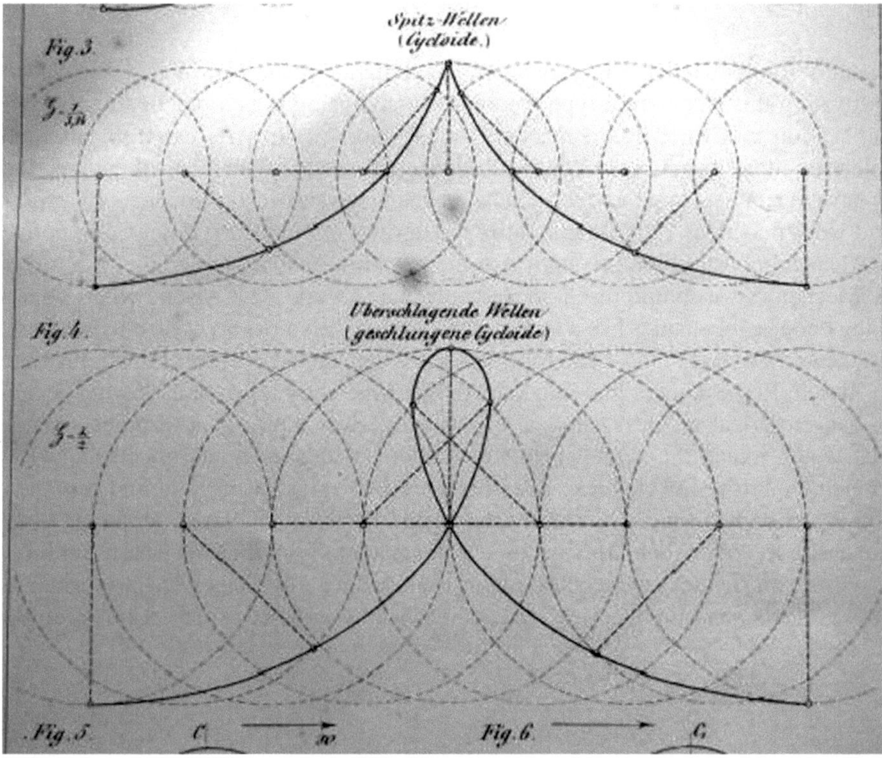

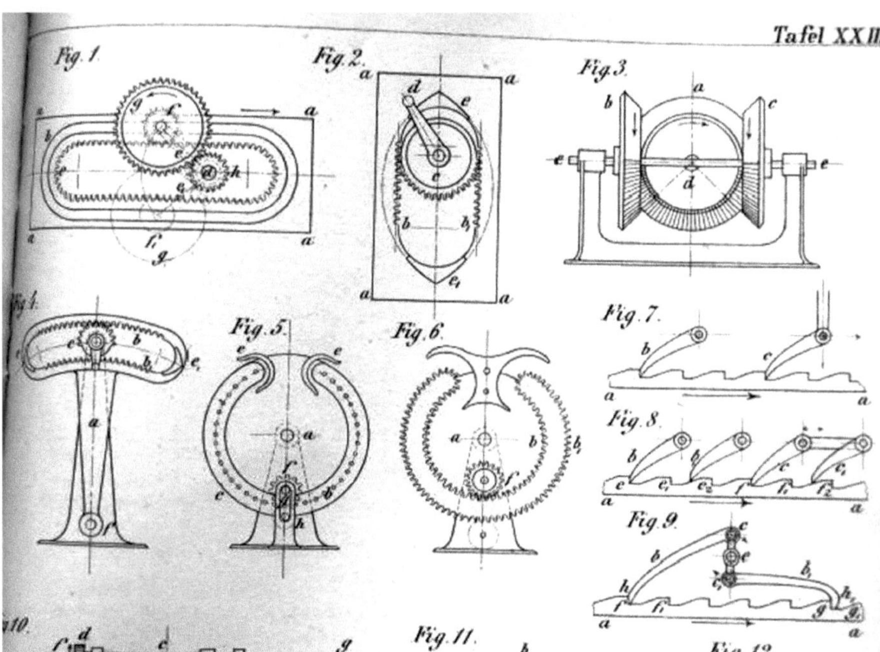

Fig. 4 (continued)

Besides Redtenbacher, the French polytechnician Artur Morin had succeeded cultivating this calculation type theoretically and using it as a practicable tool. Its publication motivated Redtenbacher, for the most part, to retreat from his high scientific ideals but it corresponded to the actual needs of mechanical engineering to have available a large variety of new mechanisms and design concepts for power and work machines fully functional in the developing industry. It helped to define mechanical-constructive orientation of the machine design.

Mechanics was omnipresent in Redtenbacher's work: "All-around where something budges, mechanics has a finger in the pie but the spirits do not stir by mechanics", could be read in the 1856 caption of a Redtenbacher portrait. During his time in Zurich, Redtenbacher had indicated that he was not interested in leaving design to half-empirical recipes. Already, his objective was aimed at a machine science that would match the achievements of modern mathematics and mechanics. The French experts of mechanics were so highly distinguished in his mind that their shining example inspired him to emulate them. He had manifested, for a long time, the desire to comprehend the machine problems totally in their highest generality.

His scientific ideas were published in 1852 in the book "Principles of Mechanics and of Mechanical Engineering" (Redtenbacher 1852a) (see Fig. 5). This became a

a

b

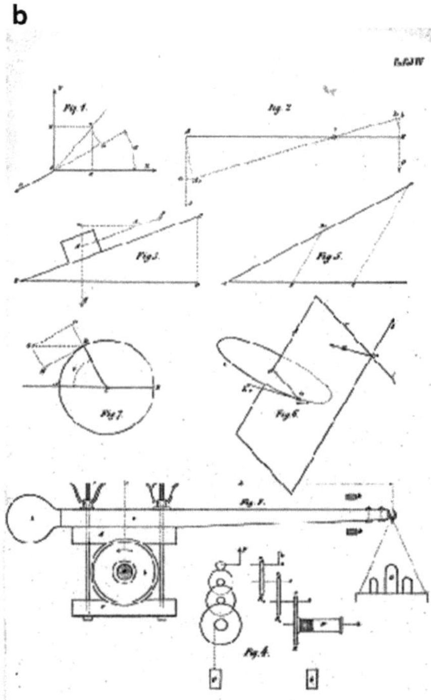

Fig. 5 (**a**) Book facing page "Principles of Mechanics and of Mechanical Engineering". (**b**) Drawings of equilibrium problems from "Principles"

principal part of his machine theory and a milestone in composing technical-scientific theories. In the book, he compellingly developed the modeling ideas for the design of a machine. With a consequent orientation to the interactions of technical function, constructional shape forming and economic feasibility, he broke new ground in design methodology. Consequently, he incorporated that vocabulary which had already heralded a specific dynamic-energetic dimension of approach under the French triumvirate Louis Marie Henry Navier, Gustave Gaspard Coriolis and Poncelet, starting with Charles A. Coulomb and his theory of simple machines during the second half of the eighteenth century. Work, energy and efficiency factor were among the main properties within the structure of categories of his machine theory, where he always tried to explain the context descriptively. By avoiding cumbersome mathematical expansions, straightforward and applicable procedures were his ultimate goal. Herein, the specific usefulness of technical coefficients outcropped, e.g., describing the strength behavior of machine elements made of metallic material or concerning the outflow of fluids from containers or reservoirs but also such geometrical properties of shape and cross-section tabularly ascertainable as moments of inertia, etc. Such comparisons gave directions to the engineer for an effective design.

Redtenbacher's approach was captured in a systematic method that was based on the machine's type of motion – changing the position and deformation of machine parts led to the determination that beneficial layouts and culminated in an analytical machine theory. Its framework was geared to a technical–economic optimal solution, the essential determining properties were then represented by a set of primary and secondary equations. With respect to the purpose of the machine and the technological process, considerations were finally given to the dependencies between the constructional elements and their technical functions. This part of Redtenbacher's procedure to design machines is the one that is based the least on scientific principles but on economic, industrial and societal issues. Here mechanical design practice totally dominates.

With the development of mass production, scientific mechanical engineering became more intensively devoted to the manufacturing processes. The temporary priority of the design process as an object of scientific examinations was emphasized by the key role which Redtenbacher attributed to technical drawing. Despite all specialization, he still considered engineering activities as an organic procedure, also involving a certain artistic creativity. The recognition that the design abilities of technicians as well the aesthetic appeal of their product were part of the process began to emerge within the evolving engineering sciences. The use of aesthetics in machine design was later embraced by Redtenbacher's student Franz Reuleaux (see, e.g., Moon 2007).

Redtenbacher saw the conscious use of engineering thinking as an expression of human power, and a chance to create a bridge between nature, technology and society and, furthermore, a basic approach for a reconciliation between the differences of the neo-humanistic and technical education. The psycho-emotional moment in the process of an original design became a central part of the inventive process of an engineer as well as the basic principles of mechanics.

Redtenbacher believed that, among the important virtues of a mechanical engineer, should be an artistic nature, including phantasy, play instinct, artistic abilities and virtuosity. Incidently, he felt that technical drawing would considerably advance keenness and clarity of thoughts of an engineer. Without it, many a trend-setting idea might be pushed ahead of its time, many a keen approach could not completely be realized. The use of descriptive geometry and technical drawing was introduced by Gaspar Monge of the Ecole Polytechnique showing again the French influence on Redtenbacher.

On the one hand, advances in the theory of machine sciences were so notable that they were beyond the means of the engineering practice, on the other hand, there were huge gaps of knowledge in such elementary problems as lubrication, tightness or fit accuracy of machine components. Increasing the speed rates and the pressure as well the enlargement of the reciprocating masses were other factors accounting for such discrepancies between theory and design rules. Only the next generation of engineering scientists could overcome the hiatus between program and reality. Nevertheless, Redtenbacher tried to verify his guiding principles during the fifties. By his essay "The Air Expansion Machine" (Redtenbacher 1852b), which was already published as a second edition in an extended form and new title "The Caloric Machine" (Redtenbacher 1853) 1 year later, he could manage a validated conformation of his conceptions from Redtenbacher (1848) in a original format. Using the example of Ericson's air expansion machine, a strange forerunner of the combustion engine and strongly debated in the literature, Redtenbacher – surely knowing the contributions of the French polytechnician Sadi Carnot – raised the question on the perfect motor. During that time, the thermal behavior of heat-power engines came more and more the vision of the engineers. Increasing the power by higher cylinder pressures and temperatures, multi-utilization of steam, minimal fuel consumption and advantageous arrangement of the thermal process were the focal points of many scientific debates.

In contrast to his student Reuleaux, Redtenbacher's contribution to machine theory therefore moved into the forefront of technical thermodynamics. Starting with fundamental considerations in which, strongly following his principles, he formed a theory of an effective hot air engine, he optimized its main parameters. However, the problem was not yet completely solved and he addressed the key question of the practicability under which the required high temperature gradient brought back down to earth all speculation to exceed the limits of the classical steam engine. With less resonance, the work "Laws of the Locomotive Construction" (Redtenbacher 1855) was addressed in 1855, even though there was an urgent need for such a book of rules. Railroad construction had been a catalyst for the growth of powerful machine factories. The novelty and complexity of this technical system required scientific principles and from March 1854, he declared in several letters that this technology, driven by the English with their long empirical knowledge, would require a scientific basis. He thereby entered the new field of interia forces, vibrations and impact loading, but was not able to cope with the complex driving system, called a locomotive, in all details. Critical comments by experts in this field could not be ignored.

His series of books was extended in 1857 by the contribution "Motion Mechanisms" (Redtenbacher 1857a), a book of detailed designs for models of kinematic mechanisms. In 1861, it was expanded by a new issue (Redtenbacher 1861a), to be collected in 1866 as one volume (see Fig. 6) which essentially presented and explained the teaching model aids collection of the mechanical engineering section constructed in the workshops of the Polytechnic School under Redtenbacher. These models (see Figs. 7 and 8) supported his lectures to the professional community from models of a simple pair of wheels or a slider crank mechanism through kinematic models of the complex non-uniformly transmitting gear boxes.

a

DIE

BEWEGUNGS-MECHANISMEN

Darstellung und Beschreibung eines Theiles der

MASCHINEN-MODELL-SAMMLUNG

der polytechnischen Schule in Carlsruhe.

von

F. REDTENBACHER

Großherzoglich Badischer Hofrath u. Director an der polytechnischen Schule in Carlsruhe.

Neue Auflage.

Mit 80 lithographirten Tafeln.

HEIDELBERG

VERLAGSBUCHHANDLUNG VON FRIEDRICH BASSERMANN.

b

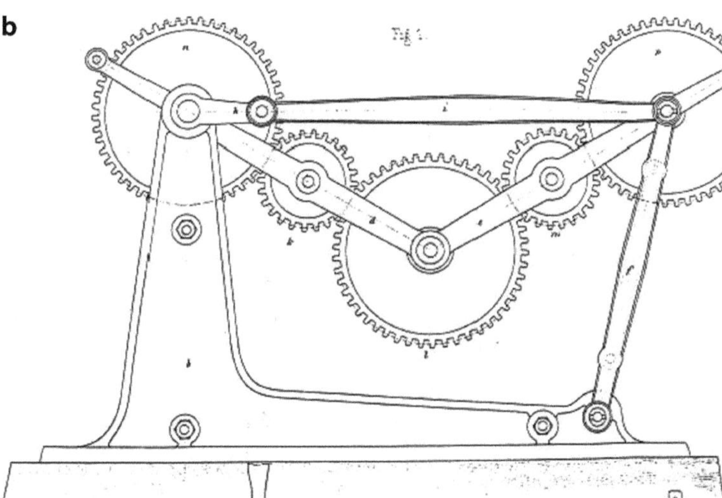

Fig. 6 (**a**) Book facing page of Redtenbacher's "Motion Mechanisms". (**b**) Drawing of gear train mechanism from "Motion Mechanisms".

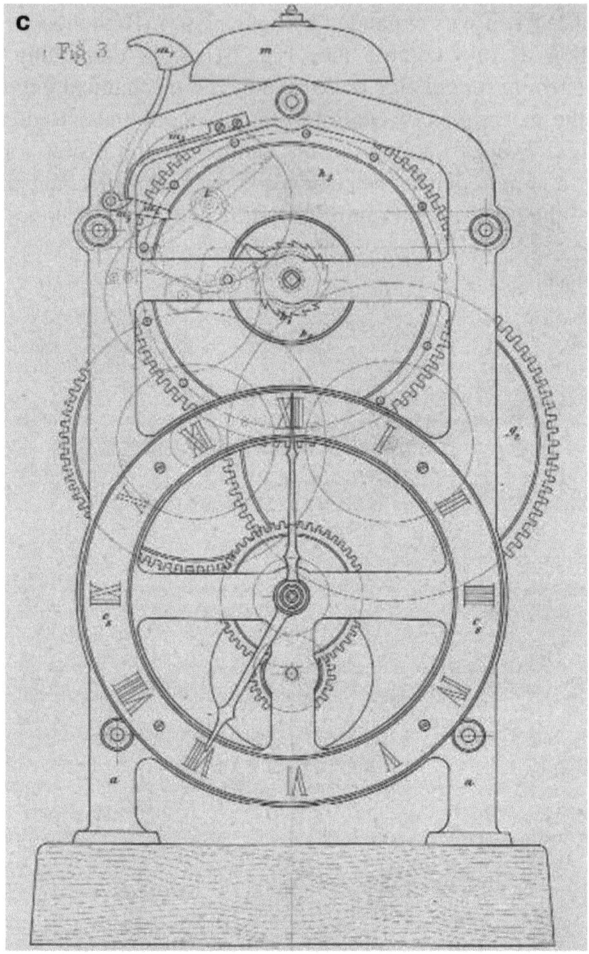

Fig. 6 (continued) (**c**) Drawing of gear train mechanism from "Motion Mechanisms"

Again and again, Redtenbacher was interested putting his explanation of the interaction between nature and technology on a consistent basis. Such wide-ranging thinking permanently inspired his research activities and led him to a complex and variable methodology. His world picture was particularly rooted in the predictability and in the logical precision of classical mechanics. Redtenbacher, who contemplated writing a history of mechanics, picked up nature-philosophical discussions and tried to translate them into a technical-scientific question. His intentions in 1857 culminated in the presentation of his opus "The System of Dynamides, Main Features of a Mechanical Physics" (Redtenbacher 1857b). What he already had implied in his "Principles", here it attained the most mature expression: an attempt

Fig. 7 Photographs of kinematic models from the Redtenbacher Model Collection of Karlsruhe University. (**a**) Reciprocating gear mechanism. (**b**) 6-part Geneva wheel intermitted mechanism. c. Stephenson valve gear mechanism

to yield a contribution to the scientific and philosophic progress of knowledge. Therefore, evidence of the theoretical influence of the flourishing technical-scientific thinking should be furnished. In the process of fermentation between different concepts of nature which loomed in the course of the accomplishment of the classical physics, Redtenbacher's contribution to continue the thoughts of Dalton Poisson and Augustin Louis Cauchy was an attempt to match the existing atomistic molecular-theoretical models with the specifics of technical-scientific styles of thinking. The center of his ideas were the so-called "Dynamides", i.e., body atoms with its enclosing aether hull, he aspired to explain graphically the connection between micro- and macro-cosmos, a venture which must fail on the mathematical explanation of complex processes of aether oscillations. By this "System of Dynamides" which was denoted a patchwork theory by Redtenbacher – obviously he appreciated the deficiencies of his constructs – he wanted to set a marker to bring forward – beyond the attempt to the self-image – the acceptance of the status of engineers. In this respect, the book had deserved broader attention from his contemporaries, especially as it dealt with important ideas in physics, e.g., such of James Clerk Maxwell. His affinity to natural-philosophical topics was expressed in 1861 a second time in his work "The Initial and the Actual Temperature of the Heavenly Bodies" (Redtenbacher 1861b). Both books are important in understanding the broadness of Redtenbacher's scientific work, but for the development of machine science, they are contributions with no impact.

Obviously, several factors came together to inspire Redtenbacher's final masterwork "Mechanical Engineering" (Redtenbacher 1862). As Redtenbacher himself explains in the preface to the first volume, the complete three volumes should summarize the essentials of all his lectures on mechanical engineering at the Polytechnic

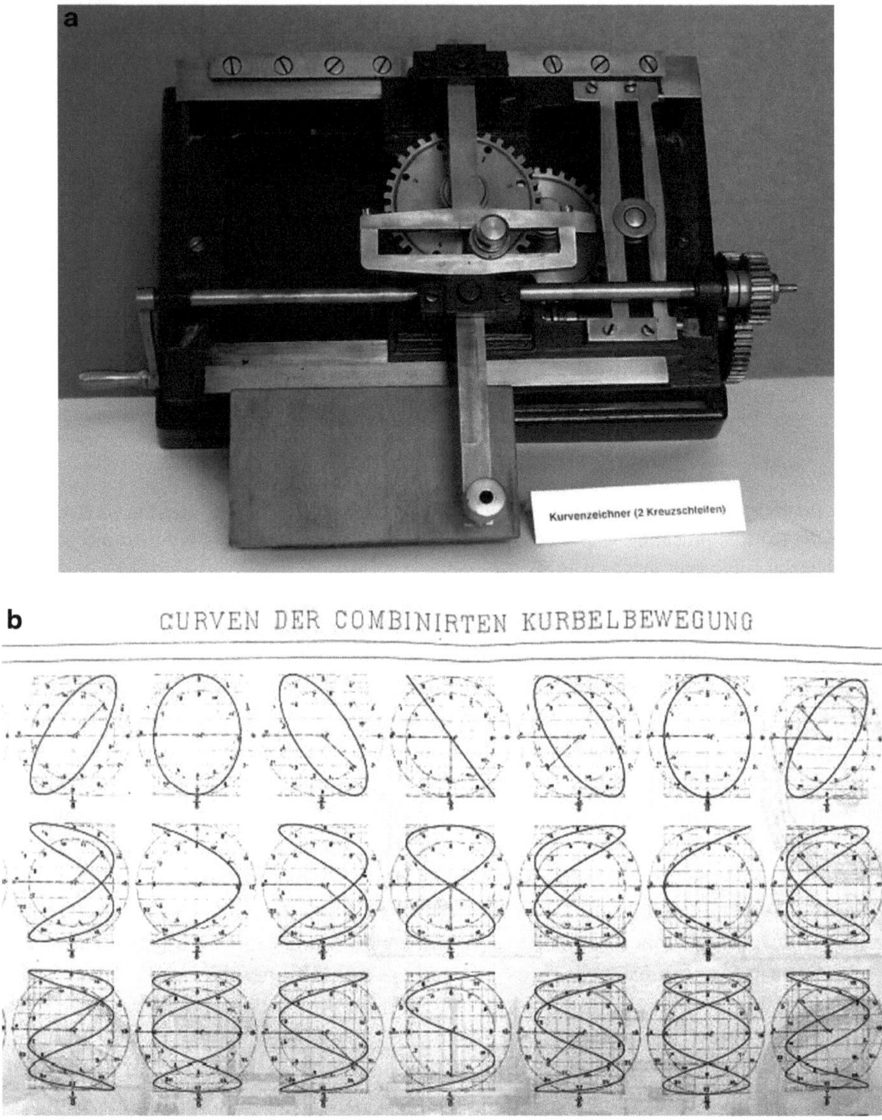

Fig. 8 (**a**) Photograph of kinematic model of a function synthesizer from the Redtenbacher Model Collection of Karlsruhe University. (**b**) Lissajous curves drawn with Redtenbacher's kinematic function generator

School of Karlsruhe. "This work composes for all students of mechanical engineering the bridge between the other works which I have published many years ago, namely the 'Principles of Mechanics and of Mechanical Engineering' and the 'Results for Mechanical Engineering'. The present work shortly called 'Mechanical Engineering' essentially is the derivation of the results from 'Principles' for application. This means that the 'Principles' are a general formalism describing what the essence of every machine is, how the motions and operations work, what conditions influence the machine and how these constraints can be traced, what ideas are important in order to develop the design to a real machine that functions in practice. With one word: The 'Principles' try to tell and to teach everything what is necessary for understanding, for design and for realization of machines, i.e., the 'Principles' are the general machine science which is taught within the first three months of engineering study. On the other hand, 'Mechanical Engineering' is intended to teach all specialities which are necessary for practical realization of all general principles and treats a series of concrete work and power machines." But Redtenbacher's desire to write this summary work, however, was urgent due to the misus of his ideas by others and the exploitation of his lectures. The final statement within the foreword to Volume 1 states explicitly "that the publisher already often was impelled to intervene against this circulation of my work and that hopefully this misapplication will not happen again in future". Only the first volume was published in Redtenbacher's lifetime. Volume 2 was ready to print when he died, volume 3 was finished according to available notes by his co-worker Josef Hart; the publication took place in 1865.

Following the intentions mentioned, the first volume spanned a wide selection of topics including strength of materials, design of machine elements, friction in machine elements, kinematics of mechanisms, clock escapements, and the design of construction cranes. His second and third volumes cover hydraulics, water wheels and turbines, locomotive design, steam engines and mining machines. The structure of the text consists of short descriptions of each topic, many formulas, some derived using differential and integral calculus, and a few line drawings at the end of each volume.

On the Influence of Redtenbacher's Works

For the evolution of mechanical engineering in central Europe from its workshop-oriented tradition into a science, Redtenbacher's three books "Results", "Principles" and "Mechanical Engineering" composed a type of comprehensive framework. To understand concrete applications, the publications "Water Wheels", "Turbines" and with some restrictions "Caloric Machines" and "Locomotives" are useful, too. To understand how Redtenbacher taught his students, explaining to them theoretical and methodical connections with the use of real models, the book "Motion Mechanisms" had additional significance. Therefore, these works will be included in the following discussion.

The influence and the renown of Redtenbacher's works were already wide spread during his lifetime. Six German editions and an additional French edition (from 1848 through 1875) of his most popular work "Results" as well three editions of "Motion Mechanisms" (between 1858 and 1866) give evidence for his wide reputation. In addition, there were two editions of "Water Wheels", "Turbines" and the "Principles" while, on the other hand, there was only one edition of the large three-volumed comprehensive work "Mechanical Engineering". The latest edition of "Results" edited by Redtenbacher's successor, Franz Grashof, in 1875 was in the library of the well-known American engineer-professor Robert Thurston of Stevens Institute and later Cornell University, i.e., Redtenbacher's work also circulated in America as well as in Germany and France. We can obviously conclude that practical handbook knowledge, rules of thumb and tabular results were most important for engineers and technicians during that time of early industrialization. But even nowadays such collections of scientific and technological facts in a short form without proofs are very important for students and practicing engineers as compendia as "Hütte" or "Dubbel" show. And a comparison with the early editions of "Hütte" in the nineteenth century verifies that it was definitively based on Redtenbacher's "Results".

The design experience represented by his precise technical drawings one can find both within the "Results" and also the tables accompanying the text of "Motion Mechanisms", and are a further step in the didactic realization of the practical-mechanic knowledge of his time. Machines and their components were visualized in all details where the machine elements and even the motion of mechanisms were displayed in their statical and strength constitution. The kinematic principles that governed machine engineering was the typical style of the French Monge school in which they demonstrated the conversion of motion and represented a catalog of mechanisms for intuitive selection of proper gear shapes for example. It was far away from a modern mathematical–geometrical analysis or even a synthesis of such units.

Redtenbacher's Kinematic Models

There is evidence that Robert Willis used such models as early as 1840 in Cambridge, as did Johann Andreas Schubert at Dresden around the same time (see, e.g., Moon 2003a, b, Mauersberger 1989b, 1997). Likewise, Redtenbacher developed one of the most extensive model collections of his day at Karlsruhe with as many a 100 different models which was only surpassed after his death by his student Franz Reuleaux at Berlin. Most important for the evolution of mechanical engineering is the fact that Redtenbacher documented the construction of these teaching models in his two monographs. This offered an opportunity for others to copy these models as did a famous commercial model-maker of the time, the Schroeder Company of Darmstadt. Unlike Reuleaux's original 800 model collection at Berlin that was destroyed in World War II, the Redtenbacher Collection is remarkably preserved at the University of Karlsruhe. Recently, has become the entire collection available on the web, appropriately alongside the Reuleaux Collection, at the website KMODDL (http://kmoddl.library.cornell.edu). Reuleaux's designs for models survived because he set Gustav Voigt to copy and make reproductions of around 350 of his models. However in competition with Voigt, I. Schröder of Darmstadt, Germany, also produced copies of Ferdinand Redtenbacher's models from (Redtenbacher, 1857a, 1861a) as well as copies of models based on Reuleaux's books. The Schröder-Redtenbacher models were quite popular because they came in two sizes, small and large. Several museums such as the Science Museum in London and the Musée des Arts et Métier in Paris purchased the larger models. A large collection of these copies of Redtenbacher models can be found in Italy at the Foundation for Science and Technology (FST) in Florence. Another large collection can be found at the University of Porto in Portugal. Cornell University has about 18 Schröder models to be found illustrated in Redenbacher's textbooks. Schröder exhibited these at many world exhibitions including the Centenial Exhibition in Philadelphia in 1876 at which Reuleaux was on the judging panel that awarded Schröder's models a prize. Thus unwittingly, Redtenbacher's influence extended beyond Karlruhe through his designs for teaching models of machines. The Redtenbacher models largely consist of kinematic mechanisms including linkages, gear systems and straightline mechanisms. However, there are cutaway models of steam-engine valve systems as well as mathematical function synthesizers.

The representation of engineering knowledge in the mid-nineteenth century began with mathematical codification in the form of coefficients, tables, dimensionless characteristic numbers, etc., and was typical for the methodical approach in the evolution phase of mechanical engineering science. In particular for the instruction of the early mechanical engineers, this type of procedure introduced drastic simplifications, and was the only practicable method. Within the conflicting goals of practicability and scientific content, Redtenbacher tended to propose a didactically-illustrative method – similar to his contemporary colleagues Julius Ludwig Weisbach at the Mining School of Freiberg in Saxony and the Scottish engineer William John Macquorn Rankine – but the scientific background had to be foremost. The benefits of such a unity of exact scientific conceptions and a realization intelligible to all were obvious: The different demands on scientific methods in research, education and engineering practice became apparent. Even in that period, the emerging of these levels of engineering-scientific activities within a scientist personality such as Redtenbacher proved its worth. Excellence in science mainly occurred where new technologies pioneered and technical problems could not be solved via conventional aids. However, it cannot be ignored that, essentially, engineering practice could be handled very well by handbooks and know-how and also workshop traditions.

Redtenbacher's posthumous fame was considerable at least in the German-speaking regions of Europe. A notable number of students alone has ensured that his name would not fall into oblivion. Already within early writings in the history of engineering science, Redtenbacher's lifework was acknowledged by well-known historians. He made history not only for his pioneering work in scientific mechanical engineering, but also as one who brought about its realization in the training of engineers. There was some continuity within the work of Gerstner (Vienna), Schubert (Dresden), Weisbach (Freiberg) and Redtenbacher (Karlsruhe) to introduce scientific methods into mechanical engineering where Redtenbacher possibly outstood and had in Grashof a coequal successor. Grashof was successful in continuing Redtenbacher's educational intentions and plans. One year after his relocation to Karlsruhe, at the annual meeting of VDI (Association of German Engineers), he delivered a lecture, "On the basic principles for the organization of polytechnic schools". Based on this speech, a new organization status for the Karlsuhe Polytechnic School was granted in 1865 as the first institution in Germany holding a university constitution with self-management and appeal proceedings to be equated in rank with the state universities.

But it is also correct to say that Redtenbacher's work is clearly less spectacular and less controversal than that of his most famous student, Reuleaux. Based on a solid disciplinary development of the scientific mechanical engineering, Reuleaux and others tried a theoretical consolidation that spurred strong pedagocical debates. The major subjects were sufficiency of theoretical models, practical relevance of theory, the role of mathematics and the variety of methods within the flourishing engineering science. Therefore, it is not surprising that Reuleaux's work received much more attention in America than Redtenbacher's although the latter's work was wider in scope than either of Reuleaux's two major works including his "Kinematics of Machinery".

Modern Interpretation of Main Contributions

To appreciate Redtenbacher's contributions to machine science, it is helpful to read the opinions of his colleagues and later educators in similar institutions.

What Redtenbacher achieved for the mechanical engineering cannot be stated more incisively than by the words of Professor Kammerer within his directorate speech discoursed in 1908 at the Technical University of Berlin, i.e., in a time when his school had become the largest and most important institution of this kind in the German empire: "…in the emergence of a mechanical engineering art from mechanical engineering craft, scientific methods had to be found to determine machine dimensions as well as economic efficiency. The leadership was firstly taken by the Karlsruhe Polytechnic School. There, in the year 1841, Ferdinand Redtenbacher started his lectureship who had got to know the mechanical engineering undertaking from his own personal experience and at the same time also governed the scientific mechanics. In place of aimless rules he placed a theory based on experience… Justifiably, he may be seen as the originator of machine science in Germany" (Keller 1909).

As another person of importance in this context, Franz Reuleaux has to be surely named as one of Redtenbacher's master students. Reuleaux was also impressed and influenced by Redtenbacher in his 2 years of study at Karlsruhe, 1850/51 and 51/52, where he had learned from Redtenbacher's books published between 1844 and 1857 and summarized in the three-volumed opus "Mechanical Engineering" in 1862–65. In his famous work of 1875 "The Kinematics of Machinery: Outlines of a Theory of Machines" (Reuleaux 1875), within the third paragraph of chapter 1, Reuleaux puts machine science in the framework which was filled out by Redtenbacher 30 years earlier in Karlsruhe. Thereby, he ascertained that the general basics of machinery, the theory of machines, machine design and kinematics, in a sense, could be extended compared to that of Ampere in order to define a machine as a whole. Reuleaux explicitly credited Redtenbacher with a key role in promoting the first three ingredients for scientification of mechanical engineering departing from the traditionals stating for example, "that in the general basics of machinery a descriptive representation was usual until the thirties and Redtenbacher first removed the stigma of indistinctness from the matter."

There is some debate among historians of machine engineering as to the relative contributions to mechanical engineering of Reuleaux and his professor Ferdinand Redtenbacher who, as noticed before, has been called one of the pioneers of German engineering education. Both men published influential books in machine design and both developed kinematic model collections for teaching. Redtenbacher's summarizing three volume work "Mechanical Engineering" followed a year after Reuleaux's own "The Constructor" (Reuleaux 1861) was published. The Redtenbacher work is wider in scope than either of Reuleaux's two major works, including his "Kinematics of Machinery" (Reuleaux 1875) of 1875/76. However, unlike his pupil, Redtenbacher did not attempt to develop a general kinematics theory of machines or mechanisms. His chapter on kinematics classified mechanisms as to the type of motion they performed and was clearly influenced by the earlier French classification schemes of Monge. Nor did Redtenbacher review the history of machines and there are not many references to the literature or machine practice.

Lasting effects emanated from Redtenbacher's main theoretical work "Principles". However, it is difficult to find his name associated with theorems, methods or even formulas in mechanical science and engineering. Examples of his theoretical paradigms showing his use of definitions and names in the sense of Newton's *Principia* were introduced in the development of machine theories and are still found in the framework of modern machine theory, albeit in a modified and extended form. In this meaning, Redtenbacher might be compared with Euler who had a similar impact on natural science. Redtenbacher reinterpreted the theorems of theoretical mechanics in such a way that they could be readily applied to problems in engineering. With regard to progress in knowledge and development of methods, Redtenbacher's "Principles" constitute a milestone in the history of mechanical engineering and are, at the same time, a credo for scientific methodology. It was not sufficient to simply mathematize problems, but the development of theoretical models took a key role in engineering science. The complexity of real machine problems and the large number of parameters within the design process preclude exact methods according to the ideals of natural science and require a limitation to the use of essential principles. The modelling has to consider the concrete purpose of use, the function and the boundary conditions. Practical experience, specifically for mechanical engineering, is the reference field for setting up of adequate theories. Nevertheless, the developing machine sciences borrowed their methodical approaches from natural science. Failing self-contained methods this approach was legitimate at the time. Deductive methods and the search for closed-form solutions were the long-term objective of all attempts for scientification by Redtenbacher and his students. At the same time, it was intuitively clear to them that the path for generalization could only be followed inductively, studying the specific features of the individual case.

How should such a theory of machine science finally look, what factors had to be taken into consideration within the modelling? In general, according to the principles of Galilei, the essentials had to be distinguished from the non-essentials. For this purpose, practical judgment and experience were required. However, the main parameters for an analytical machine theory were fixed; mostly there were key economic constraints on designing a machine as related to function, weight, power and costs. In his concept of a machine theory where one had to find efficient design proportions and the choice of appropriate machine elements and materials, Redtenbacher began with design parameters which he described as functions of geometric secondary quantities. The solution of a specific dimensioning problem then arose from maximizing or minimizing the main design parameters by step-by-step elimination from the governing system of equations.

Some fruitful approaches were created but there were some limits to such a strictly composed deductive theory. Often the economic solution could not be technically realized; due to the complexity of the machine problem, simplifications had to be introduced which diminished the applicability. Furthermore, mathematical solutions existed which avoided a technical interpretation. Altogether, Redtenbacher therefore stuck to approximate approaches. The basic principle of transfer of proven traditions onto new circumstances and, in parallel, the balancing of possible factors and disturbances, was in accordance with the requirements of that time. It is to Redtenbacher's credit, that he used the "method of ratios" borrowed from architecture for mechanical engineering and cultivated it. Initiated by practical problems, Weisbach and Redtenbacher maintained the introduction and the use of technical parameters. These had the advantage of being measurable and could be calculated. The tabulation of such coefficients provided a basis to consolidate the structure of the concept of machine science and, by the permanent usage in reference books, it became the prevalent tool of the working engineer. Such practicable parameters also conveyed their theoretical background. Redtenbacher committed himself to establish non-dimensional parameters which had the advantage of being independent of specific measurements. Furthermore, such ratios were easily committed to memory and formed the intensive thinking of engineers for the correct design rules. Such a simplified machine calculation began in the first place with the geometry of motion of the mechanisms (later object of kinematics and gearings). The second step in the design process was the consideration of the strength of materials which resulted in the specification of design rules for machine elements.

Although Redtenbacher, who had a distinct appreciation for the theoretical formulation, attached great importance to their verification, he continually tried to improve them. With a view to approaching the high aims of his "Principles", at least partially, he formulated modified specific theories for special machines. For corresponding optimization problems, he even discarded the "exclusive computing method". Consequently, Redtenbacher proposed a deductive-axiomatic procedure, e.g., treating the design of hydraulic machines where he structured the calculation as follows:

1. Listing the requirements
2. Definition of the variables
3. Development of theory
4. Formulation of constraint equations maximizing a certain effect
5. Derivation of dimensioning rules
6. Consideration of constraints (noise, losses)
7. Generation of general laws for construction and operation of the machine by comparison with experiments
8. Selection of material and cost calculation

Taking into account the imperfection of the introduced ratios which were based on static principles, Redtenbacher again and again evaluated the pros and cons of approximate methods. But ultimately he had only one essential criterion of his scientific theory which was to accurately describe real conditions in a machine. The necessary simplifications of theoretical formulations led to the practical realization that these models were not precise but which were consciously accepted. But Redtenbacher did not stop with a new theory, he tried to improve it by error estimations and reflections on the neglected effects. Such a procedure sharpened his ability to deal with existing technical solutions and to judge and to estimate trends of mechanical engineering. Even today, such a procedure seems to be very modern and shows how strong his methodical thinking influenced the later theory of mechanisms and machines. But an explicit example may be difficult to find.

Contemporary "mechanism and machine theory" was originally defined within the English-speaking area as a conglomerate of machine-scientific disciplines focused on machine kinematics and only included dynamics and design in recent years. Willis seemed to advance the view of machine theory as largely kinematics, see, e.g., Moon (2003a, b)). Rankine, however, see Rankine (1887), seems to have pursued a wider vision of the theory of machines. Redtenbacher in his time, pursued by all means a very modern concept of machine science looking much more from a holistic perspective. Traditionally, constructive and technological aspects were approximate and one tried to evolute the established machine engineering into a machine science. To realize it, Redtenbacher and his students brought together kinematics and gearings as Reuleaux did, as well as strength of materials with an intensive study of the machine elements as did Bach and fluid dynamics/thermodynamics as did Weisbach, Zeuner and Stodola.

Acknowledgements The authors thank the Department of Mechanical Engineering at the University of Karlsruhe (TH), in particular the Institute of Combustion Engines, for providing access to Redtenbacher's kinematic models.

References

Bader J (ed) (1858) Die Residenzstadt Karlsruhe. Festgabe der Stadt Karlsruhe zur 34, Naturforscher-versammlung, Karlsruhe, pp 123–154, 158–163

Fuchs WP (1959) Die geschichtliche Gestalt Ferdinand Redtenbachers. Zeitschrift für die Geschichte des Oberrheins 107:205–222

Keller K (1909) Festrede. In: Ferdinand Redtenbacher, Bericht über die Feier seines 100. Geburtstages ander Großh. Technischen Hochschule zu Karlruhe. Offprint of G. Braunschen Hof-buchdruckerei in Karlsruhe, Karlsruhe, pp 4–22

Mauersberger K (1980) Die Herausbildung der Technischen Mechanik und ihr Anteil bei der Verwissen-schaftlichung des Maschinenwesens. Dresdner Beiträge zur Geschichte der Technikwissenschaften, TU Dresden (2):1–52

Mauersberger K (1985) Vergleichende Betrachtung zur Herausbildung der Technischen Mechanik und Maschinenlehre in England, Frankreich und Deutschland bis zur Mitte des 19. Jahrhunderts. NTM – Schriftenreihe für Geschichte der Naturwissenschaften, Technik und Medizin 22(2):25–32

Mauersberger K (1989a) Jacob Ferdinand Redtenbacher (1809–1863). Ein entscheidender Beitrag zur Ver-wissenschaftlichung des Maschinenbaus. In: Buchheim G, Sonnemann R (eds) Lebensbilder von Ingenieurwissenschaftlern. VEB, Leipzig, pp 43–59

Mauersberger K (1989b) Die Entwicklung der technischen Hochschulen und ihrer Ausbildungskonzepte im 19. Jahrhundert. In: Dresdner Beiträge zur Geschichte der Technikwissenschaften. TU Dresden (17):18–36

Mauersberger K (1997) Von Karmarsch bis Reuleaux – Verallgemeinernde technik-wissenschaftliche Konzepte im 19. Jahrhundert. In: Banse G (ed) Allgemeine Technologie zwischen Aufklärung und Metatheorie. Johann Beckmann und die Folgen. Cotbuser Beiträge zur Philosophie der Technik und der Technik-wissenschaften. Berlin, pp 45–64

Moon FC (2003a) Robert Willis and Franz Reuleaux: pioneers in the Theory of Machines. Notes and Records of the Royal Society, London

Moon FC (2003b) Franz Reuleaux: contributions to 19th century kinematics and theory of machines. Appl Mech Rev 56(2):261–285

Moon FC (2007) The Machines of Leonardo da Vinci and Franz Reuleaux. In: Springer series history of mechanism and machine science, vol 2. Kluwer, Berlin/New York

Poncelet J-V (1838) Cours de mecanique appliquee aux machines. Metz, 1826; Theory des effects mecaniques de la Turbine Fourneyron. Metz, 1838

Rankine WJM (1887) A manual of machinery and millwork, 6th edn. Charles Griffin & Co. London

Redtenbacher F (1844) Theorie und Bau der Turbinen und Ventilatoren, Bassermann, Mannheim

Redtenbacher F (1846) Theorie und Bau der Wasserräder, Bassermann, Mannheim

Redtenbacher F (1848) Resultate für den Maschinenbau, Bassermann, Mannheim

Redtenbacher F (1852a) Principien der Mechanik und des Maschinenbaus, Bassermann, Mannheim

Redtenbacher F (1852b) Die Luftexpansionsmaschine, Bassermann, Mannheim

Redtenbacher F (1853) Die calorische Maschine, Bassermann, Mannheim

Redtenbacher F (1855) Die Gesetze des Lokomotivbaues, Bassermann, Mannheim

Redtenbacher F (1857a) Die Bewegungsmechanismen, Bassermann, Mannheim

Redtenbacher F (1857b) Das Dynamiden-System. Grundzüge einer Mechanischen Physik, Bassermann, Mannheim

Redtenbacher F (1861a) Die Bewegungsmechanismen, neue Folge. Bassermann, Mannheim

Redtenbacher F (1861b) Die ursprüngliche und jetzige Temperatur der Weltkörper. Bassermann, Mannheim

Redtenbacher F (1862–1865) Der Maschinenbau. 3 Bände, Bassermann, Mannheim

Redtenbacher R (ed) (1879) Ferdinand Redtenbacher – Biographische Skizze. In: Geistige Bedeutung der Me-chanik und Geschichtliche Skizze der Entdeckung ihrer Prinzipien (Lecture given in fall 1859 by Ferdinand Redtenbacher). München, pp 9–74

Reuleaux F (1861) Der Constructeur, ein Handbuch zum Gebrauch beim Maschinen-Entwerfen. Vieweg und Sohn, Braunschweig (English Transaltion: Suplee HH (trans) (1893) The constructor, a handbook of machine design. Philadelphia

Reuleaux F (1875) Theoretische Kinematik, Grundzüge einer Theorie des Maschinenwesens. Vieweg und Sohn, Braunschweig (English translation: Kennedy ABW (trans) (1876) Kinematics of machinery: outlines of a theory of machines. Macmillan, London

Schnabel F (1925) Die Anfänge des technischen Hochschulwesens. Karlsruhe 1925 (from "Festschrift anläss-lich des 100-jährigen Bestehens der TH Karlsruhe"), in particular pp 46–50

Schnabel F (1938) Ferdinand Redtenbacher. Blätter für Geschichte der Technik (4):66–71

Wauer J, Moon FC, Mauersberger K (2009) Ferdinand Redtenbacher (1809–1863): pioneer in scientific machine engineering. In: mechanism and machine theory 44, 1607–1626

Wang Zheng (1571–1644)

Baichun Zhang and Miao Tian

Abstract Wang Zheng was a Chinese official interested in machine design from a young age. In the 1620s, he became familiar with European missionaries and the knowledge they brought to China. Together with Johannes Schreck Terrenz (1576–1630), a German Jesuit missionary, he compiled *Yuanxi Qiqi Tushuo Luzui (A Record of the Best Illustrations and Descriptions of Extraordinary Devices of the Far West)* to introduce European mechanics and machines to China. This remained the most authoritative book on mechanics and machine design in China until the 1860s. Wang Zheng published two further books on machine design and constructed several devices himself. In 1644, he committed suicide because of the fall of the Ming Dynasty.

Biographical Notes

Wang Zheng (Fig. 1), who styled himself Liangpu, was born on May 12, 1571, in Jingyang County, Shanxi Province, China. His father, Wang Yingxuan, was an erudite village teacher with an interest in ancient books, skilled in calculation, methods for preserving health, geography, and astrology. His mother came from a family of scholar-officials (Wang Zheng 1987a, pp. 252–258).

Wang learned from his father from a very young age. In 1577, he went to live with his maternal grandfather and began to study with his uncle, Zhang Jian, a brother of his mother. Zhang, who held various official positions during his lifetime, was proficient in the Confucian classics and history, well acquainted with military knowledge, and had extensively studied arcane knowledge such as the arts

B. Zhang (✉) and M. Tian
Institute for History of Natural Science, Chinese Academy of Sciences, 137 Chaonei Street, Beijing 100010, China
e-mail: zhang-office@ihns.ac.cn; miaotian17@hotmail.com

M. Ceccarelli (ed.), *Distinguished Figures in Mechanism and Machine Science: Their Contributions and Legacies, Part 2*, History of Mechanism and Machine Science 7,
DOI 10.1007/978-90-481-2346-9_13,

Fig. 1 Wang Zheng Portrait (anonymous painter, modern imitation, preserved in Wang Kejü's home)

of astrology and divination. He was also good at making devices. Zhang exerted great influence on Wang, who later highly praised his morality and profound learning, claiming that his skill in crossbow and chariot-making was remarkable (Wang 1987a, p. 208).

In 1587, Wang became a student in the county school. In 1594, he successfully passed the imperial examinations at the provincial level. Over the following years, he took the highest imperial examination ten times, finally succeeding in 1622. Soon after, he was appointed an official in charge of juridical affairs in Guangping Prefecture. At the end of the autumn of 1622, he reached Guangping to take up the post. He was an able and dutiful official. According to his own account, during 1 year he inspected the army and its armaments and led the reform of the military, compiling the *Bing Yue* (Soldiers' Rules); he quelled local unrest and released innocent persons falsely accused of belonging to a rebellious sect, the White Lotus Society; and he ordered the construction of canals and embankments and designed water-gates for the Qinghe river. With the death of his step-mother in March, 1624, he left Guangping for his hometown to observe the prescribed 3-year period of mourning.[1]

[1]After the death of a parent, a son should observe a 27-month period of mourning at home, eschewing all social intercourse. If an official, he must leave his job.

At the beginning of 1627, he was appointed an official in charge of juridical affairs in Yangzhou Prefecture. He arrived at his post in July of 1627. During his tenure of office, besides his juridical duties, he reconstructed floodgates, improved irrigation systems, introduced military methods, built Confucian temples and painted.

In 1628, he again had to leave his post and return to his hometown to observe the prescribed period of mourning after the death of his father. In 1629, he organized local militia to fight against insurgents in his hometown and neighboring areas. In 1631, he was promoted to supervisor of military affairs. However, he was soon removed from his post due to a rebellion by low-ranking military officers and returned to his hometown. From then on, he invested his efforts in serving his hometown and writing. In 1643, the town was overrun by one of the rebelling armies. Refusing to serve them, he committed suicide.

As a result of this act, he was highly regarded as a loyalist and martyr by his contemporaries. However, subsequently he became famous as a member of the first generation of Chinese converts to Catholicism, and for his achievements in the history of the transmission of knowledge between Europe and China, especially machine design and mechanics.

There is no solid evidence to tell us exactly when Wang converted to Catholicism. He was attracted to it when he met the Spanish Jesuit missionary, Diego de Pantoja (1571–1618), in 1616. During 1625 and 1626, Wang helped the Jesuit missionary Nicolas Trigault (1577–1628) to compile and publish the first Chinese phonetic dictionary using transcriptions in the Latin alphabet. Interestingly, in his preface to this book, he stressed that, through the study of Western writing, he had come to understand why the ancient Chinese classic, the *Yi Jing* (*Book of Changes*), was the origin of all written characters. In 1634, he established a Catholic society, "Ren Hui" (Benevolence Society), framing its "Rules" and building a church.

Influenced by his uncle Zhang Jian, from his youth he took an interest in ingenious mechanisms and devices, especially those for military use. In 1622, he presented a memorial to the throne to introduce several military devices, and became known as an expert in machine design and military affairs. Between 1623 and 1626, he used his devices for various purposes such as dredging rivers, flood defense, and military defense. All the machines constructed by him were concerned directly with farming, daily life, or military affairs.

Thus it was that he was able to appreciate the machines that he saw brought by missionaries to China and recorded in their books. Inspired by the mechanisms of European clocks, he invented a weight-driven quadricycle in imitation of an ancient device he had sought to reconstruct for many years (Zhang Baichun 1996). In 1626, he wrote *Xinzhi Zhuqi Tushuo* (*Illustrations and Descriptions of Several Newly-built Devices*), which included eight kinds of machines. In 1627, together with the German Jesuit misssionary Johann Terrenz (1576–1630), he composed *Yuanxi Qiqi Tushuo Luzui* (*A Record of the Best Illustrations and Descriptions of Extraordinary Devices of the Far West*), which systemically introduced European mechanics and machines to China. Terrenz interpreted the Western texts in spoken Chinese while he illustrated Terrenz' interpretations and formulated them in written Chinese. After that, he used European technology to make more devices, playing a central role in the transmission of knowledge about Western machines to China.

In his old age, he was still interested in constructing devices based on Western knowledge and Chinese technology. At the end of 1640 or the beginning of 1641, he integrated his old writings and new ideas into a new book entitled "*E La Ji Ya You Zao Zhuqi Tushuo* (*Illustrations and Descriptions of the Devices Created by Gratia*)". This book described 24 kinds of devices, all of which he thought practical. Unfortunately, as the book was lost during the twentieth century, little is known of the specific content.

Wang lived the ordinary life of a low ranking scholar-official. His interest in machine design and the transmission of European mechanics was not widely shared by his contemporaries, though his endeavors were in accordance with the Confucian doctrine of serving society (Jami 1999, Huang Yi-Lung 2005). However, converting to Catholicism was definitely unusual in the seventeenth century. Nevertheless, an open attitude toward non-Confucian knowledge was not so unusual among Chinese scholars at that time. In fact, both his father and uncle, Zhang Jian, were interested in non-orthodox Confucian thought and knowledge, and he was open to knowledge from different traditions from early on in his life (Zhang Baichun and Tian Miao 2006). In his preface to *Yuanxi Qiqi Tushuo Luzui*, he claimed: "[As for] learning, [I] originally do not ask whether [it is] fine or coarse, [but] always expect [it] to be of benefit to the world; [as for] a person, I also do not ask whether [he is] a Chinese or a Westerner, [but] always expect him not to disobey Heaven" (Terrenz and Wang Zheng 1830). In his various works, he tried to incorporate Catholic ideas into Confucianism. We believe that this was an effort to legitimize Western knowledge and Catholicism in China. His satisfaction with the fact that Western knowledge and religion did not violate Confucian tradition was the basis for his adoption of Western knowledge and for his conversion to Catholicism. In fact, he believed that both European knowledge and thought were in accordance with Confucianism. Even though he was interested in Daoism, and converted to Catholicism, he remained a Confucian scholar (Tian Miao and Zhang Baichun 2007). In his *Liang Li Lüe* (*An Outline of Two Administrations*), he mentions that whenever "I arrived at a county, I would first pay homage at the temple of Confucius. I would examine the sacrificial utensils and books. Obeying the Sage, esteeming (his) *Dao* and respecting (his) learning are the main acts of righteousness of a country" (Wang Zheng 1987b, p. 27). His suicide in 1644, displaying such loyalty to the Ming emperor, is proof of his life-long commitment to Confucianism (Zhang Pengfen 1830). Loyalty toward the emperor was one of the prime principles of Confucianism, while suicide was strictly forbidden by Catholicism.

List of Wang Zheng's (Main) Works

Xinzhi Zhuqi Tushuo (新制诸器图说, *Illustrations and Descriptions of Several Newly-built Devices*), written in 1626, first published in 1628.
Yuanxi Qiqi Tushuo Luzui (远西奇器图说录最, *A Record of the Best Illustrations and Descriptions of Extraordinary Devices of the Far West*, written in 1627, first published in 1628.

Liang Li Lüe (两理略, *An Outline of Two Administrations*), written in 1637.
E La Ji Ya You Zao Zhuqi Tushuo (额辣济亚牖造诸器图说, *Illustrations and Descriptiions of the Devices Created by Gratia*), written in the 1640s.

Review of Wang Zheng's Main Works on Mechanism Design

In his *Xinzhi Zhuqi Tushuo* (*Illustrations and Descriptions of Newly-built Devices*, hereafter *Zhuqi Tushuo*), Wang Zhang illustrated and described nine devices, including a single-cylinder siphon-pump, a flume-beamed swape, a crank-operated man-powered mill, a wind-driven mill, a weight-driven geared mill, a weight-driven quadricycle, a combined clock, a mechanical cable plough, and a multiple crossbow (Wang Zheng 1830a, b, pp. 1–21). The single-cylinder siphon-pump (Fig. 2), which consists of a pipe and a single-cylinder pump, may be used to lift water.

The flume-beamed swape (Fig. 3) is a long water-lifting flume. The crank-operated man-powered, mill is not a particularly special mill, but he emphasized its two gear wheels. The wind-driven mill was drawn by him on the basis of a mill described by Nicolas Trigault. He also imitated the transmission and power mechanisms of European clocks, namely combined gear wheels and driving weights, with millstones or quadricycles in order to design a weight-driven geared mill and a weight-driven quadricycle (Fig. 4). For his combined clock, he made a copy of the transmission and power mechanisms of a European clock and combined them with a drum, a bell, and puppets (Fig. 5). The mechanical cable plough (Fig. 6) is driven by a windlass with handspikes. The multiple crossbow is a traditional Chinese device, which he tried to assemble on the basis of an unearthed trigger.[2]

Yuanxi Qiqi Tushuo Luzui (远西奇器图说录最, *A Record of the Best Illustrations and Descriptions of Extraordinary Devices of the Far West*, hereafter *Qiqi Tushuo*, see Fig. 7) is the first monograph on Western mechanics and machines in Chinese (Terrenz and Wang Zheng 1830), most of which was new for the Chinese of the seventeenth century. Following the introduction to the art of force or science of weights, there are three chapters that selectively expound Western mechanics and machines from Archimedean times to the early seventeenth century. The first chapter, which consists of 61 sections, is called *Explanations of Weight*. It discusses such topics as weight, center of gravity, geometrical center, specific gravity, buoyancy, etc. The second chapter of 92 sections, *Explanations of Devices*, discusses principles and calculations concerning simple machines such as the balance, steelyard, lever, pulley, wheel, screw, etc. The third chapter consists of illustrations and explanations of 54 Western machines, including devices to hoist and move heavy objects, water-lifting devices, wind-mills, water-mills, wood-sawing machines, and so on. This chapter also describes such mechanisms as the worm wheel and the ratchet wheel.

[2] The trigger unearthed from the ground tells us that there was no cross-bow trigger available for Wang Zheng.

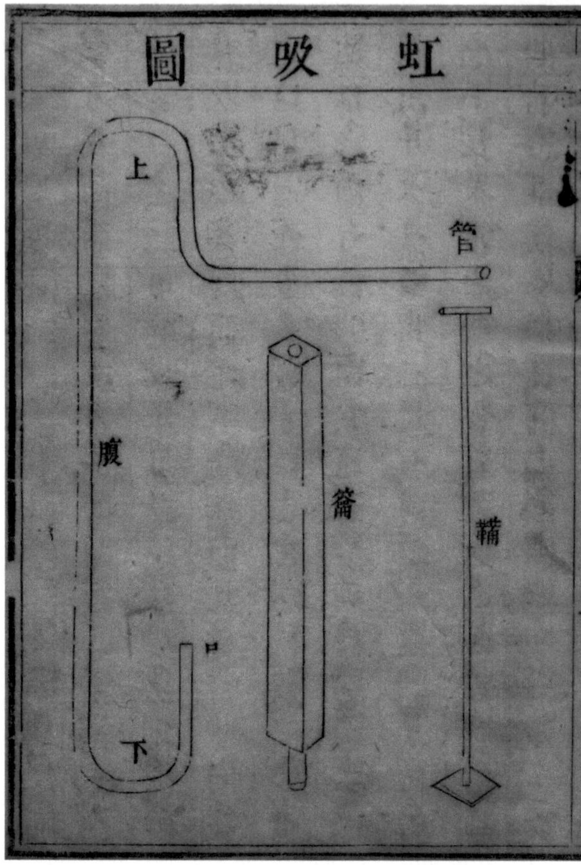

Fig. 2 Single-cylinder siphon-pump (*Xinzhi Zhuqi Tushuo*, Lailutang edition)

The first and second chapters of *Qiqi Tushuo* are probably derived mainly from Simon Steven's *Hypomnemata Mathematica... Mauritius, Princeps Auraicus, Comes Nassoviac...* (1608), Guidobaldo del Monte's *Mechanicorum liber* (1577) (Damerow and Schoepflin 2006). All the mathematical proofs in the Western sources were left out by Terrrenz and Wang. The third chapter is obviously derived from Agostino Ramelli's *Le Diverse et Artificiose Machine del Capitano* (1588), Faustus Verantius' *Machinae Novae Fausti Verantii Siceni, cum Declaratione Latina, Italica, Hispanica, Gallica et Germanica* (written *c*. 1595), Jacques Besson's *Théatre de Instruments Mathématiques et Mécaniques* (1578) and Vittorio Zonca's *Novo Teatro di Machini e Edificii* (1607). Terrenz and Wang integrated knowledge from these books into a new system of mechanical knowledge and machines (Tian Miao and Zhang Baichun 2007).

The most outstanding invention Wang made is the design of a weight-driven quadricycle (Fig. 4), which he called a Self-motion Vehicle. He made a great effort to reconstruct the famous transporting device known as the Wooden Ox and Gliding Horse, created by Zhuge Liang (181–234). He was unsuccessful until he learned

Fig. 3 Flume-beamed swape (*Xinzhi Zhuqi Tushuo*, Lailutang edition)

about the mechanism of mechanical clocks transmitted from Europe (Wang Zheng 1987b, pp. 81–82). The weight-driven quadricycle is the result of his application of such a mechanism.

According to Zhang Pengfen, it was believed in Wang's hometown that he made wooden models for farming and other purposes, such as making fire and operating bellows. It was also believed that he used his weight-driven quadricycle to help with binding and transporting stalks (Zhang Pengfen 1830). However, modern historians no longer believe such accounts.

On the Circulation of Wang Zheng's Works

Qiqi Tushuo is the most important publication that Wang Zheng was involved with. *Zhuqi Tushuo* was published attached to it. *E La Ji Ya You Zao Zhuqi Tushuo* was never published, remaining in manuscript form, and is now lost.

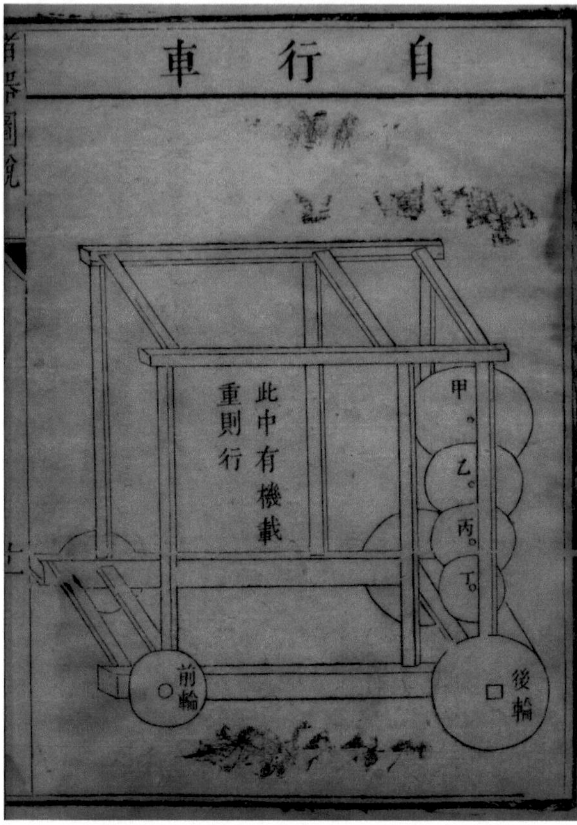

Fig. 4 Weight-driven quadricycle (*Xinzhi Zhuqi Tushuo*, Lailutang edition)

In 1628, Wu Weizhong, one of Wang's friends, printed a woodblock edition of *Qiqi Tushuo* with *Zhuqi Tushuo* in Yangzhou. Subsequently, Wang Yingkui copied the blocks and reprinted the two works in order to make them more widely available. Wang Yingkui's edition introduced some new errors and simplified a few of the illustrations. On the basis of this edition, Wu Huaigu reprinted them again in 1631. The content of this edition is the same as Wang Yingkui's, except that all the Western characters used as markers are replaced by Chinese ones.

Later, new editions of the two books based on these were reprinted in such collections as *Gujin Tushu Jicheng* (*GJTSJC, Collection of Ancient Chinese Books*, 1728), *Siku Quanshu* (*SKQS, Complete Collection in Four Treasuries*, 1781–1782), *Shoushange Congshu* (*Shoushange Book Series*, 1844) and by the Lailutang printing house (Zhang Baichun et al. 2006). *GJTSJC* and *SKQS* were compiled on imperial order. The editors of *GJTSJC* left out the first and second chapters, while the editors of *SKQS* added a short introduction to *Qiqi Tushuo*, remarking that Johann Terrenz and Wang's explanations of the art of force (science of weight) "exaggerated the marvellousness of these methods. Most of them are preposterous and

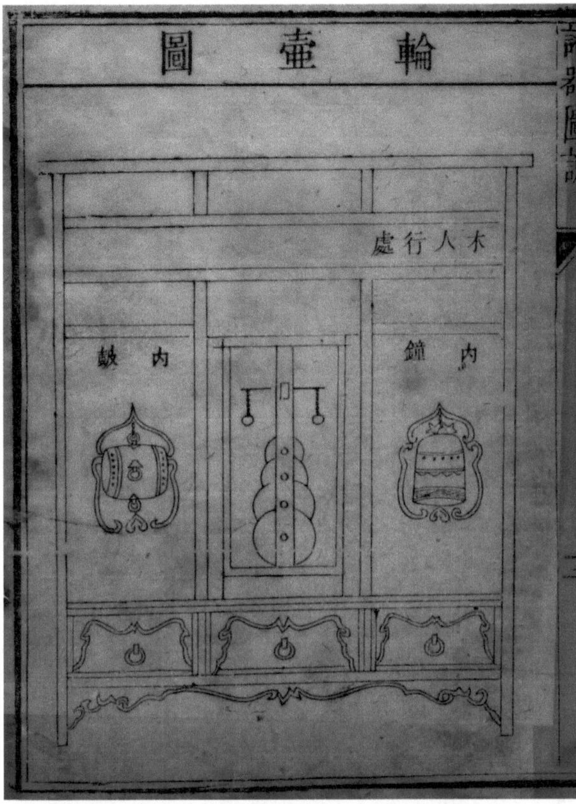

Fig. 5 European-styled clock (*Xinzhi Zhuqi Tushuo*, Lailutang edition)

excessive, not worthy of investigation. However, the skill of the manufacture of the machines is, in fact, the finest of any age" (Ji et al. 1983). This paragraph demonstrates the attitude of some Chinese scholars towards theoretical mechanics.

Shoushange Congshu and *GJTSJC* were repeatedly copied by other publishers so that *Qiqi Tushuo* and *Zhuqi Tushuo* were widely transmitted in the nineteenth and twentieth centuries not only throughout China, but also to Japan and Korea. The publishers and editors of new editions selected and changed their content based on their interests in Western knowledge about mechanics and machines. The majority of the errors in the first edition were repeated in subsequent ones. Though some revisers improved the content or added new misunderstanding, all the editions played a role in the acceptance and transmission of western scientific knowledge in China before the twentieth century.

Prior to the 1860s, *Qiqi Tushuo* was the most authoritative book on mechanics and machine design in China. Its content was repeatedly cited and discussed by later Jesuit missionaries and Chinese Scholars (Tian Miao and Zhang Baichun 2006a, Tian Miao and Zhang Baichun 2006b). Xue Fengzuo (1600–1680),

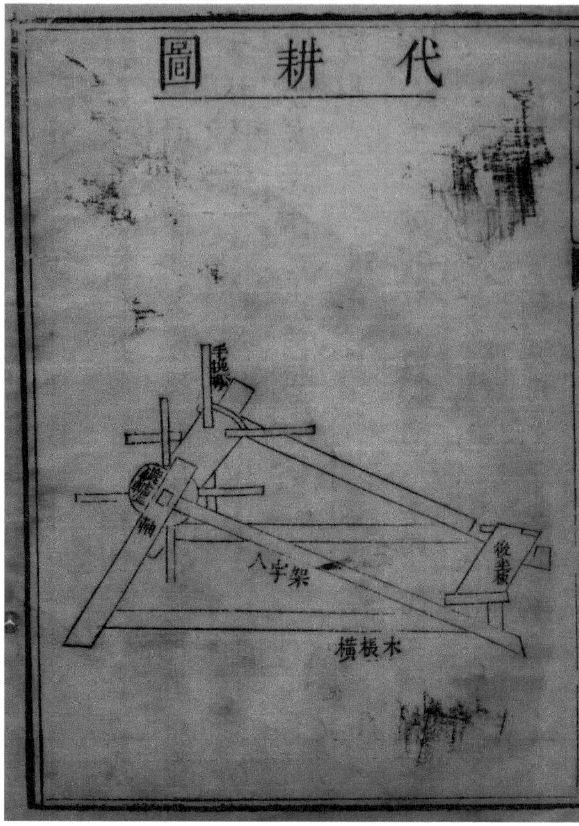

Fig. 6 Mechanical cable plough (*Xinzhi Zhuqi Tushuo*, Lailutang edition)

an astronomer and mathematician, completed a book entitled *Zhongxue* (*The science of weights*) in 1664 based on the two works (Tian Miao and Zhang Baichun 2006a). Xue had a good understanding of Western mechanical knowledge and abridged and summarized their original texts and illustrations, though he completely left out the second chapter of *Qiqi Tushuo.* He thought that the science of weights and mathematics would help people to understand the purpose and principles of machine construction, though some of the illustrations he redrew were of poor quality and he deleted some important mechanisms.

In the 1680s, another mathematician, Mei Wending (1633–1721), made a study of *Qiqi Tushuo*, contributing a commentary to its first chapter and supplementing its third chapter with some illustrations and explanations of traditional Chinese machines (Tian Miao and Zhang Baichun 2006b). He also tried to provide mathematical proofs for the mechanical knowledge in the first chapter of *Qiqi Tushuo.* A Jesuit missionary, Ferdinand Verbiest (1623–1688), compiled a book in Chinese entitled *Qiong Li Xue* in 1683, which includes mechanical theory from *Qiqi Tushuo* rather than knowledge about machines.

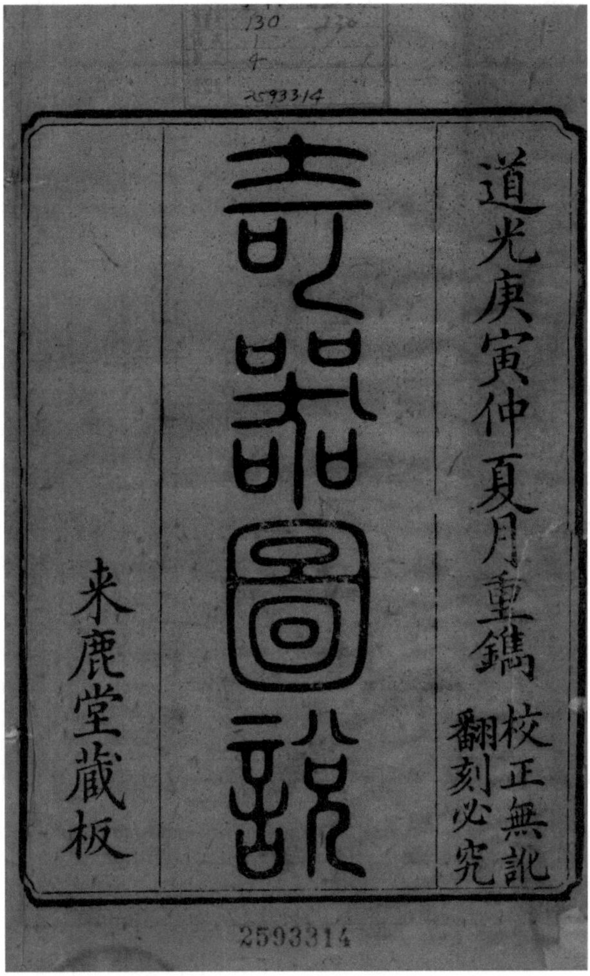

Fig. 7 The Cover of *Yuanxi Qiqi Tushuo Luzui* (Lailutang edition)

Although *Qiqi Tushuo* and *Zhuqi Tushuo* were circulated among Chinese scholars and other readers, we have not found any Chinese scholar from the eighteenth to mid-nineteenth century who made a further study of mechanics and mechanism design on their basis.

Modern Interpretation of Wang Zheng's Main Contributions

The weight-driven quadricycle was Wang Zheng's most important design (Fig. 4). A model of this kind of quadricycle could run for more than 10 m. Its mechanism was based on a system of gear wheels. Such gear wheels enable a crank-operated man-powered mill to be turned quickly and the design was used in water-powered, gear-driven mills in ancient China. Using it to make a transportation device was Wang's own invention. He surmised that a big weight-driven quadricycle could run for three *li* (one li is approximately equal to 0.5 km). However, in order to make a quadricycle or a mill run over and over again, the operator has to intermittently lift a weight that provides the power. In fact, his designs are feasible in principle, but they cannot be used to construct a large functioning quadricycle or mill (Liu Xianzhou 1958).

His main contribution to mechanism design is his role in the earliest transmission of Western knowledge about mechanics and machines into China. He also used his skill in Chinese painting to create a traditional Chinese pictorial representation of Western machines, for instance changing all the machine operators in the illustrations from European figures to Chinese ones.

His work laid the foundation for the further introduction of modern mechanics and knowledge about Western devices into China prior to the twentieth century. From the 1840s, the door to China was pushed open and Western military and other related technology was introduced, including mechanical knowledge. Mechanical terms used in *Qiqi Tushuo* had an influence on such books as *Zhong Xue* (Science of Weights, 1859) by Li Shanlan and Joseph Edkins (1823–1905), which was a Chinese translation of *An Elementary Treatise on Mechanics* by W. Whewell (1794–1866), as well as some textbooks on machines. When the Republic of China's Ministry of Education standardized technical terms, Liu Xianzhou adopted terms from *Qiqi Tushuo* when compiling a dictionary on mechanical engineering.

Acknowledgements The article is the result of the research of the Partner Group of the Max Planck Institute for the History of Science at the Chinese Academy of Sciences' Institute for the History of Natural Science. The authors wish to thank Mr. Wang Keju, Mr. John Moffett, Dr. Philip Cho, the librarians of the Institute for the History of Natural Sciences, and Mr. Urs Schoepflin, the curator of the library of the Max Planck Institute for the History of Science. Wang Keju is a descendent of Wang Zheng, and he provided the portrait of his ancestor and source materials. John Moffett, Philip Cho and Ms. Wang Lihua polished our English draft Urs Schoepflin helped the authors to find and also to digitalize relevant materials. This paper has also profited from the authors' discussions with Prof. Jürgen Renn, Dr. Matthias Schemmel, and Dr. Peter Damorow.

References

Damerow P, Schoepflin U (2006) Western sources of the *Qiqi Tushuo*. In: Zhang Baichun, Renn J (eds) Transformation and transmission: Chinese mechanical knowledge and the Jesuit intervention. Preprint 313 of Max-Planck-Institut für Wissenschaftsgeschichte, pp 89–93

Huang Yi-Lung (2005) Confucianized catholic: a case study on Wang Zheng. In: Huang Yi-Lung (ed) Two-headed snakes: the first generation of Catholic in the end of the Ming Dynasty and the beginning of the Qing Dynasty. Guoli Tsinghua daxue Press, Xizhu, pp 131–174 (in Chinese)

黄一农. 儒家化的天主教徒: 以王徵为例. 见：黄一农. 两头蛇：明末清初的第一代天主教徒. 上海: 上海世纪出版股份有限公司, 上海古籍出版社, 2006年. 第130–174页

Jami C (1999) "European Science in China" or "Western Learning": representations of cross-cultural transmission 1600–1800. Science in Context 12:413–434. Catherine Jami has analyzed the efforts Chinese converts and scholars who admired Western learning made in legitimizing Western knowledge and Catholicism in China in her paper

Ji Y, Lu X, Sun S (1983) An abstract of *Qiqi Tushuo*, 1781 or 1782, p 2, In: Wen Yuan Ge, Beijing (ed) Siku Quanshu (Siku Book Series), vol 842, pp 407–408 (in Chinese)

纪昀，陆锡熊，孙士毅．《奇器图说》提要．见：王志远誊录．林皋绘图．景印文渊阁四库全书・子部148·谱录类. 第842册. 台北：商务印书馆，1983年. 第407–408页

Liu Xianzhou (1958) Wang Zheng and Chinese first book on mechanical engineering. Chinese J Mech Eng 6(3):148–162 (in Chinese)

刘仙洲．王徵与我国第一部机械工程学，机械工程学报，1958年，第6卷，第3期，第148–162页

Terrenz J, Wang Zheng (1830) A record of the best illustrations and descriptions of extraordinary devices of the far west. Yangzhou, Lailu Tang Press, Chengdu, reprinted in 1830 (in Chinese)

邓玉函口授．王徵译绘．远西奇器图说录最．武位中校．张鹏翂梓．成都：来鹿堂刻本．1830年

Tian Miao, Zhang Baichun (2006a) Xue Fengzuo's reconstruction of mechanical knowledge from *Yuanxi Qiqi Tushuo Luzui*. J Harbin Inst Technol (Soc Sci edition) 8(6):1–8 (in Chinese)

田淼, 张柏春. 薛凤祚对《远西奇器图说录最》所述力学知识的重构. 哈尔滨工业大学学报（社会科学版），2008年, 第8卷，第6期. 第1–8页

Tian Miao, Zhang Baichun (2006b) Research of Mei Wending's commentaries on the *Yuanxi Qiqi Tushuo Luzi.* Chinese J Hist Sci Technol 27:330–339 (in Chinese)

田淼, 张柏春. 梅文鼎《远西奇器图说录最》注之研究. 中国科技史杂志，2006年, 第27卷，第4期. 第330–339页

Tian Miao, Zhang Baichun (2007) Wang Zheng who transmitted European science, technology and culture to China. J Harbin Inst Technol (Soc Sci Edition) 9(6):1–10 (in Chinese)

田淼, 张柏春. 王徵——欧洲科学技术与文化的传播者. 哈尔滨工业大学学报(社会科学版), 2007年，第9卷, 第6期. 第1–10页

Wang Zheng (1987a) Wang Zheng's posthumous writings. In: Li Zhiqin (ed). Shaanxi Renmin Press, Xian (in Chinese)

王徵著，李之勤辑. 王徵遗著. 西安：陕西人民出版社.1987年

Wang Zheng (1987b) An outline of two administrations, 1637. In: Li Z (ed) Wang Zheng's posthumous writings. Shaanxi Renmin Press, Xian, pp 12–131 (in Chinese)

王徵. 两理略. 见: 王徵著, 李之勤辑. 1987年. 第12–131页

Wang Zheng (1830a) Illustrations and descriptions of several newly-built devices. Lailu Tang Press, Chengdu, reprinted in 1830 (in Chinese)

王徵．新制奇器图说．武位中校．张鹏翂梓．成都：来鹿堂刻本. 1830年

Wang Zheng (1830b) Wang Zheng's preface to a record of the best illustrations and descriptions of extraordinary devices of the far west. In: Terrenz J, Wang Zheng (eds) (in chinese)

王徵．远西奇器图说录最（序）．第9页. 见：邓玉函口授．王徵译绘. 1830年

Zhang Baichun (1996) A new study on *Yuanxi Qiqi Tushuo Luzui* written by Wang Zheng and Johann Terrenz. J Dialect Nat 18(1):45–51 (in Chinese)

张柏春. 王徵与邓玉函《远西奇器图说录最》新探. 自然辩证法通讯，1996年, 第18卷，第1期. 第45-51页.

Zhang Baichun, Tian Miao, Liu Qiang (2006) The evolution of editions of *Yuanxi Qiqi Tushuo Luzi* and *Xinzhi Zhuqi Tushuo.* Chinese J Hist Sci Technol 27:115–136 (in Chinese)

张柏春, 田淼, 刘蔷．《远西奇器图说录最》与《新制诸器图说》版本之流变. 中国科技史杂志. 2006年. 第27卷, 第2期. 第115–136页

Zhang Baichun, Tian Miao (2006) Wang Zheng and the transmission of Western mechanical knowledge to China. In: Zhang Baichun and Renn J (eds) Transformation and transmission: Chinese mechanical knowledge and the Jesuit intervention. Preprint 313 of Max-Planck-Institut für Wissenschaftsgeschichte, pp 75–88

Zhang Pengfen (1830). Preface to a record of the best illustrations and descriptions of extraordinary devices of the far west, p. 9. In: Terrenz J, Wang Z (eds) (in Chinese)

张鹏翂. 重刊王忠节公奇器图说序.第9页. 见：邓玉函口授. 王徵译绘. 1830年

Index

MIX
Papier aus verantwortungsvollen Quellen
Paper from responsible sources
FSC® C105338

If you have any concerns about our products,
you can contact us on
ProductSafety@springernature.com

In case Publisher is established outside the EU,
the EU authorized representative is:
Springer Nature Customer Service Center GmbH
Europaplatz 3, 69115 Heidelberg, Germany

Printed by Libri Plureos GmbH
in Hamburg, Germany